Monographs in Computer Science

Editors

David Gries
Fred B. Schneider

Springer
New York
Berlin
Heidelberg
Barcelona
Budapest
Hong Kong
London
Milan
Paris
Santa Clara
Singapore
Tokyo

Monographs in Computer Science

Abadi and Cardelli, **A Theory of Objects**

Brzozowski and Seger, **Asynchronous Circuits**

Selig, **Geometrical Methods in Robotics**

Nielson [editor], **ML with Concurrency**

Castillo, Gutiérrez, and Hadi, **Expert Systems and Probabilistic Network Models**

Enrique Castillo
José Manuel Gutiérrez
Ali S. Hadi

Expert Systems and Probabilistic Network Models

With 250 Figures

Enrique Castillo
Cantabria University
39005 Santander, Spain
E-mail: castie@ccaix3.unican.es

José Manuel Gutiérrez
Cantabria University
39005 Santander, Spain
E-mail: gutierjm@ccaix3.unican.es

Ali S. Hadi
Cornell University
358 Ives Hall
Ithaca, NY 14853-3901
USA

Series Editors:
David Gries
Department of Computer Science
Cornell University
Upson Hall
Ithaca, NY 14853-7501
USA

Fred B. Schneider
Department of Computer Science
Cornell University
Upson Hall
Ithaca, NY 14853-7501
USA

Library of Congress Catologing-in-Publication Data
Castillo, Enrique.
   Expert systems and probabilistic network models / Enrique
Castillo, José Manuel Gutiérrez, Ali S. Hadi.
      p. cm. − (Monographs in computer science)
   Includes bibliographical references and index.
   ISBN 0-387-94858-9 (alk. paper)
   1. Expert systems (Computer science)   2. Probabilities.
 I. Gutiérrez, José Manuel. II. Hadi, Ali S. III. Title.
 IV. Series.
 QA76.76.E95C378 1997
 006.3'3–dc20   96-33161

Printed on acid-free paper.

© 1997 Springer-Verlag New York, Inc.
All rights reserved. This work may not be translated or copied in whole or in part without the written permission of the publisher (Springer-Verlag New York, Inc., 175 Fifth Avenue, New York, NY 10010, USA), except for brief excerpts in connection with reviews or scholarly analysis. Use in connection with any form of information storage and retrieval, electronic adaptation, computer software, or by similar or dissimilar methodology now known or hereafter developed is forbidden.
The use of general descriptive names, trade names, trademarks, etc., in this publication, even if the former are not especially identified, is not to be taken as a sign that such names, as understood by the Trade Marks and Merchandise Marks Act, may accordingly be used freely by anyone.

Production managed by Lesley Poliner; manufacturing supervised by Johanna Tschebull.
Photocomposed using the authors' LaTeX files.
Printed and bound by Braun-Brumfield, Inc., Ann Arbor, MI.
Printed in the United States of America.

9 8 7 6 5 4 3 2 1

ISBN 0-387-94858-9 Springer-Verlag New York Berlin Heidelberg   SPIN 10524048

To all the people of the former Yugoslavia with the hope that they will live together in peace and be friends, as are the authors, despite the differences in our religions, languages, and national origins.

Preface

The artificial intelligence area in general and the expert systems and probabilistic network models in particular have seen a great surge of research activity during the last decade. Because of the multidisciplinary nature of the field, the research has been scattered in professional journals in many fields such as computer science, engineering, mathematics, probability, and statistics. This book collects, organizes, and summarizes these research works in what we hope to be a clear presentation. Every effort has been made to keep the treatment of the subject as up-to-date as possible. Actually, some of the material presented in the book is yet to be published in the literature. See, for example, the material in Chapter 12 and some of the material in Chapters 7 and 11.

The book is intended for students and research workers from many fields such as computer science; engineering and manufacturing; medical and pharmaceutical sciences; mathematical, statistical, and decision sciences; business and management; economics and social sciences; etc. For this reason, we assumed no previous background in the subject matter of the book. The reader, however, is assumed to have some background in probability and statistics and to be familiar with some matrix notation (see, e.g., Hadi (1996)). In a few instances, we give some programs in *Mathematica* to perform the calculations. For a full understanding of these programs some knowledge of *Mathematica* is needed.

The book can be used as a reference or consulting book and as a textbook in upper-division undergraduate courses or in graduate-level courses. The book contains numerous illustrative examples and end-of-chapter exercises. We have also developed some computer programs to implement the various algorithms and methodologies presented in this book. The current version of these programs, together with a brief User's Guide, can be obtained from the World Wide Web site http://ccaix3.unican.es/~AIGroup. We have used these programs to do the examples and we encourage the reader to use them to solve some of the exercises. The computer programs can also help research workers and professionals apply the methodology to their own fields of study. Actually, we have used these programs to analyze some real-life applications (case studies) in Chapter 12. We therefore encourage the reader to use and explore the capabilities of these programs. It is suggested that the reader repeat the computations in the examples and solve the exercises at the end of the chapters using these programs. We hope that making such programs available will facilitate the learning of the material presented in this book. Finally, the extensive bibliography included at the end of the book can also serve as a basis for additional research.

Although some theory is present in the book, the emphasis is on applications rather than on theory. For this reason, the proofs of many theorems are left out, numerous examples are used to illustrate the concepts and theory, and the mathematical level is kept to a minimum.

The book is organized as follows. Chapter 1 is an introductory chapter, which among other things, gives some motivating examples, describes the components and development of an expert system, and surveys other related areas of artificial intelligence. Chapters 2 and 3 describe the main two types of expert systems: rule-based and probabilistic expert systems. Although the two types of expert systems are introduced separately, rule-based expert system can be thought of as a special case of the more powerful probabilistic expert system.

It is argued in Chapters 1–3 that two of the most important and complex components of expert systems are the coherence control and the inference engine. These are perhaps the two weakest links in almost all current expert systems, the former because it has appeared relatively recently and many of the existing expert systems do not have it, and the latter because of its complexity. In Chapters 1–3 we show how these subsystems can be implemented in rule-based and probability-based expert systems and how the probability assignment must be done in order to avoid inconsistencies. For example, the automatic updating of knowledge and the automatic elimination of object values are important for maintaining the coherence of the system. Chapters 5–10 are mainly devoted to the details of such implementations.

The materials in Chapter 5 and beyond require some concepts of graph theory. Since we expect that some of the readers may not be familiar with these concepts, Chapter 4 presents these concepts. This chapter is an essential prerequisite for understanding the topics covered in the remaining chapters. Building probabilistic models, which are needed for the knowledge base of a probabilistic expert system, is presented in Chapters 5–7. In particular, the independence and conditional independence concepts, which are useful for defining the internal structure of probabilistic network models and for knowing whether or not some variables or sets of variables have information about other variables, are discussed in Chapter 5. As mentioned in Chapter 4, graphs are essential tools for building probabilistic and other models used in expert systems. Chapter 6 presents the Markov and Bayesian network models as two of the most widely used graphical network models. Chapter 7 extends graphically specified models to more powerful models such as models specified by multiple graphs, models specified by input lists, multifactorized probabilistic models, and conditionally specified probabilistic models.

Chapters 8 and 9 present the most commonly used exact and approximate methods for the propagation of evidence, respectively. Chapter 10 introduces symbolic propagation, which is perhaps one of the most recent advances in evidence propagation. Chapter 11 deals with the problem of learning Bayesian network models from data. Finally, Chapter 12 includes several examples of applications (case studies).

Many of our colleagues and students have read earlier versions of this manuscript and have provided us with valuable comments and suggestions. Their contributions have given rise to the current substantially improved version. In particular, we acknowledge the help of the following (in alphabetical order): Noha Adly, Remco Bouckaert, Federico Ceballos, Jong Wang Chow, Javier Díez, Dan Geiger, Joseph Halpern, Judea Pearl, Julius Reiner, Milan Studený, and Jana Zvárová.

Enrique Castillo
Jose Manuel Gutiérrez
Ali S. Hadi

Contents

Preface		**vii**
1 Introduction		**1**
1.1	Introduction	1
1.2	What Is an Expert System?	2
1.3	Motivating Examples	3
1.4	Why Expert Systems?	7
1.5	Types of Expert System	8
1.6	Components of an Expert System	10
1.7	Developing an Expert System	14
1.8	Other Areas of AI	16
1.9	Concluding Remarks	20
2 Rule-Based Expert Systems		**21**
2.1	Introduction	21
2.2	The Knowledge Base	22
2.3	The Inference Engine	28
2.4	Coherence Control	48
2.5	Explaining Conclusions	52
2.6	Some Applications	53
2.7	Introducing Uncertainty	65
	Exercises	65
3 Probabilistic Expert Systems		**69**
3.1	Introduction	69
3.2	Some Concepts in Probability Theory	71
3.3	Generalized Rules	85
3.4	Introducing Probabilistic Expert Systems	86
3.5	The Knowledge Base	91
3.6	The Inference Engine	102
3.7	Coherence Control	104

	3.8	Comparing Rule-Based and Probabilistic Expert Systems	106
		Exercises	108

4 Some Concepts of Graphs — 113
- 4.1 Introduction — 113
- 4.2 Basic Concepts and Definitions — 114
- 4.3 Characteristics of Undirected Graphs — 118
- 4.4 Characteristics of Directed Graphs — 122
- 4.5 Triangulated Graphs — 129
- 4.6 Cluster Graphs — 139
- 4.7 Representation of Graphs — 144
- 4.8 Some Useful Graph Algorithms — 158
- Exercises — 172

5 Building Probabilistic Models — 175
- 5.1 Introduction — 175
- 5.2 Graph Separation — 177
- 5.3 Some Properties of Conditional Independence — 184
- 5.4 Special Types of Input Lists — 192
- 5.5 Factorizations of the JPD — 195
- 5.6 Constructing the JPD — 200
- Appendix to Chapter 5 — 204
- Exercises — 206

6 Graphically Specified Models — 211
- 6.1 Introduction — 211
- 6.2 Some Definitions and Questions — 213
- 6.3 Undirected Graph Dependency Models — 218
- 6.4 Directed Graph Dependency Models — 237
- 6.5 Independence Equivalent Graphical Models — 252
- 6.6 Expressiveness of Graphical Models — 259
- Exercises — 262

7 Extending Graphically Specified Models — 267
- 7.1 Introduction — 267
- 7.2 Models Specified by Multiple Graphs — 269
- 7.3 Models Specified by Input Lists — 275
- 7.4 Multifactorized Probabilistic Models — 279
- 7.5 Multifactorized Multinomial Models — 279
- 7.6 Multifactorized Normal Models — 292
- 7.7 Conditionally Specified Probabilistic Models — 298
- Exercises — 311

8 Exact Propagation in Probabilistic Network Models — 317
- 8.1 Introduction — 317

8.2	Propagation of Evidence	318
8.3	Propagation in Polytrees	321
8.4	Propagation in Multiply-Connected Networks	342
8.5	Conditioning Method	342
8.6	Clustering Methods	351
8.7	Propagation Using Join Trees	366
8.8	Goal-Oriented Propagation	377
8.9	Exact Propagation in Gaussian Networks	382
	Exercises	387

9 Approximate Propagation Methods — 393

9.1	Introduction	393
9.2	Intuitive Basis of Simulation Methods	394
9.3	General Frame for Simulation Methods	400
9.4	Acceptance-Rejection Sampling Method	406
9.5	Uniform Sampling Method	409
9.6	The Likelihood Weighing Sampling Method	411
9.7	Backward-Forward Sampling Method	413
9.8	Markov Sampling Method	415
9.9	Systematic Sampling Method	419
9.10	Maximum Probability Search Method	429
9.11	Complexity Analysis	439
	Exercises	440

10 Symbolic Propagation of Evidence — 443

10.1	Introduction	443
10.2	Notation and Basic Framework	445
10.3	Automatic Generation of Symbolic Code	447
10.4	Algebraic Structure of Probabilities	454
10.5	Symbolic Propagation Through Numeric Computations	455
10.6	Goal-Oriented Symbolic Propagation	464
10.7	Symbolic Treatment of Random Evidence	470
10.8	Sensitivity Analysis	472
10.9	Symbolic Propagation in Gaussian Bayesian Networks	474
	Exercises	478

11 Learning Bayesian Networks — 481

11.1	Introduction	481
11.2	Measuring the Quality of a Bayesian Network Model	484
11.3	Bayesian Quality Measures	486
11.4	Bayesian Measures for Multinomial Networks	490
11.5	Bayesian Measures for Multinormal Networks	499
11.6	Minimum Description Length Measures	506
11.7	Information Measures	509
11.8	Further Analyses of Quality Measures	509

11.9 Bayesian Network Search Algorithms 511
11.10 The Case of Incomplete Data 513
    Appendix to Chapter 11: Bayesian Statistics 515
    Exercises 525

12 Case Studies   529
12.1 Introduction 529
12.2 Pressure Tank System 530
12.3 Power Distribution System 542
12.4 Damage of Concrete Structures 550
12.5 Damage of Concrete Structures: The Gaussian Model . . . 562
    Exercises 567

List of Notation   **573**

References   **581**

Index   **597**

Chapter 1
Introduction

1.1 Introduction

Not so long ago, it was generally believed that problems such as theorem proving, speech and pattern recognitions, game playing (e.g., chess and backgammon), and highly complex deterministic and stochastic systems can only be tackled by humans because their formulations and solutions require some abilities that are found only in humans (e.g., the ability to think, observe, memorize, learn, see, smell, etc.). However, intensive research during the last three decades or so by researchers from several fields shows that many of these problems can actually be formulated and solved by machines.

The broad field that is now referred to as *artificial intelligence* (AI) deals with these problems, which at first seemed to be impossible, intractable, and difficult to formulate and solve using computers. A. Barr and E. A. Feigenbaum, two of the pioneers in AI research, define AI as follows (see Barr and Feigenbaum (1981), page 4):

> *Artificial intelligence is the part of Computer Science concerned with designing intelligent computer systems, that is, systems that exhibit the characteristics we associate with intelligence in human behavior—understanding language, learning, reasoning, solving problems, and so on.*

Nowadays, the field of AI consists of several subareas such as expert systems, automatic theorem proving, automatic game playing, pattern and speech recognition, natural language processing, artificial vision, robotics, neural networks, etc. This book is concerned with *expert systems*. Although expert systems are one of several research areas in the field of AI, most if not all other areas of AI have an expert system component built into them.

This chapter introduces expert systems. We start with some definitions of expert systems in Section 1.2. Section 1.3 gives some motivating examples of application of expert systems in various fields. These examples show the importance and wide applicability of expert systems in practice. Several reasons for using expert systems are outlined in Section 1.4. The main types of expert systems are presented in Section 1.5. Section 1.6 discusses the general structure of expert systems and their main components. The various steps involved in the design, development, and implementation of expert systems are given in Section 1.7. Finally, Section 1.8 gives a brief mention of other research areas of AI and provides the interested reader with some of the related references, journals, and World Wide Web (WWW) sites.

1.2 What Is an Expert System?

Several definitions of expert systems are found in the literature. For example, Stevens (1984), page 40, gives the following definition:

> *Expert Systems are machines that think and reason as an expert would in a particular domain. For example, a medical-diagnoses expert system would request as input the patient's symptoms, test results, and other relevant facts; using these as pointers, it would search its data base for information that might lead to the identification of the illness. [...] A true Expert System not only performs the traditional computer functions of handling large amounts of data, but it also manipulates that data so the output is a meaningful answer to a less than fully specified question.*

While this is still a reasonable definition of an expert system, several other definitions have evolved over the years due to the rapid development of technology (see, for example, Castillo and Alvarez (1991) and Durkin (1994)). The gist of these definitions can be summarized as follows:

Definition 1.1 Expert system. *An expert system can be broadly defined as a computer system (hardware and software) that simulates human experts in a given area of specialization.*

As such, an expert system should be able to process and memorize information, learn and reason in both deterministic and uncertain situations, communicate with humans and/or other expert systems, make appropriate decisions, and explain why these decisions have been made. One can also think of an expert system as a *consultant* that can provide help to (or in some cases completely substitute) the human experts with a reasonable degree of reliability.

Applications of expert systems in many fields of study have grown very rapidly during the last decade or so (see, for example, Quinlan (1987, 1989)). Also, Durkin (1994) examines some 2,500 expert systems and classifies them according to several criteria such as areas of applications, tasks performed, etc. As can be seen in Figure 1.1, the business, manufacturing, and medical fields continue to be the dominant areas in which expert systems are used. The following section gives a few motivating examples of applications of expert systems in several areas of applications.

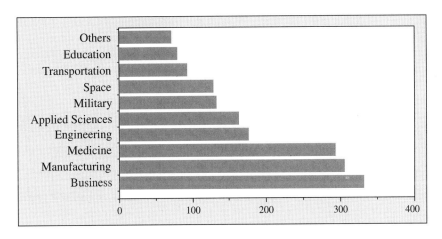

FIGURE 1.1. Fields of applications of expert systems. Adapted from Durkin (1994) and Castillo, Gutiérrez, and Hadi (1995a).

1.3 Motivating Examples

Expert systems have many applications. In this section we give a few motivating examples of the kind of problems that can be solved by expert systems. Other practical examples are given throughout the book.

Example 1.1 Banking transactions. Not too long ago, to be able to do banking transactions such as depositing or withdrawing money, one had to go to the bank during normal working hours. Nowadays, these and other transactions can be done at any time of day or night using automatic teller machines (ATMs) which are examples of expert systems. In fact, it is now possible to do these transaction from the convenience of your home by communicating with an expert system by telephone. ■

Example 1.2 Traffic control. Traffic control is one of the important applications of expert systems. Not too long ago the flow of traffic in city streets used to be controlled by putting a traffic officer in each intersection to manually operate the traffic lights. Nowadays, expert systems are used to automatically operate the traffic lights and regulate the flow of traffic in city streets and railroads (e.g., subway and train traffic). Examples of such systems are given in Section 2.6.1 and in the exercises of Chapter 2. ■

Example 1.3 Scheduling problems. Expert systems can also be used to solve complicated scheduling problems so that certain goals are optimized, for example, scheduling and assigning rooms for final examinations at a large university so that the following objectives are achieved:

- Eliminate conflict in room assignment: Only one examination is held in the same room at the same time.

- Sufficient seats: An assigned room must contain at least two seats per student.

- Minimize conflict in time: Minimize the number of students who have to take more than one examination at the same time.

- Eliminate hardship: No student should have more than two examinations in a 24-hour period.

- Minimize the number of examinations held in evening time.

Other examples of scheduling problems that can be solved by expert systems are scheduling doctors and nurses in large hospitals, scheduling shifts in a large 24-hour factory, and scheduling buses to handle rush hours or holiday traffic. ■

Example 1.4 Medical diagnosis. One of the most important applications of expert systems is in the medical field, where they can be used to answer the following questions:

1. How can one collect, organize, store, update, and retrieve medical information (e.g., records of patients) in an efficient, timely fashion? For example, suppose a doctor in a medical center is interested in information about a certain disease (D) and three associated symptoms (S_1, S_2, and S_3). An expert system can be used to search the database and extract and organize the desired information. This information can be summarized in tables such as the one given in Table 1.1 or in graphs like the one in Figure 1.2.

2. How can one learn from experience? That is, how is the knowledge of medical doctors updated as the number of diagnosed patients increases?

3. Given that a patient has a set of observed symptoms, how does one decide which disease(s) the patient is most likely to have?

4. What are the relationships among a set of (usually unobservable) diseases and a set of (observable) symptoms? In other words, what model(s) can be used to describe the relationships among the symptoms and diseases?

5. Given that the known set of symptoms is not sufficient to diagnose the disease(s) with certain degrees of certainty, what additional information should be obtained (e.g., additional symptoms to be identified or additional medical tests to be performed)?

6. What is the value of each of these additional pieces of information? In other words, what is the contribution of each of the additional symptoms or tests towards making a diagnostic decision? ■

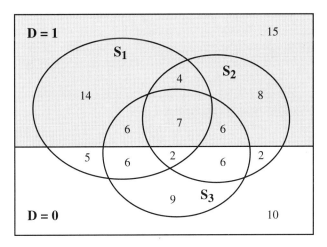

FIGURE 1.2. A graphical representation of the frequency distribution of one disease (D) and three binary symptoms (S_1, S_2, and S_3) in a medical database.

Example 1.5 Secret agents. Albert, Kathy, Nancy, and Tom are secret agents, each is currently in one of four countries: Egypt, France, Japan, and Spain. We do not know who is in which country. So we asked for information and we received the following four fax messages:

D	S_1	S_2	S_3	Frequency
1	1	1	1	7
1	1	1	0	4
1	1	0	1	6
1	1	0	0	14
1	0	1	1	6
1	0	1	0	8
1	0	0	1	0
1	0	0	0	15
0	1	1	1	2
0	1	1	0	0
0	1	0	1	6
0	1	0	0	5
0	0	1	1	6
0	0	1	0	2
0	0	0	1	9
0	0	0	0	10

TABLE 1.1. A tabular representation of the frequency distribution of one disease (D) and three binary symptoms (S_1, S_2, and S_3) in a medical database (1 represents the presence and 0 represents the absence of the indicated disease or symptom).

- From France: Kathy is in Spain.
- From Spain: Albert is in France.
- From Egypt: Nancy is in Egypt.
- From Japan: Nancy is in France.

We do not know who sent which message, but we know that Tom lies (a double agent?) and the other three agents tell the truth.

The question we wish to answer is, Which agent is in which country? This is a logic problem, and even though it involves simple statements, its solution is not immediately apparent. An expert system that solves this problem is given in Section 2.6.2. ∎

Example 1.6 Prisoners' problem. Just for fun and entertainment, here is another problem that can be solved by an expert system. Daniel Coleman, a friend of one of the authors, once mentioned the following problem during lunchtime. Consider a prison in which there are n inmates. The eye color of each inmate is either *black* or *not black*. There are no facilities in the prison by which an inmate can determine his own eye color. For example, there are no mirrors and the inmates are not allowed to communicate with each other. The king visits the prison and meets with all of the inmates. The king says to the inmates: "Starting today, you will meet here once every day and each one of you will have the opportunity to observe the eye colors of the others. At the end of each meeting, any one who can logically determine his own eye color will be set free. For your information, at least one of you has black eyes."

Given, for example, the fact that all inmates have black eyes, the question is, How many days will it take each one of them to determine his own eye color and hence be set free?

This is also a logic problem. It is not easy to solve even when the number of inmates n is as small as 3. The problem gets complicated as n gets large. Here we give the answer to the problem and let you think about why the answer is correct, but in Section 2.6.3, we shall develop an expert system to solve this problem for the simple case with $n = 2$. The solution can then be extended to more than two prisoners, although it becomes more complicated as n increases. The answer to the prisoners' problem is, all the inmates will meet every day and observe each others eyes and none of them will be able to determine his own eye color until the end of the nth day, at which time every one of them will be able to determine his own eye color. Hence they will all be set free at the end of the nth day's meeting. ∎

1.4 Why Expert Systems?

The initial development or acquisition of an expert system is usually expensive, but the maintenance and marginal cost of their repeated use is relatively low. Furthermore, the gains in terms of money, time, and accuracy resulting from using expert systems are very high, and amortization is very fast. However, before developing or acquiring an expert system a feasibility and cost-benefit study has to be made. There are several reasons for using expert systems. Chiefly among them are the following:

1. With the help of an expert system, personnel with little expertise can solve problems that require expert knowledge. This is also an important factor in cases where human experts are in short supply. In addition, the number of people with access to the knowledge increases.

2. The knowledge of several human experts can be combined together, which gives rise to a more reliable expert system, a system that is based on the collective wisdom of several experts, rather than on the experience of a single expert.

3. Expert systems can answer questions and solve problems much faster than the human expert. Thus, expert systems are invaluable in cases where time is a critical factor.

4. In some cases complexity of the problem prevents the human expert from reaching a solution. In other cases the solutions obtained by human experts are unreliable. Due to the capabilities of computers of processing a huge number of complex operations in a quick and accurate way, expert systems can provide both fast and reliable answers in situations where the human experts cannot.

5. Expert systems can be used to perform monotonous operations and others that are boring or uncomfortable to humans. Indeed, expert systems (e.g., an unmanned airplane or a spacecraft) may be the only viable option in a situation where the task to be performed may jeopardize a human life.

6. Substantial savings can be achieved from using expert systems.

The use of an expert system is especially recommended in the following situations:

- When the knowledge is difficult to acquire or is based on rules that can only be learned through experience.

- When continual improvement in knowledge is essential and/or when the problem is subject to rapidly changing legal rules and codes.

- When human experts are either expensive or difficult to find.

- When the users' knowledge of the subject matter is limited.

1.5 Types of Expert System

The problems that expert systems can deal with can be classified into two types: mainly deterministic and mainly stochastic problems. For example, although Example 1.1 (banking) and Example 1.2 (traffic control) can conceivably contain some elements of uncertainty, they are largely deterministic problems. On the other hand, in the medical field (see Example 1.4) the relationships among symptoms and diseases are known only with a certain degree of uncertainty (the presence of a set of symptoms does not always imply the presence of a disease). These kinds of problems may also include some deterministic elements, but they are largely stochastic problems.

Consequently, expert systems can be classified into two main types according to the nature of the problems they are designed to solve: deterministic and stochastic expert systems.

Deterministic problems can be formulated using a set of rules that relates several well-defined objects. Expert systems that deal with deterministic problems are known as *rule-based expert systems* because they draw their conclusions based on a set of rules using a *logical reasoning* mechanism. Chapter 2 is devoted to rule-based expert systems.

In stochastic or uncertain situations it is necessary to introduce some means for dealing with uncertainty. For example, some expert systems use the same structure of rule-based expert systems, but introducing some measure associated with the uncertainty of rules and their premises. Some propagation formulas can then be used to calculate the uncertainty associated with conclusions. Several uncertainty measures have been proposed during the last decades. Examples of these measures include *certainty factors*, used in expert systems shells such as the MYCIN expert system (see Buchanan and Shortliffe (1984)); *fuzzy logic* (see, for example, Zadeh (1983) and Buckley, Siler, and Tucker (1986)); and the Dempster and Shafer *theory of evidence* (see Shafer (1976)).

Another intuitive measure of uncertainty is *probability*, where a *joint probability distribution* of a set of variables is used to describe the relationships among the variables, and conclusions are drawn using certain well known probability formulas. This is the case with the PROSPECTOR expert system (see Duda, Gaschnig, and Hart (1980)), which uses Bayes' theorem for mineral exploring.

Expert systems that use probability as a measure of uncertainty are known as *probabilistic* expert systems and the reasoning strategy they use is known as *probabilistic reasoning*, or *probabilistic inference*. This book is devoted to probabilistic expert systems. Other related books that also provide a general introduction to other uncertainty measures are Buchanan and Shortliffe (1984), Waterman (1985), Pearl (1988), Jackson (1990), Neapolitan (1990), Castillo and Alvarez (1991), Durkin (1994) and Jensen (1996).

At the early stages of probabilistic expert systems several obstacles, due to the difficulties encountered in defining the joint probability distribution of the variables, have slowed down their development. With the introduction of *probabilistic network models*, these obstacles have been largely overcome and probabilistic expert systems have made a spectacular comeback during the last two decades or so. These network models, which include Markov and Bayesian networks, are based on a graphical representation of the relationships among the variables. This representation leads not only to efficient ways for defining the joint probability distribution but also to efficient propagation algorithms that are used to draw conclusions. Examples of such expert system shells are the HUGIN expert system (see Andersen et al. (1989)) and *X-pert Nets*,[1] which was written by the authors.

---

[1] This and other expert system shells can be obtained from the WWW site http://ccaix3.unican.es/~AIGroup.

1.6 Components of an Expert System

The definitions of expert systems given in Section 1.2 are perhaps best understood when one examines the main components of expert systems. These components are shown schematically in Figure 1.3 and are explained below.

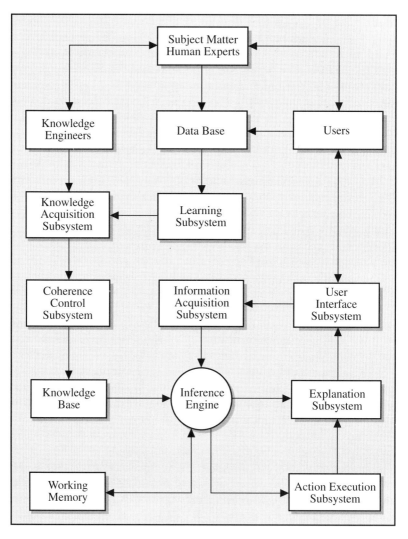

FIGURE 1.3. Typical components of an expert system, where the arrows represent the flow of information.

1.6.1 The Human Component

An expert system is usually the product of collaborative work of the *subject-matter human experts* and the *knowledge engineers*, with the *users* in mind. The human experts provide the knowledge base in the subject-matter area, and the knowledge engineers translate this knowledge into a language that the expert system can understand. The collaboration of the subject-matter specialists, the knowledge engineers, and the users is perhaps the most important element in the development of an expert system. This step requires an enormous amount of time and effort due to the different languages that the collaborating parties use and to the different experiences they have.

1.6.2 The Knowledge Base

The subject-matter specialists are responsible for providing the knowledge engineers with an ordered and structured knowledge base and a well-defined and well-explained set of relationships. This structured way of thinking may require the human experts to rethink, reorganize, and restructure the knowledge base, and as a result, the subject-matter specialists may become even more expert in their own fields.

Knowledge can be either *abstract* or *concrete*. Abstract knowledge refers to statements of general validity such as rules, probability distributions, etc. Concrete knowledge refers to information related to a particular application. For example, in medical diagnosis, the symptoms and diseases and relationships among them form the abstract knowledge, whereas particular symptoms of a given patient form the concrete knowledge. While abstract knowledge is permanent, concrete knowledge is ephemeral, i.e., it does not form a permanent part of the system and is destroyed after its use. The abstract knowledge is stored in the knowledge base, and the concrete knowledge is stored in the *working memory*. All procedures of the different systems and subsystems that are of a transient character are also stored in the working memory.

1.6.3 Knowledge Acquisition Subsystem

The knowledge acquisition subsystem controls the flow of new knowledge from the human experts to the knowledge base. It determines what new knowledge is needed or whether the received knowledge is indeed new, i.e., whether or not it is included in the knowledge base and if necessary, transmits them to the knowledge base.

1.6.4 Coherence Control

The coherence control subsystem has appeared in expert systems only recently. It is, however, an essential component of an expert system. This subsystem controls the consistency of the knowledge base and prevents any incoherent knowledge from reaching the knowledge base. In complex situations even an expert person can give inconsistent statements. Thus, without a coherence control subsystem, contradictory knowledge can creep into the system, giving an unsatisfactory performance. It is also common, especially in systems with uncertainty propagation mechanisms, to reach absurd and conflicting conclusions such as, for example, situations where the system produces probability values larger than one or smaller than zero. Thus, the coherence control subsystem checks and informs the human experts about inconsistencies. Furthermore, when the human experts are asked to provide new information, this subsystem advises them about the constraints that the new information has to satisfy. In this way, it helps the human experts to give accurate information.

1.6.5 The Inference Engine

The inference engine is the heart of every expert system. The main purpose of this component is to draw conclusions by applying the abstract knowledge to the concrete knowledge. For example, in medical diagnosis the symptoms of a given patient (concrete knowledge) are analyzed in the light of the symptoms of all diseases (abstract knowledge).

The conclusions drawn by the inference engine can be based on either *deterministic knowledge* or *probabilistic knowledge*. As one might expect, dealing with uncertain (probabilistic) situations may be considerably more difficult than dealing with certain (deterministic) ones. In many cases some facts (concrete knowledge) are not known with absolute certainty. For example, think of a patient who is not sure about his symptoms. It is also possible to work with abstract knowledge of a nondeterministic type, i.e., where random or fuzzy information is present. The inference engine is also responsible for the propagation of uncertain knowledge. Actually, in probability-based expert systems, uncertainty propagation is the main task of the inference engine; it enables it to draw conclusions under uncertainty. This task is so complex that it is probably the weakest element in almost all current expert systems. For this reason a major portion of this book is devoted to uncertainty propagation.

1.6.6 The Information Acquisition Subsystem

If the initial knowledge is very limited and conclusions cannot be reached, the inference engine utilizes the *information acquisition subsystem* in order to obtain the required knowledge and resume the inference process until conclusions can be reached. In some cases, the user can provide the required information. A *user interface* subsystem is needed for this and other purposes. Also, any information provided by the user must be checked for consistency before it is entered into the working memory.

1.6.7 User Interface

The user interface subsystem is the liaison between the expert system and the user. Thus, in order for an expert system to be an effective tool, it must incorporate efficient mechanisms to display and retrieve information in an easy way. Examples of information to be displayed are the conclusions drawn by the inference engine, the reasons for such conclusions, and an explanation for the actions taken by the expert system. On the other hand, when no conclusion can be reached by the inference engine due, for example, to a lack of information, the user interface provides a vehicle for obtaining the needed information from the user. Consequently, an inadequate implementation of the user interface that does not facilitate this process would undermine the quality of the expert system. Another reason for the importance of the user interface component is that users commonly evaluate expert (and other) systems based on the quality of the user interface rather than on that of the expert system itself, although one should not judge a book by its cover. Readers who are interested in the design of user interfaces are referred to the books by Shneiderman (1987) and Brown and Cunningham (1989).

1.6.8 The Action Execution Subsystem

The *action execution subsystem* is the component that enables the expert system to take actions. These actions are based on the conclusions drawn by the inference engine. As examples, an expert system designed to analyze railway traffic can decide to delay or stop some trains in order to optimize the overall traffic, or a system for controlling a nuclear power plant can open or close certain valves, move bars, etc., to avoid an accident. Explanations of these actions can be provided to the user by the *explanation subsystem*.

1.6.9 The Explanation Subsystem

The user may demand an explanation of the conclusions drawn or of the actions taken by the expert system. Thus, an explanation subsystem is needed to explain the process followed by the inference engine or by the action execution subsystem. For example, if an automatic teller machine decided to swallow the user's card (an action), the machine can display a message (an explanation) like the following:

> *Sorry, password still incorrect after three trials.*
> *We withheld your card, for your protection.*
> *Please contact your bank during regular office hours for help.*

In many domains of applications, explanations of the conclusions are necessary due to the risks associated with the actions to be executed. For example, in the field of medical diagnosis, doctors are ultimately responsible

14   1. Introduction

for the diagnoses made, regardless of the technical tools used to draw the conclusions. In these situations, without an explanation subsystem, doctors may not be able to explain the reasons for the diagnosis to their patients.

1.6.10   The Learning Subsystem

One of the main features of an expert system is the ability to learn. We shall differentiate between structural and parametric learning. By *structural learning* we refer to some aspects related to the structure of knowledge (rules, probability distributions, etc.). Thus, discovery of a new relevant symptom for a given disease or including a new rule in the knowledge base are examples of structural learning. By *parametric learning* we refer to estimating the parameters needed to construct the knowledge base. So, estimation of frequencies or probabilities associated with symptoms or diseases is an example of parametric learning.

Another feature of expert systems is their ability to gain *experience* based on available *data*. These data can be collected by both experts and nonexperts and can be used by the *knowledge acquisition subsystem* and by the *learning subsystem*.

It can be seen from the above components that expert systems can perform various tasks. These tasks include, but are not limited to, the following:

- Acquisition of knowledge and the verification of its coherence; hence the expert system can help the human experts in giving coherent knowledge.
- Storing (memorizing) knowledge.
- Asking for new knowledge when needed.
- Learning from the knowledge base and from the available data.
- Making inference and reasoning in both deterministic and uncertain situations.
- Explaining conclusions reached or actions taken.
- Communicating with human experts and nonexperts and with other expert systems.

1.7   Developing an Expert System

Weiss and Kulikowski (1984) outline the following steps for the design and implementation of an expert system (see also Hayes-Roth, Waterman, and Lenat (1983), Luger and Stubblefield (1989), and Figure 1.4):

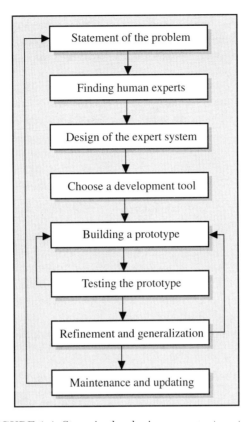

FIGURE 1.4. Steps in developing an expert system.

1. **Statement of the problem**. The first step in any project is usually the definition of the problem to be solved. Since the main goal of an expert system is to answer questions and to solve problems, this step is perhaps the most important one in the development of an expert system. If the problem is ill-defined, the system is expected to give erroneous answers.

2. **Finding human experts who can solve the problem**. In some cases, however, data bases can play the role of the human expert.

3. **Design of the expert system**. This step includes designing the structures for knowledge storage, the inference engine, the explanation subsystem, the user interface, etc.

4. **Choosing the development tool, shell, or programming language**. A decision has to be made between developing a specially designed expert system, a shell, a tool, or a programming language. If a tool or shell satisfying all design requirements exists, it should be

16   1. Introduction

used, not only for financial reasons but also due to reliability implications. Commercial shells and tools are subject to quality controls to which specially designed home programs are not.

5. **Developing and testing of a prototype**. If the prototype does not pass the desired checks, the previous steps (with the appropriate modifications) have to be repeated until a satisfactory prototype is obtained.

6. **Refinement and generalization**. In this step faults are corrected and new possibilities, apart from the initial design, are included.

7. **Maintenance and updating**. In this step user's complaints and problems must be taken into consideration, as well as correction of bugs and errors, updating of the product with new advances, etc.

All these steps influence the quality of the resulting expert systems, which must always be evaluated accordingly. For the evaluation of expert systems we refer the reader to O'Keefe, Balci, and Smith (1987), Chandrasekaran (1988), and Preece (1990).

1.8   Other Areas of AI

In this section we give a brief panoramic description of the scope and domain of some areas of AI other than expert systems. Since this book is devoted exclusively to expert systems, we provide the interested reader with some related references to other areas of AI. We should keep in mind, however, that this is not an exhaustive account of all areas of AI and that AI is a rapidly developing field, and new branches are continually emerging to deal with new situations in this ever-growing science.

There are several books that provide a general overview of most of the topics included in AI. The multivolume *Handbook of Artificial Intelligence* edited by Barr and Feigenbaum (1981, 1982) (volumes 1 and 2) and by Cohen and Feigenbaum (1982) (volume 3), and the *Encyclopedia of Artificial Intelligence*, edited by Shapiro (1987) contain detailed discussions of the various topics of AI. There are also several other books that cover the areas of AI. To mention only a few: Charniak and McDermott (1985), Rich and Knight (1991), Winston (1992), Ginsberg (1993), Russell and Norvig (1995).

As a consequence of the intensive research in the area of AI, there is also an increasing number of journals that publish articles in and related to the field of AI, journals such as *Applied Artificial Intelligence, Applied Intelligence, Artificial Intelligence, Artificial Intelligence Magazine. International Journal of Intelligent Systems,* On the other hand, journals like *Artificial*

Intelligence in Medicine, Biocybernetics and Biomedical Engineering, Cybernetics and Systems, Fuzzy Sets and Systems, IEEE Expert, IEEE Transactions on Systems, Man and Cybernetics, International Journal for Artificial Intelligence in Engineering, International Journal of Approximate Reasoning, International Journal of Computer Vision, International Journal of Expert Systems, Machine Learning, Networks, Neural Networks, and *Pattern Recognition Letters* are specialized in a single topic or in a given domain of application.[2]

1.8.1 Knowledge Representation

There are many different sources of information, or knowledge, involved in AI. The field of knowledge representation is concerned with the mechanisms for representing and manipulating this information. The resulting representation schemes should allow for efficient searching or inference mechanisms to operate on them. For example, in some cases the information can be represented by objects (or variables) and by logical rules (expressing relationships among the objects). Thus, this representation is suitable for their manipulation using logic analysis. This is the knowledge representation mechanism used, for example, in rule-based expert systems (Chapter 2). For a general overview of the different methodologies of knowledge representation see, for example Bachman, Levesque, and Reiter (1991), Bench-Capon (1990), and the *Proceedings of the International Conference on Principles of Knowledge Representation and Reasoning* (KR-89, 91, 92, and 94) published by Morgan and Kaufmann Publishers.

1.8.2 Planning

When dealing with complex problems, it becomes important to divide a large task into smaller parts that are easier to handle. Planning methods analyze different strategies for decomposing a given problem, solving each of its parts, and arriving at a final solution. The interaction among the parts will depend on the degree of decomposability of the problem. On the other hand, the beginning of parallel computing able to perform several different tasks simultaneously raises new problems that require some specialized planning strategies. In this situation, the goal is to divide tasks in a suitable form to solve many of the parts simultaneously. The readings edited by Allen, Hendler, and Tate (1990), give a general description of this field. On the other hand, the collection of papers edited by Bond and Gasser (1988) is devoted to *parallel reasoning*, also known as *distributed reasoning*.

---

[2] A list containing most of the journals in the field of AI can be obtained from the WWW site "http://ai.iit.nrc.ca/ai_journals.html"; see also "http://www.bus.orst.edu/faculty/brownc/aies/journals.htm."

18   1. Introduction

1.8.3 Automatic Theorem Proving

The ability to make logical deductions was thought for many years to be a task that can be performed only by humans. The research conducted in the 1960s in the area of automatic theorem proving showed that this task could also be carried out by programmable machines. Such machines would need not only to modify the existing knowledge, but also to create new conclusions. First, *theorem provers* have been successfully applied in several fields of mathematics such as logic, geometry, etc. The field of mathematics is a natural area for this methodology because of the existence of deductive mechanisms and of an extensive mass of knowledge. However, theorem provers can also be adapted to solve problems arising in other deductive areas with these two characteristics. General introductions of this subject are given in Wos et al. (1984) and Bundy (1983), which contains the Prolog code for a simple theorem-proving program. More recent references are Newborn (1994), Almulla (1995) and the references therein.

1.8.4 Automatic Game Playing

Automatic game playing is an example of one of the oldest and most fascinating areas of AI (see, for example, Newell, Shaw, and Simon (1963)). Computer games (such as chess, backgammon, and playing cards) have seen a massive development in recent years. For example, chess programs are now able to compete with and defeat well-known chess masters. Automatic game playing requires a deep theoretical study and has several applications in other areas such as *search methods, optimization*, etc. A good discussion of this field including historical references is found in Levy (1988).

1.8.5 Pattern Recognition

Pattern recognition is concerned with different classification techniques for identifying the subgroups, or clusters, with some common characteristics within a given group. The degree of association of any of the objects with any of the groups provides a way to infer conclusions. Thus, the algorithms developed in this area are useful tools for dealing with several problems in many domains such as image recognition, signal recognition, diagnoses of equipment failure, process control, etc. See Sing-Tze (1984) and Niemann (1990) for a general introduction to this field, and Patrick and Fattu (1984) for a discussion from a statistical point of view.

1.8.6 Speech Recognition

Speech is by far the main medium of communication used by humans. Speech recognition deals with the problem of processing spoken language and capturing the different semantic elements forming the speech. The problems associated with the different pronunciations and voice tones are the main obstacles that this discipline has to face. A general introduction to speech recognition is given in Rabiner and Juang (1993).

1.8.7 Natural Language Processing

An objective of natural language processing (NLP) is to extract as much semantic information as possible from a written text. With the increasing use of computers in transcribing information, written language is gaining an important role as a medium for communication. Because speech recognition is inherently a more difficult problem, an efficient processing of written language is necessary. The reverse problem of language processing is *language generating*, that is, to provide computers with capabilities to generate natural language sentences rather than cryptic messages. The combination of these two tasks would allow, for example, for the possibility of automatic translation between texts written in different languages. This is known as *computer assisted language translation*. A classic reference to this field is Schank and Abelson (1977). Also, descriptions of language processing and language generation are found in Allen (1995) and McKeown (1985), respectively.

1.8.8 Artificial Vision

One goal of artificial vision is the possibility of using computers to automatically recognize and locate different objects in three dimensions. Many other areas of AI such as knowledge representation, pattern recognition, and neural networks play an essential role in artificial vision. The significant technical advances in the last decade have already been applied in several commercial vision systems used in manufacturing, inspection and guidance tasks, etc. For a general introduction to this area, see Fischler and Firschein (1987) and Shapiro and Rosenfeld (1992).

1.8.9 Robotics

Robotics is one of the most popular and useful areas of AI. Robots combine mechanical components, sensors, and computers that enable them to deal with real objects and perform many tasks in a precise, fast, and tireless way. Thus, we may think of robots as computers that interact with the real world. A general survey of the robotics area is presented in McKerrow (1991), while Jones and Flynn (1993) is devoted to practical applications.

1.8.10 Neural Networks

Neural networks were created with the aim of reproducing the functions of the human brain. Network architectures with a large number of weighted connections among several layers of processors were introduced as a resemblance of the neuronal structure of the human brain. The information contained in a neural network is encoded in the structure of the network and in the weights of the connections. Then, for a particular situation, the weights of the connections have to be adjusted to reproduce a given output. This learning task is achieved using a learn-by-analogy training process, that is, the model is trained to reproduce the outputs of a set of training signals with the aim of encoding in such a way the structure of the phenomena. The appearance of fast computers on which to simulate large networks and the discovery of powerful learning algorithms have been the keystone of the quick development of these area. For a general introduction to this field see, for example, Freeman and Skapura (1991) and Lisboa (1992).

1.9 Concluding Remarks

From the brief description of the various areas of AI in this chapter, one can see that these areas are interrelated. For example, robotics utilizes other AI areas such as automatic vision and pattern and speech recognition. The AI area as a whole is highly interdisciplinary. For example, expert systems require various concepts from computer science, mathematical logic, graph theory, probability, and statistics. Thus, working in this field requires the collaboration of many researchers in different areas of specialization.

Chapter 2
Rule-Based Expert Systems

2.1 Introduction

In our daily living, we encounter many complex situations governed by deterministic rules: traffic control mechanisms, security systems, bank transactions, etc. Rule-based expert systems are an efficient tool to deal with these problems. Deterministic rules are the simplest of the methodologies used in expert systems. The knowledge base contains the set of rules defining the problem, and the inference engine draws conclusions applying classic logic to these rules. For a general introduction to rule-based expert systems, see, for example, Buchanan and Shortliffe (1984), Castillo and Alvarez (1991), Durkin (1994), Hayes-Roth (1985), Waterman (1985), and also the readings edited by García and Chien (1991). A practical approach is also given in the book of Pedersen (1989), which includes several algorithms.

This chapter is organized as follows. Section 2.2 describes the knowledge base of the rule-based expert systems and gives a definition and examples of rules, which constitute the core of the knowledge base. We then discuss how the inference engine operates (Section 2.3), how the coherence control subsystem works (Section 2.4), and how conclusions reached by the inference engine are explained (Section 2.5). Section 2.6, gives some examples of applications. Finally, Section 2.7 points to some limitations of rule-based expert systems.

22   2. Rule-Based Expert Systems

Object	Set of Possible Values
Score	{0, 1, ..., 200}
Grade	{A,B,C,D,F}
GPA	[0,4.0]
Rank	{0, 1, ..., 100}
Admit	{yes, pending, no}
Notify	{yes, no}

TABLE 2.1. An example of objects and their sets of possible values.

2.2 The Knowledge Base

There are two different types of knowledge in expert systems: *concrete knowledge* and *abstract knowledge*. Concrete knowledge is the evidence or facts that are known or given in a particular situation. This type of knowledge is dynamic, that is, it can change from one application to another. For this reason this type of knowledge is stored in the working memory. The information stored in the working memory is not permanent in nature.

In deterministic situations, the relationships among a set of objects can be represented by a set of rules. Abstract knowledge consists of a set of objects and a set of rules that governs the relationships among the objects. Abstract knowledge is stored in the knowledge base. The information stored in the knowledge base is static and permanent in nature, that is, it does not change from one application to another.

To give an intuitive idea of what a rule is, suppose that we have a set of *objects* and for simplicity, assume that each object can have one and only one of a set of possible values. Examples of objects and their values are given in Table 2.1. The following are a few examples of rules:

Rule 1: If Score > 180, then Grade = A.
Rule 2: If Rank > 80 and GPA > 3.5, then Admit = yes and Notify = yes.
Rule 3: If Rank < 60 or GPA < 2.75, then Admit = no and Notify = yes.

Each of the above rules relates two or more objects and consists of the following parts:

- The *premise* of the rule, which is the *logical expression* between the keywords *if* and *then*. The premise can contain one or more object-value statements connected by the logical operators *and*, *or*, or *not*. For example, the premise of Rule 1 consists of only one object-value statement, whereas each of the premises of Rules 2 and 3 consists of two object-value statements connected by a logical operator.

- The *conclusion* of the rule, which is the logical expression after the keyword *then*.

The above examples facilitate the following definition of a rule.

Definition 2.1 Rule. *A rule is a logical statement that relates two or more objects and includes two parts, the premise and the conclusion. Each of these parts consists of a logical expression with one or more object-value statements connected by the logical operators* and, or, *or* not.

A rule is usually written as "If *premise*, then *conclusion*." In general, both the premise and the conclusion of a rule can contain multiple object-value statements. A logical expression that contains only one object-value statement is referred to as a *simple logical expression*; otherwise, the expression is called a *complex logical expression*. For example, the logical expressions in both the premise and conclusion of Rule 1 are simple, whereas the logical expressions in the premises and conclusions of Rules 2 and 3 are complex. Correspondingly, a rule that contains only simple logical expressions is called a *simple rule*; otherwise, it is called a *complex rule*. For example, Rule 1 is simple but Rule 2 and Rule 3 are complex.

Example 2.1 Automatic teller machine. As an example of a deterministic problem that can be formulated using a set of rules, consider a situation where a user (e.g., client) wishes to withdraw money from his bank account using an automatic teller machine (ATM). Once the user inserts his card into the ATM, the machine attempts to read and verify it. If the card is not verified (e.g., because it is unreadable), the ATM returns the card to the user with a message to that effect. Otherwise the ATM will then ask the user to give his personal identification number (PIN). If the number is incorrect, the user is given a certain number of chances (trials) to give the correct PIN. If the PIN is correct, the ATM prompts the user for the desired amount. For a payment to be authorized, the amount cannot exceed a certain allowable daily limit and there also must be sufficient funds in the user's account.

Here we have seven objects, and each object can have one and only one of some possible values. Table 2.2 lists these objects and their associated possible values.

Figure 2.1 shows seven rules governing the strategy that the ATM would follow whenever a user attempts to withdraw money from the machine. In Rule 1, for example, the premise consists of six object-value statements connected by the logical operator *and* indicating that the premise is true if all six statements are true. Thus, Rule 1 relates the object *Payment* (in the conclusion) to the other six objects. Following Rule 1, the action to be taken by the ATM is to give the money to the user if the card is verified, the date has not expired, PIN is correct, the number of trials for giving the correct PIN is not exceeded, the amount requested does not exceed the

24   2. Rule-Based Expert Systems

Object	Set of Possible Values
Card	{verified, not verified}
Date	{expired, not expired}
PIN	{correct, incorrect}
Trials	{exceeded, not exceeded}
Balance	{sufficient, insufficient}
Limit	{exceeded, not exceeded}
Payment	{authorized, not authorized}

TABLE 2.2. Objects and their possible values for the automatic teller machine example.

Name	American	Politician	Woman
Barbara Jordan	yes	yes	yes
Bill Clinton	yes	yes	no
Barbara Walters	yes	no	yes
Mohammed Ali	yes	no	no
Margaret Thatcher	no	yes	yes
Anwar El-Sadat	no	yes	no
Marie Curie	no	no	yes
Pablo Picasso	no	no	no

TABLE 2.3. A database showing four objects and their values for the famous people example.

balance, and the amount requested does not exceed the daily limit. The logical expressions in each of the remaining rules in Figure 2.1 has only one statement. Note that Rule 1 indicates when to permit payment and the remaining six rules, when to refuse payment.   ∎

Example 2.2 Famous people. Suppose we have a database consisting of N individuals. For each individual, the database contains four attributes: name, gender, nationality, and profession. Suppose the database shows only whether or not a person is an American, a politician, and/or a woman. Each of the last three attributes is binary (has exactly two possible values). In this case, the database can contain at most $2^3 = 8$ disjoint subsets. These subsets are depicted in Figure 2.2. The figure also shows the name of one person in each subset. Table 2.3 gives an example of a database that contains $N = 8$ famous people. Here we have four objects: *Name*, *American*, *Politician*, and *Woman*. The first object can have one of N possible values (the names of the people) and each of the last three objects can assume either the value *yes* or the value *no*.

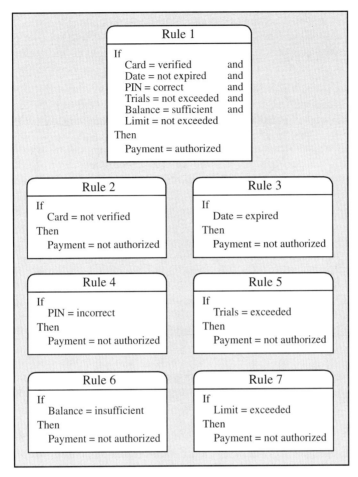

FIGURE 2.1. Examples of rules for withdrawing money from an automatic teller machine.

From Table 2.3 one can construct a rule for each person giving a total of eight rules. For example, the following rule is associated with President Clinton:

- Rule 1: If $Name = Clinton$, then $American = yes$ and $Politician = yes$ and $Woman = no$.

The remaining seven rules can be constructed in a similar way. ∎

We shall use Examples 2.1 and 2.2 later in this chapter to illustrate several concepts related to rule-based expert systems.

26   2. Rule-Based Expert Systems

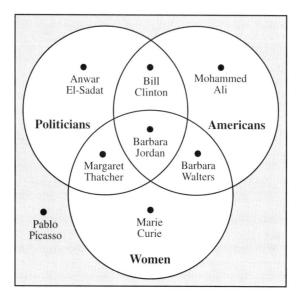

FIGURE 2.2. An example of a database with three binary attributes that divide the population into eight disjoint subsets.

Some systems impose certain restrictions on rules. For example:

- The logical operator *or* is not permitted in the premise, and

- The conclusion consists of only one simple logical expression.

There are good reasons for imposing these restrictions. First, rules that satisfy these restrictions are easy to deal with when writing computer programs. Second, the above two restrictions cause no loss of generality because a more general rule can be replaced by two or more rules satisfying these restrictions. This is called *rule substitution*. Therefore, the set of rules initially specified by the subject-matter human expert may require further rule substitutions in order for the rules to comply with the above restrictions.

Table 2.4 gives examples of rule substitution. Note that each rule in the first column can be substituted by the corresponding set of rules in the second column and that all of the rules in the second column satisfy the above conditions. For example, the first complex rule in Table 2.4:

- Rule 1: If A or B, then C,

can be replaced by the two simple rules:

- Rule 1a: If A, then C.

- Rule 1b: If B, then C.

Rule	Equivalent Rules
If A or B, then C	If A, then C If B, then C
If $\overline{A \text{ or } B}$, then C	If $\bar{A}$ and $\bar{B}$, then C
If $\overline{A \text{ and } B}$, then C	If $\bar{A}$, then C If $\bar{B}$, then C
If $(A$ or $B)$ and C, then D	If A and C, then D If B and C, then D
If $\overline{(A \text{ or } B)}$ and C, then D	If $\bar{A}$ and $\bar{B}$ and C, then D
If $\overline{A \text{ and } B}$ and C, then D	If $\bar{A}$ and C, then D If $\bar{B}$ and C, then D
If A, then B and C	If A, then B If A, then C
If A, then B or C	If A and $\bar{B}$, then C If A and $\bar{C}$, then B
If A, then $\overline{B \text{ and } C}$	If A and B, then $\bar{C}$ If A and C, then $\bar{B}$
If A, then $\overline{B \text{ or } C}$	If A, then $\bar{B}$ If A, then $\bar{C}$

TABLE 2.4. Examples of rule Substitution: The rules in the first column are equivalent to the rules in the second column. Note that in the first six examples the substitutions are applied to the premise and in the last four they are applied to the conclusion.

As another example, Table 2.5 shows that

- Rule 2: If $\overline{A \text{ or } B}$, then C,

can be replaced by the rule

- Rule 2: If $\bar{A}$ and $\bar{B}$, then C,

where $\bar{A}$ means *not* A. Table 2.5 is called a *truth table*.

28   2. Rule-Based Expert Systems

A	B	$\bar{A}$	$\bar{B}$	$\overline{A \text{ or } B}$	$\bar{A}$ and $\bar{B}$
T	T	F	F	F	F
T	F	F	T	F	F
F	T	T	F	F	F
F	F	T	T	T	T

TABLE 2.5. A truth table showing that the logical expressions $\overline{A \text{ or } B}$ and $\bar{A}$ and $\bar{B}$ are equivalent.

2.3 The Inference Engine

As we have mentioned in the previous section, there are two types of knowledge: concrete knowledge (facts or evidence) and abstract knowledge (the set of rules stored in the knowledge base). The inference engine uses both the abstract knowledge and the concrete knowledge to obtain new conclusions or facts. For example, if the premise of a rule is true, then the conclusion of the rule must also be true. The initial concrete knowledge is then augmented by the new conclusions. Thus, both the initial facts and concluded facts form the concrete knowledge at a given instant in time.

Conclusions can be classified into two types: *simple* and *mixed*. Simple conclusions are those resulting from a single rule. Mixed conclusions result from more than one rule. In order to obtain conclusions, expert systems utilize different types of inference and control strategies (see, for example, Castillo and Alvarez (1991), Durkin (1994), Shapiro (1987), Waterman (1985)). In the rest of this section we discuss the following inference strategies used by the inference engine to reach simple and mixed conclusions:

- Modus Ponens.

- Modus Tollens.

- Resolution Mechanism.

- Rule Chaining.

- Goal-Oriented Rule Chaining.

- Compiling Rules.

The first two strategies are used to reach simple conclusions and the last four are used to reach mixed conclusions.

Note, however, that none of the above strategies, if implemented alone, will lead to all logically possible conclusions. Thus, several inference strategies have to be implemented in the system in order for the inference engine to obtain as many conclusions as possible.

2.3.1 Modus Ponens and Modus Tollens

The *Modus Ponens* is perhaps the most commonly used strategy. It is used to reach simple conclusions. It examines the premise of a rule, and if the premise is true, the conclusion becomes true. To illustrate, suppose we have the rule, "If A is true, then B is true" and we also know that "A is true." Then, as shown in Figure 2.3, the Modus Ponens strategy concludes that "B is true." This strategy, which sounds trivial due to its familiarity, is the basis for a large number of rule-based expert systems.

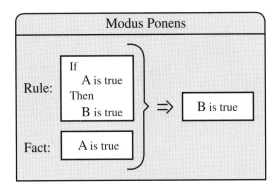

FIGURE 2.3. An illustration of the Modus Ponens strategy.

The *Modus Tollens* strategy is also used to reach simple conclusions. It examines the conclusion and if it is false, the premise becomes false. For example, suppose again that we have the rule, "If A is true, then B is true" but now we know that "B is false." Then, using the Modus Ponens strategy we cannot reach a conclusion, but as shown in Figure 2.4, the Modus Tollens strategy concludes that "A is false." Although very simple and with many useful applications, the Modus Tollens strategy is less commonly used than the Modus Ponens strategy.

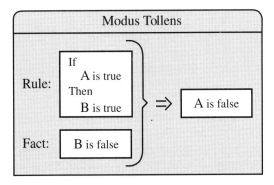

FIGURE 2.4. An illustration of the Modus Tollens strategy.

Thus, Modus Ponens is a strategy that moves forward from the premise to the conclusion of a rule, whereas the Modus Tollens strategy moves backward from the conclusion to the premise. The two strategies should not be viewed as alternative but rather complementary inference strategies. The Modus Ponens strategy needs some information about the objects in the premise of a rule to conclude, while the Modus Tollens strategy needs information about the objects in the conclusion. In fact, given an inference engine that only uses the Modus Ponens strategy, incorporating also the Modus Tollens strategy can be thought of as effectively expanding the knowledge base by adding additional rules as the following example illustrates.

Example 2.3 Modus Tollens strategy expands the knowledge base. Suppose the knowledge base consists of only Rule 1, which is shown in Figure 2.5. We can use the Modus Tollens strategy to "reverse" Rule 1 and obtain some conclusion when information about the objects in the conclusion is given. Then, applying the Modus Tollens strategy to the rule "If A, then B" is equivalent to applying the Modus Ponens strategy to the rule "If $\bar{B}$, then $\bar{A}$." In the case of Rule 1, using the equivalence

$$\overline{A = T \text{ and } B = T} \Leftrightarrow \overline{A = F} \text{ or } \overline{B = F},$$

we get Rule 1b, which is shown in Figure 2.6. Thus, using both Modus Ponens and Modus Tollens strategies when the knowledge base contains only Rule 1 is equivalent to using the Modus Ponens strategy when the knowledge base contains both Rule 1 and Rule 1b. ∎

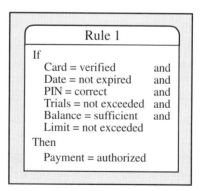

FIGURE 2.5. Rule 1 taken from Figure 2.1.

Furthermore, the performance of the inference engine depends on the set of rules in its knowledge base. There are situations where the inference engine can conclude using a set of rules, but cannot conclude using another (though logically equivalent) set of rules. Here is an illustrative example.

2.3 The Inference Engine   31

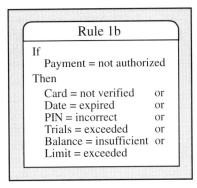

FIGURE 2.6. Rule 1b can be derived from Rule 1 using the Modus Tollens strategy.

Example 2.4 Inference with two equivalent sets of rules. Suppose again that we have two inference engines: Engine E_1, whose knowledge base contains the seven rules in Figure 2.1, and Engine E_2, whose knowledge base contains the seven rules shown in Figure 2.7. Note that the two sets of rules are logically equivalent. Suppose further that the value of *PIN* is known to be *incorrect*. If both E_1 and E_2 use only the Modus Ponens strategy, then E_1 will be able to conclude that *Payment = not authorized* (because of Rule 4), but E_2 will not conclude. Thus, using only the Modus Ponens strategy some logically derivable conclusions may not be obtained. On the other hand, if both engines use the Modus Tollens strategy, then both will conclude. ∎

2.3.2 Resolution Mechanism

The Modus Ponens and Modus Tollens strategies can be used to obtain simple conclusions. On the other hand, mixed conclusions, which are based on two or more rules, are obtained using the so-called *resolution mechanism* strategy. This strategy consists of the following steps:

1. Rules are substituted by equivalent logical expressions.

2. These logical expressions are then combined into one logical expression.

3. This logical expression is used to obtain the conclusion.

These steps are illustrated using the following examples.

Example 2.5 Resolution mechanism 1. Suppose we have two rules:

- Rule 1: If A is true, then B is true.
- Rule 2: If B is true, then C is true.

32   2. Rule-Based Expert Systems

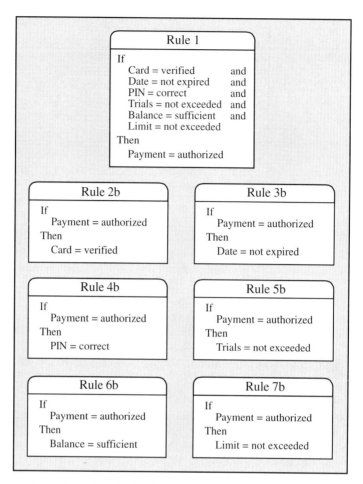

FIGURE 2.7. A set of rules (logically equivalent to the set of rules in Figure 2.1).

The first step in the resolution mechanism strategy is to substitute each of the two rules with an equivalent logical expression. This can be done as follows (refer to Figure 2.8):

- Rule 1 is equivalent to the logical expression: "A is false or B is true." A proof of this equivalence is shown in the truth table in Table 2.6.

- Similarly, Rule 2 is equivalent to the logical expression: "B is false or C is true."

The second step is to combine the above two expressions into one as follows: The logical expressions "A is false or B is true" and "B is false or C is true" imply the expression "A is false or C is true." A proof of

2.3 The Inference Engine   33

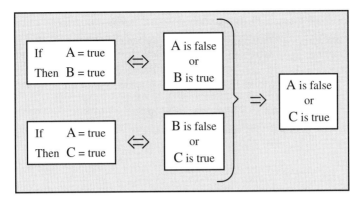

FIGURE 2.8. An example illustrating the resolution mechanism strategy.

A	B	$\bar{A}$	If A, then B	$\bar{A}$ or B
T	T	F	T	T
T	F	F	F	F
F	T	T	T	T
F	F	T	T	T

TABLE 2.6. A truth table showing that the rule "If A is true, then B is true" is equivalent to the logical expression "A is false or B is true."

A	B	C	$\bar{A}$ or B	$\bar{B}$ or C	($\bar{A}$ or B) and ($\bar{B}$ or C)	$\bar{A}$ or C
T	T	T	T	T	T	T
T	T	F	T	F	F	F
T	F	T	F	T	F	T
T	F	F	F	T	F	F
F	T	T	T	T	T	T
F	T	F	T	F	F	T
F	F	T	T	T	T	T
F	F	F	T	T	T	T

TABLE 2.7. A truth table showing that the logical expressions "A is false or B is true" and "B is false or C is true" imply the logical expression "A is false or C is true."

this equivalence is shown in Table 2.7. This last expression is then used in the third step to obtain the conclusion. The above steps are illustrated in Figure 2.8. ■

34   2. Rule-Based Expert Systems

Example 2.6 Resolution mechanism 2. Consider again the ATM example with the added object *Explain*, which can take one of the values {*yes, no*}, indicating whether or not explanation of an action taken by the ATM is needed. Let us now apply the resolution mechanism strategy to the evidence $PIN = incorrect$ and the following two rules:

- If $PIN = incorrect$ then $Payment = not\ authorized$.
- If $Payment = not\ authorized$ then $Explain = yes$.

As illustrated in Figure 2.9, the resolution mechanism strategy leads to the conclusion $Explain = yes$. In effect, following the indicated steps, we have

1. The two rules are substituted by the equivalent expressions:
   - $PIN = correct$ or $Payment = not\ authorized$
   - $Payment$ is $authorized$ or $Explain = yes$

2. The above two expressions are combined in the indicated form to give the expression $PIN = correct$ or $Explain = yes$, and

3. This last expression is combined with the evidence $PIN = incorrect$, and the mixed conclusion, $Explain = yes$, is obtained.   ∎

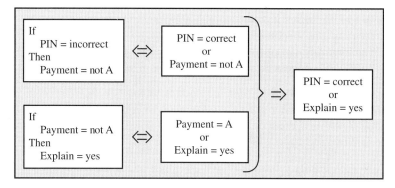

FIGURE 2.9. The resolution mechanism strategy applied to ATM example.

It is important to point out that the resolution mechanism strategy does not always lead to conclusions. We may not know the truthfulness or falsity of certain expressions. If this occurs, the expert system, or more precisely, its inference engine, must make a decision between:

- Abandon the rule because of the impossibility of any conclusion, or

- Ask the user, via the demand information subsystem, about the truthfulness or falsity of one or several expressions in order to be able to resume the inference process until a conclusion is drawn.

2.3.3 Rule Chaining

Another inference strategy that can be used to obtain mixed conclusions is called *rule chaining*. This strategy can be used when the premises of some rules coincide with the conclusions of some other rules. When rules are chained, facts can be used to conclude new facts. This is done repeatedly by executing the rules until no new conclusions can be obtained. The time it takes for the execution process to terminate clearly depends on the known facts and on the order in which the rules are executed. The rule chaining strategy is given in the following algorithm:

Algorithm 2.1 Rule Chaining.

- **Input:** A knowledge base (objects and rules) and some initial facts.
- **Output:** The set of concluded facts.

1. Assign objects to their values as given by the facts or evidence.
2. Execute each rule in the knowledge base and conclude new facts if possible.
3. Repeat Step 2 until no new facts can be obtained. ∎

This algorithm can be implemented in many ways. One way starts with rules whose premises have known values. These rules should conclude and their conclusions become facts. These concluded facts are added to the concrete facts, if any, and the process continues until no new facts can be obtained. This process is illustrated below by two examples.

Example 2.7 Rule chaining 1. Figure 2.10 shows an example of six rules relating 13 objects A through M. The relationships among these objects implied by the six rules can be represented graphically as shown in Figure 2.11, where each object is represented by a node. The arcs represent the connection between the objects in the premise of a rule and the object in its conclusion. Notice that the premises of some rules coincide with the conclusions of some others. For example, the conclusions of Rules 1 and 2 (objects C and G) are the premises of Rule 4.

Suppose now that the objects A, B, D, E, F, H, and I are known to be *true* and the other six objects have unknown values. Figure 2.12 distinguishes between objects with known values (the facts) and objects with unknown values. In this case, the rule chaining algorithm proceeds as follows:

- Rule 1 concludes that $C = true$.
- Rule 2 concludes that $G = true$.
- Rule 3 concludes that $J = true$.

36   2. Rule-Based Expert Systems

- Rule 4 concludes that $K = true$.
- Rule 5 concludes that $L = true$.
- Rule 6 concludes that $M = true$.

Since no more conclusions can be obtained, the process stops. This process is illustrated in Figure 2.12, where the numbers inside the nodes indicate the order in which the facts are concluded. ∎

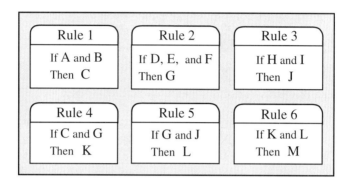

FIGURE 2.10. An example of a set of six rules relating 13 objects.

Example 2.8 Rule chaining 2. Consider again the six rules in Figure 2.10 and suppose now that the facts $H = true$, $I = true$, $K = true$, and $M = false$ are given. This is illustrated in Figure 2.13, where the objects with known values (the facts) are shaded and the goal object is circled. Lets suppose first that the inference engine uses both the Modus Ponens and Modus Tollens strategies. Then, applying Algorithm 2.1, we obtain

1. Rule 3 concludes that $J = true$ (Modus Ponens).
2. Rule 6 concludes (Modus Tollens) that $K = false$ or $L = false$, but since $K = true$ we must have $L = false$.
3. Rule 5 concludes (Modus Tollens) that $G = false$ or $J = false$, but since $J = true$ we must have $G = false$.

Then, the conclusion $G = false$ is obtained. However, if the inference engine only uses the Modus Ponens strategy, the algorithm will stop in Step 1, and no conclusion will be reached for object G. This is another example that illustrates the usefulness of the Modus Tollens strategy. ∎

Note that the rule chaining strategy clearly differentiates between the working memory and the knowledge base. The working memory contains data arising during the consultation period. Premises of rules in the data base are compared with the contents of the working memory and when conclusions are drawn they are also passed to the working memory.

2.3 The Inference Engine   37

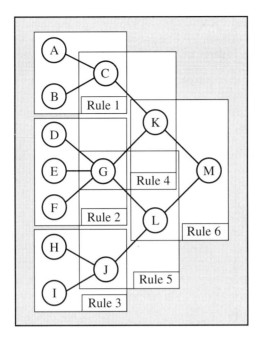

FIGURE 2.11. A graphical representation of the relationships among the six rules in Figure 2.10.

2.3.4 Goal-Oriented Rule Chaining

The goal-oriented rule chaining algorithm requires the user to first select an initial goal object; then the algorithm navigates the rules in search for a conclusion for the goal object. If no conclusion can be reached with the current information, then the algorithm asks the user for new information regarding only those elements that are relevant to obtain a conclusion for the goal object.

Some authors refer to rule chaining and goal-oriented rule chaining algorithms as the *forward chaining* and *backward chaining* algorithms, respectively. But this terminology can be confusing because both of these algorithms can actually use either or both of the Modus Ponens strategy (forward) and Modus Tollens strategy (backward) strategies.

The goal-oriented rule chaining algorithm is described below.

Algorithm 2.2 Goal-Oriented Rule Chaining.

- **Input:** A knowledge base (objects and rules), some initial facts, and a goal object.

- **Output:** The concluded fact for the goal object.

38   2. Rule-Based Expert Systems

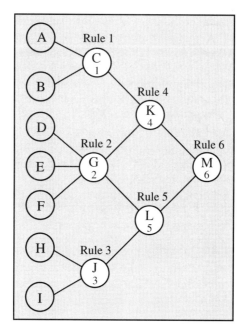

FIGURE 2.12. An example illustrating the rule chaining strategy. The nodes with known values are shaded and the numbers inside the nodes indicate the order in which the facts are concluded.

1. Assign objects to their values as given by the facts, if any. Flag all assigned objects. If the initial goal object is flagged, go to Step 7; otherwise:

    (a) Designate the *initial* goal object as the *current* goal object.
    (b) Flag the current goal object.
    (c) Let $PreviousGoals = \phi$, where ϕ is the empty set.
    (d) Designate all rules as active (executable).
    (e) Go to Step 2.

2. Find an active rule that includes the current goal object but not any of the objects in *PreviousGoals*. If a rule is found, go to Step 3; otherwise, go to Step 5.

3. Execute the rule for the current goal object. If it concludes, assign the concluded value to the current goal object and go to Step 6; otherwise go to Step 4.

4. If all objects in the rule are flagged, declare the rule as *inactive* and go to Step 2; otherwise:

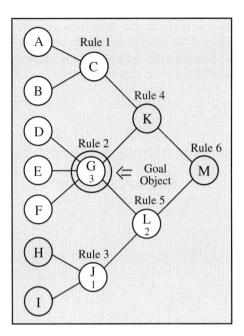

FIGURE 2.13. Another example illustrating the rule chaining algorithm. The nodes with known values are shaded, the goal object is circled, and the numbers inside the nodes indicate the order in which the facts are concluded.

(a) Add the current goal object to *PreviousGoals*.

(b) Designate one of the nonflagged objects in the rule as the current goal object.

(c) Flag the current goal object.

(d) Go to Step 2.

5. If the current goal object is the same as the initial goal object, go to Step 7; otherwise, ask the user for a value for the current goal object. If no value is given, go to Step 6; otherwise assign the object to the given value and go to Step 6.

6. If the current goal object is the same as the initial goal object, go to Step 7; otherwise, designate the previous goal object as the current goal object, eliminate it from *PreviousGoals*, and go to Step 2.

7. Return the value of the goal object if known. ■

The goal-oriented rule chaining algorithm is illustrated below by some examples.

Example 2.9 Goal-oriented rule chaining. Consider the six rules given in Figures 2.10 and 2.11. Suppose that node M is selected as the goal

40   2. Rule-Based Expert Systems

object and that objects D, E, F, and L are known to be true. These nodes are shaded in Figure 2.14. The steps of the goal-oriented rule chaining Algorithm 2.2 are illustrated in Figure 2.14, where the number inside a node indicates the order in which the node is visited. These steps are

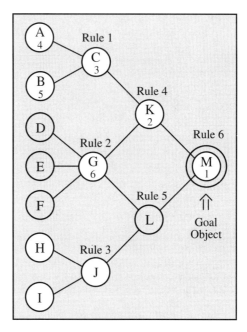

FIGURE 2.14. An example illustrating goal-oriented rule chaining algorithm. The nodes with known values are shaded, the goal object is circled, and the number inside a node indicates the order in which the node is visited.

- Step 1: The objects D, E, F, and L are assigned the value *true* and flagged. Since the goal object M is not flagged, then
  - Object M is designated as the *current* goal object.
  - Object M is flagged. Thus, $FlaggedObjects = \{D, E, F, L, M\}$.
  - $PreviousGoals = \phi$.
  - All six rules are active. Thus, $ActiveRules = \{1, 2, 3, 4, 5, 6\}$.
  - We go to Step 2.
- Step 2. We look for an active rule that includes the current goal object M. Rule 6 is found, so we go to Step 3.
- Step 3. Rule 6 cannot conclude because the value of object K is unknown. So we go to Step 4.
- Step 4. Object K is not flagged. Then

- $PreviousGoals = \{M\}$.
  - Object K is chosen as the current goal object.
  - Object K is flagged. Thus, $FlaggedObjects = \{D, E, F, L, M, K\}$.
  - Go to Step 2.

- Step 2. We look for an active rule that includes the current goal object K but not the previous goal object M. Rule 4 is found, so we go to Step 3.

- Step 3. Rule 4 cannot conclude because the values of objects C and G are unknown. Thus, we go to Step 4.

- Step 4. Objects C and G are not flagged. Then
  - $PreviousGoals = \{M, K\}$.
  - One of the nonflagged objects C and G is chosen as the current goal object. Suppose C is chosen.
  - Object C is flagged.
    Thus, $FlaggedObjects = \{D, E, F, L, M, K, C\}$.
  - Go to Step 2.

- Step 2. We look for an active rule that includes the current goal object C but not the previous goal objects $\{M, K\}$. Rule 1 is found, so we go to Step 3.

- Step 3. Rule 1 cannot conclude because the values of objects A and B are unknown. So we go to Step 4.

- Step 4. Objects A and B are not flagged. Then
  - $PreviousGoals = \{M, K, C\}$.
  - One of the nonflagged objects A and B is chosen as the current goal object. Suppose A is chosen.
  - Object A is flagged.
    Thus, $FlaggedObjects = \{D, E, F, L, M, K, C, A\}$.
  - Go to Step 2.

- Step 2. We look for an active rule that includes the current goal object A but not the previous goal objects $\{M, K, C\}$. No rule is found, so we go to Step 5.

- Step 5. Since the current goal object A is different from the initial goal object M, we ask the user for a value for the current goal object A. Suppose that A is given to be true, then we let $A = true$ and go to Step 6.

42   2. Rule-Based Expert Systems

- Step 6. The current goal object A is not the same as the initial goal object M. Then the previous goal object C is now designated as the current goal object and is eliminated from $PreviousGoals$. Thus $PreviousGoals = \{M, K\}$ and we now go to Step 2.

- Step 2. We look for an active rule that includes the current goal object C but not the previous goal objects $\{M, K\}$. Rule 1 is found, so we go to Step 3.

- Step 3. Rule 1 cannot conclude because object B is unknown. So, we go to Step 4.

- Step 4. Objects B is not flagged. Then
  - $PreviousGoals = \{M, K, C\}$.
  - The only nonflagged object B is chosen as the current goal object.
  - Object B is flagged.
    Thus, $FlaggedObjects = \{D, E, F, L, M, K, C, A, B\}$.
  - Go to Step 2.

- Step 2. We look for an active rule that includes the current goal object B but not the previous goal objects $\{M, K, C\}$. No rule is found, so we go to Step 5.

- Step 5. Since the current goal object B is different from the initial goal object M, we ask the user for a value for the current goal object B. Suppose that B is given to be true, then we let $B = true$ and go to Step 6.

- Step 6. The current goal object B is not the same as the initial goal object M. Then the previous goal object C is now designated as the current goal object and is eliminated from $PreviousGoals$. Thus $PreviousGoals = \{M, K\}$ and we now go to Step 2.

- Step 2. We look for an active rule that includes the current goal object C but not the previous goal objects $\{M, K\}$. Rule 1 is found, so we go to Step 3.

- Step 3. Since $A = true$ and $B = true$, then $C = true$ by Rule 1. We now go to Step 6.

- Step 6. The current goal object C is not the same as the initial goal object M. Then the previous goal object K is now designated as the current goal object and is eliminated from $PreviousGoals$. Thus $PreviousGoals = \{M\}$ and we now go to Step 2.

- Step 2. We look for an active rule that includes the current goal object K but not the previous goal objects $\{M\}$. Rule 4 is found, so we go to Step 3.

- Step 3. Rule 4 cannot conclude because object G is unknown. So we go to Step 4.

- Step 4. Object G is not flagged. Then
    - $PreviousGoals = \{M, K\}$.
    - The only nonflagged object G is chosen as the current goal object.
    - Object G is flagged.
      Thus, $FlaggedObjects = \{D, E, F, L, M, K, C, A, B, G\}$.
    - Go to Step 2.

- Step 2. We look for an active rule that includes the current goal object G but not the previous goal objects $\{M, K\}$. Rule 2 is found, so we go to Step 3.

- Step 3. Since $D = true$, $E = true$, and $F = true$, then $G = true$ by Rule 2. We now go to Step 6.

- Step 6. The current goal object G is not the same as the initial goal object M. Then the previous goal object K is now designated as the current goal object and is eliminated from $PreviousGoals$. Thus $PreviousGoals = \{M\}$ and we now go to Step 2.

- Step 2. We look for an active rule that includes the current goal object K but not the previous goal objects $\{M\}$. Rule 4 is found, so we go to Step 3.

- Step 3. Since $C = true$ and $G = true$, then $K = true$ by Rule 4. We now go to Step 6.

- Step 6. The current goal object K is not the same as the initial goal object M. Then, the previous goal object M is now designated as the current goal object and is eliminated from $PreviousGoals$. Thus $PreviousGoals = \phi$ and we now go to Step 2.

- Step 2. We look for an active rule that includes the current goal object M. Rule 6 is found, so we go to Step 3.

- Step 3. Since $K = true$ and $L = true$, then $M = true$ by Rule 6. We now go to Step 6.

44   2. Rule-Based Expert Systems

- Step 6. The current goal object M is the same as the initial goal object. Then we go to Step 7.
- Step 7. The algorithm returns the value $M = true$. ∎

Note that although objects H, I, and J have unknown values, the goal-oriented rule chaining algorithm was still able to conclude a value for the goal object M. The reason for this is that knowing object L renders the knowledge of objects H, I, and J to be irrelevant to the knowledge of the goal object M. ∎

A rule chaining strategy is utilized in data-oriented problems in which some facts (e.g., symptoms) are known and some conclusions (e.g., diseases) are looked for. On the other hand, a goal-oriented rule chaining strategy is utilized in goal-oriented problems in which some goals (diseases) are given and the necessary facts (symptoms) for them to be possible are desired.

Example 2.10 Goal-oriented rule chaining without Modus Tollens. Consider the six rules given in Figures 2.10 and 2.11. Suppose now that node J is selected as the goal object and that the following facts are given: $G = true$ and $L = false$. This is illustrated in Figure 2.15, where the objects with known values (the facts) are shaded and the goal object is circled. First suppose that the inference engine uses only the Modus Ponens strategy. In this case, the steps of Algorithm 2.2 are as follows:

- Step 1: The objects G and L are flagged and assigned the values $G = true$ and $L = false$. Since the goal object J is not flagged, then
    - Object J is designated as the *current* goal object.
    - Object J is flagged. Thus, $FlaggedObjects = \{G, L, J\}$.
    - $PreviousGoals = \phi$.
    - All six rules are active. Thus, $ActiveRules = \{1, 2, 3, 4, 5, 6\}$.
    - We go to Step 2.

- Step 2. We look for an active rule that includes the current goal object J. Since we are using only the Modus Ponens strategy, Rule 3 is found (it is the only rule in which the goal object J is the conclusion). So we go to Step 3.

- Step 3. Rule 3 cannot conclude because the values of objects H and I are unknown. So, we go to Step 4.

- Step 4. Objects H and I are not flagged. Then
    - $PreviousGoals = \{J\}$.
    - One of the nonflagged objects H and I is chosen as the current goal object. Suppose H is chosen.

2.3 The Inference Engine 45

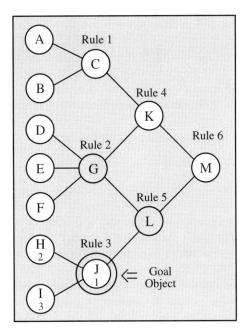

FIGURE 2.15. An example illustrating the goal-oriented rule chaining algorithm without the Modus Tollens strategy. The nodes with known values are shaded, the goal object is circled, and the number inside a node indicates the order in which the node is visited.

- Object H is flagged. Thus, $FlaggedObjects = \{G, L, J, H\}$.
- Go to Step 2.

- Step 2. We look for an active rule that includes the current goal object H but not the previous goal object J. No rule is found, so we go to Step 5.

- Step 5. Since the current goal object H is different from the initial goal object J, we ask the user for a value for the current goal object H. Suppose that no value is given for object H, so we go to Step 6.

- Step 6. The current goal object H is not the same as the initial goal object J. Then, the previous goal object J is now designated as the current goal object and is eliminated from $PreviousGoals$. Thus $PreviousGoals = \phi$ and we now go to Step 2.

- Step 2. We look for an active rule that includes the current goal object J. Rule 3 is found, so we go to Step 3.

- Step 3. Rule 3 cannot conclude because the values of objects H and I are unknown. So we go to Step 4.

- Step 4. Object I is not flagged. Then
    - $PreviousGoals = \{J\}$.
    - The only nonflagged object I is chosen as the current goal object.
    - Object I is flagged. Thus, $FlaggedObjects = \{G, L, J, H, I\}$.
    - Go to Step 2.

- Step 2. We look for an active rule that includes the current goal object I but not the previous goal object J. No rule is found, so we go to Step 5.

- Step 5. Since the current goal object I is different from the initial goal object J, we ask the user for a value for the current goal object I. Suppose that no value is given for object I, so we go to Step 6.

- Step 6. The current goal object I is not the same as the initial goal object J. Then, the previous goal object J is now designated as the current goal object and is eliminated from $PreviousGoals$. Thus $PreviousGoals = \phi$ and we now go to Step 2.

- Step 2. We look for an active rule that includes the current goal object J. Rule 3 is found, so we go to Step 3.

- Step 3. Rule 3 cannot conclude because the values of objects H and I are unknown. So we go to Step 4.

- Step 4. All objects in Rule 3 are flagged, then Rule 3 is declared inactive. Thus, $ActiveRules = \{1, 2, 4, 5, 6\}$. We go to Step 2.

- Step 2. We look for an active rule that includes the current goal object J. Since Rule 3 is declared inactive and we are not using the Modus Tollens strategy, no rule is found and we go to Step 5.

- Step 5. Since the current goal object J is the same as the initial goal object, we go to Step 7.

- Step 7. The inference engine cannot conclude a value for the goal object J. ■

Example 2.11 Goal-oriented rule chaining with Modus Tollens. In Example 2.10, the inference engine uses only the Modus Ponens strategy. Now let us consider the same setting as in Example 2.10 except that the inference engine uses both the Modus Ponens and Modus Tollens strategies. The steps of Algorithm 2.2 in this case are as follows:

- Step 1: The objects G and L are flagged and assigned the values $G = true$ and $L = false$. Since the goal object J is not flagged, then
    - Object J is designated as the *current* goal object.
    - Object J is flagged. Thus, $FlaggedObjects = \{G, L, J\}$.
    - $PreviousGoals = \phi$.
    - All six rules are active. Thus, $ActiveRules = \{1, 2, 3, 4, 5, 6\}$.
    - We go to Step 2.

- Step 2. We look for an active rule that includes the current goal object J. Since we are using both strategies, Rules 3 and 5 include the object J. Suppose that Rule 5 is chosen. We now go to Step 3. (If Rule 3 is chosen, the algorithm will take a larger number of steps to terminate.)

- Step 3. Since we are using the Modus Tollens strategy, Rule 5 concludes that $J = false$ (because $G = true$ and $L = false$). So we go to Step 6.

- Step 6. The current goal object J is the same as the initial goal object, so we go to Step 7.

- Step 7. Return $J = false$. ■

2.3.5 Compiling Rules

Another way of dealing with chained rules is to start with a set of data (information) and try to reach some goal objects. This is called *compiling rules*. When both data and goals are previously determined, rules can be compiled, that is, we write the goal objects as a function of data and obtain the so-called *goal equations*. Compiling rules is best explained by an example.

Example 2.12 Compiling rules. Consider the set of six rules in Figure 2.11 and suppose that values for the objects A, B, D, E, F, H, and I are known and the remaining objects, C, G, J, K, L, and M, are the goal objects. Let the symbol $\wedge$ denote the logical operator *and*; then using the six rules one can derive the following goal equations:

- Rule 1 implies $C = A \wedge B$.
- Rule 2 implies $G = D \wedge E \wedge F$.
- Rule 3 implies $J = H \wedge I$.
- Rule 4 implies $K = C \wedge G = (A \wedge B) \wedge (D \wedge E \wedge F)$.

- Rule 5 implies $L = G \wedge J = (D \wedge E \wedge F) \wedge (H \wedge I)$.
- Rule 6 implies $M = K \wedge L = A \wedge B \wedge D \wedge E \wedge F \wedge H \wedge I$.

The first three of these equations are equivalent to the first three rules. The last three goal equations are equivalent to the following rules, respectively:

- Rule 4a: *If A and B and D and E and F, then G.*
- Rule 5a: *If D and E and F and H and I, then J.*
- Rule 6a: *If A and B and D and E and F and H and I, then L.*

Thus, for example, if each of the the objects $\{A, B, D, E, F, H, I\}$ is known to be true, it is immediate from Rules 4a, 5a, and 6a that the objects $\{G, J, L\}$ must be true. ∎

2.4 Coherence Control

In complex situations, even expert people can give inconsistent statements (e.g., inconsistent rules and/or infeasible combinations of facts). It is therefore important to control the coherence of the knowledge during the construction of the knowledge base and during the processes of data acquisition and reasoning. If the knowledge base contains inconsistent information (e.g., rules and/or facts), the expert system is likely to perform unsatisfactorily, give absurd conclusions, and/or not perform at all.

The purpose of the coherence control subsystem is to

1. Help the user not to give inconsistent facts, for example, by supplying the user with the constraints that the information provided by the user must satisfy.

2. Prevent inconsistent or contradictory knowledge (both abstract and concrete) from creeping into the knowledge base.

Coherence control can be achieved by controlling the coherence of rules and the coherence of the facts.

2.4.1 Coherence of Rules

Definition 2.2 Coherent rules. *A set of rules is said to be coherent if there exists at least one set of values for all associated objects that produce noncontradictory conclusions.*

Accordingly, a coherent set of rules does not have to produce noncontradictory conclusions for all possible sets of values of the associated objects. That is, it is sufficient to have one set of values that produce nonconflicting conclusions.

2.4 Coherence Control

Objects		Conclusions		Contradictory
A	B	Rule 1	Rule 2	Conclusions?
T	T	$B = T$	$B = F$	Yes
T	F	$B = T$	$B = F$	Yes
F	T	—	—	No
F	F	—	—	No

TABLE 2.8. A truth table showing that Rules 1 and 2 are coherent.

Objects		Conclusions			Contradictory
A	B	Rule 1	Rule 2	Rule 3	Conclusions?
T	T	$B = T$	$B = F$	—	Yes
T	F	$B = T$	$B = F$	—	Yes
F	T	—	—	$B = T$	No
F	F	—	—	$B = T$	Yes

TABLE 2.9. A truth table showing that Rules 1–3 are coherent.

Example 2.13 Incoherent set of rules. Consider the following four rules that relate two binary $\{T, F\}$ objects A and B:

- Rule 1: If $A = T$, then $B = T$.
- Rule 2: If $A = T$, then $B = F$.
- Rule 3: If $A = F$, then $B = T$.
- Rule 4: If $A = F$, then $B = F$.

Then, one can reach the following conclusions:

1. Rules 1–2 are coherent because as shown in Table 2.8, for $A = F$, they produce no (contradictory) conclusions.

2. Rules 1–3 are coherent because for $A = F$ and $B = T$, they produce one conclusion ($B = T$), as shown in Table 2.9.

3. Rules 1–4 are incoherent because they produce contradictory conclusions for all possible values of A and B, as can be seen from Table 2.10. ∎

Note that a set of rules may be coherent, yet some sets of object-values may produce inconsistent conclusions. These sets are called *infeasible values*. For example, Rules 1–2 are coherent, yet they produce inconsistent conclusions in all cases where $A = T$. Then the coherence control subsystem should

50   2. Rule-Based Expert Systems

Objects		Conclusions				Contradictory
A	B	Rule 1	Rule 2	Rule 3	Rule 4	Conclusions?
T	T	$B = T$	$B = F$	–	–	Yes
T	F	$B = T$	$B = F$	–	–	Yes
F	T	–	–	$B = T$	$B = F$	Yes
F	F	–	–	$B = T$	$B = F$	Yes

TABLE 2.10. A truth table showing that Rules 1–4 are incoherent.

automatically eliminate the value T from the list of values of object A, thus permitting the user to select only feasible values for the objects.

Definition 2.3 Infeasible value. *A value a for object A is said to be infeasible if the conclusions reached by setting $A = a$ contradict any combination of values for the rest of the nodes.*

Thus, any infeasible value must be eliminated from the lists of object-values in order to prevent the inference engine from reaching inconsistent conclusions.

Example 2.14 Infeasible values. Consider the set of four rules in Example 2.13. Then the inference engine will conclude the following:

1. The first two rules imply that $A \neq T$, because $A = T$ always leads to inconsistent conclusions. Therefore, the value $A = T$ should be automatically eliminated from the list of feasible values for A. Since A is binary, then $A = F$ (the only possible value).

2. The first three rules imply that $A = F$ and $B = T$. Therefore, the value $B = F$ should also be automatically eliminated from the list of feasible values for B.

3. The first four rules imply that $A \neq T$, $A \neq F$, $B \neq T$, and $B \neq F$. Therefore, the values $\{T, F\}$ are eliminated from the list of values of A and B, leaving all object-value lists empty, which implies that the four rules are incoherent. ■

Note that checking the coherence of rules is done only once after each rule is introduced, and all infeasible values should be automatically eliminated from the lists of object-values once they are detected.

The set of rules that forms the abstract knowledge must be coherent, otherwise the system may reach erroneous conclusions. Thus before adding any rule to the knowledge base, the rule must be checked for consistency with the existing rules in the knowledge base. If the rule was found to be consistent with existing rules, it is added to the knowledge base; otherwise, it is returned to the human experts for corrective action.

Example 2.15 Coherence of rules. Suppose we have four objects: $A \in \{0, 1\}$, $B \in \{0, 1\}$, $C \in \{0, 1, 2\}$ and $D \in \{0, 1\}$. Consider the following rules:

- Rule 1: If $A = 0$ and $B = 0$, then $C = 0$.
- Rule 2: If $A = 0$ and $D = 0$, then $C = 1$.
- Rule 3: If $A = 0$ and $B = 0$, then $C = 1$.
- Rule 4: If $A = 0$, then $B = 0$.
- Rule 5: If $B = 0$, then $A = 1$.

Suppose that we wish to add the last three rules to a knowledge base that currently contains the first two rules. Then Rules 1 and 3 are inconsistent because they have the same premise, yet they have different conclusions. Hence, Rule 3 must be rejected and the human expert should be informed of the reason for the rejection. The human expert will then correct the rule in question and/or correct existing rules if they happen to be incorrect. Rule 4 will enter in the knowledge base because it is consistent with Rules 1 and 2. Rule 5 is inconsistent with Rule 4. Thus both rules must first be checked for consistency before being entered into the knowledge base. ∎

2.4.2 Coherence of Facts

The set of concrete knowledge (the data or evidences that are supplied by the users) must also be consistent with itself and with the set of rules in the knowledge base. Thus, the system must not accept facts that contradict the set of rules and/or the current set of facts. For example, with a knowledge base that consists of the first two rules in Example 2.15, the system must not accept the set of facts $A = 0$, $B = 0$, and $C = 1$ because they contradict Rule 1.

The system must also check whether or not a feasible solution exists and inform the user accordingly. If in the above example we try to give the information $A = 0$, $B = 0$, and $D = 0$, the system must detect that there is no value of C that is consistent with the knowledge base. Note that before knowing the values of the objects, a feasible solution exists. For example, $A = 0$, $B = 0$, $C = 0$, $D = 1$ (these facts do not contradict the knowledge base). Thus, inconsistency comes from the fact that rules and facts are inconsistent.

Coherence of facts can be achieved by the following strategies:

1. Eliminate all infeasible values (those that contradict the set of rules and/or the current set of facts) of objects once they are detected. When asking the user for information about values for a set of objects, the expert system should accept only the values of each object

that are consistent with the rules and with the previous knowledge. Consider, for example, the knowledge base in Example 2.15 and suppose that the expert system has been given the information $A = 0$ and $C = 1$; then the system must know that $B \ne 0$. Thus, this value should be eliminated from the list of feasible values for object B.

2. The inference engine must check that the known facts do not contradict the set of rules. In the above situation, for example, the system must not accept the set of facts $A = 1$, $B = 1$, and $C = 2$. If the system does not eliminate infeasible values, then the user can give contradictory evidences such as $Payment = authorized$ and $PIN = incorrect$ in the ATM Example 2.1. Thus, once the first evidence, $Payment = authorized$, is given, the system must select only the values for PIN that do not contradict rule conclusions.

3. Provide the user with a list of only objects that have not been previously assigned values.

4. For each of the objects, display and accept only the feasible values of the object.

5. Continually update the knowledge base whenever a fact is given or a conclusion is reached and eliminate values that contradict the new facts. The inference engine derives all conclusions by examining, and possibly concluding, the rules as soon as a single item of information is given to the system. Note that giving several items of information simultaneously can lead to inconsistency in the database. For example, given $A = 0$, we cannot give the combined information $B = 0$ and $C = 1$. In this case the order of the information affects the possible future inputs of the system that are compatible, that is, after giving $A = 0$ we can give either $B = 0$ or $C = 1$, but these two options put different constraints on the future inputs to the system.

   Continual and automatic updating of knowledge is important because failing to perform the updating implies the possibility for contradictory evidences to coexist in the knowledge base. Thus, as soon as an item of knowledge is incorporated into the database, knowledge should immediately be updated.

Thus, automatic elimination of infeasible values and continual updating of knowledge ensure the coherence of the knowledge base.

2.5 Explaining Conclusions

As indicated in Chapter 1, conclusions alone are not enough to satisfy the users of expert systems. Normally, users expect the system to give some kind of explanation indicating why the conclusions are drawn. During the process of rule execution, active rules form the basis of the explanation mechanism, which is regulated by the explanation subsystem.

In rule-based expert systems, it is easy to provide explanations of the drawn conclusions. The inference engine draws its conclusions based on a set of rules and it knows which rule gives rise to which conclusions. Thus, the system may provide the user with the list of concluded facts together with the rules that have been used to reach each conclusion.

Example 2.16 Explaining conclusions. Consider the six rules given in Figures 2.10 and 2.11. As in Example 2.7, suppose that the objects A, B, D, E, F, H, and I are known to be *true* and the other six objects have unknown values. Then, applying Algorithm 2.1 and examining the rules that have been executed, the expert system may provide the following explanation of the conclusions reached:

1. Facts given:

$$A = true, \quad B = true, \quad D = true, \quad E = true,$$
$$F = true, \quad H = true, \quad I = true.$$

2. Conclusions reached and explanations:

    - $C = true$, based on Rule 1.
    - $G = true$, based on Rule 2.
    - $J = true$, based on Rule 3.
    - $K = true$, based on Rule 4.
    - $L = true$, based on Rule 5.
    - $M = true$, based on Rule 6. ■

2.6 Some Applications

In this section we develop rule-based expert systems to answer some of the motivating examples given in Section 1.3.

2.6.1 Train Traffic Control

Real-life traffic control systems are necessarily complex. Here we give a somewhat esoteric example just to illustrate how to design a rule-based expert system to solve a very simple traffic control problem. Figure 2.16 shows a section of a railroad where several trains can circulate in both directions. There are five stations, $S_1, \ldots, S_5$, and 14 traffic lights, eight

54   2. Rule-Based Expert Systems

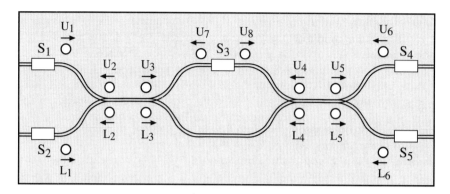

FIGURE 2.16. A section of a railroad with five stations.

Object	Value
U_1 to U_8	$\{green, red\}$
L_1 to L_6	$\{green, red\}$
S_1 to S_5	$\{free, occupied\}$

TABLE 2.11. Objects and their possible values for the train traffic control example.

in the upper part of the diagram, $U_1, \ldots, U_8$, and six in the lower part, $L_1, \ldots, L_6$. All objects and their possible values are shown in Table 2.11.

The objective is to design a set of rules needed to avoid collision. These rules are derived as follows:

1. If the traffic light U_1 is $green$, then a train in station S_1 can leave the station and no train can leave station S_2, so L_1 has to be red. The same is true for stations S_4 and S_5. This gives the first two rules in Table 2.12. Note that if the inference engine uses the Modus Tollens strategy, these rules also guarantee that when the light in one of the stations in the lower part is green, the light in the corresponding station in the upper part will be red. That is, the first two rules in Table 2.12 imply the following two rules:

    - Rule 1a: If $L_1 = green$, then $U_1 = red$.
    - Rule 2a: If $L_6 = green$, then $U_6 = red$.

2. If station S_1 is occupied, then the light U_2 must be red to prevent a train from entering the occupied station. Similarly for the other four stations. This yields the six additional rules (Rules 3–8) in Table 2.12.

3. If both U_3 and L_3 lights are red, then no train can leave station S_1. The same condition holds for lights U_5 and L_5. Thus, we have the rule

- Rule 9: If (U_3 = red and L_3 = red) or (U_5 = red or L_5 = red), then U_1 = red.

The five rules associated with the other five stations can be derived in a similar way. All six rules are shown in Table 2.12 as Rules 9–14.

4. To avoid a collision of trains coming from stations $S_1 - S_2$ and $S_4 - S_5$, the following rules are necessary:

- Rule 15: If U_3 = green, then U_4 = red
- Rule 16: If L_3 = green, then L_4 = red.

5. To prevent each of the lights in the upper part and the corresponding light in the lower part from being simultaneously green, it is necessary to include Rules 17–20 in Table 2.12.

6. Finally, to avoid a collision between a train in station S_3 with a train in the other four stations, we impose the last two rules in Table 2.12.

In order to maintain coherence of facts, it is necessary to automatically update the knowledge as soon as a new fact is given or concluded.

Let us now consider an example to illustrate the performance of the expert system whose knowledge base consists of the objects given in Table 2.11 and the set of rules given in Table 2.12.

Example 2.17 Train traffic control. In this example, we shall use the *X-pert Rules* shell. We first need to write a text file containing a description of the above knowledge base. This file is then read by *X-pert Rules*. Since *X-pert Rules* does not allow the logical operator *or* in the premise of the rules, we first need to replace Rules 9–12 by the following equivalent set of rules (see rule substitution in Table 2.4):

- Rule 9a: If (U_3 = red and L_3 = red), then U_1 = red.
- Rule 9b: If (U_5 = red and L_5 = red), then U_1 = red.
- Rule 10a: If (U_3 = red and L_3 = red), then L_1 = red.
- Rule 10b: If (U_5 = red and L_5 = red), then L_1 = red.
- Rule 11a: If (U_2 = red and L_2 = red), then U_6 = red.
- Rule 11b: If (U_4 = red and L_4 = red), then U_6 = red.

Rule	Premise	Conclusion
Rule 1	$U_1 = green$	$L_1 = red$
Rule 2	$U_6 = green$	$L_6 = red$
Rule 3	$S_1 = occupied$	$U_2 = red$
Rule 4	$S_2 = occupied$	$L_2 = red$
Rule 5	$S_3 = occupied$	$U_3 = red$
Rule 6	$S_3 = occupied$	$U_4 = red$
Rule 7	$S_4 = occupied$	$U_5 = red$
Rule 8	$S_5 = occupied$	$L_5 = red$
Rule 9	$(U_3 = red$ and $L_3 = red)$ or $(U_5 = red$ and $L_5 = red)$	$U_1 = red$
Rule 10	$(U_3 = red$ and $L_3 = red)$ or $(U_5 = red$ and $L_5 = red)$	$L_1 = red$
Rule 11	$(U_2 = red$ and $L_2 = red)$ or $(U_4 = red$ and $L_4 = red)$	$U_6 = red$
Rule 12	$(U_2 = red$ and $L_2 = red)$ or $(U_4 = red$ and $L_4 = red)$	$L_6 = red$
Rule 13	$U_2 = red$ and $L_2 = red$	$U_7 = red$
Rule 14	$U_5 = red$ and $L_5 = red$	$U_8 = red$
Rule 15	$U_3 = green$	$U_4 = red$
Rule 16	$L_3 = green$	$L_4 = red$
Rule 17	$U_2 = green$	$L_2 = red$
Rule 18	$U_3 = green$	$L_3 = red$
Rule 19	$U_4 = green$	$L_4 = red$
Rule 20	$U_5 = green$	$L_5 = red$
Rule 21	$U_1 = green$ or $L_1 = green$	$U_7 = red$
Rule 22	$U_6 = green$ or $L_6 = green$	$U_8 = red$

TABLE 2.12. Rules for the train traffic control example.

- Rule 12a: If $(U_2 = red$ and $L_2 = red)$, then $L_6 = red$.

- Rule 12b: If $(U_4 = red$ and $L_4 = red)$, then $L_6 = red$.

Thus, we create a text file "TrafficControl.txt," which contains all objects and the 26 rules.[1] Suppose that initially we have trains waiting in stations S_1, S_2, and S_3 as indicated in Figure 2.17. The following is an interactive session using *X-pert Rules* after reading the text file "TrafficCon-

---

[1]The knowledge base file "TrafficControl.txt" and the rule-based expert system shell *X-pert Rules* can be obtained from the World Wide Web site http://ccaix3.unican.es/~AIGroup.

trol.txt". We first specify the given facts: $S_1 = occupied$, $S_3 = occupied$, and $S_5 = occupied$. Then we get the following (given) facts and conclusions (concluded facts):

1. Facts:

    - $S_1 = occupied$.
    - $S_3 = occupied$.
    - $S_5 = occupied$.

2. Conclusions:

    - $U_2 = red$ (based on Rule 3).
    - $U_3 = red$ (based on Rule 5).
    - $U_4 = red$ (based on Rule 6).
    - $L_5 = red$ (based on Rule 8).

Thus, four lights are set to red to avoid collisions with the trains waiting in the stations. All other objects have unknown values. Figure 2.17 displays this information graphically.

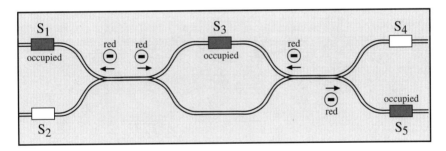

FIGURE 2.17. New conclusions resulting from the facts $S_1 = S_3 = S_5 = occupied$.

Now, suppose we wish to allow the train in S_1 to leave this section of the railroad in the east direction. Thus, we let $U_1 = green$. Then we have the following updated facts and conclusions:

1. Facts:

    - $S_1 = occupied$.
    - $S_3 = occupied$.
    - $S_5 = occupied$.
    - $U_1 = green$.

58   2. Rule-Based Expert Systems

2. Conclusions:
- $U_2 = red$ (based on Rule 3).
- $U_3 = red$ (based on Rule 5).
- $U_4 = red$ (based on Rule 6).
- $L_5 = red$ (based on Rule 8).
- $L_1 = red$ (based on Rule 1).
- $U_7 = red$ (based on Rule 21).
- $L_3 = green$ (based on Rule 9a).
- $U_5 = green$ (based on Rule 9b).
- $L_4 = red$ (based on Rule 16).
- $S_4 \neq occupied$ (based on rule Rule 7).
- $S_4 = free$ (it is the only possible value).
- $U_6 = red$ (based on rule Rule 11b).
- $L_6 = red$ (based on rule Rule 12b).

Figure 2.18 displays the resulting conclusions. Note that the train currently in station S_1 can now leave and reach station S_4. This path is indicated in Figure 2.18. ∎

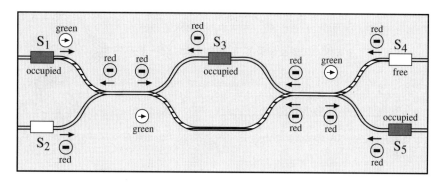

FIGURE 2.18. New conclusions resulting from the facts $S_1 = S_3 = S_5 = occupied$ and $U_1 = green$.

2.6.2 Secret Agents

We now return to the secret agents' problem introduced in Example 1.5, where each of the four secret agents, Albert, Kathy, Nancy, and Tom, is currently in one of four countries, Egypt, France, Japan, and Spain. The following fax messages were received from the agents:

- From France: Kathy is in Spain.

- From Spain: Albert is in France.

- From Egypt: Nancy is in Egypt.

- From Japan: Nancy is in France.

We do not know who sent which message, but we know that Tom lies (a double agent?) and the other three agents tell the truth. The mystery we try to solve is, Who is in which country?

We design an expert system to solve this problem. We have four objects: Albert, Kathy, Nancy, Tom. Each object can take one of four values: Egypt, France, Japan, and Spain. Since Tom is the only one who lies, a fax sent by him is considered to be false. This creates two rules for each message:

1. The message from France (Kathy is in Spain):

    - Rule 1: If Tom is in France, then Kathy is not in Spain.
    - Rule 2: If Tom is not in France, then Kathy is in Spain.

2. The message from Spain (Albert is in France):

    - Rule 3: If Tom is in Spain, then Albert is not in France.
    - Rule 4: If Tom is not in Spain, then Albert is in France.

3. The message from Egypt (Nancy is in Egypt):

    - Rule 5: If Tom is in Egypt, then Nancy is not in Egypt.
    - Rule 6: If Tom is not in Egypt, then Nancy is in Egypt.

4. The message from Japan (Nancy is in France):

    - Rule 7: If Tom is in Japan, then Nancy is not in France.
    - Rule 8: If Tom is not in Japan, then Nancy is in France.

Using only these eight rules, let us now try some possible values for Tom's country:

1. Tom is possibly in Egypt. If Tom is assigned the value Egypt, we obtain the following conclusions:

    - Kathy is in Spain by Rule 2.
    - Albert is in France by Rule 4.
    - Nancy is not in Egypt by Rule 5.
    - Nancy is in France by Rule 8.

We see that Albert and Nancy are both in France, contradicting the information that only one agent can be in each country (but the set of eight rules does not contain this information). We therefore conclude that Egypt is an infeasible value for the object Tom, that is, Tom cannot be in Egypt.

2. Tom is possibly in Japan. If Tom is assigned the value Japan, we obtain the following conclusions:

   - Kathy is in Spain by Rule 2.
   - Albert is in France by Rule 4.
   - Nancy is in Egypt by Rule 6.

   There is no contradiction here, which means that Japan is a feasible value for the object Tom.

With the above eight rules, the inference engine cannot conclude in which country each agent is, because the rules do not contain the information "only one agent can be in each country." We now consider this fact and derive an extra set of rules that accounts for this information.

Since each country can be occupied by exactly one agent, suppose that a given agent is in a given country. Then we need three rules to guarantee that none of the other three agents is in this country. Since we have four agents, then for each country we have a total of 12 rules (3 rules × 4 agents). However, using the Modus Tollens strategy, only six rules are needed for any country because the other six rules are redundant. For example, for the country Egypt we have the rules:

- Rule 9: If Albert is in Egypt, then Kathy is not in Egypt.
- Rule 10: If Albert is in Egypt, then Nancy is not in Egypt.
- Rule 11: If Albert is in Egypt, then Tom is not in Egypt.
- Rule 12: If Kathy is in Egypt, then Nancy is not in Egypt.
- Rule 13: If Kathy is in Egypt, then Tom is not in Egypt.
- Rule 14: If Nancy is in Egypt, then Tom is not in Egypt.

Note that there is a redundant set of six rules to the above set of six rules. For example, the rule:

- Rule 14a: If Tom is in Egypt, then Nancy is not in Egypt,

is equivalent (Modus Tollens) to Rule 14. Therefore, we need only six rules for each country.

The sets of six rules for each of the other three countries are generated in a similar way. We therefore have a total of 24 additional rules representing

2.6 Some Applications   61

the fact that exactly one agent can be in each country. Table 2.13 shows all of the 32 rules needed to solve the problem. An examination of the rules in Table 2.13, however, shows that Rules 5 and 14 are redundant (by Modus Tollens), hence one of them can be omitted.

Now, using the 32 rules, the inference engine will be able to determine which agent is in which country without any additional information. Upon the execution of the rules, the following conclusions will emerge:

- Albert is in France.

- Kathy is in Spain.

- Nancy is in Egypt.

- Tom is in Japan.

Note that the inference uses the Modus Tollens strategy and automatically eliminates all infeasible values (as described in Section 2.4.2) to reach the above conclusions. The above solution is the only feasible combination of values for the objects. For example, Figure 2.19 shows the conclusions reached by the *X-pert Rules* computer program with the knowledge base given in Table 2.13.

2.6.3 The Prisoners' Problem

We now return to the prisoners' problem introduced in Example 1.6. Let X_i denote the name of the ith inmate and c_i denote his eye color. Thus, c_i can be either *black* or *not black*. Let d_{ij} be defined as follows:

$$d_{ij} = \begin{cases} 1, & \text{if the } i\text{th inmate leaves on the } j\text{th day,} \\ 0, & \text{if the } i\text{th inmate does not leave on the } j\text{th day,} \end{cases}$$

Here we consider the simple case where there are only $n = 2$ inmates. The case where $n = 3$ is left as an exercise for the reader. The four possible combinations of eye colors are shown in the second and third columns of Table 2.14, where $c_i = 1$ means black and $c_i = 0$ means not black. The remaining columns show the possible values of d_{ij} for each case, for $i = 1, 2$ and $j = 1, 2, 3$. Note that both inmates have already left the prison by the third day.

Let us now analyze each of the four possible combinations of eye colors:

- The first case: $c_1 = c_2 = 0$, is clearly impossible because the King has indicated that at least one inmate has a black eye.

- The second case: $c_1 = 0$ and $c_2 = 1$. The first inmate, X_1, finds that the second, X_2, has black eyes and hence he cannot determine his eye color at the end of the first day. Therefore $d_{11} = 0$. The second inmate, X_2, observes that X_1 does not have black eyes. Therefore,

Rule	Premise	Conclusion
Rule 1	Tom is in France	Kathy is not in Spain
Rule 2	Tom is not in France	Kathy is in Spain
Rule 3	Tom is in Spain	Albert is not in France
Rule 4	Tom is not in Spain	Albert is in France
Rule 5	Tom is in Egypt	Nancy is not in Egypt
Rule 6	Tom is not in Egypt	Nancy is in Egypt
Rule 7	Tom is in Japan	Nancy is not in France
Rule 8	Tom is not in Japan	Nancy is in France
Rule 9	Albert is in Egypt	Kathy is not in Egypt
Rule 10	Albert is in Egypt	Nancy is not in Egypt
Rule 11	Albert is in Egypt	Tom is not in Egypt
Rule 12	Kathy is in Egypt	Nancy is not in Egypt
Rule 13	Kathy is in Egypt	Tom is not in Egypt
Rule 14	Nancy is in Egypt	Tom is not in Egypt
Rule 15	Albert is in France	Kathy is not in France
Rule 16	Albert is in France	Nancy is not in France
Rule 17	Albert is in France	Tom is not in France
Rule 18	Kathy is in France	Nancy is not in France
Rule 19	Kathy is in France	Tom is not in France
Rule 20	Nancy is in France	Tom is not in France
Rule 21	Albert is in Japan	Kathy is not in Japan
Rule 22	Albert is in Japan	Nancy is not in Japan
Rule 23	Albert is in Japan	Tom is not in Japan
Rule 24	Kathy is in Japan	Nancy is not in Japan
Rule 25	Kathy is in Japan	Tom is not in Japan
Rule 26	Nancy is in Japan	Tom is not in Japan
Rule 27	Albert is in Spain	Kathy is not in Spain
Rule 28	Albert is in Spain	Nancy is not in Spain
Rule 29	Albert is in Spain	Tom is not in Spain
Rule 30	Kathy is in Spain	Nancy is not in Spain
Rule 31	Kathy is in Spain	Tom is not in Spain
Rule 32	Nancy is in Spain	Tom is not in Spain

TABLE 2.13. Knowledge base for the secret agents' problem.

he concludes that he is the only one with black eyes and leaves the prison the first day. Therefore, $d_{21} = 1$. The second day X_1 comes to the meeting and finds himself alone. He then concludes that he does not have black eyes. Therefore, he leaves the prison the second day, $d_{12} = 1$.

2.6 Some Applications   63

> No Facts given.
> Conclusions:
>   Albert cannot be in_Spain.
>   Albert cannot be in_Egypt.
>   Albert cannot be in_Japan.
>   Albert is in_France.
>   (It is the only possible value).
>   Kathy cannot be in_France.
>   Kathy cannot be in_Egypt.
>   Kathy cannot be in_Japan.
>   Kathy is in_Spain.
>   (It is the only possible value).
>   Nancy cannot be in_France.
>   Nancy cannot be in_Spain.
>   Nancy cannot be in_Japan.
>   Nancy is in_Egypt.
>   (It is the only possible value).
>   Tom cannot be in_France.
>   Tom cannot be in_Spain.
>   Tom cannot be in_Egypt.
>   Tom is in_Japan.
>   (It is the only possible value).

FIGURE 2.19. Session with the *X-pert Rules* computer program.

Case	c_1	c_2	Day 1		Day 2		Day 3	
			d_{11}	d_{21}	d_{12}	d_{22}	d_{13}	d_{23}
1	0	0	—	—	—	—	—	—
2	0	1	0	1	1	—	—	—
3	1	0	1	0	—	1	—	—
4	1	1	0	0	1	1	—	—

TABLE 2.14. All possible combinations of events for the prisoners' problem with $n = 2$.

- The third case: $c_1 = 1$ and $c_2 = 0$ is the reverse of the second case, that is, $d_{11} = 1$, $d_{21} = 0$, and $d_{22} = 1$.

- The fourth and last case: On the first day each of the two inmates observes that the other has black eyes. Thus, neither one can determine his eye color. Therefore, neither one can leave on the first day and

$d_{11} = d_{21} = 0$. They both show up the second day. Therefore, each can conclude that he has black eyes and leaves, that is, $d_{12} = d_{22} = 1$.

We therefore can conclude that given the fact that both inmates have black eyes, it will take two days for each to determine his own eye color (see Table 2.14), as claimed in Example 1.6.

Let us now develop an expert system to solve the problem for the case $n = 2$.

- *The Knowledge Base:* The knowledge base consists of the objects shown in Table 2.14, the multinomial object *Error*, whose value indicates an error message to be printed when an error occurs, and the following set of rules:

  - Rule 0: If $c_1 = 0$ and $c_2 = 0$, then $Error = 1$.
  - Rule 1: If $c_2 = 1$, then $d_{11} = 0$.
  - Rule 2: If $c_2 = 0$, then $d_{11} = 1$.
  - Rule 3: If $c_1 = 1$, then $d_{21} = 0$.
  - Rule 4: If $c_1 = 0$, then $d_{21} = 1$.
  - Rule 5: If $c_2 = 1$ and $d_{21} = 0$, then $d_{12} = 1$.
  - Rule 6: If $c_2 = 1$ and $d_{21} = 1$, then $d_{12} = 0$.
  - Rule 7: If $c_1 = 1$ and $d_{11} = 0$, then $d_{22} = 1$.
  - Rule 8: If $c_1 = 1$ and $d_{11} = 1$, then $d_{22} = 0$.
  - Rule 9: If $c_2 = 1$ and $d_{21} = 1$, then $d_{13} = 0$.
  - Rule 10: If $c_1 = 1$ and $d_{11} = 0$, then $d_{23} = 0$.

- *The Inference Engine:* When the above rules are executed after the facts $c_1 = c_2 = 1$, the inference engine will give the correct answer not only for the case where all inmates have black eyes but also when only one of them has black eyes.

- *The Coherence Control Subsystem:* Suppose that the user indicates that the facts are $c_1 = 0$ and $c_2 = 0$. This is clearly inconsistent with the king's statement that at least one of the inmates has black eyes. If this fact is given by the user, then Rule 0 will activate the object *Error* and assign it the value 1. This would signal the execution subsystem to abort execution and to print the error message number 1. This error message may read: *These facts are inconsistent with the king's statement*. Note that without Rule 0, the inference engine would conclude that $d_{11} = d_{21} = 1$, that is, both inmates would conclude that they have black eyes and would leave after the first day, which is clearly not the right solution.

The above rules can be generalized to the cases where $n > 2$, but clearly, the number of rules increases as n increases.

2.7 Introducing Uncertainty

The rule-based expert systems described in this chapter can be applied only to deterministic situations. There are numerous practical cases, however, that involve uncertainty. For example, in medical diagnosis of Example 1.4, the presence of some symptoms does not always imply the existence of a given disease, even though it may be strong evidence for the existence of the disease. Thus, it is useful to extend classical logic to deal with uncertainty. Several uncertainty measures have been introduced to deal with uncertainty. Castillo and Alvarez (1990, 1991) describe the application of these measures to enhance rule-based expert systems. On the other hand Johnson and Keravnou (1988) describe some expert system prototypes based on uncertain logic. Chapter 3 describes in detail *probabilistic expert systems*, which effectively deal with uncertainty.

Exercises

2.1 In Example 2.3, we use two binary objects A and B and give an example in which the Modus Tollens strategy expands the knowledge base. Give a similar example using nonbinary objects. For example, when A and B can take values $\{0, 1, 2\}$.

2.2 Show that the two sets of rules in Figures 2.1 and 2.7 are logically equivalent.

2.3 At some point in Example 2.11, we looked for an active rule that includes the current goal object J. Two rules, Rules 3 and 5, were found and Rule 5 was chosen. Complete the steps of the algorithm if Rule 3 had been chosen instead of Rule 5.

2.4 Consider an intersection of two two-way streets as indicated in Figure 2.20, where the allowed turns are shown. Let us call $T_1 - T_3$, $R_1 - R_3$, $B_1 - B_3$, and $L_1 - L_3$ the traffic lights associated with the lanes. Define a set of rules to regulate the intersection in such a way that no collision occurs.

2.5 Consider the train line with six stations given in Figure 2.21. Complete the set of rules given in Section 2.6.1 to handle the new station S_6.

2.6 Consider the train line in Figure 2.22 with four stations $\{S_1, \ldots, S_4\}$. Design a traffic control system by selecting adequate traffic lights. Obtain a set of rules to guarantee that no collision occurs.

66 2. Rule-Based Expert Systems

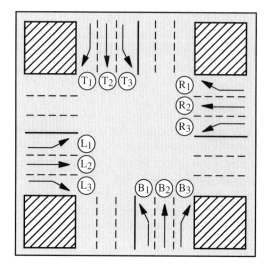

FIGURE 2.20. Intersection showing the allowable turns.

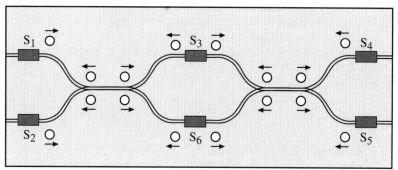

FIGURE 2.21. A train line with six stations.

2.7 Suppose we have the six-rules knowledge base given in Example 2.7. Following the process given in Examples 2.9 and 2.10, apply a goal-oriented rule chaining algorithm to conclude a value for the goal object shown in Figures 2.23(a) and 2.23(b). Gray shaded nodes indicate that the object has been assigned a value. The values are shown next to the corresponding nodes. What would be the conclusions with an inference engine including only the Modus Ponens strategy?

2.8 Design a rule-based expert system incorporating all the rules needed for playing "Tic-Tac-Toe." Two players take turns and put one of their pieces on a board with 9 squares (3 × 3), see Figure 2.24. The winner is the first player who puts 3 pieces in a column (Figure 2.24(a), a row (Figure 2.24(b)), and diagonal (Figure 2.24(c)). Consider the following strategies:

2.7 Introducing Uncertainty   67

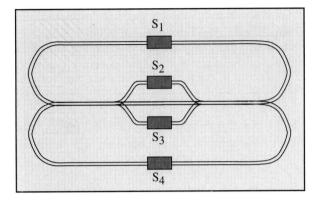

FIGURE 2.22. A train line with four stations.

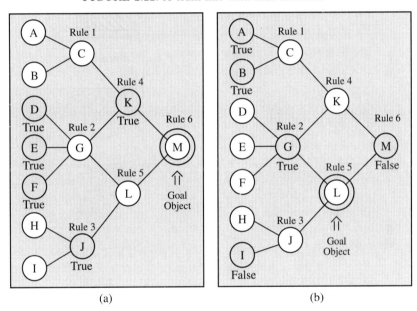

FIGURE 2.23. Initial facts and goal objects for a goal-oriented rule chaining algorithm.

- Defense strategy: Define the rules needed to prevent the other player from winning.
- Attack strategy: Add the set of rules that defines a strategy for winning.

2.9 Design a rule-based expert system to classify animals or plants based on a minimal set of characteristics. Follow these steps:

- Decide the set of animals or plants to be classified.

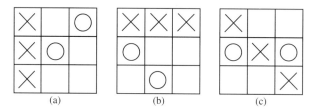

FIGURE 2.24. Tic-Tac-Toe boards: Three examples in which the player with "X" is the winner.

- Choose differential characteristics.
- Define rules for identifying each animal or plant.
- Remove unnecessary characteristics.
- Write the rules in the system.
- Perform an exhaustive check of the system.
- Redesign the system accordingly.

2.10 In the secret agent example of Section 2.6.2, what conclusion can be drawn using only the first eight rules in Table 2.13 when (a) France is given as a possible value for the object Tom and (b) Spain is given as a possible value for the object Tom.

2.11 Generate a set of rules to solve the prisoners' problem with $n = 3$ inmates (see Section 2.6.3). In this case, there are eight possible combinations and they are listed in the second to fourth columns of Table 2.15. Follow the same steps shown in Section 2.6.3 to solve the problem. Check that if all three inmates have black eyes, it will take them three days for each to determine his own eye color, as claimed in Example 1.6.

c_1	c_2	c_3	d_{11}	d_{21}	d_{31}	d_{12}	d_{22}	d_{32}	d_{13}	d_{23}	d_{33}
				Day 1			Day 2			Day 3	
0	0	0	—	—	—	—	—	—	—	—	—
0	0	1	0	0	1	1	1	—	—	—	—
0	1	0	0	1	0	1	—	1	—	—	—
0	1	1	0	0	0	0	1	1	1	—	—
1	0	0	1	0	0	—	1	1	—	—	—
1	0	1	0	0	0	1	0	1	—	1	—
1	1	0	0	0	0	1	1	0	—	—	1
1	1	1	0	0	0	0	0	0	1	1	1

TABLE 2.15. Combinations of events for the prisoners' problem with $n = 3$.

Chapter 3
Probabilistic Expert Systems

3.1 Introduction

Deterministic rule-based expert systems, introduced in Chapter 2, do not deal with uncertainties because objects and rules are treated deterministically. In most practical applications, however, uncertainty is the rule not the exception. For example, a question that often arises in medical diagnosis is: Given that a patient has a set of symptoms, which disease is the patient most likely to have? This situation involves some degree of uncertainty because

- The facts (concrete knowledge) may not be accurate. For example, a patient may not be sure whether or not he had a fever last night. Thus, there is a degree of uncertainty in the information associated with each patient (subjectivity, imprecision, lack of information, errors, missing data, etc.).

- The abstract knowledge is not deterministic. For example, the relationships among diseases and symptoms are not deterministic because the same group of symptoms may be associated with different diseases. In fact, it is not uncommon finding two patients with the same symptoms but different diseases.

Thus, a need for expert systems that deal with uncertain situations is clear. This chapter describes one type of expert system that effectively deals with uncertainties. These are the *probabilistic expert systems*, also known as *probability-based expert systems*.

Probability measure was selected to deal with uncertainty in the first expert systems (see Cheeseman (1985) or Castillo and Alvarez (1991)). But unfortunately, some problems were rapidly encountered due to the use of some independence assumptions to reduce computational complexity. As a result, at the very early stages of expert systems, probability measure was considered to be impractical. Most criticisms of probability methods were based on the huge number of parameters required, the impossibility of their precise assessment or estimation, or the unrealistic assumptions of independence.

Consequently, several alternative uncertainty measures, as certainty factors, beliefs, plausibilities, necessities or possibilities, have been suggested in the literature to deal with situations where uncertainty is present (see, for example, Shafer (1976), Zadeh (1983), Buchanan and Shortliffe (1984), Yager et al. (1987), and Almond (1995)).

However, with the appearance of probabilistic network models (mainly Bayesian and Markov networks, which are presented in Chapter 6), probability has made a spectacular comeback, and it is today one of the most intuitive and accepted measures of uncertainty. Lindley (1987), for example, states:

> "The only satisfactory description of uncertainty is probability. By this it is meant that every uncertainty statement must be in the form of a probability; that several uncertainties must be combined using the rules of probability, and that the calculation of probabilities is adequate to handle all situations involving uncertainty. In particular, alternative descriptions of uncertainty are unnecessary."

This chapter introduces probabilistic expert systems, which are based on probability as a measure of uncertainty. We describe in detail their main components (e.g., knowledge base, inference engine, and coherence control) and compare them with rule-based expert systems. Section 3.2 is a brief discussion of some concepts in probability theory that are needed for understanding the material in this and other chapters. Section 3.3 defines and discusses generalized rules as an attempt to extend rule-based expert systems to deal with uncertain situations. Then, keeping with our treatment of rule-based expert systems, we examine the structure of the knowledge base, the inference engine, and coherence control of probabilistic expert systems. In particular, Section 3.4 introduces probabilistic expert systems by an example. Section 3.5 describes the knowledge base and presents several models to describe the relationships among a set of variables of interest. In Section 3.6, we discuss the inference engine. The problem of coherence control is presented in Section 3.7. Finally, in Section 3.8 we end with a comparison of rule-based and probabilistic expert systems.

3.2 Some Concepts in Probability Theory

In this section we introduce the following background material in probability theory, which we use in the remainder of this chapter:

- Probability measure.
- Probability distributions.
- Dependence and independence.
- Bayes' theorem.
- Types of errors.

Readers who are familiar with these concepts can skip this section and go directly to Section 3.3. On the other hand, the material presented in this section is a bare minimum. For more concepts and results, the interested reader can consult any of the standard books on probability theory and statistics, for example, DeGroot (1987), Durrett (1991), Hogg (1993), and Billingsley (1995).

3.2.1 Probability Measure

To measure uncertainty we start with a given frame of discernment S, in which all mutually exclusive and collectively exhaustive outcomes of a given experiment, or elements of a given set, are included. The set S is also referred to as the *sample space*. Then, the aim is to assign to every subset of S a real value measuring the degree of uncertainty about its occurrence. In order to obtain measures with clear physical and practical meanings, some general and intuitive properties are used to define a class of measures known as *probability measures*.

Definition 3.1 Probability measure. *A function p mapping any subset $A \subseteq S$ into the interval $[0, 1]$ is called a probability measure if it satisfies the following axioms:*

- **Axiom 1 (Boundary):** $p(S) = 1$.
- **Axiom 2 (Additivity):** *For any infinite sequence, $A_1, A_2, \ldots$, of disjoint subsets of S, then*

$$p\left(\bigcup_{i=1}^{\infty} A_i\right) = \sum_{i=1}^{\infty} p(A_i). \tag{3.1}$$

Axiom 1 states that despite our degree of uncertainty, at least one element in the universal set S will occur (that is, the set S is exhaustive). Axiom 2 is an aggregation formula that can be used to compute the probability of a union of disjoint subsets. It states that the uncertainty of a given subset is the sum of the uncertainties of its disjoint parts. Note that this property also holds for finite sequences.

From the above axioms, many interesting properties of the probability measure can be derived. For example:

- **Property 1 (Boundary):** $p(\phi) = 0$.

- **Property 2 (Monotonicity):** If $A \subseteq B \subseteq S$, then $p(A) \leq p(B)$.

- **Property 3 (Continuity-Consistency):** For every increasing sequence $A_1 \subseteq A_2 \subseteq \ldots$ or decreasing sequence $A_1 \supseteq A_2 \supseteq \ldots$ of subsets of S we have

$$\lim_{i \to \infty} p(A_i) = p(\lim_{i \to \infty} A_i).$$

- **Property 4 (Inclusion-Exclusion):** Given any pair of subsets A and B of S, the following equality always holds:

$$p(A \cup B) = p(A) + p(B) - p(A \cap B). \tag{3.2}$$

Property 1 states that the evidence associated with a complete lack of information is defined to be zero. Property 2 shows that the evidence of the membership of an element in a set must be at least as great as the evidence that the element belongs to any of its subsets. In other words, the certainty of an element belonging to a given set A must not decrease with the addition of elements to A.

Property 3 can be viewed as a consistency or a continuity property. If we choose two sequences converging to the same subset of S, we must get the same limit uncertainty. Property 4 states that the probabilities of the sets $A, B, A \cap B$, and $A \cup B$ are not independent; they are related by (3.2).

A classical example illustrating these axioms is rolling a six-sided fair die. Here the sample space is $S = \{1, 2, 3, 4, 5, 6\}$, that is, the possible outcomes of the die. Let $p(A)$ denote the probability that event A occurs. Then, for example, we have $p(S) = 1$, $p(\{1\}) = 1/6$, $p(\{3\}) = 1/6$, and $p(\{1,3\}) = p(\{1\}) + p(\{3\}) = 1/3$.

3.2.2 Probability Distributions

Let $\{X_1, \ldots, X_n\}$ be a set of discrete random variables and $\{x_1, \ldots, x_n\}$ be a set of their possible realizations or instantiations. Note that a variable is denoted by an uppercase letter and its realization is denoted by the corresponding lowercase letter. For example, if X_i is a binary variable, then x_i can be either 1 or 0. The following results continue to hold if the

3.2 Some Concepts in Probability Theory   73

variables are continuous, but in this case the summation symbols must be replaced by the integral symbols.

Let $p(x_1, \ldots, x_n)$ denote the *joint probability distribution*[1] (JPD) of the variables in X, that is,

$$p(x_1, \ldots, x_n) = p(X_1 = x_1, \ldots, X_n = x_n). \tag{3.3}$$

Then the *marginal probability distribution* (MPD) of the ith variable is obtained by

$$p(x_i) = p(X_i = x_i) = \sum_{x_1, \ldots, x_{i-1}, x_{i+1}, \ldots, x_n} p(x_1, \ldots, x_n). \tag{3.4}$$

Knowledge about the occurrence of an event can modify the probabilities of other events. For example, the probability that a patient has a given disease can change after the results of a blood test become available. Thus, each time new information becomes available, the probabilities of events may change. This leads to the concept of *conditional probability*.

Definition 3.2 Conditional probability. *Let X and Y be two disjoint subsets of variables such that $p(y) > 0$. Then, the conditional probability distribution (CPD) of X given $Y = y$ is given by*

$$p(X = x | Y = y) = p(x|y) = \frac{p(x, y)}{p(y)}. \tag{3.5}$$

Equation (3.5) implies that the JPD of X and Y can be written as

$$p(x, y) = p(y) p(x|y). \tag{3.6}$$

One particular case of (3.5) is obtained when X is a single variable and Y is a subset of variables. In this case, (3.5) becomes

$$p(x_i | x_1, \ldots, x_k) = \frac{p(x_i, x_1, \ldots, x_k)}{p(x_1, \ldots, x_k)}$$

$$= \frac{p(x_i, x_1, \ldots, x_k)}{\sum_{x_i} p(x_i, x_1, \ldots, x_k)}, \tag{3.7}$$

which is the CPD of the ith variable, X_i, given a subset of variables $\{X_1, \ldots, X_k\}$. The sum in the denominator of (3.7) is taken over all possible values of X_i. Note that both the marginal probability in (3.4) and

---

[1] When the variables are discrete, $p(x_1, \ldots, x_n)$ is often referred to as a *probability mass function*, and when the variables are continuous, it is referred to as a *probability density function*. But for simplicity, we shall refer to it as a *joint probability distribution* of the variables.

the conditional probability in (3.7) still apply if the single variable X_i is replaced by a subset of variables as long as all variables are disjoint. Note also that in (3.7) if the set $\{X_1, \ldots, X_k\}$ is replaced by the empty set ϕ, then (3.7) reduces to $p(x_i)$. Thus, one may think of the marginal probability as a special case of the conditional probability.

3.2.3 Dependence and Independence

Definition 3.3 Independence of two variables. *Let X and Y be two disjoint subsets of the set of random variables $\{X_1, \ldots, X_n\}$. Then X is said to be independent of Y if and only if*

$$p(x|y) = p(x), \qquad (3.8)$$

for all possible values x and y of X and Y; otherwise X is said to be dependent on Y.

Note that if x and y are possible values of X and Y, then $p(x) > 0$ and $p(y) > 0$. Thus, the condition $p(y) > 0$ is natural in the sense that no $Y = y$ can be observed if the condition does not hold.

Equation (3.8) means that if X is independent of Y, then our knowledge of Y does not affect our knowledge about X, that is, Y has no information about X. Also, if X is independent of Y, we can then combine (3.5) and (3.8) and obtain $p(x, y)/p(y) = p(x)$, which implies

$$p(x, y) = p(x)p(y). \qquad (3.9)$$

Equation (3.9) indicates that if X is independent of Y, then the JPD of X and Y is equal to the product of their marginals. Actually, (3.9) provides a definition of independence equivalent to that in (3.8).

One important property of the independence relation is its *symmetry*, that is, if X is independent of Y, then Y is independent of X. This is because

$$p(y|x) = \frac{p(x, y)}{p(x)} = \frac{p(x)p(y)}{p(x)} = p(y). \qquad (3.10)$$

Because of the symmetry property, we say that X and Y are *independent* or *mutually independent*. The practical implication of symmetry is that if knowledge of Y is relevant (irrelevant) to X, then knowledge of X is relevant (irrelevant) to Y.

The concepts of dependence and independence of two random variables can be extended to the case of more than two random variables as follows:

Definition 3.4 Independence of a set of variables. *The random variables $\{X_1, \ldots, X_m\}$ are said to be independent if and only if*

$$p(x_1, \ldots, x_m) = \prod_{i=1}^{m} p(x_i), \qquad (3.11)$$

for all possible values $x_1, \ldots, x_m$ of $X_1, \ldots, X_m$; otherwise they are said to be dependent.

In other words, $\{X_1, \ldots, X_m\}$ are said to be independent if and only if their JPD is equal to the product of their MPD. Note that (3.11) is a generalization of (3.9).

Note also that if $X_1, \ldots, X_m$ are conditionally independent of each other given another subset $Y_1, \ldots, Y_n$, then

$$p(x_1, \ldots, x_m | y_1, \ldots, y_n) = \prod_{i=1}^{m} p(x_i | y_1, \ldots, y_n). \tag{3.12}$$

An important implication of independence is that it is not worthwhile gathering information about independent (irrelevant) variables. That is, independence means irrelevance.

Example 3.1 Four variables. Consider the following characteristics (variables and their possible values) of people in a given population:

- $Sex = \{man, woman\}$

- $Smoking = \{smoker\ (s), non\text{-}smoker\ (\bar{s})\}$

- $MaritalStatus = \{married\ (m), not\ married\ (\bar{m})\}$

- $Pregnancy = \{pregnant\ (p), not\ pregnant\ (\bar{p})\}$

The JPD of these four variables is given in Table 3.1 Thus, for example, 50% of the people in the population are women, and

$$\frac{0.01 + 0.04 + 0.01 + 0.10}{(0.01 + 0.04 + 0.01 + 0.10) + (0.00 + 0.02 + 0.00 + 0.07)} = 64\%$$

of the smokers are women.

Let A denote a person chosen at random from this population. Without knowing whether the person is a smoker, the probability that the person is a woman is $p(A = woman) = 0.50$. But if we know that the person is a smoker, this probability changes from 0.50 to $p(A = woman | A = s) = 0.64$. Therefore, we have $p(A = woman | A = s) \neq p(A = woman)$; hence, the variables Sex and $Smoking$ are dependent.

Suppose now we know that the person is pregnant. Then we have

$$p(A = woman | A = p) = 1 \neq p(A = woman) = 0.50;$$

hence, the variables Sex and $Pregnancy$ are dependent. Thus, the two variables $Smoking$ and $Pregnancy$ contain relevant information about the variable Sex. However, the event "the person is pregnant" contains much

		man		woman	
		s	$\bar{s}$	s	$\bar{s}$
m	p	0.00	0.00	0.01	0.05
	$\bar{p}$	0.02	0.18	0.04	0.10
$\bar{m}$	p	0.00	0.00	0.01	0.01
	$\bar{p}$	0.07	0.23	0.10	0.18

TABLE 3.1. The JPD of four variables: *Sex* (*man, woman*), *Smoking* ($s, \bar{s}$), *Marital Status* ($m, \bar{m}$), and *Pregnancy* ($p, \bar{p}$).

x	y	z	$p(x,y,z)$
0	0	0	0.12
0	0	1	0.18
0	1	0	0.04
0	1	1	0.16
1	0	0	0.09
1	0	1	0.21
1	1	0	0.02
1	1	1	0.18

TABLE 3.2. The JPD of three binary variables.

more information about *Sex* than the event "the person is a smoker." This can be measured by the ratio

$$\frac{p(A = woman|A = p)}{p(A = woman|A = s)} = \frac{1}{0.64} > 1.$$

On the other hand, the variable *Marital Status* does not contain relevant information about the variable *Sex* and vice versa. This can be seen from Table 3.1, where the joint probabilities are equal to the product of the marginal probabilities for all possible values of the two variables. Thus, the variables *Sex* and *Marital Status* are independent. ∎

Example 3.2 Probability distributions. Consider the JPD of three binary variables X, Y, and Z given in Table 3.2. Then we have:

- The marginal probability distributions of X, Y, and Z are shown in Table 3.3. For example, the marginal probability distribution of X is computed by

$$p(X = 0) = \sum_{y=0}^{1}\sum_{z=0}^{1} p(0, y, z) = 0.12 + 0.18 + 0.04 + 0.16 = 0.5,$$

$$p(X=1) = \sum_{y=0}^{1}\sum_{z=0}^{1} p(1,y,z) = 0.09 + 0.21 + 0.02 + 0.18 = 0.5.$$

- The pairwise JPDs are given in Table 3.4. For example, the JPD of X and Y is computed by

$$p(X=0, Y=0) = \sum_{z=0}^{1} p(0,0,z) = 0.12 + 0.18 = 0.3,$$

$$p(X=0, Y=1) = \sum_{z=0}^{1} p(0,1,z) = 0.04 + 0.16 = 0.2,$$

$$p(X=1, Y=0) = \sum_{z=0}^{1} p(1,0,z) = 0.09 + 0.21 = 0.3,$$

$$p(X=1, Y=1) = \sum_{z=0}^{1} p(1,1,z) = 0.18 + 0.02 = 0.2.$$

- The conditional probability distributions of one variable given the other are shown in Table 3.5. For example, the conditional probability distribution of X given Y is computed by

$$p(X=0|Y=0) = \frac{p(X=0, Y=0)}{p(Y=0)} = \frac{0.3}{0.6} = 0.5,$$

$$p(X=0|Y=1) = \frac{p(X=0, Y=1)}{p(Y=1)} = \frac{0.2}{0.4} = 0.5,$$

$$p(X=1|Y=0) = \frac{p(X=1, Y=0)}{p(Y=0)} = \frac{0.3}{0.6} = 0.5,$$

$$p(X=1|Y=1) = \frac{p(X=1, Y=1)}{p(Y=1)} = \frac{0.2}{0.4} = 0.5.$$

From the above results we see that $p(x,y) = p(x)p(y)$ for all values x and y, hence X and Y are independent. Note that this independence can also be checked with the equivalent definition of independence $p(x|y) = p(x)$. However, we observe that $p(x,z) \neq p(x)p(z)$ for some (in this case all) values x and z. Hence X and Z are dependent. Similarly, it can be shown that Y and Z are dependent. ∎

The concepts of dependence and independence deal with two subsets of variables. Now we turn to a generalization of the concept of independence when more than two sets of variables are involved.

78   3. Probabilistic Expert Systems

x	$p(x)$
0	0.5
1	0.5

y	$p(y)$
0	0.6
1	0.4

z	$p(z)$
0	0.27
1	0.73

TABLE 3.3. Marginal probability distributions.

x	y	$p(x,y)$
0	0	0.3
0	1	0.2
1	0	0.3
1	1	0.2

x	z	$p(x,z)$
0	0	0.16
0	1	0.34
1	0	0.11
1	1	0.39

y	z	$p(y,z)$
0	0	0.21
0	1	0.39
1	0	0.06
1	1	0.34

TABLE 3.4. Pairwise joint probability distributions.

y	x	$p(x\|y)$
0	0	0.5
0	1	0.5
1	0	0.5
1	1	0.5

z	x	$p(x\|z)$
0	0	16/27
0	1	11/27
1	0	34/73
1	1	39/73

z	y	$p(y\|z)$
0	0	21/27
0	1	6/27
1	0	39/73
1	1	34/73

TABLE 3.5. Conditional probability distributions of one variable given another.

Definition 3.5 Conditional dependence and independence.
Let X, Y and Z be three disjoint sets of variables, then X is said to be conditionally independent of Y given Z, if and only if

$$p(x|z,y) = p(x|z), \quad (3.13)$$

for all possible values x, y, and z of X, Y, and Z; otherwise X and Y are said to be conditionally dependent given Z.

When X and Y are conditionally independent given Z, we write $I(X,Y|Z)$. The statement $I(X,Y|Z)$ is referred to as a *conditional independence statement* (CIS). Similarly, when X and Y are conditionally dependent given Z, we write $D(X,Y|Z)$, which is called a *conditional dependence statement*. We sometimes write $I(X,Y|Z)_p$ or $D(X,Y|Z)_p$ to signify that the statement is derived from or implied by the probabilistic model associated with probability p (the JPD).

3.2 Some Concepts in Probability Theory

The definition of conditional independence conveys the idea that once Z is known, knowing Y can no longer influence the probability of X. In other words, if Z is already known, knowledge of Y does not add any new information about X.

An alternative but equivalent definition of conditional independence is given by

$$p(x,y|z) = p(x|z)p(y|z). \qquad (3.14)$$

The equivalence of (3.13) and (3.14) can be demonstrated in a manner similar to that in (3.9).

Note that (unconditional) independence can be treated as a particular case of conditional independence. For example, we can write $I(X,Y|\phi)$, to mean that X and Y are unconditionally independent, where ϕ is the empty set. Note, however, that X and Y can be unconditionally independent but conditionally dependent given Z, that is, the CIS $I(X,Y|\phi)$ and $D(X,Y|Z)$ can hold simultaneously.

Example 3.3 Conditional dependence and independence. Consider the JPD of the three binary variables X, Y, and Z shown in Table 3.2. In Example 3.2 it is determined whether any two variables are (unconditionally) independent. We have the following CIS:

$$I(X,Y|\phi), \quad D(X,Z|\phi), \quad \text{and } D(Y,Z|\phi).$$

For example, to determine whether X and Y are independent, we need to check whether $p(x,y) = p(x)p(y)$ for all possible values x and y.

We can also determine whether any two variables are conditionally independent given the third variable. Let us see, for example, whether X and Y are conditionally independent given Z. We need to check whether $p(x|y,z) = p(x,y,z)/p(y,z) = p(x|z)$ for all possible values x, y, and z. We compute

$$p(x|y,z) = \frac{p(x,y,z)}{p(y,z)},$$

$$p(x|z) = \frac{p(x,z)}{p(z)}.$$

These distributions are shown in Table 3.6, from which we see that $p(x|y,z) \neq p(x|z)$, and hence $D(X,Y|Z)$. Thus, the JPD in Table 3.2 implies that X and Y are unconditionally independent, $I(X,Y|\phi)$, yet they are conditionally dependent given Z, $D(X,Y|Z)$. ∎

3.2.4 Bayes' Theorem

A well-known and useful formula in probability theory is derived as follows. Using (3.5) and (3.7), we have

80   3. Probabilistic Expert Systems

y	z	x	$p(x\|y,z)$
0	0	0	$12/21 \approx 0.571$
0	0	1	$9/21 \approx 0.429$
0	1	0	$18/39 \approx 0.462$
0	1	1	$21/39 \approx 0.538$
1	0	0	$4/6 \approx 0.667$
1	0	1	$2/6 \approx 0.333$
1	1	0	$16/34 \approx 0.471$
1	1	1	$18/34 \approx 0.529$

z	x	$p(x\|z)$
0	0	$16/27 \approx 0.593$
0	1	$11/27 \approx 0.407$
1	0	$34/73 \approx 0.466$
1	1	$39/73 \approx 0.534$

TABLE 3.6. Probability distributions derived from the JPD in Table 3.2.

$$p(x_i|x_1,\ldots,x_k) = \frac{p(x_i,x_1,\ldots,x_k)}{\sum_{x_i} p(x_i,x_1,\ldots,x_k)}$$

$$= \frac{p(x_i)p(x_1,\ldots,x_k|x_i)}{\sum_{x_i} p(x_i)p(x_1,\ldots,x_k|x_i)}. \quad (3.15)$$

Equation (3.15) is known as *Bayes' theorem*.

To illustrate the use of Bayes' theorem, suppose that a patient can be either healthy (has no disease) or have one of $m-1$ diseases $\{D_1,\ldots,D_{m-1}\}$. For simplicity of notation, let D be a random variable that can take one of m possible values, $\{d_1,\ldots,d_m\}$, where $D = d_i$ means the patient has disease D_i, and $D = d_m$ means the patient has no disease. Suppose also we have n associated symptoms $\{S_1,\ldots,S_n\}$. Now, given that a patient has a set of symptoms $\{s_1,\ldots,s_k\}$, we wish to compute the probability that the patient has the disease D_i, that is, $D = d_i$. Then, using Bayes' theorem, we obtain

$$p(d_i|s_1,\ldots,s_k) = \frac{p(d_i)p(s_1,\ldots,s_k|d_i)}{\sum_{d_i} p(d_i)p(s_1,\ldots,s_k|d_i)}. \quad (3.16)$$

We make the following remarks about (3.16):

- The probability $p(d_i)$ is called the *marginal*, *prior*, or *initial* probability of the disease $D = d_i$ because it can be obtained *before* knowing any symptoms.

- The probability $p(d_i|s_1,\ldots,s_k)$ is the *posterior* or *conditional* probability of the disease $D = d_i$ because it is computed *after* knowing the symptoms $S_1 = s_1,\ldots,S_k = s_k$.

- The probability $p(s_1,\ldots,s_k|d_i)$ is referred to as the *likelihood* that a patient with the disease $D = d_i$ will show the symptom $S_1 = s_1,\ldots,S_k = s_k$.

3.2 Some Concepts in Probability Theory

Thus, we can use Bayes' theorem to update the posterior probability distribution using both the prior and the likelihood, as illustrated in the following example:

Example 3.4 Gastric adenocarcinoma. A medical center has a data set consisting of medical records of $N = 1,000$ patients. These records are summarized graphically in Figure 3.1. There are 700 patients (the shaded region) who have the disease *gastric adenocarcinoma* (G), and 300 who do not have this disease (we consider *being healthy* as another possible value for disease D). Three symptoms, *pain* (P), *weight loss* (W), and *vomiting* (V), are thought to be associated with this disease. Accordingly, when a new patient arrives at the medical center, there is a $700/1,000 = 70\%$ chance that the patient has gastric adenocarcinoma. This is the initial, or prior, probability because it is calculated based on the initial information, that is, before knowing any information about the patient.

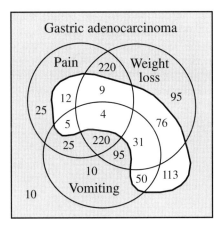

FIGURE 3.1. The number of patients in a medical center classified by one disease (gastric adenocarcinoma) and three symptoms (pain, vomiting, and weight loss).

For simplicity of notation, we use g to indicate that the disease is present and $\bar{g}$ to indicate that the disease is absent. Similar notations are used for the symptoms. Then we can also make the following statements:

- Prior probability: 440 of 1,000 patients vomit. Thus, $p(v) = card(v)/N = 440/1,000 = 0.44$, where $card(v)$ denotes the number of patients in the database who vomit. This means that 44% of the patients vomit.

- Likelihood probability: 50% of the patients who have the disease vomit, because $p(v|g) = card(v,g)/card(g) = 350/700 = 0.5$, whereas only 30% of those patients who do not have the disease vomit, because $p(v|\bar{g}) = card(v,\bar{g})/card(\bar{g}) = 90/300 = 0.3$.

- Likelihood probability: 45% of patients who have the disease vomit and lose weight, $p(v, w|g) = card(v, w, g)/card(g) = 315/700 = 0.45$, whereas only 12% of those who do not have the disease vomit and lose weight, $p(v, w|\bar{g}) = card(v, w, \bar{g})/card(\bar{g}) = 35/300 \approx 0.12$.

Because the initial probability that the patient has gastric adenocarcinoma, $p(g) = 0.7$, is not high enough to make a diagnostic decision (note that making a decision at this moment implies a probability 0.3 of misdiagnosis), the doctor decides to examine the patient in order to obtain more information. Suppose that the results of the examination show that the patient has the symptoms vomiting and weight loss ($V = v$ and $W = w$). Now, given this evidence (the patient has these symptoms), what is the probability that the patient has the disease? This posterior probability can be obtained from the prior and likelihood probabilities by applying Bayes' theorem in two steps as follows:

- After observing that $V = v$ the posterior probability is

$$p(g|v) = \frac{p(g)p(v|g)}{p(g)p(v|g) + p(\bar{g})p(v|\bar{g})}$$

$$= \frac{0.7 \times 0.5}{(0.7 \times 0.5) + (0.3 \times 0.3)} = 0.795.$$

- After observing that $V = v$ and $W = w$ the posterior probability is

$$p(g|v, w) = \frac{p(g)p(v, w|g)}{p(g)p(v, w|g) + p(\bar{g})p(v, w|\bar{g})}$$

$$= \frac{0.7 \times 0.45}{(0.7 \times 0.45) + (0.3 \times 0.12)} = 0.9. \quad (3.17)$$

Note that when applying Bayes' theorem at a given step, the posterior probability computed at a given step is the same as the prior probability in the next step. For example, the posterior probability, which has been calculated in the first step above, can be used as the prior probability in the second step, that is,

$$p(g|v, w) = \frac{p(g|v)p(w|g, v)}{p(g|v)p(w|g, v) + p(\bar{g}|v)p(w|\bar{g}, v)}$$

$$= \frac{0.795 \times 0.9}{(0.795 \times 0.9) + (0.205 \times 0.389)} = 0.9,$$

which gives the same answer as in (3.17). Notice also how the probability changes after we observe the evidences. The probability of having the disease was initially 0.7, and then it increases to 0.795, and then to 0.9

after observing the cumulative evidence $V = v$ and $W = w$, respectively. At the end of the last step, the patient has a 0.9 probability of having the disease. This probability may be high enough (as compared to the prior probability of only 0.7) for the doctor to diagnose the patient as having the disease. However, it would be convenient to observe some new evidence before making a diagnosis. ∎

3.2.5 Types of Errors

Symptoms are observable but diseases are not. But since diseases and symptoms are related, medical doctors usually use the symptoms as a basis for diagnosing diseases. A difficulty with this approach, however, is that the relationships among symptoms and diseases are not perfect. For example, the same symptom can be caused by different diseases. By studying the relationships among symptoms and diseases medical doctors gain more knowledge and experience, and hence they become more able to diagnose diseases with a high degree of certainty.

However, it should be recognized that when making decisions in uncertain situations, these decisions can be incorrect. There are two possible types of errors in uncertain situations:

- A false positive decision, also known as type I error, and
- A false negative decision, also known as type II error.

In a medical diagnosis situation, for example, the possible errors are

- *Type I Error:* A patient does not have the disease but the doctor concludes that the patient has the disease.
- *Type II Error:* A patient has the disease but the doctor concludes that the patient does not have the disease.

These types of errors are illustrated in Table 3.7. In actuality (true state of nature), a patient can either have or not have a disease. The doctor has to make a decision regarding whether a patient has a disease. This decision is correct if it matches the true state of nature; otherwise the decision is wrong. Thus, when making a diagnosis, the doctor is subject to making one of the above two errors depending on the true state of nature.

In some situations, however, the consequences of one error can be far more damaging than the consequences of the other. For example, if the suspected disease is cancer, one can argue that Type II error is more serious than Type I error. It is true that if the patient does not have cancer but the doctor concludes otherwise, the patient would suffer psychologically and possibly physically (due to the effect of treatment or of an operation). On the other hand, if the patient actually has cancer and the doctor concludes otherwise, this error could lead to death.

Doctor's decision	State of nature	
	Yes	No
Yes	Correct decision	Incorrect decision (Type I)
No	Incorrect decision (Type II)	Correct decision

TABLE 3.7. The doctor is subject to making one of the above two errors depending on the true state of nature.

Ideally, the doctor would like to keep the probabilities of making the above errors to a minimum, but the relative risks associated with the two types of errors have to be taken into consideration when making a diagnosis. To illustrate, suppose that a new patient with an unknown disease comes to the medical center. After an examination by a doctor, it is determined that the patient has k symptoms, $s_1, s_2, \ldots, s_k$. The question that both the doctor and the patient need to know is, Given these symptoms, which disease is the patient most likely to have? The answer to this question can be obtained by computing the posterior probabilities of $D = d$ for each disease $d = d_i$ given the symptoms $s_1, s_2, \ldots, s_k$, that is, $p(d_i|s_1, s_2, \ldots, s_k)$. This probability can be computed using (3.16). Thus, given that the patient has the symptoms $s_1, s_2, \ldots, s_k$, the doctor may conclude that the disease the patient is most likely to have is the one with $max_i\{p(d_i|s_1, s_2, \ldots, s_k)\}$. If $max_i\{p(d_i|s_1, s_2, \ldots, s_k)\}$ is close enough to 1, the doctor may decide that the patient has the corresponding disease. Otherwise, additional examination and identification of new symptoms (evidence) may be necessary.

Equation (3.16) can still be used to compute the new conditional probability for each disease given all the accumulated symptoms (information) as we have done in Example 3.4. This process has to be repeated, adding more evidence, until $max_i\{p(d_i|s_1, s_2, \ldots, s_k)\}$ gets close enough to 1. When this occurs, the medical doctor can make a decision and terminate the process of diagnosis. How *close* is *close enough to* 1 is usually decided by the doctor, depending on the risks associated with erroneous decisions.

We therefore need to measure the consequences of our decisions. One way to do this is to use the so called *utility function*. A utility function assigns a utility value to every possible decision. Let X be a random variable with probability distribution $p(x)$ and let $u(x)$ be a utility value assigned to decision x. Then the expected value of this utility is given by

$$E[u] = \sum_x u(x)p(x).$$

We can assign different utility functions $u_i(x); i = 1, \ldots, q$ to different decisions and decide in favor of the decision maximizing the utility.

3.3 Generalized Rules

The probability measure discussed in Section 3.2.1 can be used to measure uncertainty and to extend rule-based expert systems to handle situations that involve uncertainty. One way of introducing uncertainty in rule-based expert systems is to use *generalized rules*. For example, given the deterministic rule

- Rule 1: If A is true, then B is true,

we may introduce some uncertainty by associating a probability to this statement

- Rule 2: If A is true, then the probability that B is true is $p(b) = \theta$,

where $0 \leq \theta \leq 1$ is a measure of the uncertainty of B. Clearly, Rule 1 is a special case of Rule 2 because it is obtained from Rule 2 by setting $\theta = 1$ (certainty). But when $0 < \theta < 1$ (uncertainty), Rule 1 is no longer appropriate. Therefore, we may think of Rule 2 as a generalized rule. Thus, the value of θ determines the level of implication as follows (see Figure 3.2):

- *Strong implication* ($\theta = 1$): In classical logic, the one we have used in rule-based expert systems (Modus Ponens and Modus Tollens), if the premise of a rule is true, its conclusion must also be true. Thus, given the rule

    If A is true, then B is true,

    we can say that A implies B with probability 1. This is illustrated in Figure 3.2(a).

- *Weak implication* ($0 < \theta < 1$): The above rule can be seen in a generalized sense: If A sometimes implies B, we say that A implies B with probability $p(B = true | A = true)$. This case is shown in Figure 3.2(b).

- *No implication* ($\theta = 0$): The case where A does not imply B can be thought of as A implies B with probability 0. This is illustrated in Figure 3.2(c).

Using generalized rules requires uncertainty measures for both objects and rules, together with aggregation formulas to combine the uncertainty of the objects in the premises with that of the rules to obtain the uncertainty of the objects in the conclusions. Note that now every statement (fact) must be accompanied by a given measure of uncertainty and that when combining several uncertain facts, we must give conclusions their corresponding measure of uncertainty.

One of the first expert systems that used probability measure is PROSPECTOR, an expert system for mineral exploration (Duda, Hart, and

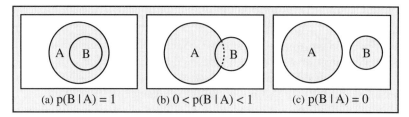

FIGURE 3.2. Examples of uncertain implications: (a) A implies B with probability 1, (b) A implies B with probability θ, where $0 < \theta < 1$, and (c) A implies B with probability 0.

Nilsson (1976), Duda, Gaschnig, and Hart (1980)). In addition to the rules forming the knowledge base, prior probabilities are associated with some of the objects forming the model, and some conditional probabilities are associated with the rules. Thus, when new evidence is observed, some probabilistic method for the propagation of probabilities has to be used in the model.

Thus, one can deal with uncertainty using generalized rules or similar schemes. However, these models are not without problems. When combining several uncertain facts, we must give conclusions their corresponding measure of uncertainty. In order to propagate the uncertainties of the observed evidence, it is necessary to make ad hoc adjustments and assumptions of conditional independence that may not be justified (see Neapolitan (1990), Chapter 4). This is the case, for example, of the method of odds likelihood ratios (Duda, Hart, and Nilsson (1976)) developed for propagating probabilities in the PROSPECTOR expert system, or the certainty factor propagation method used in MYCIN (see Buchanan and Shortliffe (1984)).

An alternative way of using probability measure is to describe the relationships among the objects (variables) by a JPD. We refer to expert systems that use the JPD of the set of variables as the basis for making inferences as *probabilistic expert systems*. In the remainder of this chapter, we introduce probabilistic expert systems, describe their components, and compare them with rule-based expert systems.

3.4 Introducing Probabilistic Expert Systems

The core of the knowledge base in rule-based expert systems is the set of rules that describe the relationships among the set of objects (variables). In probabilistic expert systems the relationships among the variables are described by their JPD. Thus, the JPD forms what we call the abstract knowledge or the knowledge of general application. To facilitate the discussion we use an example from the medical diagnosis area (symptoms and diseases), but the concepts described clearly apply to many other fields of applications. In fact, medical diagnoses is one of the areas where expert systems have found more application (see Section 1.2), and as we shall see in the next section, some models of probabilistic expert systems were developed to deal with problems with the special "symptoms-disease" structure.

Example 3.5 Medical diagnosis. Suppose that we have a database with information about N patients and that a patient can have one and only one of m given diseases, $d_1, \ldots, d_m$, as shown in Figure 3.3 for $m = 5$ diseases. Assume also that a patient can show none, one, or more of n symptoms $S_1, \ldots, S_n$, as illustrated in Figure 3.4 for $n = 3$ symptoms. For simplicity, let us assume that the random variable disease D takes as values diseases $d_1, \ldots, d_m$ and that symptoms are binary variables, so each takes the value 1 if present and the value 0 if absent. Note that any random variable in the set $\{D, S_1, \ldots, S_n\}$ partitions the universal set of patients into a class of disjoint and exhaustive sets. Then, combining diseases and symptoms, each patient can be classified into one and only one region such as those shown in Figure 3.5, where Figures 3.3 and 3.4 are superimposed. For example, the black dot in Figure 3.5 represents a patient who has disease d_4 and who shows all three symptoms: S_1, S_2, and S_3.

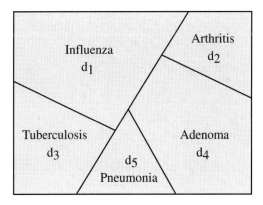

FIGURE 3.3. A graphical representation of a population of patients classified by five mutually exclusive diseases d_1-d_5.

The objects, or variables, here are the disease D and symptoms $S_1, \ldots, S_n$. The JPD of the variables $(D, S_1, \ldots, S_n)$ is given by the frequencies, that is, the number of patients in the various regions of the diagram in Figure 3.5. Continuing with the notation introduced in Section 3.2.2, a variable is represented by an uppercase letter and the corresponding lowercase letter represents one of its possible values (realizations). In this example, the disease D is assumed to take m possible values and the symptoms are assumed to be binary. In other words, the possible values of D are $d_1, \ldots, d_m$,

88   3. Probabilistic Expert Systems

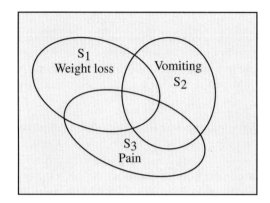

FIGURE 3.4. A graphical representation of a population of patients classified by three symptoms S_1–S_3.

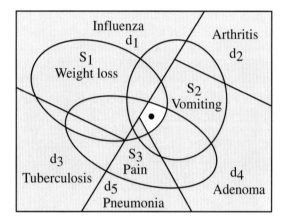

FIGURE 3.5. A graphical representation of a population of patients classified by five mutually exclusive diseases d_1–d_5 and three symptoms S_1–S_3.

and the possible values of the jth variable S_j are either 1 (present) or 0 (absent). ∎

The probabilities associated with a disease d can be estimated by

$$p(D = d) \approx card(D = d)/N, \qquad (3.18)$$

where N is the total number of patients in the database and $card(D = d)$ is the number of patients with $D = d$. For example,

- Disease d_1 present: $p(D = d_1) \approx card(D = d_1)/N$,
- Disease d_1 absent: $p(D \neq d_1) \approx card(D \neq d_1)/N$.

A problem that often arises in medical diagnosis is that we only observe a subset of the symptoms, and based on the observed symptoms, we wish to

3.4 Introducing Probabilistic Expert Systems

Disease d	$p(d\|s_1,\ldots,s_k)$
d_1	0.2
d_2	0.1
d_3	0.8 ← most likely
d_4	0.4
d_5	0.0 ← least likely
d_6	0.7
$\vdots$	$\vdots$

TABLE 3.8. Conditional probabilities of all diseases d_i, given the set of symptoms $S_1 = s_1, \ldots, S_k = s_k$.

diagnose the disease(s) causing the symptoms with a reasonable degree of certainty. In other words, we need to address the following question: Given that a patient has a subset of k symptoms $S_1 = s_1, \ldots, S_k = s_k$, which disease is the patient most likely to have? Thus, the problem is to compute the probability that a patient has disease d_i, given a set of values $s_1, \ldots, s_k$ for the symptoms $S_1, \ldots, S_k$. In other words, for $i = 1, \ldots, m$, we wish to compute the conditional probabilities $p(D = d_i | S_1 = s_1, \ldots, S_k = s_k)$. You may think of this as a generalized classification problem: A patient can be classified into one or more groups (diseases). For example, we may obtain the probabilities shown in Table 3.8.

Probabilistic expert systems can be used to solve these and other problems. For example:

1. Expert systems can memorize information. One can store and retrieve information from the database. An example of such a database is given in Table 3.9, where it is assumed that diseases and symptoms are categorical variables (binary or multinomial). For instance, Table 3.10 may represent the information of a database with ten patients for the diagnosis problem with five binary diseases and three binary symptoms introduced in Example 3.5.

2. Expert systems can count or compute the absolute and relative frequencies for any subset of the variables from the database and use them to calculate the conditional probabilities $p(d_i|s_1,\ldots,s_k)$ using the well-known conditional probability formula[2]

$$p(d_i|s_1,\ldots,s_k) = \frac{p(d_i,s_1,\ldots,s_k)}{p(s_1,\ldots,s_k)}. \tag{3.19}$$

---

[2]For notational simplicity we write $p(D = d_i|S_1 = s_1,\ldots,S_k = s_k)$ as $p(d_i|s_1,\ldots,s_k)$.

Patient	Disease d	Symptoms		
		s_1	$\ldots$	s_n
1	d_m	1	$\ldots$	1
2	d_1	0	$\ldots$	0
3	d_3	1	$\ldots$	0
$\vdots$	$\ddots$	$\vdots$	$\ddots$	$\vdots$
N	d_m	1	$\ldots$	1

TABLE 3.9. An example of a database of N patients and their corresponding diseases and symptoms.

Patient	Disease D	Symptoms		
		S_1	S_2	S_3
1	d_5	1	1	1
2	d_2	1	0	1
3	d_3	1	1	0
4	d_5	0	0	1
5	d_3	0	1	0
6	d_1	1	1	0
7	d_1	1	1	1
8	d_3	1	0	0
9	d_1	1	1	1
10	d_5	1	0	1

TABLE 3.10. An example of a database of ten patients for the diagnosis problem of Example 3.5.

This probability can be estimated by

$$\frac{card(d_i, s_1, \ldots, s_k)}{card(s_1, \ldots, s_k)}, \qquad (3.20)$$

where $card(d_i, s_1, \ldots, s_k)$ is the frequency of the patients in the database who have the indicated set of values of the variables. For example, given the database with ten patients in Table 3.10, we can calculate the frequencies associated with any combination of values of symptoms and diseases by counting the number of cases that match this evidence in the database. For example, $card(D \ne d_1|S_1 = 1, S_2 = 1) = 2$ because there are two patients (patients 1 and 3) that do not present disease d_1 but present symptoms S_1 and S_2. Similarly, $card(D = d_1|S_1 = 1, S_2 = 1) = 3$, $card(S_1 = 1, S_2 = 1) = 5$, etc. Then, using (3.5) the conditional probabilities associated with a given disease and a set of symptoms can be calculated. For example:

$$p(D \neq d_1|S_1 = 1, S_2 = 1) \approx \frac{card(D \neq d_1|S_1 = 1, S_2 = 1)}{card(S_1 = 1, S_2 = 1)} = \frac{2}{5} = 0.4,$$

$$p(D = d_1|S_1 = 1, S_2 = 1) \approx \frac{card(D = d_1|S_1 = 1, S_2 = 1)}{card(S_1 = 1, S_2 = 1)} = \frac{3}{5} = 0.6.$$

3. Expert systems can learn from experience. Once a new patient is examined and diagnosed, the new information is added to the database and the frequencies changed accordingly. For example, if a new patient presenting symptoms $S_1 = 1, S_2 = 1$, and $S_3 = 0$ is known to have the disease d_1, we can update this new information in the probabilistic expert system by including the case as a new entry in the database shown in Table 3.10.

4. Expert systems can make (or help human experts make) decisions such as

   - Do we have enough information to diagnose the disease?
   - Do we need further tests? If yes, which test will provide the most information about the suspected disease?

In the following three sections we describe the three main components of probabilistic expert systems.

3.5 The Knowledge Base

As we have seen in Chapter 2, the knowledge base of a rule-based expert system consists of a set of objects (variables) and a set of rules. The knowledge base of a probabilistic expert system consists of a set of variables, $\{X_1, \ldots, X_n\}$, and a JPD over all the variables $p(x_1, \ldots, x_n)$. Thus, to construct the knowledge base of a probabilistic expert system, we need to specify the JPD of the variables.

The most general model is based on a direct specification of the JPD. That is, a numerical value (parameter) is associated with every possible combination of values. Unfortunately, the direct specification of a JPD involves a huge number of parameters. For example, with n binary variables, the most general JPD has 2^n parameters (the probabilities $p(x_1, \ldots, x_n)$ for every possible realization $\{x_1, \ldots, x_n\}$ of the variables), a number so large that no existing computer can store it even for n as small as 50. This was one of the early criticisms of using probability in expert systems. In most practical situations, however, many subsets of the variables can be independent or conditionally independent. In such cases, simplifications of the most general model can be obtained by exploiting the independency

92   3. Probabilistic Expert Systems

structure among the variables. This may result in significant reductions in the number of parameters. In this section, we discuss the following examples of such simplifications:

1. The Dependent Symptoms Model (DSM).

2. The Independent Symptoms Model (ISM).

3. The Independent Relevant Symptoms Model (IRSM).

4. The Dependent Relevant Symptoms Model (DRSM).

These four models, however, are ad hoc models that apply mainly in the medical field (see Castillo and Alvarez (1991)). More general and powerful probabilistic models (e.g., Markov network models, Bayesian network models, and conditionally specified models) are presented in Chapters 6 and 7. These models can be used in medical as well as other practical applications.

To present the above four models we use the medical diagnosis problem introduced in Section 3.2.4, where we have n symptoms $S_1, \ldots, S_n$, and a random variable D that can take one of m possible values $d_1, \ldots, d_m$. We wish to diagnose the presence of a disease given a set of symptoms $s_1, \ldots, s_k$. The JPD of the disease and symptoms is $p(d, s_1, \ldots, s_n)$.

As we mentioned above, the most general form of this JPD depends on an infeasibly large number of parameters. To reduce the number of parameters, one can impose certain assumptions (constraints) among them. The models presented in the following subsections are examples of such imposed assumptions. In all four models, the diseases are assumed to be independent given the symptoms.

3.5.1 The Dependent Symptoms Model

In this model, it is assumed that the symptoms are dependent but the diseases are independent of each other given the symptoms. The DSM is illustrated graphically in Figure 3.6, where every symptom is connected to every other symptom and to every possible value of D (indicating dependence).

The JPD for the DSM can then be written as

$$p(d_i, s_1, \ldots, s_n) = p(s_1, \ldots, s_n)p(d_i|s_1, \ldots, s_n). \qquad (3.21)$$

Note that this equation is obtained using (3.6) with $X = \{D\}$ and $Y = \{S_1, \ldots, S_n\}$. Now, $p(d_i|s_1, \ldots, s_n)$ can be expressed as

$$p(d_i|s_1, \ldots, s_n) = \frac{p(d_i, s_1, \ldots, s_n)}{p(s_1, \ldots, s_n)} \qquad (3.22)$$

$$= \frac{p(d_i)p(s_1, \ldots, s_n|d_i)}{p(s_1, \ldots, s_n)} \qquad (3.23)$$

$$\propto p(d_i)p(s_1, \ldots, s_n|d_i). \qquad (3.24)$$

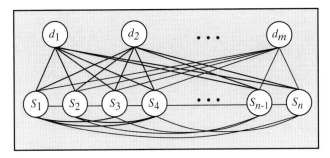

FIGURE 3.6. A graphical illustration of the dependent symptoms model.

The first of the above equations follows from (3.5), and the second is obtained by applying (3.6). The proportionality follows because $p(s_1, \ldots, s_n)$ is a normalizing constant.

Note that (3.24) only includes prior and likelihood probabilities (conditional probabilities of the symptoms given one of the diseases) whose values can be estimated using the objective information given by the frequencies of diseases and symptoms in the population. From (3.24) we see that the parameters needed for the knowledge base of the DSM are

- The marginal probabilities $p(d_i)$, for all possible values of D.

- The likelihood probabilities $p(s_1, \ldots, s_n | d_i)$, for all possible combinations of values of symptoms and diseases.

For example, for m possible diseases and n binary symptoms, the marginal probability distribution of D, $p(d_i)$, depends on $m-1$ parameters (because the m parameters must add up to one). Thus, we need to specify $m-1$ parameters for the marginal probability distribution of D. For the likelihood probabilities $p(s_1, \ldots, s_n | d_i)$, we need to specify $(2^n - 1)$ parameters[3] for each possible value of D, a total of $m(2^n - 1)$ parameters. Thus, the DSM requires a total of $m - 1 + m(2^n - 1) = m2^n - 1$ parameters.

Example 3.6 The DSM. To illustrate the DSM consider the data in Example 3.4, which are given in Figure 3.1. The gastric adenocarcinoma is the only disease of interest here. Thus, the variable D takes two possible values, g (when a patient has gastric adenocarcinoma) and $\bar{g}$ (when a patient does not have gastric adenocarcinoma). There are three binary symptoms, P, V, and W. It is sometimes convenient to use 1 and 0 to indicate the presence and absence of the symptom, respectively. To specify the DSM, we need the marginal probability distribution $p(d_i)$ and the conditional probability

---

[3]Note that for n binary symptoms there are 2^n parameters (one parameter for each possible combination of the symptoms). However, these parameters must add up to one; hence we have only $2^n - 1$ free parameters for each possible value of D.

d	p(d)
$\bar{g}$	0.3
g	0.7

			p(p,v,w|d)	
p	v	w	d = g	d = $\bar{g}$
0	0	0	0.014	0.377
0	0	1	0.136	0.253
0	1	0	0.014	0.167
0	1	1	0.136	0.103
1	0	0	0.036	0.040
1	0	1	0.314	0.030
1	1	0	0.036	0.017
1	1	1	0.314	0.013

TABLE 3.11. Probability distributions required for the specification of the DSM.

distribution of the symptoms given the disease, $p(p, v, w|d_i)$. These probability distributions are extracted from Figure 3.1 and tabulated in Table 3.11.

Using (3.24) and the probability distributions in Table 3.11, we can compute the probability of the disease given any combination of symptoms. These probabilities are given in Table 3.12. For example, the conditional probability distribution of the disease given that all three symptoms are present is calculated as follows:

$$p(\bar{g}|p, v, w) \propto p(\bar{g})p(p, v, w|\bar{g}) = 0.3 \times 0.013 = 0.0039,$$
$$p(g|p, v, w) \propto p(g)p(p, v, w|g) = 0.7 \times 0.314 = 0.2198.$$

Dividing by the normalizing constant $0.2198 + 0.0039 = 0.2237$, we obtain

$$p(\bar{g}|p, v, w) = 0.0039/0.2237 = 0.02,$$
$$p(g|p, v, w) = 0.2198/0.2237 = 0.98,$$

which are given in the last row of Table 3.12. ∎

The main problem with the DSM is the extremely large number of parameters required. Clearly, specifying the frequencies for all these combinations becomes increasingly difficult and then impossible as the number of diseases and symptoms increases. For example, with 100 diseases and 200 symptoms (which is not an unrealistic situation), the number of frequencies (parameters) needed is larger than 10^{62}, so big that no existing computer can store it.

The above discussion assumes binary symptoms (symptoms having only two possible options such as fever, no fever; pain, no pain; etc.). The difficulties with the DSM are magnified even further in cases where symptoms have multiple options or levels such as high fever, medium fever, low fever, and no fever.

p	v	w	$d=g$	$d=\bar{g}$
0	0	0	0.08	0.92
0	0	1	0.56	0.44
0	1	0	0.17	0.83
0	1	1	0.75	0.25
1	0	0	0.68	0.32
1	0	1	0.96	0.04
1	1	0	0.83	0.17
1	1	1	0.98	0.02

TABLE 3.12. The DSM: The $p(d|p,v,w)$ for $d=\bar{g}$ and $d=g$.

3.5.2 The Independent Symptoms Model

Due to the infeasibility of the DSM in many practical situations, further simplification of the model is necessary. A possible simplification is to assume that for a given disease, the symptoms are conditionally independent of each other. The resultant model is called the *independent symptoms model* (ISM). The ISM is illustrated in Figure 3.7, where the symptoms are not linked, indicating independence.

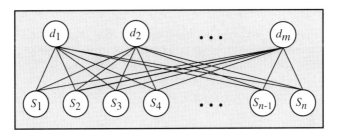

FIGURE 3.7. A graphical illustration of the independent symptoms model.

Because the symptoms are assumed to be conditionally independent given the disease, we have

$$p(s_1,\ldots,s_n|d_i) = \prod_{j=1}^{n} p(s_j|d_i). \qquad (3.25)$$

Thus, we can write the CPD of disease D given the symptoms $s_1,\ldots,s_n$ as

$$p(d_i|s_1,\ldots,s_n) = \frac{p(d_i)p(s_1,\ldots,s_n|d_i)}{p(s_1,\ldots,s_n)}$$

$$= \frac{p(d_i)\prod_{j=1}^{n} p(s_j|d_i)}{p(s_1,\ldots,s_n)} \qquad (3.26)$$

96   3. Probabilistic Expert Systems

$$\propto p(d_i) \prod_{j=1}^{n} p(s_j|d_i). \qquad (3.27)$$

Substituting (3.26) in (3.21), we obtain the ISM.

Equation (3.26) shows how the independence assumption modifies the probabilities of all diseases when new symptoms become known. Thus, the initial probability of disease d_i is $p(d_i)$, but after knowing symptom s_j, for $j = 1, \ldots, k$, it becomes proportional to $p(s_j|d_i)$. Note that each new symptom leads to a new factor. Note also that $p(s_1, \ldots, s_n)$ in the denominator of (3.26) is a normalizing constant and need not be explicitly computed.

From (3.27), the parameters needed for the knowledge base of the ISM are

- The marginal probabilities $p(d_i)$, for all possible values of disease D.

- The conditional probabilities $p(s_j|d_i)$, for all possible values of symptom S_j and disease D.

Thus, with the independent symptoms assumptions, the number of parameters is considerably reduced. With m possible diseases and n binary symptoms, the total number of parameters is $m(n+1) - 1$. For example, with $m = 100$ diseases and $n = 200$ symptoms, we have 20,099 parameters in the ISM as opposed to more than 10^{62} parameters for the DSM.

Example 3.7 The ISM. To illustrate the ISM, we use the records of two medical centers, each of which consists of $N = 1,000$ patients; two values of D (g and $\bar{g}$); and three symptoms, P, V, and W. The data are summarized in Table 3.13. Notice that the data for Medical Center 1 are the same as the data in Figure 3.1, but given here in a tabular instead of graphical form.

To specify the ISM, we need the marginal probability distribution, $p(d_i)$, of the disease and the conditional probability distribution of each symptom given the disease, $p(p|d_i), p(v|d_i)$, and $p(w|d_i)$. These probability distributions are extracted from Table 3.13 and given in Table 3.14. Note that only 7 parameters are free. One interesting aspect of the two data sets is that although they are quite different, they lead to identical probability distributions as given in Table 3.14.

The conditional probability distribution of D given various combinations of the symptoms for the two medical centers are given in Table 3.15. Note that $p(\bar{g}|p, v, w) = 1 - p(g|p, v, w)$. The true values are calculated directly from Table 3.13 using the definition of conditional probability in (3.5). The values in the columns labeled ISM are calculated using the formula for the ISM in (3.27). For example, for Medical Center 1, the true value of $p(g|p, v, w)$ is calculated by

$$p(g|p, v, w) = \frac{p(g, p, v, w)}{p(p, v, w)} = \frac{220}{220 + 4} = 0.98.$$

3.5 The Knowledge Base   97

Medical Center 1						
		g		$\bar{g}$		
		p	$\bar{p}$	p	$\bar{p}$	Total
v	w	220	95	4	31	350
	$\bar{w}$	25	10	5	50	90
$\bar{v}$	w	220	95	9	76	400
	$\bar{w}$	25	10	12	113	160
Total		490	210	30	270	1,000
Medical Center 2						
		g		$\bar{g}$		
		p	$\bar{p}$	p	$\bar{p}$	Total
v	w	140	210	0	0	350
	$\bar{w}$	0	0	30	60	90
$\bar{v}$	w	280	0	0	120	400
	$\bar{w}$	70	0	0	90	160
Total		490	210	30	270	1,000

TABLE 3.13. The number of patients classified by a disease G and three symptoms, P, V, and W in two medical centers.

d	$p(d)$
$\bar{g}$	0.3
g	0.7

d	p	$p(p\|d)$
$\bar{g}$	0	0.9
$\bar{g}$	1	0.1
g	0	0.3
g	1	0.7

d	v	$p(v\|d)$
$\bar{g}$	0	0.7
$\bar{g}$	1	0.3
g	0	0.5
g	1	0.5

d	w	$p(w\|d)$
$\bar{g}$	0	0.6
$\bar{g}$	1	0.4
g	0	0.1
g	1	0.9

TABLE 3.14. Probability distributions required for the specification of the ISM.

The value of $p(g|p, v, w)$ according to the ISM is computed using (3.27) as follows:

$$p(g|p, v, w) \propto p(g)p(p|g)p(v|g)p(w|g) = 0.7 \times 0.7 \times 0.5 \times 0.9 = 0.2205,$$
$$p(\bar{g}|p, v, w) \propto p(\bar{g})p(p|\bar{g})p(v|\bar{g})p(w|\bar{g}) = 0.3 \times 0.1 \times 0.3 \times 0.4 = 0.0036.$$

Dividing 0.2205 by the normalizing constant $0.2205 + 0.0036 = 0.2241$, we obtain $p(g|p, v, w) = 0.2205/0.2241 = 0.98$ and $p(\bar{g}|p, v, w) = 0.0036/0.2241 = 0.02$.

A comparison between the true and ISM probabilities in Table 3.15 shows that the two sets of probabilities are in close agreement for Medical Center 1 but that they are in sharp disagreement for Medical Center 2. For example, for Medical Center 2 the true value of $p(g|p, v, \bar{w})$ is 0 as compared to the ISM value of 0.82. This indicates the failure of the ISM for describing the probability distribution of the data for Medical Center 2. Note that we

			Medical Center 1		Medical Center 2	
p	v	w	True	ISM	True	ISM
0	0	0	0.08	0.08	0.00	0.08
0	0	1	0.56	0.56	0.00	0.56
0	1	0	0.17	0.18	0.00	0.18
0	1	1	0.75	0.74	1.00	0.74
1	0	0	0.68	0.66	1.00	0.66
1	0	1	0.96	0.96	1.00	0.96
1	1	0	0.83	0.82	0.00	0.82
1	1	1	0.98	0.98	1.00	0.98

TABLE 3.15. The $p(g|p, v, w)$ for the data in Table 3.13. The true values are computed using the definition of conditional probability in (3.5). The ISM values are computed using the ISM formula in (3.27). Note that $p(\bar{g}|p, v, w) = 1 - p(g|p, v, w)$.

have two data sets with the same prior probabilities and likelihoods, yet the ISM is appropriate for one and not for the other. From this example one can conclude that prior probabilities and likelihoods are not generally sufficient for specifying a probabilistic model. ∎

Example 3.7 illustrates the fact that the performance of a probabilistic expert system hinges on the correct specification of the JPD. Therefore, care must be taken when choosing a probabilistic model.

Although the assumption of independence gives rise to a large reduction in the number of parameters, the number of parameters in the ISM is still too high to be practical. We therefore need to simplify the model further.

3.5.3 Independent Relevant Symptoms Model

A further reduction of the number of parameters can be achieved by assuming that each disease has a small set of relevant symptoms. Accordingly, for each value d_i of disease D some relevant symptoms $S_1, \ldots, S_r$ (relatively few) are selected and the rest of the symptoms are assumed to be independent for that value of D. This IRSM is illustrated in Figure 3.8. Note that for d_1, the set of relevant symptoms is $\{S_1, S_2\}$; for d_2, the set of relevant symptoms is $\{S_2, S_3, S_4\}$; and so on.

For simplicity of notation, suppose that $S_1, \ldots, S_{r_i}$ are relevant to disease d_i and that the remaining symptoms $S_{r_i+1}, \ldots, S_n$ are irrelevant. According to the IRSM, $p(s_j|d_i)$ is the same for all symptoms that are irrelevant to the disease d_i. Then the CPD of disease d_i given the symptoms $s_1, \ldots, s_n$ can be written as

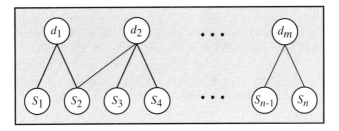

FIGURE 3.8. A graphical illustration of the independent relevant symptoms model.

$$p(d_i|s_1,\ldots,s_n) = \frac{p(d_i)p(s_1,\ldots,s_n|d_i)}{p(s_1,\ldots,s_n)}$$

$$= \frac{p(d_i)\prod_{j=1}^{r_i}p(s_j|d_i)\prod_{j=r_i+1}^{n}p(s_j|d_i)}{p(s_1,\ldots,s_n)}$$

$$= \frac{p(d_i)\prod_{j=1}^{r_i}p(s_j|d_i)\prod_{j=r_i+1}^{n}p_j}{p(s_1,\ldots,s_n)} \quad (3.28)$$

$$\propto p(d_i)\prod_{j=1}^{r_i}p(s_j|d_i)\prod_{j=r_i+1}^{n}p_j, \quad (3.29)$$

where $p_j = p(s_j|d_i)$, which is the same for all diseases to which S_j is irrelevant. Substituting (3.28) in (3.21), we obtain the IRSM.

From (3.29), we need to store the following probabilities in the knowledge base of the IRSM:

- The marginal probabilities $p(d_i)$, for all possible values of disease D.

- The conditional probabilities $p(s_j|d_i)$, for each possible value of D and each of their corresponding relevant symptoms.

- The probabilities p_j, for each possible value of D that has at least one irrelevant symptom. (This implies that $p_j = p(s_j|d_i)$ is the same for all symptoms irrelevant to d_i.)

Equation (3.28) implies that in the knowledge base we need to store the probabilities of every relevant symptom for every disease, and the same probability for all symptoms irrelevant to each value of D. Thus, for m possible diseases and n binary symptoms, the number of parameters in the IRSM is

$$m - 1 + n - a + \sum_{i=1}^{m}r_i, \quad (3.30)$$

where r_i is the number of relevant symptoms for disease d_i and a is the number of symptoms that are relevant to all diseases. The number of parameters is significantly reduced when r_i is much less than n. For example, with 100 diseases and 200 symptoms, if $r_i = 10$ for all diseases,[4] the number of parameters in the IRSM is reduced from 20,099 for the ISM to 1,299 for the IRSM.

Note that the IRSM can be obtained from the ISM by specifying further restrictions on the parameters of the ISM, because in the IRSM the probabilities $p(s_j|d_i)$ must be the same for all symptoms irrelevant to diseases d_i. The number of constraints is

$$a - n + \sum_{j=1}^{n} n_j,$$

where n_j is the number of diseases to which the symptom S_j is irrelevant. Thus, the number of parameters in the IRSM is equal to the number of parameters in the ISM, $(m(n+1) - 1)$, minus the number of constraints. This gives,

$$m(n+1) - 1 + n - a - \sum_{j=1}^{n} n_j, \qquad (3.31)$$

which is equal to the formula in (3.30).

3.5.4 The Dependent Relevant Symptoms Model

Although the IRSM reduces the number of parameters considerably, unfortunately, it is unrealistic because symptoms associated with certain diseases come in groups known as syndromes. Thus, it may be unreasonable to assume that the relevant symptoms are independent. The *dependent relevant symptoms model* (DRSM) avoids this shortcoming. The DRSM is the same as the IRSM but without requiring the relevant symptoms to be independent, given the corresponding disease. In this way, we only assume that irrelevant symptoms are independent but relevant symptoms may be dependent. Thus, one may think of the DRSM as a compromise between the DSM and IRSM. The DRSM is illustrated in Figure 3.9. By comparison with Figure 3.8, we see that in the DRSM, the symptoms relevant to each disease are connected, indicating dependence.

Suppose that $S_1, \ldots, S_{r_i}$ are relevant to disease d_i and that the remaining symptoms $S_{r_i+1}, \ldots, S_n$ are irrelevant. Then according to the DRSM, the CPD of d_i given the symptoms $s_1, \ldots, s_n$ can be written as

---

[4]Note that $r_i = 10$ for all diseases implies that $a = 0$, that is, every disease has at least one irrelevant symptom.

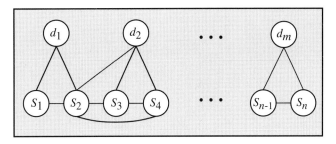

FIGURE 3.9. A graphical illustration of the dependent relevant symptoms model.

$$p(d_i|s_1,\ldots,s_n) = \frac{p(d_i)p(s_1,\ldots,s_{r_i}|d_i)\prod_{j=r_i+1}^{n}p(s_j|d_i)}{p(s_1,\ldots,s_n)}$$

$$= \frac{p(d_i)p(s_1,\ldots,s_{r_i}|d_i)\prod_{j=r_i+1}^{n}p_j}{p(s_1,\ldots,s_n)} \quad (3.32)$$

$$\propto p(d_i)p(s_1,\ldots,s_{r_i}|d_i)\prod_{j=r_i+1}^{n}p_j, \quad (3.33)$$

where $p_j = p(s_j|d_i)$, which is the same for all diseases to which S_j is irrelevant. Substituting (3.32) in (3.21), we obtain the DRSM. For this model, we need to store the following probabilities in the knowledge base:

- The marginal probabilities $p(d_i)$, for all possible values of disease D.

- The conditional probabilities $p(s_1,\ldots,s_{r_i}|d_i)$, for all possible combinations of values of disease D and its relevant symptoms $S_1,\ldots,S_{r_i}$.

- The probabilities p_j, for each possible value of D that has at least one irrelevant symptom. (As in the IRSM, this implies that $p_j = p(s_j|d_i)$ is the same for for all symptoms irrelevant to d_i.)

Accordingly, for m binary diseases and n binary symptoms, the total number of parameters in the DRSM is

$$m - 1 + n - a + \sum_{i=1}^{m}(2^{r_i} - 1) = n - 1 - a + \sum_{i=1}^{m}2^{r_i}. \quad (3.34)$$

Note that when $r_i = r$ for all d_i, (3.34) becomes $m2^r + n - 1$. Note also that if all symptoms are relevant to every disease ($a = n$ and $r_i = n$ for all D), the DRSM becomes the DSM. Table 3.16 shows a comparison of the number of parameters required for specifying the models discussed in this section in the case of $m = 100$ binary diseases, $n = 200$ binary symptoms, and $r = 10$ relevant symptoms per disease.

	Number of parameters	
Model	Formula	Value
DSM	$m2^n - 1$	$> 10^{62}$
ISM	$m(n+1) - 1$	20,099
IRSM	$m(r+1) + n - 1$	1,299
DRSM	$m2^r + n - 1$	102,599

TABLE 3.16. The number of parameters required for specifying four models in the case of $m = 100$ binary diseases, $n = 200$ binary symptoms, and $r = 10$ relevant symptoms per disease.

In the DRSM the number of parameters is greatly reduced compared to the DSM, yet the model is realistic because it considers the dependencies among the most important (relevant) symptoms for each disease. However, because of the dependence assumption, the number of parameters in the DRSM is larger than the number of parameters in the ISM and IRSM models.

An extra reduction can be obtained by dividing the set of relevant symptoms into subsets (blocks) that are assumed to be mutually independent, but symptoms inside each block are dependent.

3.5.5 Concluding Remarks

In this section we discussed four ad hoc models for describing the relationships among a set of variables. The set of parameters needed to define the knowledge base depends on the choice of model. Each of these models has its own advantages and disadvantages. However, these four models only apply in particular situations. In Chapters 6 and 7, we introduce more general probabilistic models such as Markov network models, Bayesian network models, models specified by input lists, and conditionally specified models.

Whatever model is adopted, however, the knowledge base must contain the set of variables of interest and the minimum set of parameters (probabilities or frequencies) needed to specify the JPD of the variables.

3.6 The Inference Engine

There are two types of knowledge in probabilistic expert systems:

1. The *abstract knowledge*, which consists of the set of variables and the associated set of probabilities needed to construct the JPD of the variables. This type of knowledge is stored in the knowledge base.

2. The *concrete knowledge*, which consists of a set of values of some variables (e.g., symptoms) known to be true to the user. This information is referred to as the *evidence* set. It is stored in the working memory.

The inference engine uses both the abstract and concrete knowledge to answer certain questions or queries posed by the user. Examples of such queries are:

- **Question 1:** Before a patient is examined by a doctor, which disease is the patient most likely to have? Here, no evidence is available. The patient is yet to be examined and the set of symptoms that the patient exhibits is yet to be determined. The problem is to compute the marginal (initial) probability distribution of D,

$$p(D = d_i), \quad i = 1, \ldots, m.$$

- **Question 2:** Given that a patient has a subset of symptoms $S_1 = s_1, \ldots, S_k = s_k$, which disease is the patient most likely to have? The evidence set in this case consists of the values $s_1, \ldots, s_k$. The problem at hand is to compute the CPD for each disease d_i given the evidence $s_1, \ldots, s_k$:

$$p(D = d_i | s_1, \ldots, s_k), \quad i = 1, \ldots, m.$$

The marginal probability distribution of D, $p(D = d_i)$, is also known as the *prior distribution* because it is computed before the evidence. The conditional probability distribution of d_i given a realization of the set of symptoms $p(d_i | s_1, \ldots, s_k)$ is known as the *posterior distribution* because it is computed after knowing the evidence. Note that the marginal (prior) probability distribution can be thought of as a special case of the conditional (posterior) probability distribution, where the set of observed symptoms is the empty set, ϕ.

One task of the inference engine in probabilistic expert systems is to compute the conditional probabilities of different diseases when new symptoms or data become known. The inference engine is responsible for updating the conditional probabilities:

$$p(d_i | s_1, \ldots, s_k) = \frac{p(d_i, s_1, \ldots, s_k)}{p(s_1, \ldots, s_k)}; \quad i = 1, \ldots, m, \tag{3.35}$$

for all possible subsets of the symptoms, and to decide which ones have high conditional probabilities. Normally a small number is selected and these are listed for the user (e.g., medical doctors and patients) to observe and make the appropriate decisions.

In (3.35), the role of the term $p(s_1, \ldots, s_k)$ is to act as a normalizing constant. Therefore, a decision based on the maximum of $p(d_i | s_1, \ldots, s_k)$

coincides with that based on the maximum of $p(d_i, s_1, \ldots, s_k)$. Thus, the ratios

$$R_i = \frac{p(d_i, s_1, \ldots, s_k)}{\max_i p(d_i, s_1, \ldots, s_k)}; \; i = 1, \ldots, m, \qquad (3.36)$$

provide information about the relative significance of any of the diseases.

Note that Bayes' theorem is used to compute the posterior probabilities with ease when we have only a few diseases and symptoms. But when the number of variables (diseases and/or symptoms) is large, which is usually the case in practice, more efficient models and methods are needed to compute both the initial and the posterior probabilities. These methods, which are known as *evidence or uncertainty propagation* methods, are presented in Chapters 8, 9 and 10.

3.7 Coherence Control

One of the most serious problems in expert systems is the presence of some incoherences in its knowledge base and/or working memory. There are several reasons for this. For example,

1. Human experts can provide inconsistent abstract knowledge.

2. The user can provide inconsistent concrete knowledge.

3. The inference engine does not update the facts (see Section 2.4.2).

4. There is no coherence control subsystem that prevents inconsistent knowledge from reaching the knowledge base and/or working memory.

We use the following examples to illustrate the importance of maintaining a coherent knowledge in expert systems.

Example 3.8 Two-variable constraints. Suppose we have only two binary variables, D and S. As shown in the previous section, the probabilities required for the knowledge base of any of the above introduced methods are $p(d)$, $p(s)$, $p(s|d)$. Thus the expert system starts by asking the user to provide values for $p(d)$ and $p(s)$. These values must satisfy the trivial constraints $0 \leq p(d) \leq 1$ and $0 \leq p(s) \leq 1$. Once $p(d)$ and $p(s)$ are specified and checked, the system asks the user for the values of $p(s|d)$. The system should then inform the user about the constraints that these values have to satisfy. For example, give lower and upper bounds for them. In some cases, some of the values are redundant and the expert systems should automatically assign the appropriate values without asking the user. For example,

$$p(s|D=0) + p(s|D=1) = p(s), \text{ for all } s.$$

Thus, we have

$$p(s|D=1) = p(s) - p(s|D=0). \qquad (3.37)$$

Therefore, once $p(s)$ is known, the expert system does not need to ask the user for all the values $p(s|d)$, because only two of them are necessary: $p(S=0|D=0)$ and $p(S=1|D=0)$. Furthermore, these two probabilities must add up to one. Therefore, only one of these probabilities is enough to define the knowledge base. ■

In addition to relationships among the different probabilities involved in the definition of the JPD, there are also some constraints the probabilities have to satisfy in order to be consistent. Therefore, the coherence control should be able to inform the user about the constraints for the new pieces of information. The following example illustrates this idea.

Example 3.9 Two sets constraints. Suppose we have only two sets A and B. The probabilities that can be involved in the definition of the knowledge base of a probabilistic expert system are $p(A), p(B), p(A \cup B)$, and $p(A \cap B)$. These probabilities must satisfy the following constraints:

$$\begin{aligned} & 0 \leq p(A) \leq 1, \\ & 0 \leq p(A) \leq 1, \\ & max\{0, p(A) + p(B) - 1\} \leq p(A \cap B) \leq min\{p(A), p(B)\}, \\ & max\{p(A), p(B)\} \leq p(A \cup B) \leq min\{1, p(A) + p(B)\} \end{aligned} \qquad (3.38)$$

Note that the constraint $p(A) + p(B) - 1 \leq p(A \cap B)$ is obtained as follows:

$$\begin{aligned} p(A \cap B) &= p(\overline{\overline{A \cap B}}) = p(\overline{\overline{A} \cup \overline{B}}) = 1 - p(\overline{A} \cup \overline{B}) \\ &\geq 1 - (1 - p(A) + 1 - p(B)) = p(A) + p(B) - 1. \end{aligned}$$

Thus the expert system starts by asking the user about the values of $p(A)$ and $p(B)$. These values must satisfy the first two constraints in (3.38). Once $p(A)$ and $p(B)$ are specified and checked, the knowledge acquisition subsystem asks for the values of either $p(A \cap B)$ or $p(A \cup B)$; the system should inform the user about the upper and lower bounds for these probabilities given in the last two constraints in (3.38). Otherwise, values outside the coherent intervals can be given. In that case the axioms of probability would be violated and the system will likely generate wrong conclusions. Suppose that $p(A \cap B)$ was given and checked; then $p(A \cup B)$ is automatically assigned the value

$$p(A \cup B) = p(A) + p(B) - p(A \cap B), \qquad (3.39)$$

in accordance with (3.2). ■

The reader can imagine how complex the constraints can get as the number of subsets increases. Therefore, the chance that the user will violate

106   3. Probabilistic Expert Systems

the constraints increases as the number of variables increases. In these situations it is important to have a system able to control the coherence of knowledge (Smith (1961)).

In some probabilistic models (e.g. Bayesian network models presented in Chapter 6), coherence control is not a problem because the models are coherent by construction. In other probabilistic models, however, the coherence must be controlled.

In some probabilistic models the coherence control subsystem is a necessity, not a luxury. The coherence control subsystem prevents incoherent knowledge from entering the knowledge base and/or the working memory. A method for checking the consistency of a probabilistic model is described in Chapter 7.

3.8 Comparing Rule-Based and Probabilistic Expert Systems

We conclude this chapter with a brief comparison between rule-based and probabilistic expert systems. We discuss their similarities and differences and their advantages and disadvantages. Table 3.17 gives a summary of some of the components of each type of expert system and the structure (logic or probabilistic) on which they are based.

1. **Knowledge Base:**

    The abstract knowledge of a rule-based expert system consists of the objects and the set of rules. The abstract knowledge of a probabilistic expert system consists of the probability space, which includes variables, their possible values, and their JPD. On the other hand, the concrete knowledge in both systems is the facts, that is, evidence associated with cases to be analyzed.

    The knowledge base in rule-based expert systems is easy to implement, since we need to handle only simple elements, such as objects, sets of values, premises, conclusions, and rules. However, the knowledge we can store is limited when compared with probabilistic expert systems. A shortcoming of probabilistic systems is the large number of parameters involved, which makes the specification of the probabilistic model difficult.

2. **Inference Engine:**

    In rule-based expert systems, conclusions are drawn from facts applying inference strategies such as modus ponens, modus tollens, and rule chaining. Thus, the inference engine is fast and easy to implement. In probabilistic systems the inference engine is more complicated than in the case of rule-based expert systems. The inference engine of a

3.8 Comparing Rule-Based and Probabilistic Expert Systems

	Rule-Based	Probabilistic
Knowledge base	Abstract: Objects, rules Concrete: Facts	Abstract: Variables, JPD Concrete: Facts
Inference engine	Inference strategies Rule chaining	Conditional probability evaluation methods
Explanation subsystem	Based on active rules	Based on conditional probabilities
Learning	Change in objects and rules	Change in probabilistic model

TABLE 3.17. A comparison between rule-based and probabilistic expert systems.

probabilistic expert system is based on the evaluation of conditional probabilities using one of several methods proposed for the different types of probabilistic expert systems (see Chapters 8 and 9). The degree of difficulty depends on the selected model and ranges from low, for general independence models, to high, for general dependence models.

3. **Explanation Subsystem:**

   Explanation is easy in the case of rule-based expert systems since we know which rules are active at a given moment. The inference engine knows which rules have been used (are active) in the chaining process and have contributed to obtaining conclusions and which rules were tried without success.

   In the case of probabilistic expert systems, the information about which variables influence others is encoded in the JPD. Then, explanation is based on the relative values of conditional probabilities that measure the degree of the dependencies. A comparison of conditional probabilities for different sets of conditioning evidences allows analyzing their effects and drawing conclusions.

4. **Learning Subsystem:**

   In rule-based expert systems, learning consists of incorporating new objects, new sets of feasible values for the objects, new rules or modification of existing objects, sets of values, or rules. In probabilistic expert systems, learning consists of incorporating or modifying the structure of the probability space: variables, sets of possible values, or the parameters (probability values).

Exercises

3.1 Use the JPD in Table 3.2 to compute the following CPD, for all values of x, y, and z:

(a) $p(x|y,z)$.

(b) $p(y|x,z)$.

(c) $p(z|x,y)$.

3.2 Construct a JPD of three variables X, Y, and Z from which one can conclude that X and Y are independent, X and Z are dependent, and Y and Z are dependent. Then, use the JPD to compute the following CPD, for all values of x, y, and z:

(a) $p(y|x)$.

(b) $p(x|y)$.

(c) $p(x|y,z)$.

3.3 Consider the JPD in Table 3.2.

(a) Generate all possible CISs involving the variables X, Y, and Z: $\{I(X,Y|\phi), \ldots\}$.

(b) Check which ones of these CISs are implied by the JPD in Table 3.2.

3.4 In Example 3.4 we applied Bayes' theorem to show that after observing the evidences $V = v$ and $W = w$, the posterior probability of gastric adenocarcinoma is 0.9. Complete the diagnosis problem by calculating the posterior probabilities with the additional information $P = p$. With this additional information, what is the chance that the diagnosis of the disease is wrong? How does this probability change when $P = \bar{p}$?

3.5 Use the data in Example 3.4 to compute the posterior probabilities of gastric adenocarcinoma

(a) Using Bayes' theorem.

(b) Using the definition of conditional probability and Figure 3.1.

Consider the cases given by the following sets of evidences:

(a) $V = \bar{v}$ and $W = w$.

(b) $V = v$ and $W = \bar{w}$.

(c) $V = \bar{v}$ and $W = \bar{w}$.

(d) $V = \bar{v}$, $W = \bar{w}$, and $P = \bar{p}$.

3.8 Comparing Rule-Based and Probabilistic Expert Systems

$Disease$	Relevant symptoms
D_1	S_1, S_2, S_5
D_2	S_2, S_3, S_5
D_3	S_3, S_4, S_5

TABLE 3.18. Diseases and their relevant symptoms.

3.6 Show that the two formulas in (3.30) and (3.31) for the number of parameters in the IRSM are equal.

3.7 Given a population of patients classified by five mutually exclusive diseases $D_1, \ldots, D_5$ and three binary symptoms S_1, S_2, S_3. Make appropriate assumptions for each of the four probabilistic expert system models described in Section 3.5, then determine the number of parameters required by each model.

3.8 Given a population of patients classified by three mutually exclusive diseases D_1, D_2, D_3 and five binary symptoms $S_1, \ldots, S_5$. What are the parameters that are required to specify each of the following models:

(a) The ISM.
(b) The IRSM, given that the relevant symptoms to each of the three diseases are as shown in Table 3.18.
(c) The DRSM, given that the relevant symptoms to each of the three diseases are as shown in Table 3.18.

3.9 Consider the medical diagnosis problem described in Example 3.5 and suppose we want to build an ISM probabilistic expert system for this diagnosis problem. Write a computer program to do the following:

(a) Read the prior probabilities $p(d_i)$, $i = 1, \ldots, m$, from a text file.
(b) Read the likelihoods $p(s_j|d_i)$, $i = 1, \ldots, d$; $j = 1, \ldots, n$, from a text file.
(c) Update the probabilities of diseases after symptoms are known, using the JPD of the resulting model (3.26) and Bayes' theorem (3.16).

3.10 Five different human experts have been asked to give values for the following probabilities and in the order indicated:

$$p(a), p(b), p(c), p(a,b), p(a,c), p(b,c), \text{ and } p(a,b,c).$$

The data are given in Table 3.19. Using the results in Section 3.7, determine whether the information given by each expert is coherent.

		Human Expert				
Order	Data	1	2	3	4	5
1	$p(a)$	0.8	0.8	0.5	0.5	0.6
2	$p(b)$	0.7	0.7	0.6	0.6	0.5
3	$p(c)$	0.5	0.5	0.6	0.7	0.4
4	$p(a,b)$	0.6	0.2	0.3	0.4	0.3
5	$p(a,c)$	0.4	0.2	0.3	0.2	0.2
6	$p(b,c)$	0.2	0.3	0.4	0.4	0.2
7	$p(a,b,c)$	0.1	0.2	0.2	0.2	0.1

TABLE 3.19. Five sets of probabilities provided by five human experts.

3.11 In Example 3.9, we give the constraints needed to control the coherence in cases where we have two sets A and B. Now, consider three sets A, B, and C and let

$L_1 = \max\{0, p(A) + p(B) - 1\}$,
$U_1 = \min\{p(A), p(B)\}$,
$L_2 = \max\{0, p(A) + p(C) - 1\}$,
$U_2 = \min\{p(A), p(C)\}$,
$L_3 = \max\{0, p(A \cap B) + p(A \cap C) - p(A), p(B) + p(C) - 1,$
$\quad p(A) + p(B) + p(C) - 1 - p(A \cap B) - p(A \cap C)\}$,
$U_3 = \min\{p(C), p(B), p(C) - p(A \cap C) + p(A \cap B),$
$\quad p(B) - p(A \cap B) + p(A \cap C)\}$,
$L_4 = \max\{0, p(A \cap B) + p(A \cap C) - p(A), p(A \cap B) + p(B \cap C),$
$\quad -p(B), p(A \cap C) + p(B \cap C) - p(C)\}$,
$U_4 = \min\{p(A \cap B), p(A \cap C), p(B \cap C),$
$\quad p(A) + p(B) + p(C) - p(A \cap B) - p(A \cap C) - p(B \cap C) - 1\}$.

Show that the following constraints on probabilities are needed in this case:

(a) $0 \leq p(A) \leq 1$,
(b) $0 \leq p(B) \leq 1$,
(c) $0 \leq p(C) \leq 1$,
(d) $L_1 \leq p(A \cap B) \leq U_1$,
(e) $L_2 \leq p(A \cap C) \leq U_2$,
(f) $L_3 \leq p(B \cap C) \leq U_3$,
(g) $L_4 \leq p(A \cap B \cap C) \leq U_4$.

3.12 Suppose we wish to classify four objects: kite, bird, plane, and human, based on the following binary characteristics (variables): Fly (whether the object can fly), Engine (whether the object has an engine), and Blood (whether the object has blood). You may also identify other objects that can be classified using the above variables.

3.8 Comparing Rule-Based and Probabilistic Expert Systems   111

(a) Design a rule-based and a probabilistic expert system for solving the above classification problem.

(b) Which of the two expert systems is more efficient in this case?

3.13 Design a probabilistic expert system for helping students to select a major field of study. Proceed as follows:

(a) Select a set of $m = 10$ different fields of study $\{X_1, \ldots, X_{10}\}$.

(b) Select a set of $n = 5$ appropriate indicators (abilities) $\{Y_1, \ldots Y_5\}$ that can be used to select a field of a study.

(c) Estimate the prior probabilities $p(x_i), i = 1, \ldots, 10$ by the proportion of students in each of the fields.

(d) Specify the likelihoods $p(y_j|x_i)$ for each field X_i and each indicator Y_j by choosing reasonable values. Note that the prior and likelihood probabilities constitute the abstract knowledge.

(e) Use Bayes' theorem and the formulas in this chapter to design the inference engine.

3.14 Design a rule-based expert system for helping students to select a major field of study. Proceed as follows:

(a) Select a set of $m = 10$ different fields of study $\{X_1, \ldots, X_{10}\}$.

(b) Select a set of $n = 5$ appropriate indicators (abilities) $\{Y_1, \ldots Y_5\}$ that can be used to select a field of a study.

(c) Choose a reasonable set of rules relating abilities and fields of study.

(d) Use the inference strategies and rule chaining to design the inference engine.

Compare this expert system with the one in the previous exercise. Which one is more efficient in this situation?

Chapter 4
Some Concepts of Graphs

4.1 Introduction

In this chapter we introduce some elemental concepts of graph theory that are needed in the rest of the book. Graphs are essential tools for building probabilistic and other models used in artificial intelligence and expert systems. Many of the theoretical and practical results of graph theory can be used to analyze different aspects in this field. Readers who are familiar with these concepts can skim, or even skip, the chapter and go directly to Chapter 5. Readers who wish to read more about graph theory are referred to books such as Harary (1969), Berge (1973), Bondy and Murty (1976), Golumbic (1980), Liu (1985), Ross and Wright (1988), and Biggs (1989).

We start by introducing some basic concepts and definitions in Section 4.2. Two main types of graphs, *undirected* and *directed*, and their most important properties are presented in Sections 4.3 and 4.4, respectively. A special type of graph, called a *triangulated* graph, is discussed in Section 4.5. Other types of graphs resulting from grouping nodes in an original graph, such as *clique*, *join*, and *family* graphs are presented in Section 4.6. Different forms of graph representations such as *symbolic*, *pictorial*, and *numerical* representations are presented and compared from interpretative and algorithmic points of view in Section 4.7. Finally, Section 4.8 gives a series of graph algorithms with the help of which we can obtain some information or perform some operations on graphs.

114   4. Some Concepts of Graphs

4.2 Basic Concepts and Definitions

Suppose we have a set of possibly related objects $X = \{X_1, X_2, \ldots, X_n\}$. The set X can be represented pictorially by a set of *nodes*, or *vertices*, one node for each element of X. These nodes can be connected by line segments, arcs, or arrows, which are referred to as *links* or *edges*. If there is a link between two nodes X_i and X_j we use L_{ij} to denote such a link. The set of all links is denoted by $L = \{L_{ij} \mid X_i \text{ and } X_j \text{ are linked}\}$. The sets X and L define a graph. We first give an example, then a more formal definition of a graph.

Example 4.1 Graphs. Figure 4.1 is an example of a graph that consists of seven nodes $X = \{A, B, \ldots, G\}$, each one is represented by a circle, and a set of six links,

$$L = \{L_{AB}, L_{AC}, L_{BD}, L_{CE}, L_{DF}, L_{DG}\}.$$

Each link is represented by a line segment connecting two nodes. ∎

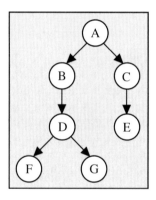

FIGURE 4.1. An example of graph or a network.

Definition 4.1 Graph or network. *A graph $G = (X, L)$ is defined by two sets X and L, where X is a finite set of nodes $X = \{X_1, X_2, \ldots, X_n\}$ and L is a set of links, that is, a subset of all possible ordered pairs of distinct nodes. The words graph and network are used synonymously in this book.*

A more general definition of a graph is possible. For example, a node can be connected to itself. However, we do not need such a generalization in this book. In the field of expert systems, graphs are used to represent a set of propositional variables (nodes), and the dependence relationships among them (links).

The links of a graph can be *directed* or *undirected*, depending on whether or not the order of the involved nodes matters.

Definition 4.2 Directed link. Let $G = (X, L)$ be a graph. When $L_{ij} \in L$ and $L_{ji} \notin L$, the link L_{ij} is called a directed link. A directed link between nodes X_i and X_j is denoted by $X_i \to X_j$.

Definition 4.3 Undirected link. Let $G = (X, L)$ be a graph. When $L_{ij} \in L$ and $L_{ji} \in L$, the link between nodes X_i and X_j is called an undirected link. An undirected link between nodes X_i and X_j is denoted by $X_i - X_j$ or $X_j - X_i$.

Definition 4.4 Directed and undirected graphs. A graph in which all the links are directed is called a directed graph and a graph in which all the links are undirected is called an undirected graph.

Thus, in a directed graph, the order of the nodes defining a link is important, whereas in an undirected graph, that order is immaterial.

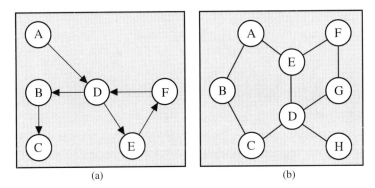

FIGURE 4.2. Examples of (a) a directed and (b) an undirected graph.

Example 4.2 Directed and undirected graphs. Examples of a directed and an undirected graph are given in Figures 4.2(a) and 4.2(b), respectively. The graph in Figure 4.2(a) is given by:

$$X = \{A, B, C, D, E, F\},$$
$$L = \{A \to D, B \to C, D \to B, F \to D, D \to E, E \to F\},$$

whereas in Figure 4.2(b), we have

$$X = \{A, B, C, D, E, F, G, H\},$$
$$L = \{A - B, B - C, C - D, D - E, E - A, E - F, F - G, G - D, D - H\}.$$

■

Definition 4.5 Adjacency set. Given a graph $G = (X, L)$ and a node X_i, the adjacency set of X_i is the set of nodes directly attainable from X_i, that is, $Adj(X_i) = \{X_j \in X \mid L_{ij} \in L\}$.

116   4. Some Concepts of Graphs

This definition provides an alternative description of a graph by specifying the set of nodes, X, and the adjacency sets for every node in X; that is, the graph (X, L) can also be represented as (X, Adj), where $X = \{X_1, \ldots, X_n\}$ is the set of nodes and $Adj = \{Adj(X_1), \ldots, Adj(X_n)\}$ is the set of all adjacency sets. As we shall see in Section 4.8, this representation of a graph is suitable for computational purposes.

Example 4.3 Adjacency sets. The directed graph given in Figure 4.2(a) has the associated adjacency sets:

$$Adj(A) = \{D\}, \quad Adj(B) = \{C\}, \quad Adj(C) = \phi,$$
$$Adj(D) = \{B, E\}, \quad Adj(E) = \{F\}, \quad Adj(F) = \{D\}.$$

On the other hand, the undirected graph in Figure 4.2(b) has the associated adjacency sets:

$$Adj(A) = \{B, E\}, \quad Adj(B) = \{A, C\},$$
$$Adj(C) = \{B, D\}, \quad Adj(D) = \{C, E, G, H\},$$
$$Adj(E) = \{A, D, F\}, \quad Adj(F) = \{E, G\},$$
$$Adj(G) = \{D, F\}, \quad Adj(H) = \{D\}.$$

Thus, the graphs shown in Figure 4.2 can be equivalently defined by (X, L) or by (X, Adj). ∎

The adjacency set of a given node X_i contains the nodes that can be directly reached from X_i. Thus, by starting at a given node and successively passing to one of the nodes in its adjacency list, we can form a *path* through the graph. As we shall see, the concept of path between two nodes plays a key role in graph theory.

Definition 4.6 Path between two nodes. *A path from node X_i to node X_j is an ordered set of nodes $(X_{i_1}, \ldots, X_{i_r})$, starting in $X_{i_1} = X_i$ and ending in $X_{i_r} = X_j$, such that there is a link from X_{i_k} to $X_{i_{k+1}}$, $k = 1, \ldots, r-1$, that is,*

$$X_{i_{k+1}} \in Adj(X_{i_k}), \quad k = 1, \ldots, r-1.$$

The length of this path is $(r-1)$, the number of links it contains.

In the case of undirected graphs, a path $(X_{i_1}, \ldots, X_{i_r})$ can also be represented as $X_{i_1} - \ldots - X_{i_r}$ indicating the undirected character of the links. Similarly, $X_{i_1} \to \ldots \to X_{i_r}$ is another way of representing a path in directed graphs.

Example 4.4 Paths. Consider the directed graph given in Figure 4.2(a). There is only one path of length 2 from D to F, $D \to E \to F$. Also, there are two different paths from A to B: $A \to D \to B$, of length 2, and $A \to D \to E \to F \to D \to B$, of length 5, but there is no path from B

4.2 Basic Concepts and Definitions   117

to A. On the other hand, there exists a path between every pair of nodes in the undirected graph of Figure 4.2(b). For example, some of the paths from A to H are

$$A - E - D - H, \text{ of length 3},$$
$$A - B - C - D - H, \text{ of length 4, and}$$
$$A - E - F - G - D - H, \text{ of length 5}.$$ ■

Note that in directed graphs, the directions of the links have to be taken into consideration when forming the path. For example, in the directed graph of Figure 4.2(a), there is a path from A to C ($A \to D \to B \to C$), but there is no path from C to A.

Definition 4.7 Closed path. *A path* $(X_{i_1}, \ldots, X_{i_r})$ *is said to be closed if it has the same starting and ending nodes, that is,* $X_{i_1} = X_{i_r}$.

Example 4.5 Closed paths. The path $D \to G \to F \to D$ in the directed graph in Figure 4.3(a) is a closed path. On the other hand, the undirected graph in Figure 4.3(b) contains several closed paths. For example, the path $A - B - C - D - E - A$ is a closed path. ■

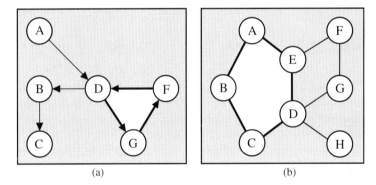

FIGURE 4.3. Examples of closed paths in (a) a directed graph and (b) an undirected graph.

Note that if a path contains a node more than once, then it contains a closed subpath. For example, the path $C - D - E - F - G - D - H$ in the graph in Figure 4.3(b) contains node D twice. Therefore, it contains a closed subpath: $D - E - F - G - D$. Then, eliminating the closed subpath from the original path, we get a shorter path, $C - D - H$, between nodes C and H.

4.3 Characteristics of Undirected Graphs

In this section we present some characteristics of undirected graphs. Characteristics of directed graphs are given in Section 4.4.

4.3.1 Some Definitions Related to Undirected Graphs

Definition 4.8 Complete graph. *An undirected graph is said to be complete if there exists a link between every pair of its nodes.*

Thus, for any set of n nodes there is only one complete graph. The complete graph of n nodes is usually denoted as K_n. Figure 4.4 shows an example of K_5.

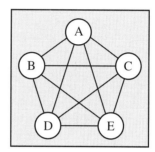

FIGURE 4.4. An example of a complete graph with five nodes.

Definition 4.9 Complete set. *A subset of nodes S of a graph G is said to be complete if there are links between every pair of nodes in S.*

It follows from the definition that any two connected nodes in a graph form a complete set. Thus, for example, the graph in Figure 4.3(b) contains no complete sets of three or more nodes. But the graph in Figure 4.5(a) contains two complete subsets of size 3: $\{D, E, G\}$ and $\{E, F, G\}$.

Among the complete sets of an undirected graph the maximal complete sets plays an important role.

Definition 4.10 Clique. *A complete set of nodes C is called a clique if it is maximal, that is, it is not a proper subset of another complete set.*

Example 4.6 Cliques. Figure 4.5(a) contains the following cliques: $C_1 = \{A, B\}$, $C_2 = \{B, C\}$, $C_3 = \{C, D\}$, $C_4 = \{D, H\}$, $C_5 = \{D, E, G\}$, $C_6 = \{E, F, G\}$ and $C_7 = \{A, E\}$. However, if we add some extra links to the graph, some of the previous maximal complete sets are no longer maximal, and the graph contains different cliques. For example, the graph in Figure 4.5(b) is obtained by adding three links to the graph in Figure 4.5(a). The sets C_1, C_2, C_3, and C_7 are no longer complete. Thus, the graph in

Figure 4.5(b) contains five cliques: $C_1 = \{A, B, D, E\}$, $C_2 = \{B, C, D\}$, $C_3 = \{D, H\}$, $C_4 = \{D, E, G\}$, and $C_5 = \{E, F, G\}$. ∎

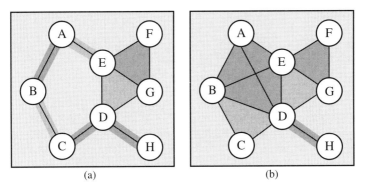

FIGURE 4.5. Examples of cliques associated with two different graphs.

Definition 4.11 Loop. *A loop is a closed path in an undirected graph.*

Example 4.7 Loop. Consider the undirected graph given in Figure 4.5(b). The closed path $A - B - C - D - E - A$ is a loop of length 5. Note that by replacing a path between two nodes within the loop by an alternative path, we obtain a different loop. For example, replacing the link $D - E$ by the path $D - G - F - E$ in the previous loop, we get the loop $A - B - C - D - G - F - E - A$ of length 7. ∎

Definition 4.12 Neighbors of a node. *The set of nodes adjacent to a node X_i in an undirected graph is referred to as the neighbors of X_i, $Nbr(X_i) = \{X_j \mid X_j \in Adj(X_i)\}$.*

Note that for undirected graphs, the adjacency set of a given node coincides with the neighbors of the node. For example, the shaded nodes $\{A, D, F\}$ in Figure 4.6 are the neighbors of node E.

Definition 4.13 Boundary of a set of nodes. *The union of the sets of the neighbors of the nodes of a given subset S, excluding the nodes in S, is called the boundary of S and is denoted by $Bnd(S)$.*

$$Bnd(S) = \left(\bigcup_{X_i \in S} Nbr(X_i) \right) \setminus S,$$

where $X \setminus S$ denotes the set of all nodes in X except those in S.

For example, the shaded nodes $\{A, C, F, G, H\}$ are the boundary of the set $\{D, E\}$ in Figure 4.7.

Note that $Nbr(X_i) = Bnd(X_i)$, where X_i is a single node in X.

120   4. Some Concepts of Graphs

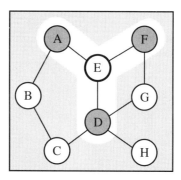

FIGURE 4.6. The neighbors of node E are $\{A, D, F\}$.

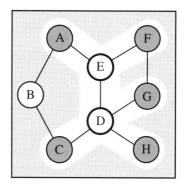

FIGURE 4.7. The boundary of the set $\{D, E\}$ is $\{A, C, F, G, H\}$.

4.3.2 Types of Undirected Graphs

In many practical situations it is important to know whether or not there is a path between two nodes in a given graph. For example, in the field of expert systems, graphs are used to represent dependence relationships among a set of variables. Thus, information about the existence of one or several paths between two nodes in the network is valuable information in order to know the dependence structure contained in the graph. From this point of view, a useful classification of graphs is based on the number of different paths existing between two nodes.

Definition 4.14 Connected undirected graphs. *An undirected graph is said to be connected if there exists at least one path between every two nodes. Otherwise, it is said to be disconnected.*

For example, the graph in Figure 4.7 is connected, because there is a path between every two nodes. However, the graphs in Figure 4.8 are disconnected, because there is no path between nodes A and F, for example. Note that at first sight, the graph in Figure 4.8(a) does not seem to be disconnected because the links cross each other. The disconnectedness of this graph is more visually apparent in Figure 4.8(b), which corresponds to the same graph as the one in Figure 4.8(a). The problem of graphical representation is analyzed in detail in Section 4.7.

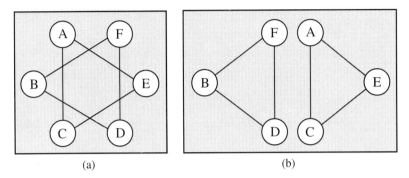

FIGURE 4.8. Two different representations of the same disconnected graph.

If a graph is disconnected, it can be partitioned into connected subgraphs, called *connected components*. For example, the disconnected graph in Figure 4.8(b) contains two connected components: $\{A, C, E\}$ and $\{B, D, F\}$. In practice, graphs are usually assumed to be connected, because each of the connected components of the disconnected graph can be thought of as a connected graph. In Section 4.8 we shall develop an algorithm for determining whether or not a graph is connected and for calculating its connected components.

The topological complexity of a graph increases as the number of different paths between two nodes increases. Thus, in addition to knowing whether or not a graph is connected, it is also important to know the number of different paths between every two nodes.

Definition 4.15 Tree. *A connected undirected graph is said to be a tree if for every pair of nodes there exists a unique path.*

It follows from this definition that a tree is a connected graph, but removing any of its links renders it disconnected. Similarly, a tree has no loops, but if we add one link anywhere, the link forms a loop.

Figure 4.9(a) shows an example of a tree. Note that removing any link will divide the tree into two disconnected trees. On the other hand, adding any link, as in Figure 4.9(b), will create a loop and the graph will no longer be a tree.

Definition 4.16 Multiply-connected graphs. *A connected undirected graph is called multiply-connected if it contains at least one pair of nodes that are joined by more than one path, or equivalently, if it contains at least one loop.*

Note that if a graph contains two different paths between two nodes, they can be combined to form a loop. Thus, the two definitions of multiply-connected graphs are equivalent. For example, the graph in Figure 4.9(b) is multiply connected, since there are two paths $D - E - G - J$ and $D - F - H - J$ joining nodes D and J. These two paths form the loop $D - E - G - J - H - F - D$.

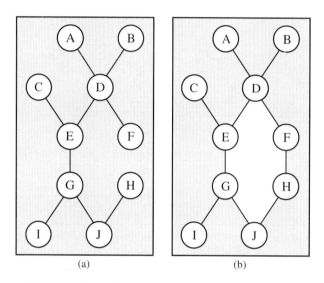

FIGURE 4.9. Examples of (a) a tree and (b) a multiply-connected graph.

The various types of undirected graphs are shown schematically in Figure 4.10.

4.4 Characteristics of Directed Graphs

In this section we discuss some characteristics of directed graphs.

4.4.1 Some Definitions Related to Directed Graphs

Definition 4.17 Parents and children. *When there is a directed link $X_i \to X_j$ from X_i to X_j, then X_i is said to be a parent of X_j, and X_j is said to be a child of X_i.*

The set of all parents of a given node X_i is denoted by Π_{X_i}. For example, in Figure 4.11, C and D are the parents of E, and G is the only child of E. In a directed graph, the set of children of a node X_i coincides with the adjacency set associated with the node.

4.4 Characteristics of Directed Graphs   123

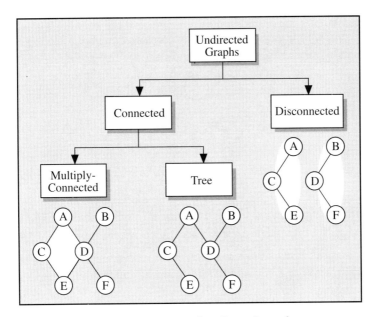

FIGURE 4.10. Types of undirected graphs.

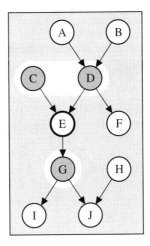

FIGURE 4.11. Parents and children of node E.

Definition 4.18 Family of a node. *The set consisting of a node and its parents is called the family of the node.*

For example, Figure 4.12 shows the families associated with all nodes in the graph. In this example there are families with one, two, or three nodes (represented with different shadow intensities). As we shall see, the different families of a graph will play an important role in next chapters, since the local structure given by the topology of the graph can be exploited to define

124   4. Some Concepts of Graphs

joint probability distributions over the nodes in the network by using local distributions defined on the families of the graph.

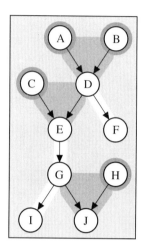

FIGURE 4.12. Families associated with the nodes of the graph.

Definition 4.19 Ancestors of a node. *A node X_j is said to be an ancestor of node X_i is there is a path from X_j to X_i.*

Definition 4.20 Ancestral set. *A set of nodes S is said to be an ancestral set if it contains the ancestors of all its nodes.*

Definition 4.21 Descendants of a node. *The descendants of a node X_i constitute the set of all nodes X_j in all directed paths emanating from X_i, except X_i.*

As an example, Figure 4.13 shows the sets of ancestors and descendants of node E.

So far, we have seen that every node in a graph can have several attributes or parameters. For example, a node has its own name, parents, children, family, ancestral set, and descendants. A node can also have a *number*, that is, given a set of nodes, we can arrange them in any order we please. The order in which the nodes are arranged, however, is usually dictated by the topology of the graph to give insight about some particular property. An *ordering*, or *numbering*, of a set of nodes $X = \{X_1, \ldots, X_n\}$ is obtained by assigning each node a unique number in $\{1, \ldots, n\}$.

Definition 4.22 Numbering. *Given a set of nodes $X = \{X_1, \ldots, X_n\}$, a numbering, α, is a bijection that assigns each number in $\{1, \ldots, n\}$ a unique node:*

$$\alpha : \{1, \ldots, n\} \longrightarrow \{X_1, \ldots, X_n\}.$$

4.4 Characteristics of Directed Graphs

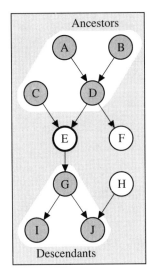

FIGURE 4.13. Ancestors and descendants of node E.

Thus, $\alpha(i)$ refers to the ith node in the numbering. A numbering can be represented as the ordered sequence of nodes $(\alpha(1), \ldots, \alpha(n))$.

A useful numbering of the nodes in a directed graph is the *ancestral numbering*.

Definition 4.23 Ancestral numbering. *A numbering of nodes in a directed graph is called ancestral if the number corresponding to any node is lower than the numbers corresponding to all of its children.*

For example, the two sets of numberings shown in Figures 4.14 are two different ancestral numberings of the same graph. Thus, ancestral numbering of a given graph is not necessarily unique. On the other hand, there are some graphs that do not admit any ancestral numbering. This problem is analyzed in detail, from theoretic and algorithmic points of view, in Section 4.7.1.

A directed graph can be easily converted to an undirected graph by dropping the directionality of the links.[1]

Definition 4.24 Undirected graph associated with a directed graph. *The undirected graph obtained by replacing every directed link in a directed graph with an undirected link is called the undirected graph associated with the directed graph.*

---

[1] Note that the reverse operation has several alternatives. For each undirected link $X_i - X_j$ we can consider two possible directed links: $X_i \to X_j$ and $X_j \to X_i$. Consequently, there are several directed graphs associated with the same undirected graph (for further discussion see Ross and Wright (1988)).

126   4. Some Concepts of Graphs

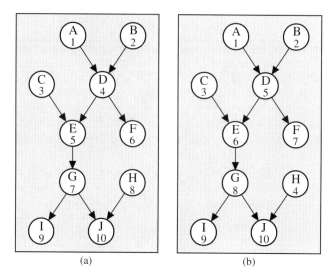

FIGURE 4.14. Two different ancestral numberings of the same directed graph.

For example, Figure 4.15(b) is the undirected graph associated with the directed graph in Figure 4.15(a).

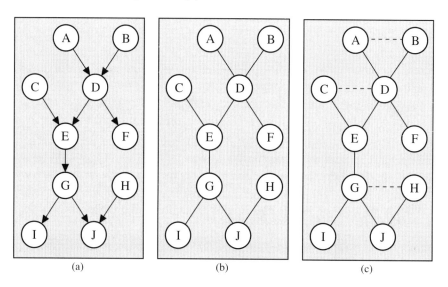

FIGURE 4.15. An example of (a) a directed graph, (b) the associated undirected graph, and (c) the associated moralized graph.

Definition 4.25 Moral graph. *The graph obtained by first joining (adding a link between) every pair of nodes with a common child in a directed*

4.4 Characteristics of Directed Graphs

graph, and then dropping the directionality of the links, is called a moral graph.

For example, Figure 4.15(c) shows the moral graph associated with the graph in Figure 4.15(a). Note that each of the pairs $\{(A,B),(C,D),(G,H)\}$ has a common child. The moral graph is obtained by adding the three links indicated by dotted lines and then ignoring the directionality of all links.

The name of *moral graph* comes from the fact that constructing the moral graph associated with a directed graph consists of joining or marrying every pair of nodes with a common child. However, this name is unfortunate because the parents of one node could themselves be the children of another node, hence, siblings. Thus, the so-called moral graph could also be immoral at the same time.

In directed graphs closed paths receive different names depending on their directed or undirected character. When the path is defined in the original directed graph it is called a *cycle*, but when the direction of the links does not matter, that is, when the path is defined in the associated undirected graph, it is called a loop (see Sec. 4.3).

Definition 4.26 Cycle. *A cycle is a closed directed path in a directed graph.*

Example 4.8 Loops and cycles. Figure 4.16(a) shows a directed graph that contains only one cycle: $D \rightarrow G \rightarrow F \rightarrow D$. The corresponding undirected graph, however, contains two loops: $D - G - F - D$ and $A - B - D - A$. ■

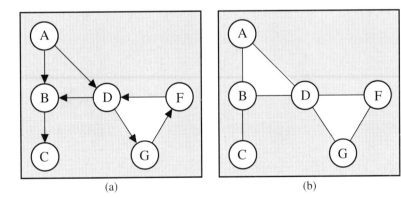

FIGURE 4.16. Loops and cycles associated with a directed graph.

4.4.2 Types of Directed Graphs

Definition 4.27 Connected directed graphs. *A directed graph is said to be connected if the associated undirected graph is connected; otherwise, it is said to be disconnected.*

Definition 4.28 Trees and multiply-connected directed graphs. *A connected directed graph is said to be a tree if the associated undirected graph is a tree. Otherwise, it is said to be multiply-connected.*

Definition 4.29 Cyclic and acyclic graphs. *A directed graph is said to be cyclic if it contains at least one cycle. Otherwise, it is called a directed acyclic graph (DAG).*

DAGs play an important role in later chapters as they are used as a basis for building an important class of probabilistic models known as *Bayesian network models*.

In directed graphs, trees are usually classified into two different types depending on the number of arrows pointing to the same node.

Definition 4.30 Simple trees and polytrees. *A directed tree is called a simple tree if every node has at most one parent. Otherwise, it is called a polytree.*

Figure 4.17 shows examples of a simple tree and a polytree. Figure 4.18 shows examples of cyclic and multiply-connected directed graphs. Various types of directed graphs are shown schematically in Figure 4.19.

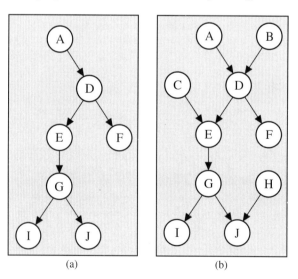

FIGURE 4.17. Examples of directed graphs: (a) a simple tree and (b) a polytree.

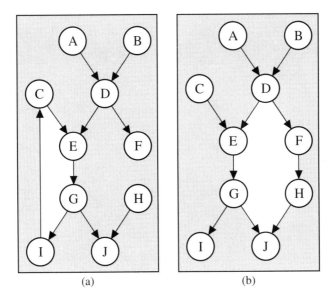

FIGURE 4.18. Examples of directed graphs: (a) a cyclic graph and (b) a multiply-connected DAG.

4.5 Triangulated Graphs

Triangulated graphs are a special type of undirected graph that has many interesting applications in several fields. For example, in Chapter 6 we shall see that this type of graph constitutes the graphical structure of a type of probabilistic network model known as *decomposable models* (Lauritzen, Speed, and Vijayan (1984)). Triangulated graphs are also referred to as *rigid circuits* (Dirac (1961)) and *chordal graphs* (Gavril (1972, 1974)).

This section introduces triangulated graphs and provides algorithms for testing whether or not a given graph is triangulated and for adding the necessary links to triangulate a graph if it is not already triangulated.

Definition 4.31 Chord of a loop. *A chord is a link between two nodes in a loop that is not contained in the loop.*

For example, in Figure 4.20, the link $E - G$ is a chord in the loop $E - F - G - D - E$. Note that a chord breaks the loop and decomposes it into two smaller loops $E - F - G - E$ and $E - G - D - E$. On the other hand, the loop $A - B - C - D - E - A$ does not have a chord.

Clearly, loops of length three cannot contain a chord. Loops of length three are often referred to as *triangles*.

Definition 4.32 Triangulated graph. *An undirected graph is said to be triangulated, or chordal, if every loop of length four or more has at least one chord.*

130   4. Some Concepts of Graphs

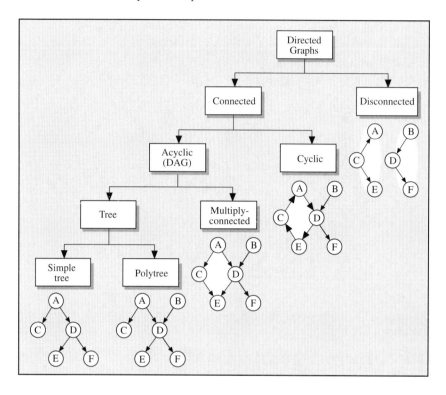

FIGURE 4.19. Types of directed graphs.

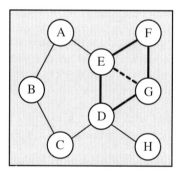

FIGURE 4.20. Example of a loop with one chord.

Example 4.9 Triangulated graphs. Figure 4.21(a) shows a triangulated graph where all loops of length four or more have at least one chord. The graph contains two loops of length four, $A - B - E - C - A$ and $B - C - E - D - B$, and one loop of length five, $A - B - D - E - C - A$, and each of them has at least one chord.

On the other hand, the graph in Figure 4.21(b) is not triangulated because a loop of length five, $A-B-C-D-E-A$, does not have a chord. ∎

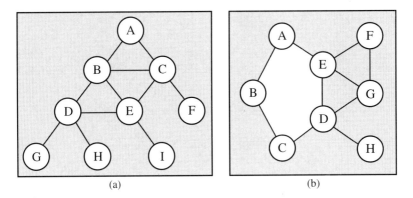

FIGURE 4.21. (a) A triangulated graph and (b) A nontriangulated graph.

If a graph is not triangulated, it can be made so by adding chords to break the loops. This process is called *filling-in* or *triangulation*. It is important to remark that triangulating a graph does not consist of dividing the graph into triangles. For example, the graph in Figure 4.21(a) is triangulated, so it does not need the addition of extra links, like those indicated by the dashed lines shown in Figure 4.22.

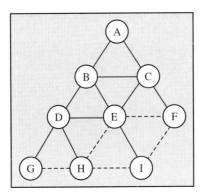

FIGURE 4.22. Triangulation does not mean dividing the graph into triangles.

Since a loop can be broken in several ways, there are several different ways to triangulate a graph. For example, the two graphs in Figure 4.23 show two different triangulations associated with the same graph in Figure 4.21(b).

When filling-in a graph, it is desirable to add as few chords as possible. A fill-in is said to be *minimal* if it contains a minimum number of chords. Note that the fill-in in Figure 4.23(a) is minimal but the one in Figure 4.23(b)

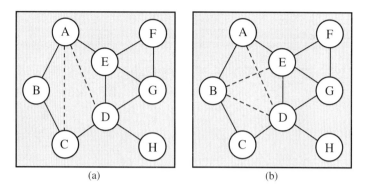

FIGURE 4.23. Two different triangulations of the same graph. The dashed lines represent the chords added to triangulate the graph.

is not, because we can remove the links $A - D$ or $B - E$ and the resulting graph is still triangulated. The problem of computing a minimal fill-in, however, is NP-complete[2] (Yannakakis (1981)). Several fast (linear-time) algorithms have been proposed for testing whether an undirected graph is triangulated and for triangulating graphs that are not triangulated (see Rose, Tarjan, and Leuker (1976), and Tarjan and Yannakakis (1984)), but they do not guarantee a minimum fill-in. We introduce below a conceptually simple algorithm called the *Maximum Cardinality Search* (see Tarjan and Yannakakis (1984)), but first we need some definitions.

Definition 4.33 Perfect numbering. *A given numbering, α, of the nodes of a graph is called a perfect numbering if the subset of nodes*

$$Bnd(\alpha(i)) \cap \{\alpha(1), \ldots, \alpha(i-1)\}$$

is complete for $i = 2, \ldots, n$.

Example 4.10 Perfect numbering. Figure 4.24(a) shows a numbering for the nodes of the graph: $\alpha(1) = A$, $\alpha(2) = B$, $\alpha(3) = C$, $\alpha(4) = E$, and so on. Let us check whether the conditions for perfect numbering are satisfied:

- For $i = 2$, $Bnd(\alpha(2)) \cap \{\alpha(1)\} = Bnd(B) \cap \{A\} = \{A, C, D, E\} \cap \{A\} = \{A\}$, which is a trivially complete set.

- For $i = 3$, $Bnd(\alpha(3)) \cap \{\alpha(1), \alpha(2)\} = \{A, B, E, F\} \cap \{A, B\} = \{A, B\}$ is complete, since the link $A - B$ is contained in the graph.

- For $i = 4$, $Bnd(\alpha(4)) \cap \{\alpha(1), \alpha(2), \alpha(3)\} = \{B, C, D, I\} \cap \{A, B, C\} = \{B, C\}$ is also complete.

---

[2] An introduction to algorithm complexity and NP-complete problems is found in the book by Garey and Johnson (1979).

Similarly, for $i = 5, \ldots, 9$, the reader can verify that

$$Bnd(\alpha(i)) \cap \{\alpha(1), \ldots, \alpha(i-1)\}$$

is complete. Thus, α is a perfect numbering. ∎

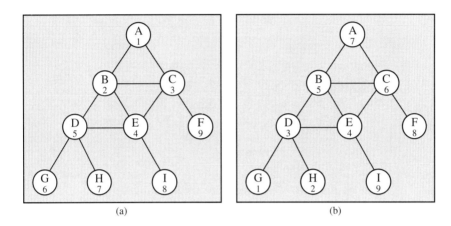

FIGURE 4.24. Two different perfect numberings of the nodes.

Note that a perfect numbering is not necessarily unique. For example, Figure 4.24(b) shows another perfect numbering for the same graph. Note also that there are graphs without any perfect numbering. For example, the graph in Figure 4.21(b), does not have any perfect numbering; the presence of large loops without chords makes it impossible to build a perfect numbering of the nodes.

The maximum cardinality search (MCS) algorithm is based on a theorem (see Fulkerson and Gross (1965), and Golumbic (1980)) relating perfect numbering and graph triangulation.

Theorem 4.1 Triangulation and perfect numbering. *An undirected graph admits a perfect numbering iff it is triangulated, where iff means if and only if.*

The MCS algorithm gives a numbering for the nodes of an undirected graph. This numbering is a perfect numbering only if the original graph is a triangulated graph.

Algorithm 4.1 Maximum Cardinality Search.

- **Input:** An undirected graph $G = (X, L)$ and an initial node X_i.
- **Output:** A numbering α of the nodes in X.

Maximum Cardinality Search

Input: A graph $G = (X, L)$ and an initial node X_i
Output: A numbering α of the nodes in X

Initial Step:
$\alpha(1) \leftarrow X_i$
$Numbered \leftarrow \{X_i\}$

Iteration Step:
for i = 2 to n
$\quad X_k \leftarrow$ choose a node X_k in $X \setminus Numbered$
$\qquad$ with maximum $|Nbr(X_k) \cap Numbered|$
$\quad \alpha(i) \leftarrow X_k$
$\quad$ add X_k to $Numbered$

FIGURE 4.25. Pseudocode for the MCS algorithm.

1. *Initialization:* Assign the first number in the numbering to the initial node X_i, that is, $\alpha(1) = X_i$.

2. Repeat the iteration step for $i = 2, \ldots, n$.

3. *Iteration i:* In the ith iteration an unnumbered node X_k with the maximum number of numbered neighbors is chosen to be numbered, that is, $\alpha(i) = X_k$. Ties are broken arbitrarily. ■

Pseudocode for the MCS algorithm is given in Figure 4.25. The following theorem provides a way of recognizing triangulated graphs using the MCS algorithm (see Tarjan (1983) and Tarjan and Yannakakis (1984)).

Theorem 4.2 MCS numbering. *Every numbering of the nodes of a triangulated graph obtained by the MCS algorithm is a perfect numbering.*

Therefore, when the numbering generated by Algorithm 4.1 is not perfect, then the graph is not triangulated. Accordingly, Algorithm 4.1 can be easily modified to check whether a given graph is triangulated. When the graph is not triangulated, the algorithm can also be used to generate a fill-in for the graph. We refer the reader to Tarjan and Yannakakis (1984) or Neapolitan (1990) for details on how to implement the algorithm for it to run in linear time $o(n + l)$, where n is the number of nodes and l is the number of links in the graph. For illustrative purposes, we introduce here the following algorithm:

4.5 Triangulated Graphs

Algorithm 4.2 Maximum Cardinality Search Fill-In.

- **Input:** An undirected graph $G = (X, L)$ and an initial node X_i.
- **Output:** A fill-in L', such that, $G' = (X, L \cup L')$ is a triangulated graph.

Initialization Steps:

1. Initially, the fill-in is empty, that is, $L' = \phi$.

2. Let $i = 1$ and assign the first number in the numbering to the initial node X_i, that is, $\alpha(1) = X_i$.

Iteration Steps:

3. An unnumbered node X_k with a maximum number of numbered neighbors is assigned label i, $\alpha(i) = X_k$.

4. If $Nbr(X_k) \cap \{\alpha(1), \ldots, \alpha(i-1)\}$ is not complete, add to L' the necessary links to make this set complete and go to Step 2; otherwise, go to Step 5.

5. If $i = n$, then stop; otherwise, let $i = i + 1$ and go to Step 3. ∎

Using Theorem 4.2, it can be shown that when the graph is triangulated, the fill-in L' resulting from Algorithm 4.2 is the empty set; otherwise, the set of links L' contains the additional links that render the graph triangulated.

Example 4.11 MCS fill-in. The graph in Figure 4.26 is not triangulated. Let us apply Algorithm 4.2 to build a fill-in for this graph. We choose, for example, node C as the initial node for this algorithm.

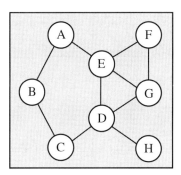

FIGURE 4.26. Nontriangulated undirected graph.

- Step 1: $L' = \phi$.
- Step 2: Let $i = 1$ and $\alpha(1) = C$.

- Step 3: Each of the nodes B and D has one labeled neighbor, and all other unlabeled nodes have zero labeled neighbors. To break the tie we choose node D and label it 2, that is, $\alpha(2) = D$.

- Step 4: Note that in this case, the previous labeled neighbors form a trivial complete set. Therefore, we do not need to include any link in L'.

- Step 5: Since $i \neq n$, we increase i by one and go to Step 3.

- Steps 3–5: Following a similar process, the nodes B and E are labeled 3 and 4, respectively.

- Steps 3–4: The node with the maximum number of labeled neighbors is A. However, as shown in Figure 4.27(a), the set of labeled neighbors of A, $\{B, E\}$, is not complete. So, we must add the link $B - E$ (see Figure 4.27(b)) to L' and start again with Step 2. Note that now $L' = \{B - E\}$.

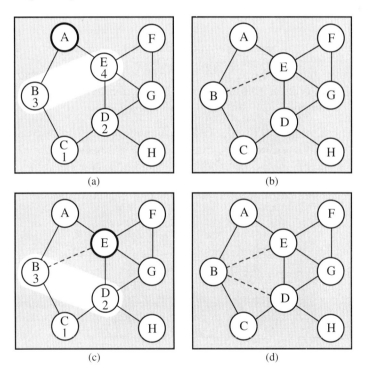

FIGURE 4.27. Perfect numbering of the nodes using the MCS algorithm.

- Steps 2 – 5: The nodes C, D, and B are labeled 1, 2, and 3, respectively.

4.5 Triangulated Graphs

- Steps 3 − 4: Node E has the maximum number of labeled neighbors, $\{B, D\}$, but this set is not complete (see Figure 4.27(c)). Therefore, we add the link $B - D$ to L' and start again with Step 2. Now, $L' = \{B - E, B - D\}$ (see Figure 4.27(d)).

- Steps 2 − 5: The nodes C, D, B, E, A, G, F, and H are labeled 1 through 8, respectively. The resulting graph $G' = (X, L \cup L')$ is now a triangulated graph, and the final numbering shown in Figure 4.28 is a perfect numbering. ∎

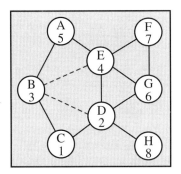

FIGURE 4.28. Perfect numbering of the nodes using the MCS algorithm.

Note that depending on the choice of the starting node and on how ties are broken, several triangulations of the same graph can be obtained. For example, the MCS algorithm can also produce the two perfect numberings shown in Figure 4.29.

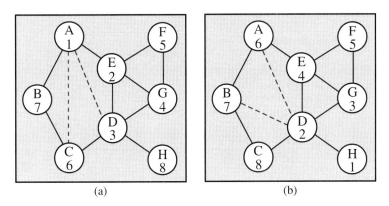

FIGURE 4.29. Two different perfect numberings for the graph in Figure 4.26.

One interesting property of triangulated graphs, which is useful in working with the so-called Markov network models (Chapters 6 and 8), is known as the *running intersection property*.

138   4. Some Concepts of Graphs

Definition 4.34 Running intersection property. *An ordering of the cliques of an undirected graph $(C_1, \ldots, C_m)$ is said to satisfy the running intersection property if the set $C_i \cap (C_1 \cup \ldots \cup C_{i-1})$ is contained in at least one of the cliques $\{C_1, \ldots, C_{i-1}\}$, for all $i = 1, \ldots, m$.*

This property states that the cliques of a graph can be ordered in such a way that the nodes that are common to a given clique and the union of the preceding cliques are also contained in at least one of the preceding cliques. An ordered sequence of cliques satisfying the running intersection property is referred to as a *chain of cliques*. Some undirected graphs have no chain of cliques, yet other undirected graphs have more than one chain of cliques. The following theorem characterizes the graphs with at least one chain of cliques.

Theorem 4.3 Chain of cliques. *An undirected graph has an associated chain of cliques iff it is triangulated.*

Next, we give an algorithm for building a chain of cliques from an undirected graph. This algorithm is based on the maximum cardinality search (MCS) algorithm. The algorithm assumes that the graph is triangulated. Otherwise, the graph can be triangulated using Algorithm 4.2.

Algorithm 4.3 Generating a Chain of Cliques.

- **Input:** A triangulated undirected graph $G = (X, L)$.
- **Output:** A chain of cliques $(C_1, \ldots, C_m)$ associated with G.

1. *Initialization:* Choose any node to serve as an initial node, then use Algorithm 4.1 to obtain a perfect numbering of the nodes, $X_1, \ldots, X_n$.

2. Calculate the cliques of the graph, C.

3. Assign to each clique the largest perfect number of its nodes.

4. Order the cliques, $(C_1, \ldots, C_m)$, in ascending order according to their assigned numbers (break ties arbitrarily). ∎

Example 4.12 Generating a chain of cliques. Let us apply Algorithm 4.3 to generate a chain of cliques associated with the triangulated graph given in Figure 4.30(a). First, we use Algorithm 4.1 to obtain a perfect numbering of the nodes. Figure 4.30(b) shows the perfect numbers obtained starting with node A as the initial node. The cliques of the graph are: $C_1 = \{A, B, C\}$, $C_2 = \{B, C, E\}$, $C_3 = \{B, D, E\}$, $C_4 = \{C, F\}$, $C_5 = \{D, G\}$, $C_6 = \{D, H\}$, and $C_7 = \{E, I\}$. Now, we assign to each clique the largest perfect number of its nodes. For example, for clique C_1, the largest perfect number of the nodes A, B, C is three, corresponding to node C. Thus, clique C_1 is assigned the number 3. The number corresponding to clique

C_2 is 4 (corresponding to node E), and so on. Note that these cliques are already ordered in ascending order according to their assigned numbers. C_1 is the clique with lowest associated number, then C_2, and so on. Therefore, $(C_1, \ldots, C_7)$ is a chain of cliques for the graph of Figure 4.30(a). ∎

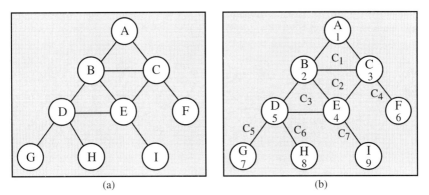

FIGURE 4.30. (a) A triangulated undirected graph and (b) its associated perfect numbering of the nodes that is needed to build an associated chain of cliques.

4.6 Cluster Graphs

Cluster graphs are formed by grouping some nodes with common characteristics in a given graph. This process is known as *clustering*, and it allows us to obtain new graphs with simple structures that essentially retain the topological properties of the original graph. In Chapters 6 and 8 we shall show several applications of cluster graphs.

Definition 4.35 Cluster. *A set of nodes of a graph is called a cluster.*

Definition 4.36 Cluster graph associated with a graph. *Given a graph $G = (X, L)$ and a set of clusters of X, $C = \{C_1, \ldots, C_m\}$, such that $X = C_1 \cup \ldots \cup C_m$, then the graph $G' = (C, L')$ is called a cluster graph of G if L' contains only links between clusters containing common nodes, that is, $(C_i, C_j) \in L' \Rightarrow C_i \cap C_j \neq \phi$.*

For a discussion of the properties of cluster graphs see Beeri et al. (1983) and Jensen (1988) and references therein.

In general, we are not interested in arbitrary cluster graphs, because we want to preserve the topological structure of the original graph as much as possible. So we shall consider special types of cluster graphs that satisfy certain desirable properties.

140   4. Some Concepts of Graphs

Definition 4.37 Clique graph. *A cluster graph is called a clique graph if its clusters are the cliques of the associated graph.*

For example, the cluster graph in Figure 4.31 is a clique graph associated with the graph in Figure 4.30(a).

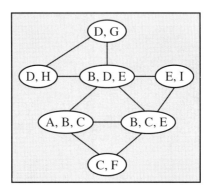

FIGURE 4.31. Cluster graph associated with the graph in Figure 4.30(a).

Definition 4.38 Join, or junction, graph. *A clique graph associated with an undirected graph is called a join or junction graph if it contains all the possible links joining two cliques with a common node.*

Note that the join graph associated with a given graph is unique. For example, the clique graph in Figure 4.31 is the join graph associated with the graph in Figure 4.30(a).

Note that in a join graph the set of clusters with a common node forms a complete set. This property guarantees that clusters with common nodes are always connected. Consequently, a join graph is highly connected even for graphs with a small number of nodes. Of interest, however, is constructing a graph with a simpler structure (e.g., a tree) that still preserves the connectedness between clusters with common nodes.

Definition 4.39 Join or junction tree. *A clique graph is called a join or junction tree if it is a tree and if every node that belongs to two clusters also belongs to every cluster in the path between them.*

Note that in a join tree there is a path between every two clusters with a common node.

Example 4.13 Join tree. The cluster tree in Figure 4.32(b) is a join tree obtained from the join graph in Figure 4.32(a) by removing four links. One can verify that the cliques contained is a path between every two clusters with common nodes also contain these common nodes. For example, the cliques $\{D, H\}$ and $\{B, D, E\}$ have a common node, D, which is also contained in the cliques of the only path between them, $\{D, H\} - \{D, G\} - \{B, D, E\}$. ∎

In Chapter 8 we shall see that several evidence propagation methods take advantage of the local structures contained in a join tree to simplify the calculations needed to update the probabilities.

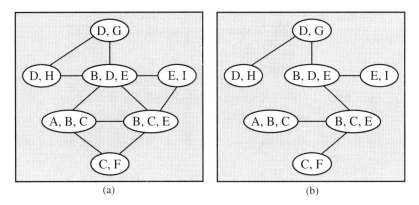

FIGURE 4.32. (a) A join graph and (b) a join tree associated with it.

The next theorem indicates whether it is possible to prune a join graph to obtain a join tree (see Jensen (1988)).

Theorem 4.4 Join tree. *An undirected graph has a join tree iff it is triangulated.*

Example 4.14 A graph without a join tree. Figure 4.33(a) shows a nontriangulated graph and its associated join graph with cliques $C_1 = \{A, B\}, C_2 = \{B, D\}, C_3 = \{C, D\}$, and $C_4 = \{A, C\}$. Note that in this situation, it is impossible to build a join tree from this graph because the graph is not triangulated. For example, if we remove the link $C_1 - C_4$ from the associated cluster graph given in Figure 4.33(b), the graph becomes a tree, but it is not a join tree because node A is contained in C_1 and C_4, but not in the two other cliques in the path $C_1 - C_2 - C_3 - C_4$. ∎

In Section 4.5 we have introduced the running intersection property, which allows us to order the cliques of a triangulated graph obtaining a chain of cliques. The next algorithm builds a join tree associated with a triangulated graph by organizing a chain of cliques into a tree structure.

Algorithm 4.4 Generating a Join Tree.

- **Input:** A triangulated undirected graph $G = (X, L)$.
- **Output:** A join tree $G' = (C, L')$ associated with G.

142   4. Some Concepts of Graphs

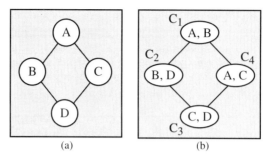

FIGURE 4.33. (a) Nontriangulated graph and (b) the associated join graph.

1. *Initialization:* Use Algorithm 4.3 to obtain a chain of cliques of graph G, $(C_1, \ldots, C_m)$.

2. For each clique $C_i \in C$, choose from $\{C_1, \ldots, C_{i-1}\}$ a clique C_k with maximum number of common nodes and add the link $C_i - C_k$ to L' (initially empty). Break ties arbitrarily. ∎

Example 4.15 Generating a join tree. Let us apply Algorithm 4.4 to generate a join tree associated with the triangulated graph given in Figure 4.30(a). In Example 4.12, we obtained the chain of cliques $C_1 = \{A, B, C\}$, $C_2 = \{B, C, E\}$, $C_3 = \{B, D, E\}$, $C_4 = \{C, F\}$, $C_5 = \{D, G\}$, $C_6 = \{D, H\}$, and $C_7 = \{E, I\}$. We can then generate a join tree by adding the necessary links to L' (initially empty) as follows:

- There are two cliques, C_2 and C_3, with maximum number of nodes in common with clique C_7. Breaking the tie, we choose clique C_3 and we add the link $C_7 - C_3$ to L'.

- C_3 and C_5 have the maximum number of nodes in common with C_6. Breaking the tie, we choose clique C_5 and we add the link $C_6 - C_5$ to L'.

- Among the set $\{C_1, C_2, C_3, C_4\}$, C_2 is the clique with maximum number of elements in common with C_5. Thus, we add the link $C_5 - C_3$ to L'.

- Proceeding in a similar way, we add the links $C_4 - C_2$, $C_3 - C_2$, and $C_2 - C_1$.

Thus, a join tree for the graph $G = (X, L)$ is given by $G' = (C, L')$. The resulting join tree is given in Figure 4.34, which is the same as the one given in Figure 4.32(b). Note that because ties are broken arbitrarily, there can be more than one join tree associated with the same undirected graph. ∎

So far, we have dealt with the problem of building cluster trees for undirected graphs. However, the concept of cluster trees can also be applied

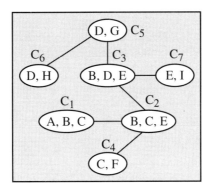

FIGURE 4.34. A join tree associated with the graph in Figure 4.30(a).

to directed graphs by working with the undirected graph associated with the directed graph. As we shall see in Chapter 8 the families of the nodes in a directed graph play an important role in evidence propagation. Also, some probabilistic models are defined through local functions defined in the families of the nodes (see Section 6.4.2). Thus, we are interested in cluster graphs in which all the families of the graph are contained in at least one cluster. This leads to the following definition.

Definition 4.40 Family tree. *A family tree of a directed graph D is a join tree of some undirected graph G associated with D in which the family of every node is contained in at least one cluster.*

Note that by taking the moral graph associated with the directed graph, we guarantee that any family will be contained in at least one clique. Thus, applying Algorithm 4.4 to any triangulated version of the moral graph we will obtain a family tree for the original directed graph. The following algorithm can then be used to generate a family tree associated with a directed graph.

Algorithm 4.5 Generating a Family Tree.

- **Input:** A directed graph $D = (X, L)$.
- **Output:** A family tree $G' = (C, L')$ associated with D.

1. Moralize the directed graph.
2. Triangulate the resulting undirected graph using Algorithm 4.2.
3. Apply Algorithm 4.4 to calculate a join tree of the resulting graph.

■

Example 4.16 Generating a family tree. Consider the directed graph given in Figure 4.35(a), where the families of the nodes are indicated by different shades: $\{A\}$, $\{B\}$, $\{C\}$, $\{A, B, D\}$, $\{C, D, E\}$, $\{D, F\}$, $\{E, G\}$, $\{F, H\}$, $\{G, I\}$, and $\{G, H, J\}$. Let us apply Algorithm 4.5 to obtain a family tree associated with this graph:

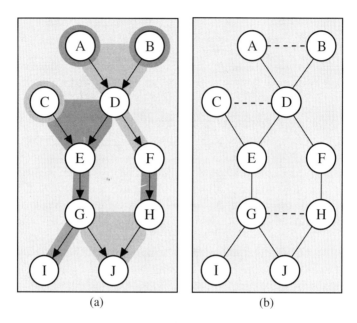

FIGURE 4.35. (a) A directed graph (with the families indicated by different shades) and (b) the corresponding moralized graph.

- The corresponding moral graph is shown in Figure 4.35(b), where three links (dashed lines) were added.

- A triangulated graph obtained using Algorithm 4.2 is shown in Figure 4.36(a), where two additional links were added.

- Finally, applying Algorithm 4.4 as in Example 4.15, we obtain the associated family tree shown in Figure 4.36(b). Note that all the families of the directed graph in Figure 4.35(a) are contained in at least one cluster of this family tree. ■

4.7 Representation of Graphs

Graphs can be represented in several different but equivalent ways, depending on the field of application. The most common and useful methods of representing graphs are:

4.7 Representation of Graphs   145

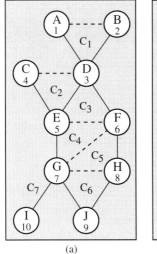
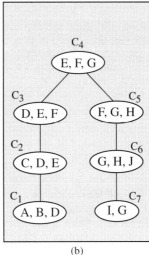

(a)   (b)

FIGURE 4.36. (a) Perfect numbering and clique ordering for a triangulation of the graph in Figure 4.35(b), and (b) a join tree associated with the graph.

- Symbolically as (X, L), where X contains the set of objects and L contains the set of links between pairs of variables, or as (X, Adj), where Adj is the list of adjacency sets.

- Pictorially as a diagram consisting of a set of nodes (one node for each variable) and a set of lines or arrows (each of which represents a link in the set L).

- Numerically, using certain types of matrices.

Each of these representations has some advantages and shortcomings. For example, the symbolic representation has the advantage of simplicity (any graph can be described by two lists), but on the other hand, it does not give direct information about the topology of the graph. The pictorial representation of a graph has visual appeal. It is easier to understand the relationships among the nodes by visually examining a diagram than by reading the links in the set L. However, if the number of nodes and links is large, pictorial representations become messy and sometimes infeasible.

In this section we present two methods for the pictorial representation of a graph and discuss the advantages and shortcomings of each (Section 4.7.1). We also show how the adjacency sets symbolic representation of a graph can be numerically described using certain types of matrices. We illustrate the usefulness of this type of graph representation in characterizing some properties of the graph (Section 4.7.2).

146   4. Some Concepts of Graphs

4.7.1 Pictorial Representations of Graphs

A graph was previously defined as a set of nodes and an associated set of links. The question to be addressed in this section is, How to draw the nodes and links of a graph, for example, on a sheet of paper or on a computer screen?

Clearly, a graph can be represented pictorially in many different ways. However, some representations are better than others in terms of simplicity, ability to show the main characteristics of the graph, and visual appearance. Pictorial representations of graphs also allow us to analyze certain properties of the graphs. For example, the type of graph that can be represented in the plane in such a way that the links do not cross each other is known as a *planar graph* and has several interesting properties. The books by Preparata and Shamos (1985) and Tamassia and Tollis (1995) offer a description of the problems associated with graph representations. We consider that a good pictorial representation of a graph is one in which

1. The diagram can be easily and quickly constructed using some algorithms.

2. The characteristics of the graphs can be easily and quickly seen from the drawn diagram. For example, Figure 4.37 shows two different diagrams for the same graph, but it is easier to see that the graph is multiply-connected from the diagram in Figure 4.37(b) than from the one in Figure 4.37(a). For another example refer back to Figure 4.8, where it can be seen that it is easier to see the disconnectedness of the graph from the diagram in Figure 4.8(b) than from the one in Figure 4.8(a).

3. The diagram is visually simple. For example, by keeping the number of crossed links to a minimum.

In this section we present two different ways of systematically drawing a given graph:

1. The *circular* representation, and

2. The *multilevel* representation.

Circular Representation of Graphs

One of the easiest ways to represent a graph pictorially is to draw its nodes on a circumference of a circle at approximately equal distances. See, for example, Figure 4.37(a). This form of representation has an important property: it ensures that no more than two nodes fall on a straight line. When three or more nodes fall on a straight line, if the links are also represented by straight lines, the links between the two outer nodes will be

4.7 Representation of Graphs   147

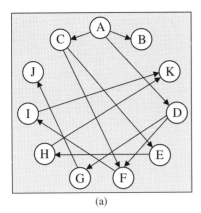 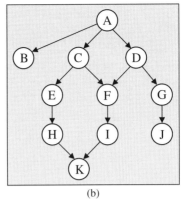

FIGURE 4.37. Two pictorial representations of the same graph: (a) circular and (b) multilevel representations.

hidden. In this sense a circular diagram is the optimal representation for highly connected graphs.

Circular representation of graphs has the following advantages:

- It is easy to construct.

- All links are visually transparent in the diagram.

- It is most suitable for complete or nearly complete graphs.

The main disadvantage of circular representation is that too many links can cross each other, which can make the diagram visually complicated or displeasing. For example, although the two diagrams in Figure 4.37 represent the same graph, there are no crossed links in diagram (b).

Multilevel Representation of Graphs

The basic idea of multilevel representation is to organize the nodes in different levels or layers in such a way that there is no link between nodes in the same level and that every node is connected to some node in the previous level. Thus, a clean and efficient graphical representation can be obtained by placing the nodes corresponding to the same level along either horizontal (as in Figure 4.38) or vertical levels (as in Figure 4.39). To develop this idea, we introduce some definitions.

Definition 4.41 Totally disconnected subset. *Given a graph (X, L), a subset of the nodes $S \subset X$ is called a totally disconnected subset if there is no link between nodes in S, that is, if $(X_i, X_j) \in L \Rightarrow X_i \notin S$ or $X_j \notin S$.*

148   4. Some Concepts of Graphs

Definition 4.42 Multilevel representation. *A multilevel representation of an undirected graph (X, L) is a partition*

$$X = \bigcup_{k=1}^{m} S_k, \qquad (4.1)$$

where the levels S_i, $i = 1, \ldots, m$, are disjoint and totally disconnected subsets of X such that

if $X_i \in S_k \Rightarrow \exists\, X_j \in S_{k-1}$ with $X_i \in Adj(X_j)$.

That is, there are no links between nodes in the same level and the nodes in one level are adjacent to at least one of the nodes in the previous level, if any.

Note that the nodes forming the first level only have to satisfy the property of being a totally disconnected set. Therefore the selection of the first level is quite arbitrary, and hence, a graph may have several multilevel representations. We refer to the nodes in the first level as the *root nodes*.

Accordingly, if we can partition the nodes as in (4.1), we can draw the nodes in a multilevel diagram (multistory building) in which the kth subset S_k occupies the kth level. For example, the graphs in Figures 4.38(a) and (b) are two different multilevel representations of the same graph. The levels associated with the representation in Figure 4.38(a) are

$$\{\{A\}, \{B, C\}, \{D, E, F\}, \{G, H, I\}\},$$

and the levels for the representation given in Figure 4.38(b) are

$$\{\{E\}, \{B, H, C\}, \{D, A, F\}, \{G, I\}\}.$$

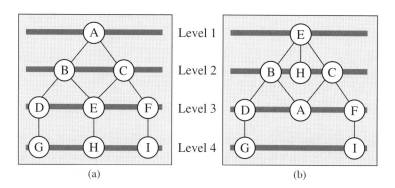

FIGURE 4.38. Two multilevel representations of the same graph. Shaded lines indicate the horizontal levels.

Note that these two representations contain a single root node. Figure 4.39 shows a different multilevel representation with root nodes $\{D, H\}$.

Some advantages of the multilevel representation are

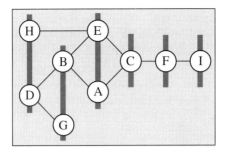

FIGURE 4.39. A vertical multilevel representation of the graph in Figure 4.38.

- It is the most suitable for trees or sparse graphs.
- It has the visual appeal of organizing the nodes in levels.

Note that we can always obtain a multilevel representation for an undirected graph by choosing any totally disconnected subset of nodes as the set of root nodes and assigning it to the first level. Thus, the second level is formed by some of the nodes adjacent to the root nodes, and so on. For example, in the representations given in Figures 4.38 the only root nodes are A and E, respectively. However, in Figure 4.39 the set of root nodes is $\{D, H\}$. This is the basic idea of the following algorithm.

Algorithm 4.6 Multilevel Representation.

- **Input:** A graph (X, Adj) of n nodes and a set of root nodes R.
- **Output:** A multilevel representation S of the graph.

1. *Initialization:* Assigned $= R$. $Level(1) = R$. $Level(k) = \phi$ for $k = 2, \ldots, n$. Set $j = 1$.

2. If $Level(j) = \phi$, stop: $S = \{Level(1), \ldots, Level(j-1)\}$ is a multilevel representation of the graph; otherwise, set $CurrentLevel = Level(j)$ and go to Step 3.

3. Select $X_k \in CurrentLevel$:

    (a) Add to $Level(j+1)$ and to $Assigned$ the elements in $Adj(X_k) \setminus Assigned$.

    (b) Add to $Level(j+1)$ the elements in $Adj(X_k) \cap Level(j)$, remove these elements from $Level(j)$ and from $CurrentLevel$ and go to Step 4.

4. Remove X_k from $CurrentLevel$. If $CurrentLevel = \phi$, set $j = j+1$ and go to Step 2; otherwise, go to Step 3. ∎

In the above algorithm, Step 3(a) adds to the level all the not-previously-assigned neighbors of the nodes in the previous level, whereas Step 3(b) avoids neighboring nodes in the same level.

Note that if the input set of root nodes R is not totally disconnected, the algorithm automatically removes nodes from the first level until a totally disconnected subset of R is obtained.

Example 4.17 Multilevel representation. Let us apply Algorithm 4.6 to obtain a multilevel representation for the graph in Figures 4.39. This graph is defined by the following adjacent sets:

$$Adj(A) = \{B, C\}, \quad Adj(B) = \{A, D, E\}, \quad Adj(C) = \{A, E, F\},$$
$$Adj(D) = \{B, G\}, \quad Adj(E) = \{B, C, H\}, \quad Adj(F) = \{C, I\},$$
$$Adj(G) = \{D\}, \quad Adj(H) = \{E\}, \quad Adj(I) = \{F\}.$$

Consider the set of root nodes $\{D, H\}$. Table 4.1 shows the process resulting from applying Algorithm 4.6. The rows in this table show the state of the corresponding variables and sets at the end of each step. Thus, we will finally obtain the multilevel representation

Level(1)={D,H},   Level(2)={B,G},
Level(3) ={A,E},   Level(4)={C},
Level(5) ={F},   Level(6)={I}.

The graphical representation associated with this partition is shown in Figure 4.39.

Note that applying Algorithm 4.6 to the graph in Figure 4.38 with root nodes A and E gives the multilevel representations shown in Figures 4.38(a) and 4.38(b), respectively. ■

In the case of directed graphs, one can consider the associated undirected graph to build a multilevel representation for a graph. However, the above process does not take into account the direction of the links. We shall see that in the case of directed acyclic graphs (DAGs), we can give a directed character to the multilevel representation in the sense that all the arrows in the graph will point in the same direction (up-to-down in horizontal representation and left-to-right in vertical representation).

Definition 4.43 Directed multilevel representation. A *multilevel representation of an undirected graph* (X, L) *is a partition*

$$X = \bigcup_{k=1}^{m} S_k, \tag{4.2}$$

where the levels S_i, $i = 1, \ldots, m$, are disjoint and totally disconnected subsets of X such that

$$X_i \in S_k \text{ and } (X_i, X_j) \in L \Rightarrow X_j \in S_r \text{ with } r > k,$$

4.7 Representation of Graphs 151

Step	X_k	j	$Level(j)$	$Level(j+1)$	$CurrentLevel$
1	—	1	$\{D,H\}$	ϕ	ϕ
2	—	1	$\{D,H\}$	ϕ	$\{D,H\}$
3(a)	D	1	$\{D,H\}$	$\{B,G\}$	$\{D,H\}$
3(b)	D	1	$\{D,H\}$	$\{B,G\}$	$\{D,H\}$
4	D	1	$\{D,H\}$	$\{B,G\}$	$\{H\}$
3(a)	H	1	$\{D,H\}$	$\{B,G,E\}$	$\{H\}$
3(b)	H	1	$\{D,H\}$	$\{B,G,E\}$	$\{H\}$
4	H	2	$\{B,G,E\}$	ϕ	ϕ
2	H	2	$\{B,G,E\}$	ϕ	$\{B,G,E\}$
3(a)	B	2	$\{B,G,E\}$	$\{A\}$	$\{B,G,E\}$
3(b)	B	2	$\{B,G\}$	$\{A,E\}$	$\{B,G\}$
4	B	2	$\{B,G\}$	$\{A,E\}$	$\{G\}$
3(a)	G	2	$\{B,G\}$	$\{A,E\}$	$\{G\}$
3(b)	G	2	$\{B,G\}$	$\{A,E\}$	$\{G\}$
4	G	3	$\{A,E\}$	ϕ	ϕ
2	G	3	$\{A,E\}$	ϕ	$\{A,E\}$
3(a)	A	3	$\{A,E\}$	$\{C\}$	$\{A,E\}$
3(b)	A	3	$\{A,E\}$	$\{C\}$	$\{A,E\}$
4	A	3	$\{A,E\}$	$\{C\}$	$\{E\}$
3(a)	E	3	$\{A,E\}$	$\{C\}$	$\{E\}$
3(b)	E	3	$\{A,E\}$	$\{C\}$	$\{E\}$
4	E	4	$\{C\}$	ϕ	ϕ
2	E	4	$\{C\}$	ϕ	$\{C\}$
3(a)	C	4	$\{C\}$	$\{F\}$	$\{C\}$
3(b)	C	4	$\{C\}$	$\{F\}$	$\{C\}$
4	C	5	$\{F\}$	ϕ	ϕ
2	C	5	$\{F\}$	ϕ	$\{F\}$
3(a)	F	5	$\{F\}$	$\{I\}$	$\{F\}$
3(b)	F	5	$\{F\}$	$\{I\}$	$\{F\}$
4	F	6	$\{I\}$	ϕ	ϕ
2	F	6	$\{I\}$	ϕ	$\{I\}$
3(a)	I	6	$\{I\}$	ϕ	$\{I\}$
3(b)	I	6	$\{I\}$	ϕ	$\{I\}$
4	I	7	$\{I\}$	ϕ	ϕ

TABLE 4.1. Steps of the multilevel representation from Algorithm 4.6.

152   4. Some Concepts of Graphs

that is, all the parents of a given node are in levels preceding the level of the node.

For example, Figure 4.40 shows two different directed multilevel representations of the graph in Figure 4.37. The levels (subsets S_k, $k = 1, \ldots, 5$) are distinguished from each other in the figure by shading. The levels in Figures 4.40(a) and (b) are

$$\{\{A\}, \{B, C, D\}, \{E, F, G\}, \{H, I, J\}, \{K\}\}$$

and

$$\{\{A, B\}, \{C, D\}, \{E, F\}, \{H, I, G\}, \{K, J\}\},$$

respectively. Both partitions satisfy (4.2).

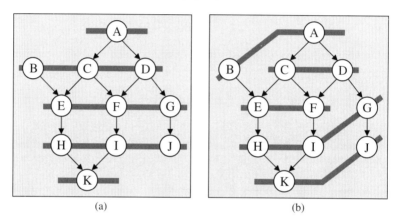

FIGURE 4.40. Two different multilevel representations of a directed graph.

One advantage of the directed multilevel type of representation is that they lead to simple and easy to interpret diagrams, since all links of the diagram are in the same direction. Figures 4.40, for example, shows a multilevel diagram with an "up-down" representation.

When trying to draw a directed multilevel representation, one may wonder whether any directed graph can be decomposed according to a directed multilevel representation. The answer is given by the following theorem.

Theorem 4.5 Directed multilevel representation. *A directed graph (X, L) admits a directed multilevel representation iff (X, L) is a directed acyclic graph (DAG).*

Note that if we use the set of nodes without parents $\{X_i \mid \Pi_{X_i} = \phi\}$ of a DAG as the set of root nodes R, Algorithm 4.6 leads to a directed multilevel representation of the DAG, as in (4.2).

The following theorem provides a useful relationship between directed multilevel representations and ancestral numbering.

Theorem 4.6 Ancestral numbering. *If $\{S_1, \ldots, S_m\}$ is a directed multilevel representation of a DAG (X, L), then any numbering α of the nodes satisfying $\alpha(X_i) > \alpha(X_j)$ for $X_i \in S_k$ and $X_j \in S_r$ with $k > r$ is an ancestral numbering of the nodes.*

Thus, using Algorithm 4.6 we can obtain an ancestral numbering for a DAG. Moreover, DAGs are the only type of directed graphs that have an ancestral numbering. The proof of this result is left as an exercise for the reader.

In the rest of this section, we develop an algorithm for partitioning any DAG as in (4.2). For this, we need the following definitions.

Definition 4.44 Ascending depth. *The ascending depth of a node X_i in a DAG, denoted by $AD(X_i)$, is the maximum of the lengths of all paths in the graph ending with X_i.*

Definition 4.45 Descending depth. *The descending depth of a node X_i in a DAG, denoted by $DD(X_i)$, is the maximum of the lengths of all paths in the graph starting with X_i.*

To calculate the ascending depth of one node it is enough to know the ascending depth of its parents. Similarly, to calculate the descending depth of one node it is enough to know the descending depth of its children. Descending depth increases in the sense of the oriented links. The reverse is true for ascending depth.

Ascending and descending depths satisfy the following properties:

- $0 \leq AD(X_i) \leq n-1$ and $0 \leq DD(X_i) \leq n-1$, where n is the number of nodes.

- If X_i has no parents, then $AD(X_i) = 0$.

- If X_i has no children, then $DD(X_i) = 0$.

By calculating the depth of all nodes in a graph, we divide the graph into levels (sets of nodes with the same depth).

Definition 4.46 Depth levels of a graph. *Given a directed graph, the kth ascending depth level, AL_k, is the subset $\{X_i \in X \mid AD(X_i) = k\}$. Similarly, the kth descending depth level, DL_k, is the subset $\{X_i \in X \mid DD(X_i) = k\}$.*

The number of nonempty levels of a graph can be calculated as a function of the length of its paths. Let m be the length of the largest path in a graph, then $L_k = \phi, \forall k > m$ and $L_k \neq \phi, \forall k \leq m$. Thus, we have a finite number of levels that define a partition of the graph satisfying Definition 4.43. This is provided in the following theorem.

154   4. Some Concepts of Graphs

Theorem 4.7 Depth levels and directed representation. *For any DAG, the sets $\{DL_k : k = 0, \ldots, m\}$ and $\{AL_k : k = 0, \ldots, m\}$ are two directed multilevel representations satisfying (4.2).*

For example, Figure 4.41 shows the ascending and descending depth levels associated with the directed graph in Figure 4.40.

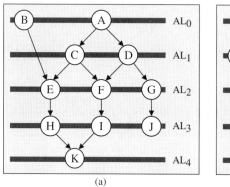
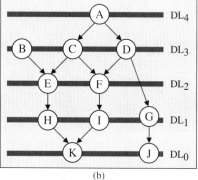

FIGURE 4.41. (a) Ascending and (b) descending depth levels of a directed graph.

We now give an algorithm for calculating the depth-levels of a directed graph. Given the iterative character of this algorithm, it also allows testing whether or not the graph is cyclic. We give the algorithm for the ascending depths. A similar algorithm for the descending depths is left as an exercise for the reader.

Algorithm 4.7 Ascending-Depth for Directed Graphs.

- **Input:** A graph (X, Adj) of n nodes.
- **Output:** The ascending depth levels $\{AL_0, \ldots, AL_k\}$ of the graph.

1. *Initialization:* Let $AD(X_i) = 0$ for all nodes X_i with no parents. If all nodes have some parents, then stop (the graph contains at least one cycle). Otherwise, let $depth = 1$ and continue with Step 2.

2. If $depth \le n$, go to Step 3; otherwise, stop (the graph contains some cycles).

3. Select one node X_i with $AD(X_i) = depth - 1$. For all nodes X_j adjacent to X_i let $AD(X_j) = depth$. Repeat this process with all nodes in the $depth - 1$ level, then go to Step 4.

4. If no node has depth equal to $depth$, then stop, the depths of all nodes have been calculated. Otherwise, add one unit to $depth$ and return to Step 2.   ■

Ascending depth levels algorithm

Input: A directed graph (X, Adj) of n nodes.
Output: The corresponding ascending-depth levels.

Initial Step:
for $k = 1$ to n **do**
    if node X_k has no parents **then** $AD(X_k) \leftarrow 0$
if $AD(X_k) \neq 0$, $k = 1, \ldots, n$
    Stop. The graph contains some cycles.

Iteration Step i:
if $i > n$ **then**
    Stop. The graph contains some cycles.
otherwise
    for all X_k such that $AD(X_k) = i - 1$ **do**
        $AD(X_r) \leftarrow i$, for all $X_r \in Adj(X_k)$
  if $AD(X_k) \neq i$, $k = 1, \ldots, n$
        Stop. All depths have been calculated.
  otherwise
        go to the iteration Step $i + 1$

FIGURE 4.42. Pseudocode for the ascending-depth algorithm for directed graphs.

Figure 4.42 shows the pseudocode for the ascending-depth algorithm. If the algorithm does not stop before Step n or stops at Step 0, then the graph contains some cycles. Otherwise, the graph is acyclic, the depths of all nodes have been calculated, and the graph contains as many depth levels as the number of iteration steps performed.

4.7.2 Numerical Representation of Graphs

Graphs can also be represented numerically using certain types of matrices. From this numerical representation several characteristics of the graph can be easily obtained. We start with some definitions:

Definition 4.47 Adjacency or incidence matrix. Let $G = (X, L)$ be a graph of n nodes and let $A = (a_{ij})$ be an $n \times n$ matrix, where

$$a_{ij} = \begin{cases} 1, & \text{if } L_{ij} \in L, \\ 0, & \text{otherwise.} \end{cases}$$

The matrix A is called the adjacency matrix associated with the graph G.

156   4. Some Concepts of Graphs

By means of this matrix and using only algebraic manipulations, some characteristics of a graph such as the number of paths joining any two nodes, whether or not the graph is connected, etc., can be obtained.

For example, Figure 4.43, shows the adjacency building process for a given graph. When $a_{ij} = 0$, there is no direct link from node X_i to node X_j. When $a_{ij} = 1$, there is a direct link between X_i and X_j; in other words, we say that X_i and X_j are adjacent nodes, hence the name adjacency matrix. The adjacency matrix is also known as the *incidence matrix* because its elements indicate whether or not there is a direct link between X_i and X_j.

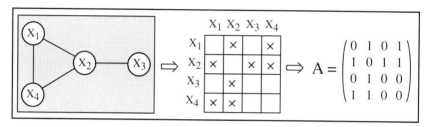

FIGURE 4.43. Building the adjacency matrix of a graph.

Clearly, the matrix A has all the information about the links of the associated graph; hence it completely characterizes the graph. For example:

- The adjacency matrix of an undirected graph is symmetric.

- Since $L_{ii} \notin L$ for all values of i, all the diagonal elements of A are zeros.

- The adjacency matrix of a complete undirected graph must have all off-diagonal elements equal to 1.

Using the adjacency matrix one can obtain several characteristics of the corresponding graph. For example, we can determine whether or not a path exists between every pair of nodes. We can also calculate the lengths of all existing paths. The following theorem shows how the adjacency matrix can be used for this task.

Theorem 4.8 Powers of the adjacency matrix. *Let A^r be the rth power of the adjacency matrix associated with the graph $G = (X, L)$. Then the ijth element of A^r is equal to t iff there exist t paths between X_i and X_j each of length r.*

Proof: The proof of Theorem 4.8 is obtained by induction as follows: It is true for $r = 1$, because $a_{ij} = 1$ if there is a path of length 1 (a link) between nodes i and j, and $a_{ij} = 0$ otherwise.

Assuming that the result is true for A^r, we prove it for A^{r+1}. We have

$$A^{r+1} = A^r A \Leftrightarrow a_{ij}^{r+1} = \sum_{k=1}^{n} a_{ik}^r a_{kj},$$

that is, if there are a_{ik}^r paths of length r between nodes X_i and X_k and there is one link between nodes X_k and X_j ($a_{kj} = 1$), then we have a_{ik}^r paths of length $(r+1)$. ■

It follows from Theorem 4.8 that

- The ijth element of A^r is zero iff there exists no path of length r between X_i and X_j.

- By calculating the successive powers of the adjacency matrix of a given graph $A, A^2, A^3, \ldots$, we can directly determine the number of paths of lengths $1, 2, 3, \ldots$ joining any pair of nodes.

We illustrate these properties by an example.

Example 4.18 Powers of the adjacency matrix. The first three powers of the adjacency matrix of the graph in Figure 4.43 are

$$A = \begin{pmatrix} 0 & 1 & 0 & 1 \\ 1 & 0 & 1 & 1 \\ 0 & 1 & 0 & 0 \\ 1 & 1 & 0 & 0 \end{pmatrix}, \quad A^2 = \begin{pmatrix} 2 & 1 & 1 & 1 \\ 1 & 3 & 0 & 1 \\ 1 & 0 & 1 & 1 \\ 1 & 1 & 1 & 2 \end{pmatrix}, \quad A^3 = \begin{pmatrix} 2 & 4 & 1 & 3 \\ 4 & 2 & 3 & 4 \\ 1 & 3 & 0 & 1 \\ 3 & 4 & 1 & 2 \end{pmatrix},$$

from which we can see, for example, that there is only one path of length 3 joining the nodes X_1 and X_3, that is, ($a_{13}^3 = 1$). In Figure 4.43, this path is $X_1 - X_4 - X_2 - X_3$. ■

The adjacency matrix can also be used to determine whether a graph is connected or disconnected. To this aim we introduce another important matrix associated with a graph, the *attainability matrix*.

Definition 4.48 Attainability matrix. *The attainability matrix,* $T = (t_{ij})$, *is defined by*

$$t_{ij} = \begin{cases} 1, & \text{if there exists a path joining nodes } X_i \text{ and } X_j \\ 0, & \text{otherwise.} \end{cases}$$

It is clear that the attainability matrix is related to the powers of the adjacency matrix. The following result gives a bound for the maximum number of powers needed to obtain the attainability matrix.

Theorem 4.9 Limited length of paths. *Given a graph with n nodes, if there exists a path from node X_i to node X_j, then there also exists a path of length $< n$ from X_i to X_j.*

The proof of the above theorem is left as an exercise for the reader. Therefore, the attainability matrix can be obtained from a finite number of powers, $A, A^2, A^3, \ldots, A^{n-1}$, of the adjacency matrix and that the number of powers needed is $n - 1$. In fact, we have

$$t_{ij} = \begin{cases} 0, & if \ a_{ij}^k = 0, \forall k < n \\ 1, & otherwise. \end{cases} \qquad (4.3)$$

In a connected graph all elements of the attainability matrix must be different from zero. Thus, from the structure of the attainability matrix we can determine whether the graph is connected, and if it is disconnected, we can also determine its connected components.

Example 4.19 Attainability matrix. Given the graph in Figure 4.44, it is possible to determine its attainability matrix by knowing the first $n = 5$ powers of its adjacency matrix. The adjacency matrix of this graph is

$$A = \begin{pmatrix} 0 & 1 & 1 & 0 & 0 & 0 \\ 1 & 0 & 1 & 0 & 0 & 0 \\ 1 & 1 & 0 & 0 & 0 & 0 \\ 0 & 0 & 0 & 0 & 1 & 1 \\ 0 & 0 & 0 & 1 & 0 & 1 \\ 0 & 0 & 0 & 1 & 1 & 0 \end{pmatrix}.$$

Calculating the first five powers of this matrix, perhaps using a computer program capable of multiplying matrices, we find the corresponding attainability matrix using (4.3):

$$T = \begin{pmatrix} 1 & 1 & 1 & 0 & 0 & 0 \\ 1 & 1 & 1 & 0 & 0 & 0 \\ 1 & 1 & 1 & 0 & 0 & 0 \\ 0 & 0 & 0 & 1 & 1 & 1 \\ 0 & 0 & 0 & 1 & 1 & 1 \\ 0 & 0 & 0 & 1 & 1 & 1 \end{pmatrix}.$$

From T, we can clearly distinguish two connected components: $\{X_1, X_2, X_3\}$ and $\{X_4, X_5, X_6\}$. These conclusions may be difficult to obtain from a graphical representation of a graph. Thus, in complex graphs, the adjacency and the attainability matrices are very useful tools, from a theoretical and computational point of view, for determining several aspects or properties of graphs. ∎

4.8 Some Useful Graph Algorithms

In the previous sections we introduced several properties of graphs. In this section we provide some graph algorithms useful for determining whether a given graph possesses some of these properties. More specifically, given a graph, we wish to

4.8 Some Useful Graph Algorithms

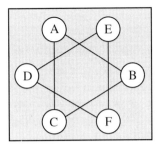

FIGURE 4.44. Example of a disconnected graph.

1. Obtain a path between two nodes.

2. Test whether the graph is connected and to find its connected components.

3. Identify loops or cycles, if present in the graph.

It is not our aim here to give optimally efficient algorithms for solving these problems, but to show, with the help of some illustrative examples, the basic ideas behind the suggested methods. Readers interested in the design of efficient algorithms can consult Cormen, Leiserson, and Rivest (1990) and Golumbic (1980). More information about graph algorithms can be obtained in Gibbons (1985), McHugh (1990), and Skiena (1990).

4.8.1 Search Methods

Many graph algorithms require a mechanism for exploring the nodes and links of a graph. For example, among other things, search algorithms can be used for obtaining a path between two nodes or looking for a loop or cycle in a graph. These methods are the basis for the construction of the algorithms introduced in this section.

The exploration of a graph starts from an initial node and consists of the specification of moving back and forth through the links, passing from a node to a neighboring node. Thus, at any stage of the process, a decision has to be made as to which node to visit next. For example, Figure 4.45 shows an example an exhaustive search technique of the graph starting at node A. Note that following the given sequence and passing from one node to a neighboring node, we can visit all the nodes in the graph in the following order: $A, B, D, G, H, C, E, I, F, J$, and K. Note also that any link is traveled at most twice: Once in the forward direction (full lines) reaching a new node and once in the backward direction (dashed lines) going back to a previously visited node.

Several heuristic search techniques have been proposed in the literature (see, for example, Rich and Knight (1991)). In this section we discuss two search methods useful for exploring a graph:

160   4. Some Concepts of Graphs

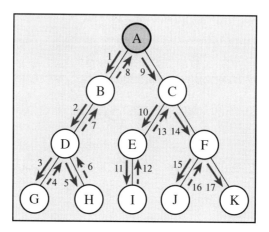

FIGURE 4.45. Example of a search process.

- **Depth-first search method:** At each step of the depth-first method one of the unvisited neighbors of the current node is chosen as the new current node (see Figure 4.46(a), where the number indicates the order in which the node is visited). In the case that no unvisited neighbor exists, the algorithm goes back to the previous current node and the search continues until all nodes have been visited.

- **Breadth-first search method:** The breadth-first method visits the graph layer by layer, starting at an initial node. First, all the neighbors of the initial node are visited. Next, the neighbors of each of those neighbors are visited, and so on (see Figure 4.46(b), where the number indicates the order in which the node is visited).

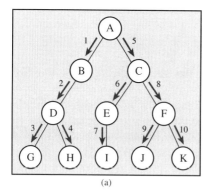
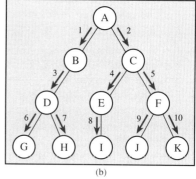

FIGURE 4.46. An illustration of the (a) depth-first and (b) breadth-first search methods. The numbers indicate the order in which the nodes are visited.

In the next sections, several algorithms are developed based on these two strategies.

4.8.2 Pathfinding Algorithms

Given a graph $G = (X, L)$, we wish to find a path from node X_i to node X_j, if such path exists. In this section we give two pathfinding algorithms based on the two search strategies introduced above. For this purpose, it is more convenient and efficient to use the representation of a graph given by its adjacency sets (see Definition 4.5). The undirected graph in Figure 4.47(a) can be represented as (X, L), where X is the set of nodes $\{A, B, C, D, E, F, G\}$ and L is the set of links $\{L_1, \ldots, L_8\}$. However, for computational purposes, a more suitable representation of this graph is given by the adjacency sets:

$$\begin{array}{lll} Adj(A) = \{B, C, D\}, & Adj(B) = \{A, E\}, & Adj(C) = \{A, F\}, \\ Adj(D) = \{A, F\}, & Adj(E) = \{B, G\}, & Adj(F) = \{C, D, G\}, \\ Adj(G) = \{E, F\}. & & \end{array} \quad (4.4)$$

Thus, $G = (X, L)$ can also be represented as $G = (X, Adj)$, where Adj is the sets of adjacent nodes given in (4.4). This representation is more efficient for searching methods because it avoids an unnecessary checking of all the links of a graph to choose the next node in a path.

The directed graph in Figure 4.47(b) has the following adjacency sets:

$$\begin{array}{lll} Adj(A) = \{B, C, D\}, & Adj(B) = \{E\}, & Adj(C) = \{F\}, \\ Adj(D) = \{F\}, & Adj(E) = \{G\}, & Adj(F) = \{G\}, \\ Adj(G) = \phi. & & \end{array} \quad (4.5)$$

Thus, the adjacency sets provide a representation independent of the directed or undirected character of a graph. Note that if we are given a directed graph in Figure 4.47(b) and we want to perform some task of undirected character (presence of loops, undirected paths, etc.), we can consider the adjacency sets given by its associated undirected graph (4.4).

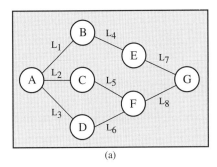
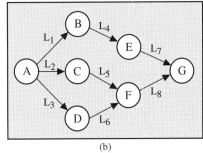

FIGURE 4.47. Examples of an (a) undirected and (b) directed graph.

Applying the search strategies introduced in the previous section, two main path-searching algorithms can be implemented: *depth-first pathfinding* and *breadth-first pathfinding*.

Algorithm 4.8 Depth-First Pathfinding Algorithm.

- **Input:** An arbitrary graph (X, Adj) and two nodes X_i and X_j.

- **Output:** A path $Path = \{X_{i_1}, \ldots, X_{i_r}\}$ from node $X_i = X_{i_1}$ to node $X_j = X_{i_r}$. If there is no such path, then $Path = \phi$.

1. *Initialization:* Let $X_k = X_i$, $Path = \{X_i\}$, and $Visited = \{X_i\}$.

2. *Iteration:* If all nodes in $Adj(X_k)$ have been visited, or we have $Adj(X_k) = \phi$, go to Step 4; otherwise, go to Step 3.

3. *Forward step:* Choose a node $X_r \in Adj(X_k)$, such that $X_r \notin Visited$, and add X_r to both $Path$ and $Visited$. If $X_r = X_j$, then return $Path$ and stop (a path has been found); otherwise, let $X_k = X_r$ and go to Step 2.

4. *Backward step:* If $X_k = X_i$, then stop (there is no path from X_i to X_j); otherwise, remove X_k from $Path$, let X_k be the last node in $Path$, and go to Step 2. ∎

Figure 4.48 shows the pseudocode for Algorithm 4.8. In each step of this algorithm, two lists are updated:

- The *Path* list, which contains the path from X_i to the last visited node.

- The *Visited* list, which contains the nodes visited so far.

Example 4.20 Depth-first path finding. Given the undirected graph in Figure 4.49(a), we wish to obtain a path between nodes A and F. The results of applying Algorithm 4.8 to this graph are given in Table 4.2. This table shows the values of the current node X_k, its unvisited adjacent nodes $Adj(X_k) \setminus Visited$, and the lists *Path* and *Visited* at the end of the indicated steps. Note that in this case no backward step is performed and the algorithm obtains the path $A - B - E - G - F$ (see Figure 4.49(a)).

Depth-First Pathfinding Algorithm

Input: A graph (X, Adj) and two nodes X_i and X_j.
Output: A path from X_i to X_j, or ϕ if there is no path.

Initial Step:

$X_k \leftarrow X_i$
$Path \leftarrow \{X_i\}$
$Visited \leftarrow \{X_i\}$

Iteration Step:

if there exists $X_r \in Adj(X_k) \setminus Visited$, **then**
    add X_r to both $Visited$ and $Path$
    if $X_r = X_j$, **then**
        stop. A path has been found.
    otherwise
        Let $X_k \leftarrow X_r$
        repeat the iteration step.
otherwise
    if $X_k = X_i$, **then**
        stop. There is not such a path.
    otherwise
        remove X_k from $Path$
        $X_k \leftarrow$ last node in $Path$
        repeat the iteration step.

FIGURE 4.48. Pseudocode for the depth-first pathfinding algorithm.

Steps	X_k	$Adj(X_k) \setminus Visited$	Visited	Path
1	A	$\{B, C, D\}$	$\{A\}$	$\{A\}$
2, 3	B	$\{E\}$	$\{A, B\}$	$\{A, B\}$
2, 3	E	$\{G\}$	$\{A, B, E\}$	$\{A, B, E\}$
2, 3	G	$\{F\}$	$\{A, B, E, G\}$	$\{A, B, E, G\}$
2, 3	F	$\{C, D, G\}$	$\{A, B, E, G, F\}$	$\{A, B, E, G, F\}$

TABLE 4.2. Steps of the depth-first pathfinding algorithm for finding a path between A and F in the undirected graph in Figure 4.47(a).

On the other hand, if we consider the directed graph given in Figure 4.49(b), the depth-first search algorithm has to use forward and backward searching, and it obtains the path $A \to C \to F$. Table 4.3 shows the steps

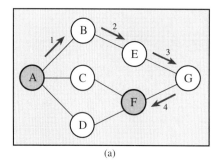 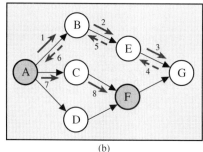

FIGURE 4.49. Steps of the depth-first pathfinding algorithm for finding a path between nodes A and F in (a) an undirected graph and (b) a directed graph.

Steps	X_k	$Adj(X_k) \setminus Visited$	Visited	Path
1	A	$\{B, C, D\}$	$\{A\}$	$\{A\}$
2, 3	B	$\{E\}$	$\{A, B\}$	$\{A, B\}$
2, 3	E	$\{G\}$	$\{A, B, E\}$	$\{A, B, E\}$
2, 3	G	ϕ	$\{A, B, E, G\}$	$\{A, B, E, G\}$
2, 4	E	ϕ	$\{A, B, E, G\}$	$\{A, B, E\}$
2, 4	B	ϕ	$\{A, B, E, G\}$	$\{A, B\}$
2, 4	A	$\{C, D\}$	$\{A, B, E, G\}$	$\{A\}$
2, 3	C	$\{F\}$	$\{A, B, E, G, C\}$	$\{A, C\}$
2, 3	F	ϕ	$\{A, B, E, G, C, F\}$	$\{A, C, F\}$

TABLE 4.3. Steps of the depth-first pathfinding algorithm for finding a path between A and F in the directed graph in Figure 4.47(b).

of the depth-first search algorithm for this example. In this case, we reach node G, but $Adj(G) = \phi$. Therefore, no path is possible and the algorithm goes back to the previous node to start searching again (see Figure 4.49(b)). ■

Note that Algorithm 4.8 always obtains simple paths, that is, paths that do not contain the same node twice (do not contain loops or cycles). The path found, however, is not necessarily the shortest path. As we shall see in Section 4.8.4, if we are interested in a closed path, this algorithm can be easily modified to find loops and cycles in a graph.

Next we consider the *breadth-first* search strategy. We present an algorithm for deciding whether or not a path exists between two given nodes, but not for finding the path. We leave the task of finding the path as an exercise for the reader.

Algorithm 4.9 Breadth-First Path-Checking Algorithm.

- **Input:** An arbitrary graph (X, Adj) and two nodes X_i and X_j.
- **Output:** Whether or not there exists a path between nodes X_i and X_j.

1. *Initialization:* Let $Visited = \phi$ and $NodesToBeVisited = \{X_i\}$.

2. *Iteration:* Select the first node, X_k, in the $NodesToBeVisited$ list, remove it from this list and add it to the $Visited$ list.

3. If $X_k = X_j$, then there is a path between X_i and X_j. Stop. Otherwise, if all neighbors of X_k have been already visited go to Step 4; otherwise, go to Step 5.

4. If no more nodes need to be visited, that is, if $NodesToBeVisited$ is empty, then there is no path between X_i and X_j. Stop. Otherwise, return to Step 2.

5. Append all unvisited nodes in $Adj(X_k)$ to the $NodesToBeVisited$ list and return to Step 2. ∎

Figure 4.50 shows the pseudocode for the breadth-first path-checking algorithm. Note that in each step of Algorithm 4.9, two lists are updated:

- The *Visited* list, which contains the nodes visited so far.
- The *NodesToBeVisited* list, which contains the queue of nodes yet to be visited.

If during this algorithm we reach node X_j, a path exists. Otherwise, after an exhaustive search of the path, the algorithm concludes that a path does not exist.

Example 4.21 Breadth-first path-checking algorithm. We now apply the breadth-first algorithm to check whether a path between nodes A and F exists in the undirected and directed graphs in Figure 4.51. In this case, the algorithm follows the same steps in both situations. Table 4.4 shows the steps of this process. The algorithm concludes at Step 3 that there is a path between the two nodes. ∎

The complexity of these algorithms is linear in the number of nodes and the number of links in the graph. The efficiency of both algorithms depends on the graph topology and the selected nodes. In general, the depth-first algorithm is more efficient than the breadth-first algorithm when the initial and final nodes are joined only by long paths (see Figure 4.52). When there is a short path between the nodes, the situation is reversed and the breadth-first algorithm becomes more efficient.

166   4. Some Concepts of Graphs

Breadth-First Search Algorithm

Input: A graph (X, Adj) and two nodes X_i and X_j.
Output: *True* if there is a path from X_i to X_j, *False* otherwise.

Initial Step:
$NodesToBeVisited \leftarrow \{X_i\}$
$Visited \leftarrow \phi$

Iteration Step:
$X_k \leftarrow$ first node in $NodesToBeVisited$
Remove X_k from $NodesToBeVisited$
Add X_k to $Visited$
if $X_k = X_j$, **then**
    Return *True* (there is a path from X_i to X_j) and **Stop**.
otherwise
    $S \leftarrow Adj(X_k) \setminus Visited$
    if $S \neq \phi$ **then**
        Add S at the beginning of $NodesToBeVisited$
        Repeat the iteration Step.
    otherwise
    if $NodesToBeVisited = \phi$, **then**
        Return *False* (there is no path from X_i to X_j) and **Stop**.
    otherwise
        Repeat the iteration Step.

FIGURE 4.50. Pseudocode for the breadth-first path-checking algorithm.

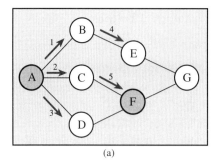
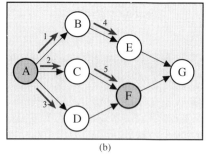

(a)   (b)

FIGURE 4.51. Steps of the breadth-first algorithm for checking whether a path between nodes A and F exists.

Steps	Node X_k	$Visited$	$NodesToBeVisited$
1	–	ϕ	$\{A\}$
2	A	$\{A\}$	$\{\}$
3,5	A	$\{A\}$	$\{B,C,D\}$
2	B	$\{A,B\}$	$\{C,D\}$
3,5	B	$\{A,B\}$	$\{C,D,E\}$
2	C	$\{A,B,C\}$	$\{D,E\}$
3,5	C	$\{A,B,C\}$	$\{D,E,F\}$
2	D	$\{A,B,C,D\}$	$\{E,F\}$
3,4	D	$\{A,B,C,D\}$	$\{E,F\}$
2	E	$\{A,B,C,D,E\}$	$\{F\}$
3,5	E	$\{A,B,C,D,E\}$	$\{F,G\}$
2	F	$\{A,B,C,D,E,F\}$	$\{G\}$

TABLE 4.4. Steps of the breadth-first path-checking algorithm and states of some elements at the end of each step.

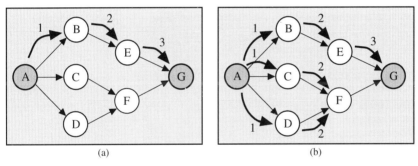

FIGURE 4.52. Finding a path between the nodes A and G with (a) depth-first and (b) breadth-first algorithms.

4.8.3 Testing Connectedness

The search methods discussed in the previous section can also be applied for testing the connectedness of a graph. The idea is to perform an exhaustive search of the graph, obtaining the set S of all nodes attainable from an initial node of the graph. If the graph is connected, then S will contain all the nodes in the graph; otherwise, the subset of nodes S will only contain the nodes in the connected component of the graph that contains the initial node.

Note that by considering the same initial and final node, $X_i = X_j$, Algorithms 4.8 and 4.9 can be used to perform an exhaustive search of the graph, that is, the list $Visited$ used at the end of both algorithms will contain all the nodes attainable from the initial node X_i.

Algorithm 4.10 Finding Connected Components of a Graph.

- **Input:** A graph (X, Adj).

- **Output:** The set of connected components C of (X, Adj).

1. *Initialization:* Let $Visited = \phi$, $C = \phi$.

2. If $X \setminus Visited = \phi$, return C and stop; otherwise, choose a node $X_i \in X \setminus Visited$ and go to Step 3.

3. Use Algorithm 4.8 or 4.9 to perform an exhaustive search of the graph (X, Adj) starting at node X_i and obtain the set S of visited nodes.

4. Add S to C. Add all the nodes in S to $Visited$. Go to Step 2. ∎

Note that if the set C contains only one connected component, then the graph is connected; otherwise, it is disconnected and C contains all the connected components of the graph.

Example 4.22 Testing connectedness. We have seen in Section 4.3.2 that the undirected graph (X, L) given in Figure 4.53(a) is disconnected. Let us apply Algorithm 4.10 to calculate all its connected components.

- Initially we set $Visited = \phi$ and $C = \phi$.

- $X \setminus Visited = X = \{A, B, C, D, E, F\}$. We choose the first of these nodes as the initial node $X_k = A$ for the first exhaustive search.

- Using Algorithm 4.8 with $X_i = X_j = A$, we obtain the set of visited nodes $S = C_1 = \{A, C, E\}$.

- Therefore, we have $C = \{C_1\}$ and $Visited = \{A, C, E\}$.

- $X \setminus Visited = \{B, D, F\}$. We take $X_k = B$.

- Using again Algorithm 4.8 with $X_i = X_j = B$, we obtain the set of visited nodes $C_2 = \{B, D, F\}$.

- Now we have $Visited = \{A, C, E, B, D, F\}$, $C = \{C_1, C_2\}$.

- Since $X \setminus Visited = \phi$, the algorithm returns C and stops.

Note that the set of connected components contains two subsets: $C_1 = \{A, C, E\}$ and $C_2 = \{B, D, F\}$. Therefore, the graph in Figure 4.53(a) is disconnected, and it contains two connected components, C_1 and C_2, as shown in Figure 4.53(b). ∎

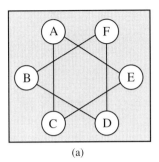 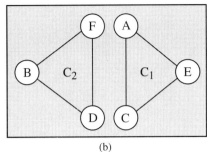

FIGURE 4.53. (a) An undirected disconnected graph and (b) its two connected components.

4.8.4 Finding Loops and Cycles

As we mentioned at the end of Example 4.20, the pathfinding algorithms can be easily modified to search for loops or cycles. In this section we modify only the depth-first search algorithm, but similar modifications of the breadth-first search algorithm can be made by the reader. Since our objective now is to find a closed path (a loop or a cycle), we can check at every step of Algorithm 4.8 whether there is some node included in the path that is also contained in the list of adjacent nodes of the current node. The resulting closed paths are loops if the graphs are undirected and cycles, if they are directed. The modified algorithm proceeds by selecting an initial arbitrary node and looking for a closed path contained in the graph until all nodes have been visited.

Algorithm 4.11 Depth-First Closed-Path Search Algorithm.

- **Input:** A graph (X, Adj).

- **Output:** A closed path, $Path$. If the graph contains no closed path, then $Path = \phi$.

1. *Initialization:* $Path = \phi$, and $Visited = \phi$.

2. If there is $X_i \in X \setminus Visited$, go to Step 3; otherwise, stop (there is no closed path in the graph).

3. Add X_i to $Visited$ and set $Path = \{X_i\}$, then take $X_k = X_i$ and $Previous = X_i$.

4. *Iteration:* If there exists $X_r \in Adj(X_k) \cap Path$, with $X_r \neq Previous$, then add X_r to $Path$ and stop (a closed path has been found); otherwise, go to Step 5.

5. If all nodes in $Adj(X_k)$ have been already visited, or $Adj(X_k) = \phi$, go to Step 7; otherwise, go to Step 6.

Step	X_k	Previous	Path	Visited
1	–	–	ϕ	ϕ
2, 3	A	A	$\{A\}$	$\{A\}$
4, 5, 6	B	A	$\{A, B\}$	$\{A, B\}$
4, 5, 6	C	B	$\{A, B, C\}$	$\{A, B, C\}$
4, 5, 7	B	B	$\{A, B\}$	$\{A, B, C\}$
5, 6	D	B	$\{A, B, D\}$	$\{A, B, C, D\}$
4	A	–	$\{A, B, D, A\}$	$\{A, B, C, D\}$

TABLE 4.5. Steps of Algorithm 4.11 for finding loops in the undirected graph in Figure 4.54(a).

6. *Forward step:* Choose some $X_r \in Adj(X_k)$, such that $X_r \notin Visited$. Set $Previous = X_k$, add X_r to both $Path$ and $Visited$, let $X_k = X_r$, and go to Step 4.

7. *Backward step:* Remove X_k from $Path$. If $X_k = X_i$, then go to Step 2; otherwise, let X_k be the last node in $Path$, and go to Step 5. ∎

The above algorithm takes an arbitrary node X_i in the graph and looks for a closed path. If no closed path is found (the algorithm returns to the initial node X_i), then if all nodes have been visited, no closed path is contained in the graph; otherwise, the algorithm takes as the new initial node one of the unvisited nodes and starts over again. Note that this property makes this algorithm suitable not only for directed and undirected graphs, but also for connected and disconnected graphs. This algorithm is illustrated by the following example.

Example 4.23 Finding loops and cycles. Consider the undirected graph in Figure 4.54(a), which contains two loops, $A - B - D - A$ and $D - G - F - D$. Suppose that we apply Algorithm 4.11 starting with node A. Table 4.5 gives the steps of the algorithm. The steps are illustrated in Figure 4.54(a). The table shows, for any step of the algorithm, the current node X_k, the associated $Previous$ node, the current $Path$, the set $Adj(X_k) \cap Path$, which is used to indicate whether a loop exists, and the $Visited$ set containing the nodes visited so far. The steps are summarized as follows: Using alphabetical ordering to break ties, we travel from A to B and from B to C. When reaching node C we cannot continue, so we go back to node B and take the only possible unvisited neighbor, D. The set $Adj(D) \cap Path$ contains node A, which is not the node previous to D. Therefore, the loop $A - B - D - A$ is found. Note that by ignoring A we can continue traveling looking for a different loop. In this way, we can obtain all the loops contained in the graph. For example, if in Step 5 we choose node G or F instead of node A, a different loop is obtained: $D - G - F - D$.

4.8 Some Useful Graph Algorithms   171

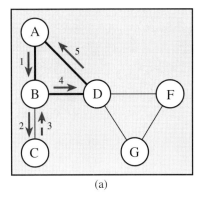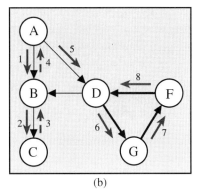

FIGURE 4.54. Steps of depth-first closed-path algorithm for (a) an undirected and (b) a directed graph.

Step	X_k	Previous	Path	Visited
1	–	–	ϕ	ϕ
2, 3	A	A	$\{A\}$	$\{A\}$
4, 5, 6	B	A	$\{A, B\}$	$\{A, B\}$
4, 5, 6	C	B	$\{A, B, C\}$	$\{A, B, C\}$
4, 5, 7	B	B	$\{A, B\}$	$\{A, B, C\}$
5, 7	A	B	$\{A\}$	$\{A, B, C\}$
5, 6	D	A	$\{A, D\}$	$\{A, B, C, D\}$
4, 5, 6	G	D	$\{A, D, G\}$	$\{A, B, C, D, G\}$
4, 5, 6	F	G	$\{A, D, G, F\}$	$\{A, B, C, D, G, F\}$
4	F	G	$\{A, D, G, F, D\}$	$\{A, B, C, D, G, F\}$

TABLE 4.6. Steps of Algorithm 4.11 for finding cycles in the directed graph in Figure 4.54(b).

Consider now the directed graph in Figure 4.54(b), which contains one cycle. Proceeding as in the previous case, and starting with the node A, we obtain the sequence of steps indicated in Table 4.6 and Figure 4.54(b). In this case, we end up with the cycle $D \to G \to F \to D$. Note that the graph in Figure 4.54(a) is the undirected graph associated with this directed graph. Thus, by changing the adjacency sets, we can obtain either the cycles in the directed graph or the loops in the associated undirected graph. ∎

172   4. Some Concepts of Graphs

Exercises

4.1 Given a connected graph $G = (X, L)$ and a link $L_{ij} \in L$, show that the following statements are equivalent:

(a) The graph $(X, L \setminus \{L_{ij}\})$ is connected.

(b) L_{ij} belongs to some loop.

4.2 Given an undirected graph G, prove that the following statements are equivalent:

(a) There exists a unique path between every pair of nodes in G.

(b) G is connected, but by removing any link it becomes disconnected.

(c) G has no loops, but if we add one link anywhere it forms one loop.

Thus, these statements provide three different but equivalent definitions of a tree.

4.3 Show that any graph having a number of links equal or larger than the number of nodes contains at least one loop.

4.4 Consider the graph in Figure 4.55:

(a) Find the ancestral and descendant sets of node C.

(b) What is the boundary of the set $\{B, C\}$ in the underlying undirected graph?

(c) Repeat the previous calculations in the graph resulting from reversing the links $A \to C$ and $C \to E$ in the graph in Figure 4.55.

(d) What can you say about the sets of ancestors and descendants of a node within the cycle $A \to B \to E \to C \to A$?

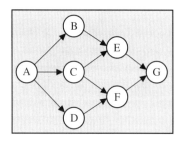

FIGURE 4.55. Example of a directed graph.

4.5 Show that a graph that contains a loop of length 4 or more without a chord cannot have a perfect numbering of the nodes. That is, show that untriangulated graphs have no perfect numberings.

4.6 Complete Example 4.10, verifying that both of the numberings shown in Figures 4.24 are perfect numberings of the given graph.

4.7 Using the MCS fill-in algorithm (Algorithm 4.2), triangulate the graph in Figure 4.55. How many different possibilities of triangulation exist? Which of them is the best? Build a family tree of the original directed graph and a join tree for the resulting triangulated graph.

4.8 Repeat the previous exercise considering the graph in Figure 4.56.

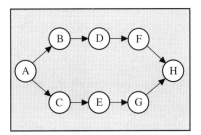

FIGURE 4.56. A directed graph.

4.9 Using the MCS fill-in algorithm (Algorithm 4.2) and choosing node F as the initial node, triangulate the graph in Figure 4.57. Follow the same steps given in Example 4.11.

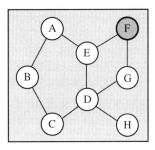

FIGURE 4.57. A nontriangulated undirected graph.

4.10 Prove that DAGs are the only type of directed graphs that have an ancestral numbering.

4.11 Given the graph with the following adjacency matrix:

$$A = \begin{pmatrix} 0 & 0 & 1 & 1 & 0 & 1 \\ 0 & 0 & 1 & 0 & 1 & 1 \\ 0 & 0 & 0 & 1 & 1 & 0 \\ 0 & 0 & 0 & 0 & 0 & 1 \\ 0 & 0 & 0 & 0 & 0 & 1 \\ 0 & 0 & 0 & 0 & 0 & 0 \end{pmatrix}.$$

- Is it an undirected or a directed graph?
- Draw the graph.
- Is it connected?
- How many paths of length 3 are there between the different pairs of nodes?

4.12 Prove Theorem 4.9.

4.13 What can we say about a graph such that its attainability matrix has zero main diagonal?

4.14 Determine the number of paths of length m joining any two nodes of the complete undirected graph with n nodes, K_n (see Definition 4.8).

4.15 Modify the breadth-first path-checking algorithm (Algorithm 4.9) to find a path between two nodes.

4.16 Apply the depth-first pathfinding algorithm (Algorithm 4.8) to find a path between nodes A and G in the graph of Figure 4.55. Proceed by searching the nodes in alphabetical order. What would happen if we eliminate the link $E \to G$ from the graph?

4.17 Use the depth-first closed-path algorithm (Algorithm 4.11) to find loops in the graph of Figure 4.55. Build the table of *Visited* nodes and *Path* for each step of the algorithm. What would happen if we eliminate the link $A \to B$ from the graph?.

4.18 Write and test a program in your preferred computer language for the following algorithms:

- Maximum cardinality search (Algorithm 4.1).
- Maximum cardinality search fill-in (Algorithm 4.2).
- MCS generating a join tree (Algorithm 4.4).
- Multilevel representation (Algorithm 4.6).
- Depth-first pathfinding algorithm (Algorithm 4.8).
- Breadth-first path-checking algorithm (Algorithm 4.9).
- Finding connected components of a graph (Algorithm 4.10).
- Depth-first closed-path search algorithm (Algorithm 4.11).

Chapter 5
Building Probabilistic Models

5.1 Introduction

We have seen in Chapter 3 that the knowledge base of a probabilistic expert system consists of a set of variables and a probabilistic model describing the relationships among them. We have also seen that all the information about the relationships among a set of variables is contained in the joint probability distribution (JPD) of the variables. Thus, the performance of a probabilistic expert system hinges on the correct specification of the JPD. Therefore, an important task for expert systems developers is to specify the JPD as accurately as possible. Human experts often collaborate to achieve this objective.

Building the knowledge base of a probabilistic expert system usually requires the following steps:

1. **Statement of the problem.** As we mentioned in Chapter 1, the first step in developing an expert system is the definition of the problem to be solved or the question to be answered. An example of a problem definition is: Given a patient with a set of symptoms, what disease is the patient likely to have? The definition of the problem is a crucial step because with an ill-defined problem the expert system is doomed to failure.

2. **Variable selection.** The next step is to select a set of variables that, according to the human experts, are relevant to the problem at hand. For example, in the medical diagnosis setup, the variables of

interest may be the set of diseases and the associated set of symptoms, which must be carefully selected. This means that symptoms with adequate discriminant powers must be chosen to differentiate the selected diseases.

3. **Collection of relevant data.** This step involves the collection of the relevant information (data) needed to construct the model that is used to answer the problem. By data we mean quantitative as well as qualitative information obtained from the human experts. The collection of data should be carefully done using experimental design techniques and/or well-designed sample surveys. The help of statisticians should be sought here because statistical methods can increase the quality of the data and ensure the validity of the statistical techniques that will be used to analyze the data and draw conclusions from the data.

4. **Construction of the JPD.** Once a relevant set of variables has been selected, the next step is to construct a JPD that describes the relationships among them. This is perhaps the most critical and difficult step in the development of an expert system:

    (a) It is critical because the accuracy of the obtained results depends on the accuracy of the specified JPD, that is, the quality of the results cannot exceed the quality of the constructed model. Wrong specification of the JPD can lead expert systems to give absurd and/or conflicting results.

    (b) The form of the JPD (that is, the dependence and independence structure among the variables) is seldom known in practice. Thus, one has to estimate or approximate the dependence structure based on a set of data. Again, the quality of the approximation cannot exceed the quality of the data used to estimate the model.

    (c) Even if the dependence structure is known, the structure may depend on an infeasibly large number of unknown parameters (see Section 3.5). Direct specification of these parameters becomes increasingly difficult and/or practically impossible as the number of variables increases. In any case, the parameters have to be either specified by the human experts or estimated from the data.

This and the next two chapters are devoted to the construction of the JPD of a set of variables of interest that is needed for the knowledge base of a probabilistic expert system. Two main approaches for building probabilistic models are

- Graphically specified models.

- Models specified by a set of conditional independence statements (CISs).

These approaches are discussed in detail in Chapters 6 and 7, respectively. In this chapter we present the necessary background material for building probabilistic models using these approaches. Section 5.2 introduces some criteria for *graph separation*. These criteria can be used to determine whether certain CISs hold in a given graph. Recall from Section 3.2.3 that a CIS is written as $I(X,Y|Z)$, where X, Y, and Z are mutually disjoint subsets of the set of variables $\{X_1,\ldots,X_n\}$. This CIS means that X and Y are conditionally independent given Z, or equivalently, Z separates X and Y. Note that all three subsets of variables involved in a CIS are assumed to be mutually disjoint. Section 5.3 introduces several properties of conditional independence. Given a list of CISs, these properties can be used to derive additional CISs from the given list and to build the JPD. Some special types of lists of CISs are given in Section 5.4. Section 5.5 discusses important ways of factorizing any JPD into products of conditional probability distributions (CPDs). Finally, in Section 5.6, we describe the steps necessary for constructing probabilistic models.

5.2 Graph Separation

Graphs are a powerful tool for describing the relationships among a set of variables $\{X_1,\ldots,X_n\}$. Thus, one way to build the JPD of the set of variables is to ask the human experts to provide a graph that they believe describes the relationships among the set of variables. A question that arises naturally here is

- **Question 5.1.** Can (directed and undirected) graphs be represented by an equivalent set of CISs in the sense that CISs derived from the graph coincide with those in the set of CISs? If yes, how to derive the list of CISs implied by the graph?

The answer to this question is yes, a graph can be represented by an equivalent set of CISs. One way to derive the list of CISs implied by a graph is to generate all possible CISs; then each CIS can be checked for validity using some graph separation criteria. These criteria depend on whether we are dealing with directed or undirected graphs.

5.2.1 Separation in Undirected Graphs

In many practical situations the structural relationship among a set of variables $\{X_1,\ldots,X_n\}$ can be represented by an undirected graph G. As discussed in Chapter 4, each variable can be represented by a node in G. If two variables are unconditionally dependent, their relationship can be represented by a path connecting their corresponding nodes in G. On the other hand, if they are unconditionally independent, the corresponding nodes should not be connected by a path.

Similarly, if the dependence between variables X and Y is indirect and mediated by a third variable Z (that is, if X and Y are conditionally dependent given Z), we display Z as a node that does not intersect the path between X and Y, i.e., Z is not a *cutset* separating X and Y. This correspondence between conditional dependence and cutset separation in undirected graphs forms the basis of the theory of *Markov fields* (Isham (1981), Lauritzen (1982), Wermuth and Lauritzen (1983)), and has been given axiomatic characterizations (Pearl and Paz (1987)).

To represent conditional independence using undirected graphs we need to define a separation criterion, which we refer to as the *U-separation criterion*, and its associated algorithm.

Definition 5.1 U-separation. *Let X, Y, and Z be three disjoint subsets of nodes in an undirected graph G. We say that Z separates X and Y iff every path between each node in X and each node in Y contains at least one node in Z. When Z separates X and Y in G, we write $I(X, Y|Z)_G$ to indicate that this CIS is derived from G; otherwise we write $D(X, Y|Z)_G$ to indicate that X and Y are conditionally dependent given Z in the graph G.*

We shall say that X is *graphically independent* of Y given Z if Z separates X and Y. Therefore, given an undirected graph, one can derive all CISs from the graph using the above U-separation criterion. This provides the answer to Question 5.1 for undirected graphs. The answer for directed graphs is given in Section 5.2.2.

Example 5.1 U-separation. Figure 5.1 illustrates the concept of U-separation with four different cases. In all cases the three subsets of interest are marked by rectangles. To distinguish among the three subsets, the box for the first subset is left unshaded, the box for the second subset is lightly shaded, and the box for the third (separating) subset is darkly shaded.

- In Figure 5.1(a), the variables A and I are conditionally independent given E, because every path between A and I contains E. Thus, $I(A, I|E)_G$.

- In Figure 5.1(b), A and I are conditionally dependent given B, because there is a path $(A - C - E - I)$ that does not contain B.

- In Figure 5.1(c), the subsets $\{A,C\}$ and $\{D,H\}$ are conditionally independent given the subset $\{B,E\}$, because every path between the two subsets contains either B or E. Therefore, we write

$$I(\{A,C\},\{D,H\}|\{B,E\})_G.$$

- Finally, in Figure 5.1(d), the subsets $\{A,C\}$ and $\{D,H\}$ are conditionally dependent given the subset $\{E,I\}$, because there is a path $(A - B - D\}$ that does not contain the variables E and I. Thus,

$$D(\{A,C\},\{D,H\}|\{E,I\})_G.$$

Following this process, we can check whether the graph satisfies any given CIS. ∎

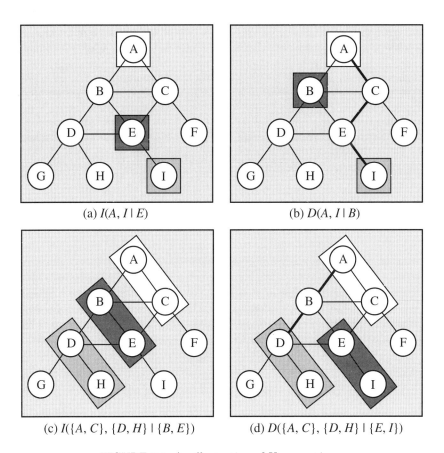

FIGURE 5.1. An illustration of U-separation.

5.2.2 Separation in Directed Graphs

To determine whether a given CIS holds in a directed graph, we use a criterion known as the *D-separation* criterion. To facilitate a definition of *D*-separation, let us consider first an example consisting of the following six variables whose relationships are depicted in Figure 5.2:

- E: Employment status
- V: Investment income
- W: Wealth
- H: Health
- C: Charitable contributions
- P: Happiness

This graph shows that employment status and the investment income are direct causes of wealth. Wealth and health influence happiness. Finally, wealth also has a direct influence on charitable contributions. One would expect, for example, wealth and health to be unconditionally independent, but conditionally dependent given happiness because once we know the outcome of happiness, wealth and health become dependent; an increase in our belief in one variable would decrease our belief in the other. To detect the independencies induced by this graph, we need a new graph separation criterion valid for directed graphs. This is the concept of *D-separation*; see Pearl (1988) and Geiger, Verma, and Pearl (1990a).

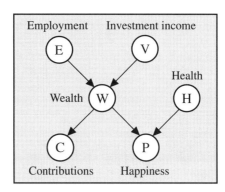

FIGURE 5.2. A directed graph used to illustrate the *D*-separation criterion.

Definition 5.2 Head-to-head node in a path. *Given a directed acyclic graph (DAG) and an undirected path $(\ldots - U - A - V - \ldots)$, then node A is said to be head-to-head in the path if it has converging path arrows (i.e., if the directed graph contains the links $U \to A$ and $V \to A$).*

Example 5.2 Head-to-head node. Node P is the only head-to-head node in the undirected path $E - W - P - H$ in Figure 5.2. Note that although node W has two converging arrows, it is not a head-to-head node in this path because the link $V \to P$ is not in this path. However, node W is a head-to-head node in the undirected path $E - W - V$. ∎

Definition 5.3 D-Separation. *Let X, Y, and Z be three disjoint subsets of nodes in a DAG D; then Z is said to D-separate X and Y, iff along every undirected path from each node in X to each node in Y there is an intermediate node A such that either*

1. *A is a head-to-head node in the path, and neither A nor its descendants are in Z, or*

2. *A is not a head-to-head node in the path and A is in Z.*

When Z D-separates X and Y in D, we write $I(X,Y|Z)_D$ to indicate that this CIS is derived from D; otherwise we write $D(X,Y|Z)_D$ to indicate that X and Y are conditionally dependent given Z in the graph D.

Therefore, if we can find any node along every undirected path satisfying one of the above two conditions, then $I(X,Y|Z)_D$; otherwise, $D(X,Y|Z)_D$. These conditions convey the idea that the inputs (parents) of any causal mechanism become dependent once the output (a child) is known. For instance, in the directed graph given in Figure 5.2, employment status and investment income are unconditionally independent, that is, $I(E,V|\phi)_D$. However, if we have some evidence about wealth, E and V become dependent, $D(E,V|W)_D$, because an increase in our belief in either cause would decrease our belief in the other.

Example 5.3 *D*-separation. Consider the directed graph in Figure 5.2. The following statements, for example, can be concluded from the graph:

- Case (a). Unconditional independence, $I(E,V|\phi)_D$: Nodes E and V are unconditionally independent because they are D-separated by ϕ. As can be seen from Figure 5.3(a), the only undirected path $E-W-V$ connecting E and V contains the head-to-head node W, and neither this node nor its descendants are in ϕ.

- Case (b). Conditional dependence, $D(E,H|P)_D$: Nodes E and H are conditionally dependent given P. It can be seen from Figure 5.3(b) that the only undirected path $E - W - P - H$ connecting E and H contains the two nodes W and P, and that neither of these nodes satisfies the conditions for D-separation. Thus, P does not D-separate E and H.

- Case (c). Conditional independence, $I(C,P|\{E,W\})_D$: Nodes C and P are conditionally independent given $\{E,W\}$ because the only undirected path $C - W - P$ connecting C and P contains a single intermediate node, W, which is not head-to-head, but it is in $\{E, W\}$, see Figure 5.3(c).

- Case (d). Conditional dependence, $D(C, \{H, P\}|E)_D$: Nodes C and $\{H, P\}$ are conditionally dependent given E, see Figure 5.3(d). Note that the undirected path $C - W - P$ between C and P contains node W, which is not a head-to-head node in this path, but it is not in $\{E\}$. ∎

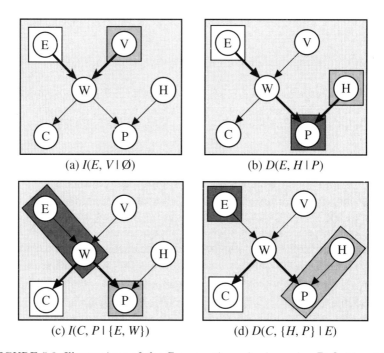

FIGURE 5.3. Illustrations of the D-separation criterion using Definition 5.3.

Thus, one can use D-separation as a criterion for representing independence models by directed graphs, and the answer to Question 5.1 for both directed and undirected graphs is now complete. The following, is an alternative definition of D-separation that is easier to apply than Definition 5.3.

Definition 5.4 D-Separation. *Let $X, Y,$ and Z be three disjoint subsets of nodes in a DAG D, then Z is said to D-separate X and Y iff Z separates X and Y in the moral graph of the smallest ancestral set[1] containing X, Y, and Z.*

---

[1] Recall from Definition 4.20, that an ancestral set is a set that contains the ancestors of all its nodes.

Lauritzen et al. (1990), who showed that Definitions 5.3 and 5.4 are equivalent, refer to Definition 5.4 as A-separation.

Note that the idea of moralization used in this definition coincides with the first of the two conditions given in Definition 5.3. If there is a head-to-head node A in a path joining X and Y, and A or some of its descendants are in Z, then A will be contained in the smallest ancestral set containing X, Y, and Z. Then, since A is a head-to-head node, even in the case that A is in Z there will be an undirected path in the moralized graph joining X and Y whose nodes will not be contained in Z. This suggests the following algorithm for D-separation:

Algorithm 5.1 D-Separation.

- **Input:** A DAG, D, and three disjoint subsets of nodes X, Y, and Z.

- **Output:** *True* if the CIS $I(X, Y|Z)$ holds in D; *False* otherwise.

1. Identify the smallest subgraph containing X, Y, and Z and their ancestral subsets.

2. Moralize the obtained subgraph.

3. Use the U-separation criterion to determine whether Z separates X and Y. ∎

Example 5.4 D-separation. Consider again the directed graph in Figure 5.2 and let us use Algorithm 5.1 to derive the four statements made in Example 5.3. These cases are illustrated in Figure 5.4, where a dashed link means that the link is removed because the corresponding nodes are not in the ancestral sets.

- Case (a). Unconditional independence, $I(E, V|\phi)_D$: In Figure 5.4(a), there is no path connecting E and V in the moral graph of the smallest subgraph containing E, V, and ϕ and their ancestral sets. Hence, $I(E, V|\phi)_D$.

- Case (b). Conditional dependence, $D(E, H|P)_D$: In Figure 5.4(b), there is a path $E - W - H$ connecting E and H in the moral graph of the smallest ancestral subgraph containing E, H, and P that does not contain any node in $\{P\}$. Hence, $D(E, H|P)_D$.

- Case (c). Conditional independence, $I(C, P|\{E, W\})_D$: There are two paths between C and P, $C - W - P$ and $C - W - H - P$, in the moral graph of the smallest subgraph containing C, E, W, and P and their ancestral sets (see Figure 5.4(c)). Both paths contain the node W, which is in the set $\{E, W\}$. Hence, $I(C, P|\{E, W\})_D$.

- Case (d). Conditional dependence, $D(C, \{H, P\}|E)_D$: In Figure 5.4(d), the path $C - W - P$ connects C and $\{H, P\}$ in the moral graph of the smallest subgraph containing C, H, P, E, and their ancestral sets. However, the path does not contain node E. Hence, $D(C, \{H, P\}|E)_D$. ∎

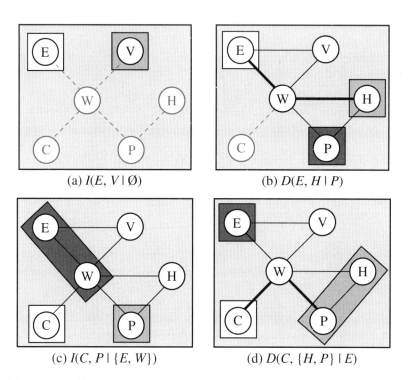

FIGURE 5.4. Illustrations of the D-separation criterion using Definition 5.4.

5.3 Some Properties of Conditional Independence

So far we have introduced three different models to define conditional independence relationships: probabilistic models, undirected graphical models, and directed graphical models. In this section, we analyze some properties of conditional independence that hold in some of these models. These properties allow us to derive additional CISs from an initial list of CISs obtained from the given model. For example, given a JPD, $p(x_1, \ldots, x_n)$, of a set of variables $\{X_1, \ldots, X_n\}$, one can obtain the complete set of CISs implied by the JPD by enumerating all possible CISs and checking to see whether each statement actually holds in the JPD. In practice, however, the JPD is often unknown, and as a result, we may only have a list of CISs provided by the human experts, that partially describes the relationships among the set of variables. This list is referred to as an *input list*.

Definition 5.5 Input list. *An input list M is a set of CISs of the form $I(X, Y|Z)$, where X, Y, and Z are three disjoint subsets of the variables $\{X_1, \ldots, X_n\}$, which means that X and Y are conditionally independent given Z.*

We first need to know whether the CISs in the initial input list imply other CISs not included in the list but are true in the given model; that is, we need to address the following question:

- **Question 5.2:** Given an input list M of CISs, how does one derive additional CISs from M that hold in the given model, if any, using certain properties of conditional independence?

We address this question in this section and give an algorithm for generating additional CISs from a given input list. For a model to satisfy the CISs in the input list and to be compatible with probability axioms, it must satisfy other CISs. The new CISs are called *derived* CISs. The derived CISs, if any, must then be confirmed by the human experts in order for the model to be consistent with reality. The set of (initial plus derived) CISs defines a model that describes the relationships among the variables. These models, referred to as *models specified by input lists*, are presented in Chapter 7.

To address the above and other questions, we need to introduce some properties of conditional independence. These properties are illustrated using undirected graphs in Figures 5.5 and 5.6, where each of the three subsets involved in an independence relation (e.g., $I(X, Y|Z)$) is contained in a rectangular box. To distinguish among the three subsets, the box for the first subset is left unshaded, the box for the second subset is lightly shaded, and the box for the third (separating) subset is darkly shaded.

The following are some properties of conditional CISs, the first four of which are shown, in the appendix to this chapter, to be satisfied by any JPD. For further discussion, see, for example, Lauritzen (1974) and Dawid (1979, 1980). In Chapter 6 (Theorems 6.1 and 6.8), we give the properties that are satisfied by undirected and directed graphical models, respectively.

1. **Symmetry:** If X is conditionally independent of Y given Z, then Y is conditionally independent of X given Z, that is,
$$I(X, Y|Z) \Leftrightarrow I(Y, X|Z). \tag{5.1}$$

This property is illustrated in Figure 5.5(a).

2. **Decomposition:** If X is conditionally independent of $Y \cup W$ given Z, then X is conditionally independent of Y given Z, and X is conditionally independent of W given Z, that is,
$$I(X, Y \cup W|Z) \Rightarrow I(X, Y|Z) \text{ and } I(X, W|Z), \tag{5.2}$$

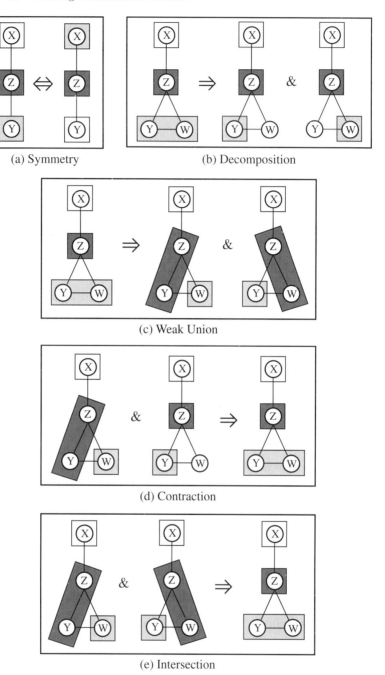

FIGURE 5.5. Graphical illustrations of some properties of CISs: (a) Symmetry, (b) Decomposition, (c) Weak union, (d) Contraction, and (e) Intersection. (The separating subset is indicated by a darkly shaded rectangle, and the other two subsets are indicated by an unshaded and a lightly shaded rectangle.)

5.3 Some Properties of Conditional Independence

where $Y \cup W$ denotes Y union W. Note that Y and W need not be disjoint sets here.

This property is illustrated in Figure 5.5(b). Note that the reverse of (5.2) is known as the *composition property*, which is not satisfied by all JPDs as the following example illustrates.

Example 5.5 Composition property violation by JPDs. Consider the set of binary random variables $\{X, Y, Z, W\}$. Two proposed JPDs of the variables are given in Table 5.1. They have been obtained by fixing numerical values for some of the parameters (those indicated with two decimal digits in Table 5.1) and calculating the remaining values in order for the JPD p_1 to violate the composition property and for p_2 to satisfy the composition property.

It can be shown that $p_1(x, y, z, w)$ satisfies the CISs $I(X, Y|Z)$ and $I(X, W|Z)$ but not the CIS $I(X, Y \cup W|Z)$; thus, the composition property does not hold. It is interesting to note that there exist no values (x, y, z, w) that satisfy the condition

$$p(x|y, w, z) = p(x|z).$$

On the contrary, the JPD $p_2(x, y, z, w)$ satisfies the CISs $I(X, Y|Z)$, $I(X, W|Z)$, and $I(X, Y \cup W|Z)$. Thus, $p_2(x, y, z, w)$ satisfies the composition property but $p_1(x, y, z, w)$ does not. ∎

3. **Weak Union:**

$$I(X, Y \cup W|Z) \Rightarrow I(X, W|Z \cup Y) \text{ and } I(X, Y|Z \cup W). \quad (5.3)$$

This property is illustrated in Figure 5.5(c). It states that learning irrelevant information Y cannot help the irrelevant information W to become relevant to X.

4. **Contraction:** If W is irrelevant to X after the learning of some irrelevant information Y, then W must have been irrelevant before we knew Y, that is,

$$I(X, W|Z \cup Y) \text{ and } I(X, Y|Z) \Rightarrow I(X, Y \cup W|Z). \quad (5.4)$$

This property is illustrated in Figure 5.5(d).

The *weak union* and *contraction* properties together mean that irrelevant information should not alter the relevance of other relevant information in the system. In other words, what was relevant remains relevant, and what was irrelevant remains irrelevant.

The above four properties hold for any JPD, but the next property is shown in the appendix to this chapter to hold for nonextreme JPDs.

188   5. Building Probabilistic Models

x	y	z	w	$p_1(x,y,z,w)$	$p_2(x,y,z,w)$
0	0	0	0	0.012105300	0.0037500
0	0	0	1	0.005263160	0.0050000
0	0	1	0	0.000971795	0.1312200
0	0	1	1	0.024838000	0.1574640
0	1	0	0	0.01	0.0087500
0	1	0	1	0.02	0.01
0	1	1	0	0.03	0.2361960
0	1	1	1	0.04	0.02
1	0	0	0	0.05	0.03
1	0	0	1	0.06	0.04
1	0	1	0	0.07	0.05
1	0	1	1	0.08	0.06
1	1	0	0	0.09	0.07
1	1	0	1	0.10	0.08
1	1	1	0	0.11	0.09
1	1	1	1	0.296822000	0.0076208

TABLE 5.1. Examples of two JPDs where $p_2(x,y,z,w)$ satisfies the composition property but $p_1(x,y,z,w)$ does not.

5. **Intersection:**

$$I(X,W|Z \cup Y) \text{ and } I(X,Y|Z \cup W) \Rightarrow I(X, Y \cup W|Z).$$

This property is illustrated in Figure 5.5(e). It states that unless Y affects X when W is known or W affects X when Y is known, neither W nor Y nor their combination can affect X.

The following four properties may be satisfied by some, but not all, JPDs.

6. **Strong Union:** If X is conditionally independent of Y given Z, then X is also conditionally independent of Y given $Z \cup W$, that is,

$$I(X,Y|Z) \Rightarrow I(X,Y|Z \cup W). \qquad (5.5)$$

This property is illustrated using the undirected graph in Figure 5.6(a). The following example shows that this property does not hold in directed graphical models.

Example 5.6 Strong union property violation by DAGs. Consider the DAG given in Figure 5.7(a). Using the D-separation criterion, we can conclude that the CIS $I(X,Y|Z)$ holds (because in the

5.3 Some Properties of Conditional Independence 189

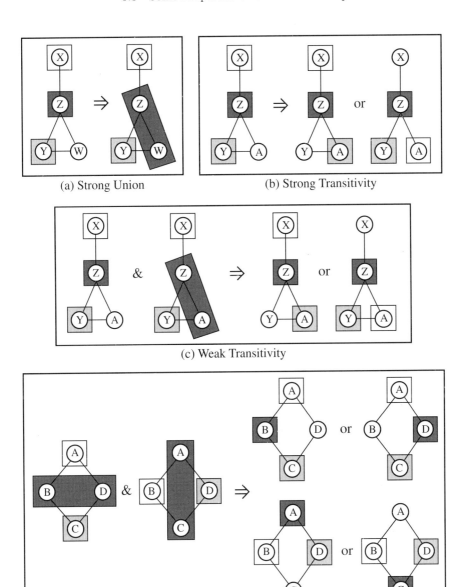

FIGURE 5.6. Graphical illustration of some properties of CISs: (a) Strong union, (b) Strong transitivity, (c) Weak transitivity, and (d) Chordality. (The separating subset is indicated by a darkly shaded rectangle and the other two subsets are indicated by an unshaded and a lightly shaded rectangle.)

190   5. Building Probabilistic Models

moral graph of the smallest subgraph containing X, Y, and Z, there is only one path between X and Y and it contains Z). However, if we increase the conditioning set by adding node W then the nodes become dependent (see Figure 5.7(b)). This is because in the moral graph of the smallest subgraph containing X, Y, W, and Z, there is a path between X and Y and it does not contain Z. Thus, we have $D(X,Y|\{Z,W\})$ and the strong union property does not hold in directed graphical models. ■

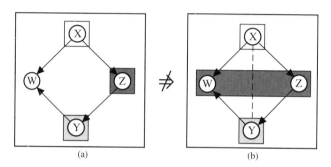

FIGURE 5.7. A graphical illustration showing that the strong union property does not hold in directed graphs.

7. **Strong Transitivity:** If X is conditionally dependent on A given Z and A is conditionally dependent on Y given Z, then X is conditionally dependent on Y given Z, that is,

$$D(X,A|Z) \text{ and } D(A,Y|Z) \Rightarrow D(X,Y|Z),$$

or equivalently,

$$I(X,Y|Z) \Rightarrow I(X,A|Z) \text{ or } I(A,Y|Z), \tag{5.6}$$

where A is a single variable.

The strong transitivity property states that two variables are dependent if there exists another variable A that depends on both. This property is illustrated in Figure 5.6(b).

8. **Weak Transitivity:** If X and A are conditionally dependent given Z, and Y and A are conditionally dependent given Z, then either X and Y are conditionally dependent given Z, or X and Y are conditionally dependent given $Z \cup A$, that is,

$$D(X,A|Z) \text{ and } D(A,Y|Z) \Rightarrow D(X,Y|Z) \text{ or } D(X,Y|Z \cup A),$$

5.3 Some Properties of Conditional Independence

or equivalently,

$$I(X,Y|Z) \text{ and } I(X,Y|Z \cup A) \Rightarrow I(X,A|Z) \text{ or } I(A,Y|Z), \quad (5.7)$$

where A is a single variable. This is illustrated in Figure 5.6(c).

9. **Chordality:** If A and C are conditionally dependent given B, and A and C are conditionally dependent given D, then A and C are conditionally dependent given $B \cup D$, or B and D are conditionally dependent given $A \cup C$, that is,

$$D(A,C|B) \text{ and } D(A,C|D) \Rightarrow D(A,C|B \cup D) \text{ or } D(B,D|A \cup C),$$

or equivalently,

$$I(A,C|B \cup D) \text{ and } I(B,D|A \cup C) \Rightarrow I(A,C|B) \text{ or } I(A,C|D), \quad (5.8)$$

where $A, B, C,$ and D are single variables. This property is illustrated in Figure 5.6(d).

We conclude this section with the following remarks:

1. Strong union (SU) implies weak union:

$$I(X, Y \cup W|Z) \overset{SU}{\Rightarrow} I(X, Y \cup W|Z \cup W) \Rightarrow I(X, Y|Z \cup W).$$

2. Strong transitivity implies weak transitivity.

3. Strong union and intersection (IN) imply contraction:

$$I(X,Y|Z) \overset{SU}{\Rightarrow} \left. \begin{array}{l} I(X,Y|Z \cup W) \\ I(X,W|Z \cup Y) \end{array} \right\} \overset{IN}{\Rightarrow} I(X, Y \cup W|Z).$$

4. Strong union and intersection also imply composition:

$$\left. \begin{array}{l} I(X,Y|Z) \overset{SU}{\Rightarrow} I(X,Y|Z \cup W) \\ I(X,W|Z) \overset{SU}{\Rightarrow} I(X,W|Z \cup Y) \end{array} \right\} \overset{IN}{\Rightarrow} I(X, Y \cup W|Z).$$

The above properties are used in the next section to derive additional CISs from a given input list of CISs and to define some special types of input lists.

5.4 Special Types of Input Lists

Now that we have presented some properties of conditional independence, we are ready to answer Question 5.2, which is stated above but repeated here for convenience:

- **Question 5.2:** Given an input list M of CISs, how does one derive additional CISs from M that hold in the given model, if any, using certain properties of conditional independence?

Note that so far, we have not required the input lists provided by the human experts to satisfy any conditions other than for the subsets within each CIS to be disjoint. When we impose further conditions, such as that the list has to satisfy a given set of properties, we obtain special types of input lists, some of which are defined below.

Definition 5.6 Graphoid. *A graphoid is a set of CISs that is closed under the properties of symmetry, decomposition, weak union, contraction, and intersection.*

Definition 5.7 Semigraphoid. *A semigraphoid is a set of CISs that is closed under the properties of symmetry, decomposition, weak union, and contraction.*

Thus, a graphoid must satisfy the first five properties, whereas a semigraphoid must satisfy only the first four properties (see Pearl and Paz (1987) and Geiger (1990)).

Definition 5.8 Probabilistic input list. *An input list M is said to be a probabilistic input list if it contains all and only those CISs derived from a JPD $p(x_1, \ldots, x_n)$.*

Definition 5.9 Nonextreme probabilistic input list. *A nonextreme probabilistic input list is a probabilistic input list obtained from a nonextreme, or positive, JPD, that is, $p(x_1, \ldots, x_n)$ is restricted to the open interval $(0, 1)$.*

Since all JPDs satisfy the first four properties of conditional independence, then all probabilistic input lists are semigraphoids. But because the intersection property is satisfied only by CISs derived from a nonextreme JPD, then all nonextreme JPDs are graphoids.

Definition 5.10 Input list compatible with a JPD. *An input list M is said to be compatible with a JPD $p(x_1, \ldots, x_n)$ if all CISs in M are satisfied by $p(x_1, \ldots, x_n)$.*

Note that a compatible input list is a list that can be derived from a JPD $p(x_1, \ldots, x_n)$. Unlike probabilistic input lists, it does not have to be complete, that is, it does not have to contain all CISs that can be derived from $p(x_1, \ldots, x_n)$. Thus, not all compatible input lists are probabilistic input lists.

If an input list M is compatible with a JPD $p(x_1, \ldots, x_n)$, then the smallest semigraphoid generated by M must be compatible with $p(x_1, \ldots, x_n)$. This is because all CISs derived from a JPD must satisfy the first four properties of conditional independence. Thus, one interesting practical problem consists of generating the smallest graphoid or semigraphoid that contains a given input list M. The following algorithm can be used to achieve this objective:

Algorithm 5.2 Generating the minimal graphoid.

- **Input:** An initial and finite input list M.

- **Output:** The minimal graphoid containing M.

1. Generate new CISs by applying the symmetry, decomposition, weak union, contraction, and intersection properties to the set of all initial CISs in M. Augment M by the newly generated CISs.

2. Repeat Step 1 until no new CISs are generated. The resulting set is the desired graphoid. ∎

A semigraphoid can also be generated using the above algorithm if in Step 1 we do not use the intersection property. The following example illustrates Algorithm 5.2.

Example 5.7 Generating graphoids. Suppose that we have a set of four variables $\{X_1, X_2, X_3, X_4\}$ and that a human expert gives the following input list:

$$M = \{I(X_1, X_2|X_3),\ I(X_1, X_4|X_2),\ I(X_1, X_4|\{X_2, X_3\})\}. \qquad (5.9)$$

Table 5.2 shows the initial and derived CISs needed for the input list to be a semigraphoid and a graphoid, respectively. The new CISs are generated using a computer program, called *X-pert Maps*,[2] which implements Algorithm 5.2. Table 5.2 also shows the CISs that are used to derive the new CISs. ∎

Thus, for any input list M compatible with a JPD $p(x_1, \ldots, x_n)$, the first five properties of conditional independence can be used to augment M with new CISs derived from M. Both the initial and derived CISs are still compatible with $p(x_1, \ldots, x_n)$. This raises the following question:

---

[2] The program *X-Pert Maps* can be obtained from the World Wide Web site http://ccaix3.unican.es/~AIGroup.

Initial Input List
$M = \{I(X_1, X_2\|X_3),\ I(X_1, X_4\|X_2),\ I(X_1, X_4\|X_2X_3)\}$

CISs Defining a Semigraphoid		
Property	Derived CIS	Derived From
Symmetry	$I(X_2, X_1\|X_3)$	$I(X_1, X_2\|X_3)$
Symmetry	$I(X_4, X_1\|X_2)$	$I(X_1, X_4\|X_2)$
Symmetry	$I(X_4, X_1\|X_2X_3)$	$I(X_1, X_4\|X_2X_3)$
Contraction	$I(X_1, X_2X_4\|X_3)$	$I(X_1, X_2\|X_3)$ and $I(X_1, X_4\|X_2X_3)$
Symmetry	$I(X_2X_4, X_1\|X_3)$	$I(X_1, X_2X_4\|X_3)$
Weak Union	$I(X_1, X_2\|X_3X_4)$	$I(X_1, X_2X_4\|X_3)$
Symmetry	$I(X_2, X_1\|X_3X_4)$	$I(X_1, X_2\|X_3X_4)$
Decomposition	$I(X_1, X_4\|X_3)$	$I(X_1, X_2X_4\|X_3)$
Symmetry	$I(X_4, X_1\|X_3)$	$I(X_1, X_4\|X_3)$

Additional CISs Defining a Graphoid		
Property	Derived CIS	Derived From
Intersection	$I(X_1, X_2X_4\|\phi)$	$I(X_1, X_2\|X_3X_4)$ and $I(X_1, X_4\|X_2)$
Symmetry	$I(X_2X_4, X_1\|\phi)$	$I(X_1, X_2X_4\|\phi)$
Decomposition	$I(X_1, X_2\|\phi)$	$I(X_1, X_2X_4\|\phi)$
Symmetry	$I(X_2, X_1\|\phi)$	$I(X_1, X_2\|\phi)$
Weak Union	$I(X_1, X_2\|X_4)$	$I(X_1, X_2X_4\|\phi)$
Symmetry	$I(X_2, X_1\|X_4)$	$I(X_1, X_2\|X_4)$
Decomposition	$I(X_1, X_4\|\phi)$	$I(X_1, X_2X_4\|\phi)$
Symmetry	$I(X_4, X_1\|\phi)$	$I(X_1, X_4\|\phi)$

TABLE 5.2. The smallest semigraphoid and graphoid generated by the input list M in (5.9) using Algorithm 5.2.

- **Question 5.3.** Does the application of the above properties lead to a probabilistic input list? In other words, is the set of CISs complete, in the sense that there exists a JPD $p(x_1, \ldots, x_n)$ such that

$$p(x|y, z) = p(x|z) \Leftrightarrow I(X, Y|Z)$$

for every CIS $I(X, Y|Z)$ in M?

Pearl and Paz (1987) (see Pearl, (1988) p. 88) first conjectured that the set of the first four properties (symmetry, decomposition, weak union, and contraction) is complete. However, this conjecture was refuted by Studený (1989) by first finding a property that cannot be derived from the above four properties and then showing that there is no finite complete set of properties characterizing probabilistic models (Studený (1992)).

We conclude this section with the following definition.

Definition 5.11 Dependency model. *Any model M of a set of variables $\{X_1,\ldots,X_n\}$ from which we can determine whether $I(X,Y|Z)$ is true, for all possible triplets of disjoint subsets X, Y, and Z, is called a dependency model.*

Thus, from a dependency model we can conclude which CISs $I(X,Y|Z)$ are true and which are false. Accordingly, a graph (using the corresponding separation criterion), a JPD (with the definition of conditional independency), and various types of input lists introduced in this section (semigraphoids, graphoids, probabilistic input lists, etc) are dependency models. As we shall see in the next chapters, the qualitative structure of a probabilistic model can be represented by a dependency model that provides a way to factorize the corresponding JPD. In the next section, we introduce some concepts about factorizations of a JPD.

5.5 Factorizations of the JPD

Any JPD of a set of random variables can be defined in terms of a set of smaller CPDs. In this section we discuss some ways of factorizing JPDs.

Definition 5.12 Factorizing the JPD by potentials. *Let $C_1,\ldots,C_m$ be subsets of a set of variables $X = \{X_1,\ldots,X_n\}$. If the JPD of $X_1,\ldots,X_n$ can be written as a product of m nonnegative functions Ψ_i $(i = i,\ldots,m)$, that is,*

$$p(x_1,\ldots,x_n) = \prod_{i=1}^{m} \Psi_i(c_i), \tag{5.10}$$

where c_i is a realization of C_i, we say that (5.10) is a factorization of the JPD. The functions Ψ_i are called factor potentials of the JPD.

Examples of such factorizations are given in Chapter 6. Note that the subsets $C_1,\ldots,C_m$ are not necessarily disjoint and that Ψ_i is not necessarily a probability function. When we restrict the Ψ_i to be probability functions, we obtain special types of factorizations, some of which are discussed below.

Let $\{Y_1,\ldots,Y_m\}$ be a partition (a class of mutually exclusive and collectively exhaustive subsets) of $\{X_1,\ldots,X_n\}$. An important class of factorizations of the JPD, known as the *chain rule factorizations*, is defined below.

Definition 5.13 Chain rule factorizations of the JPD. *Any JPD of a set of ordered variables $\{X_1,\ldots,X_n\}$ can be expressed as a product of m CPDs of the form*

$$p(x_1,\ldots,x_n) = \prod_{i=1}^{m} p(y_i|b_i), \tag{5.11}$$

or equivalently,

$$p(x_1,\ldots,x_n) = \prod_{i=1}^{m} p(y_i|a_i), \tag{5.12}$$

where $B_i = \{Y_1,\ldots,Y_{i-1}\}$ is the set of variables before Y_i and $A_i = \{Y_{i+1}, \ldots, Y_n\}$ is the set of variables after Y_i. Note that a_i and b_i are realizations of A_i and B_i, respectively.

When Y_i consists of only a single variable for all i, then $m = n$ and the set $\{Y_1,\ldots,Y_n\}$ is simply a permutation of the set $\{X_1,\ldots,X_n\}$. In this case, (5.11) and (5.12) become

$$p(x_1,\ldots,x_n) = \prod_{i=1}^{n} p(y_i|b_i) \tag{5.13}$$

and

$$p(x_1,\ldots,x_n) = \prod_{i=1}^{n} p(y_i|a_i), \tag{5.14}$$

respectively. Equations (5.11) and (5.12) are called chain rule factorizations of the JPD $p(x_1,\ldots,x_n)$, whereas Equations (5.13) and (5.14) are called canonical chain rule factorizations of the JPD, $p(x_1,\ldots,x_n)$.

Example 5.8 Chain rule. Consider a case of four variables $\{X_1,\ldots,X_4\}$ partitioned as $Y_1 = \{X_1\}$, $Y_2 = \{X_2\}$, $Y_3 = \{X_3\}$, $Y_4 = \{X_4\}$. Then (5.13) and (5.14) give the following equivalent chain rule factorizations of the JPD:

$$p(x_1,\ldots,x_4) = p(x_1)p(x_2|x_1)p(x_3|x_1,x_2)p(x_4|x_1,x_2,x_3) \tag{5.15}$$

and

$$p(x_1,\ldots,x_4) = p(x_1|x_2,x_3,x_4)p(x_2|x_3,x_4)p(x_3|x_4)p(x_4). \tag{5.16}$$

This implies that the JPD can be expressed as a product of four CPDs. Note that chain rule factorizations are not unique because one can apply the chain rule to different partitions of $\{X_1,\ldots,X_4\}$ and obtain different chain rule factorizations. For example, the following are two different but equivalent chain rule factorizations associated with the same JPD, obtained from different partitions of $\{X_1,\ldots,X_4\}$. For example:

- The partition $Y_1 = \{X_1\}$, $Y_2 = \{X_2, X_3\}$, and $Y_3 = \{X_4\}$ gives

$$p(x_1,\ldots,x_4) = p(x_1)p(x_2,x_3|x_1)p(x_4|x_1,x_2,x_3).$$

- The partition $Y_1 = \{X_1, X_4\}$ and $Y_2 = \{X_2, X_3\}$ gives

$$p(x_1,\ldots,x_4) = p(x_1,x_4)p(x_2,x_3|x_1,x_4).$$

∎

We have seen in Section 3.5 that the number of parameters in a given probabilistic model can be reduced by imposing certain constraints (assumptions). For example, the various models presented in Section 3.5 are obtained by imposing certain general conditional independence assumptions. To see how the specification of a CIS gives rise to a parameter reduction, we write the JPD as a product of a set of CPDs in a chain rule factorization. As the following example illustrates, a factorization of the JPD in terms of CPDs is a convenient way to compute the constraints on the parameters imposed by a CIS.

Example 5.9 Imposing constraints. Consider the four variables in Example 5.8 and suppose that the human experts provide two CISs:

$$I(X_3, X_1|X_2) \text{ and } I(X_4, \{X_1, X_3\}|X_2). \tag{5.17}$$

We wish to compute the constraints among the parameters of the JPD imposed by these CISs. The first of these statements implies

$$p(x_3|x_1, x_2) = p(x_3|x_2), \tag{5.18}$$

and the second statement implies

$$p(x_4|x_1, x_2, x_3) = p(x_4|x_2). \tag{5.19}$$

Note that the general form of the JPD is not a suitable representation for calculating the constraints given by (5.18) and (5.19). However, by substituting these two equalities in (5.15), we obtain

$$p(x_1, \ldots, x_4) = p(x_1)p(x_2|x_1)p(x_3|x_2)p(x_4|x_2). \tag{5.20}$$

Assuming that the variables are binary, then the JPD in (5.15) depends on $2^4 - 1 = 15$ free (unconstrained) parameters.[3] On the other hand, the JPD in (5.20) depends on seven parameters ($p(x_1)$ depends on one parameter and each of the other three CPDs depends on two parameters). Therefore, the two CISs in (5.17) give rise to a reduction in the number of parameters from 15 to 7. ■

Definition 5.14 Canonical CPD. *Let $U_i \subset X = \{X_1, \ldots, X_n\}$. A CPD $p(x_i|u_i)$ is said to be a canonical CPD if X_i is a single variable not included in U_i.*

The following theorem, given by Gelman and Speed (1993), guarantees the existence of a set of canonical CPDs from any set of CPDs not in canonical form.

---

[3] Actually there are sixteen parameters but they must add up to one, hence there are only fifteen unconstrained parameters.

Theorem 5.1 Existence of canonical forms. Let $X = \{X_1, \ldots, X_n\}$ and suppose that a set $P = \{p(u_1|v_1), \ldots, p(u_m|v_m)\}$ of marginal and/or conditional probability distributions is given, where U_i and V_i are disjoint subsets of X, such that $U_i \neq \phi$ and V_i can be empty (for the case of marginals). Then from P one can obtain equivalent representations such that each new set U_i contains a single variable in X.

Proof: From $p(u_i|v_i)$, using the chain rule, one can obtain as many new conditionals as the number of variables in U_i, that is, the set

$$\{p(x_j|c_{ij}, v_i) \; \forall X_j \in U_i\}, \tag{5.21}$$

where $C_{ij} = \{X_r \,|\, X_r \subset U_i, \, r < j\}$. ∎

The following algorithm converts a given set P of CPDs to an equivalent canonical form.

Algorithm 5.3 Converts a Set of CPDs to Canonical Form.

- **Input:** A set $P = \{p(u_i|v_i), \, i = 1, \ldots, m\}$ of m CPDs, where U_i and V_i are disjoint subsets of X.

- **Output:** An equivalent set P^* in canonical form.

1. Initializations: Set $P^* = \phi$ and $i = 1$.

2. Set $j = 1$, $S_i = U_i \cup V_i$, and L = number of variables in U_i.

3. Remove one of the variables in U_i from S_i, say X_ℓ, then add $p(x_\ell|s_i)$ to P^*.

4. If $j < L$, increase j by 1 and go to Step 3; otherwise, go to Step 5.

5. If $i < m$, increase i by 1 and go to Step 2; otherwise return P^*. ∎

Example 5.10 Suppose that $X = \{A, B, C, D\}$ and that the set of conditional probabilities $P = \{p(a, b|c), p(a, c, d|b)\}$ is given. In the notation of Algorithm 5.3, the sets U_i and V_i are

$$U_1 = \{A, B\}, \quad V_1 = \{C\},$$
$$U_2 = \{A, C, D\}, \quad V_2 = \{B\}.$$

To convert the two CPDs in P to their corresponding canonical representations, we use Algorithm 5.3 and obtain

$$\begin{aligned} p(a, b|c) &= p(a|b, c)p(b|c), \\ p(a, c, d|b) &= p(a|c, d, b)p(c|d, b)p(d|b). \end{aligned} \tag{5.22}$$

Thus, we obtain the canonical representation

$$P^* = \{p(a|b, c)p(b|c); p(a|c, d, b)p(c|d, b)p(d|b)\}. \tag{5.23}$$

```
Canonical[P_List]:= Module[{U,V,S,l,PCan},
    PCan={};
    Do[U=P[[i,1]]; (* First element of the i-th pair *)
        V=P[[i,2]];
        S=Join[U,V];
        l=Length[U];
        Do[S=Drop[S,1]; (* Removes the last element *)
            AppendTo[PCan,{{U[[j]]},S}]
          ,{j,1,l}]
      ,{i,1,Length[P]}];
    Return[PCan]
]
```

FIGURE 5.8. A *Mathematica* program for converting a given set P of CPDs to canonical form.

Figure 5.8 shows a *Mathematica* program for converting a given set P of CPDs to a canonical form. Given as input a list of pairs $\{U, V\}$, the program returns a list in canonical form. For example, given the CPDs in (5.22) the following *Mathematica* command obtains the corresponding canonical forms shown in (5.23):

In:=Canonical[List[{{A,B},{C}},{{A,C,D},{B}}]]

Out:=List[{{A},{B,C}},{{B},{C}},{{A},{C,D,B}},
          {{C},{D,B}},{{D},{B}}]   ∎

Definition 5.15 Standard canonical CPD. *Let $\{Y_1, \ldots, Y_n\}$ be a permutation of a set of variables $X = \{X_1, \ldots, X_n\}$. A CPD $p(y_i|s_i)$ is said to be a standard canonical CPD if Y_i is a single variable and S_i contains either all variables with indices below i or all variables with indices above i, that is, either $S_i = \{Y_1, \ldots, Y_{i-1}\}$ or $S_i = \{Y_{i+1}, \ldots, Y_n\}$.*

For example, if $Y = \{Y_1, Y_2, Y_3, Y_4\}$, $p(y_1)$ and $p(y_3|y_1, y_2)$ are standard canonical CPDs, but $p(y_2|y_1, y_3)$ and $p(y_1|y_3, y_4)$ are not.

Definition 5.16 Standard canonical representation of the JPD. *Let $\{Y_1, \ldots, Y_n\}$ be a permutation of a set of variables $X = \{X_1, \ldots, X_n\}$. Then the JPD $p(x)$ can be expressed as a product of n standard canonical CPDs of the form*

$$p(x) = \prod_{i=1}^{n} p(y_i|b_i), \qquad (5.24)$$

where $B_i = \{Y_1, \ldots, Y_{i-1}\}$; or equivalently,

$$p(x) = \prod_{i=1}^{n} p(y_i|a_i), \qquad (5.25)$$

where $A_i = \{Y_{i+1}, \ldots, Y_n\}$. Equations (5.24) and (5.25) are called standard canonical representations, or standard canonical factors of the JPD. The terms $p(y_i|b_i)$ and $p(y_i|a_i)$ are called standard canonical components, or factors.

For example, in (5.24) and (5.25) two standard canonical representations of the JPD $p(x_1, \ldots, x_4)$, are given. Like canonical forms, standard canonical representations are not unique because one can apply the chain rule to different permutations of X and obtain different standard canonical forms.

The practical consequences of the existence of a canonical representation for any given set P of CPDs are

1. Any set P of CPDs not expressed in canonical form can always be reexpressed in canonical form.

2. Any JPD can be reexpressed as a product of standard canonical CPDs using the chain rule. Thus standard canonical forms of a JPD always exist.

3. Only CPDs of single variables are needed to specify the JPD of all variables, that is, U_i is a single variable for all i.

    The main advantages of this representation are

    - The probability assessment is greatly simplified because direct assessment of the JPD is difficult. Human experts are required to give only the CPDs of single variables given some other variables. These CPDs are usually much smaller than the JPD.

    - The computer implementations are also simplified because only one type of conditional is required.

4. Standard canonical forms make it easy to determine whether a given set of CPDs are consistent and whether it uniquely defines a JPD (see Chapter 7).

5.6 Constructing the JPD

Although the problem of constructing a JPD of the set of variables is difficult, it can be greatly simplified if the JPD can be factorized as a product of smaller CPDs. This factorization is always possible in cases where the model contains several subsets of variables that are independent and/or conditionally independent. To find the desired factorization, we first need to specify the *dependence structure* of the model. The dependence structure of a model refers to the *qualitative structure* of the relationships among the variables. For example, we need to specify which variables are independent and/or conditionally independent of others and which are not. The dependence structure, and hence the desired factorization, can be obtained in one of several ways:

1. **Graphically specified models:** The human experts construct a graph or a set of graphs that describe the relationships among the variables. A complete list of CISs can then be derived from the graphs using graph separation criteria. The resultant models are referred to as *graphically specified models*. Examples of such models are *Markov network models*, and *Bayesian network models*. These models are discussed in detail in Chapters 6 and 7. The tasks of verifying the graphs, understanding their implications, and modifying them appropriately are best be done through the understanding of the dependence and independence relationships that exist among the set of variables.

2. **Models specified by input lists:** Graphs are easy and powerful tools for describing the dependence structure of a probabilistic model. The problem with using graphs to describe probabilistic models, however, is that not all JPDs can be represented by a graph (see Section 6.2). An alternative to graphically specified models is to ask the human experts to provide a list M of CISs that describes the relationships among the variables. The list of CISs, which is called an *input list*, reflects the opinions of the human experts regarding the relationships among the variables. Each of the CISs in an input list indicates which variables contain relevant information about others, and whether or not the knowledge of some variables makes others irrelevant in making inference about a given set of variables. This initial input list can then be completed by deriving from it additional CISs that satisfy any desired number of conditional independence properties. The completed input list can then be used to express the JPD as a product of CPDs. The resultant model is referred to as a *model specified by input lists*. A detailed discussion of these models is given in Chapter 7.

3. **Conditionally specified models:** As an alternative to graphs and input lists, the human experts can provide a set

$$P = \{p_1(u_1|v_1), \ldots, p_m(u_m|v_m)\}$$

of marginal and/or conditional probability distributions from which the JPD can be derived. This set, however, must satisfy certain compatibility and uniqueness conditions. Chapter 7 gives a detailed discussion of how to check the given set for compatibility and uniqueness and how to obtain the corresponding JPD.

Thus, an advantage of using either graphs or input lists to construct the JPD is that they usually lead to specification of the JPD using a set of CPDs the product of which gives the JPD. These CPDs are often much smaller than the JPD itself, which makes the process of constructing the JPD easier than the direct specification of the JPD. This divide-and-conquer way of breaking up the JPD into a product of smaller CPDs is discussed further in Chapters 6 and 7.

Once the factorization (qualitative structure) of the JPD is defined using either a graph or an input list, the JPD is obtained as a product of a set of CPDs. These CPDs, which provide the quantitative information needed to specify the JPD, may depend on sets of parameters. These parameters have to be either specified by the human experts or estimated from the data.

Thus, if the structural form is unknown, which is often the case in practice, both the structural form and the parameters need to be estimated from the available data. This problem, which is referred to as *learning*, is addressed in Chapter 11.

To summarize, the construction of the JPD can be done in two steps:

1. Factorization of the JPD as a product of CPDs. This factorization can be obtained in one of three ways:

   (a) Using graphs and obtaining graphically specified models (Chapter 6).

   (b) Using input lists of CISs and obtaining models specified by input lists (Chapter 7).

   (c) Using a set of CPDs and obtaining conditionally specified models (Chapter 7).

2. Estimation of the parameters of each of the CPDs required to specify the JPD.

The above way of constructing the JPD is illustrated in Figure 5.9. A solid arrow from rectangle A to rectangle B means that every member of A is also a member of B, whereas a dashed arrow means that some, but not necessarily all, members of A can be found in B. The easiest way to construct a JPD is perhaps to start with a graph that is believed by the human experts to describe the dependence structure of the variables. One can then use the graph to obtain a factorization of the JPD. Alternatively, one can start with an input list of CISs that according to the human experts, describes the relationships among the variables. From this input list, one can then obtain a factorization of the JPD in terms of smaller CPDs. Thus, the factorization is obtained based on either a graph or an input list of CISs.

The factorization of the JPD will suggest the parameters needed to specify the JPD. Once these parameters are estimated, the JPD is obtained as a product of the CPDs.

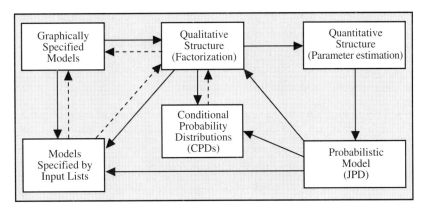

FIGURE 5.9. Ways of building JPDs and the relationships among them.

On the other hand, if the JPD is known (which is rarely the case in practice, however), one can obtain several factorizations using the *chain rule* as discussed in Section 5.5. Also, one can obtain the set of CISs directly from the JPD by enumerating all possible CISs and checking to see whether each statement actually holds in the JPD. From the list of valid CISs, one can obtain a factorization of the JPD.

The above way of constructing the JPD raises several questions.

- **Question 5.4:** Can any input list be represented by a graph in the sense that CISs derived from the input list coincide with those derived from the graph?

Although a graph can be represented by an equivalent set of CISs, not every input list of CISs can be represented by a graph. For this reason, in Figure 5.9 we have a solid arrow from the box representing graphically specified models to the box representing models specified by input lists, and a dotted arrow in the opposite direction. Further discussion and illustration of this fact are provided in Chapter 6 for both directed and undirected graphs.

- **Question 5.5:** How does one obtain a JPD that portrays the independencies found in a directed or an undirected graph?

- **Question 5.6:** How does one obtain a JPD that portrays the independencies found in an input list?

The answers to Questions 5.5 and 5.6 are given in Chapters 6 and 7, respectively.

204   5. Building Probabilistic Models

Unfortunately, graphs are not always able to reproduce conditional independencies implied by all input lists or by all JPDs. Therefore, it is important to characterize the class of JPD that can be reproduced by graphs or input lists. This raises the following questions:

- **Question 5.7:** What is the class of JPD that can be represented by graphs?

- **Question 5.8:** What is the class of input lists that can be represented by graphs?

- **Question 5.9:** What is the set of CPDs needed to specify the JPD and what are the associated sets of parameters that need to be quantified?

The answers to these questions are given in Chapters 6 and 7. There we shall find that although every graph has an equivalent factorization of the JPD, not all JPD (factorizations) can be represented by a graph. Hence, in Figure 5.9 there is a solid arrow from graphically specified models to factorized models and a dotted arrow in the opposite direction. Similarly, we shall find that every JPD defines a set of CISs, but not all sets of CISs determine a factorization of the JPD. Consequently, in Figure 5.9 there are solid arrows from JPD and factorized models to models specified by input lists and a dotted arrow from models specified by input lists to factorized models.

From the above discussion and from Figure 5.9, we see that a JPD can be constructed in three ways:

- Graph → Factorized model → Parameter estimation → JPD.

- Input list → Factorized models → Parameter estimation → JPD.

- A set of CPDs → Factorized models → Parameter estimation → JPD.

We shall see in Chapters 6 and 7 that it is easier to start with a graph but that the most general way is to start with an input list of CISs.

Appendix to Chapter 5

In this appendix we prove that some properties of conditional independence are satisfied by JPDs.

5.7.1 Proof of the Symmetry Property

Given that $I(X, Y|Z)$ holds in the JPD $p(x, y, z)$, then

$$p(x|y, z) = p(x|z) \Leftrightarrow p(x, y|z) = p(x|z)p(y|z). \qquad (5.26)$$

Let us now show that $I(Y, X|Z)$ also holds. Provided that $p(x, z) > 0$, we have

$$p(y|x, z) = \frac{p(x, y|z)}{p(x|z)} = \frac{p(x|z)p(y|z)}{p(x|z)} = p(y|z) \Rightarrow I(Y, X|Z),$$

where the second equality follows from (5.26). ∎

5.7.2 Proof of the Decomposition Property.

Given that $I(X, Y \cup W|Z)$ holds in the JPD $p(x, y, w, z)$, then

$$p(x|z, y, w) = p(x|z). \qquad (5.27)$$

We first show that $I(X, Y|Z)$ also holds. We have

$$\begin{aligned}p(x|z, y) &= \sum_v p(x, v|z, y) \\ &= \sum_v p(x|z, y, v) p(v|z, y),\end{aligned}$$

where $V = W \setminus Y$ means the set W excluding the elements in Y. Applying (5.27) we get

$$\begin{aligned}p(x|z, y) &= \sum_v p(x|z) p(v|z, y) \\ &= p(x|z) \sum_v p(v|z, y) \\ &= p(x|z).\end{aligned}$$

The last equality follows because

$$\sum_v p(v|z, y) = 1,$$

that is, the sum of the probabilities over all possible values of a random variable is one. Thus, we have $p(x|z, y) = p(x|z)$, and hence $I(X, Y|Z)$. Similarly, it can be shown that the CIS $I(X, W|Z)$ also holds. ∎

5.7.3 Proof of the Weak Union Property

Given that $I(X, Y \cup W|Z)$ holds in the JPD $p(x, y, w, z)$, then

$$p(x|z, y, w) = p(x|z). \qquad (5.28)$$

We first show that this CIS implies $I(X, W|Z \cup Y)$. The decomposition property applied to $I(X, Y \cup W|Z)$ implies $I(X, Y|Z)$, that is,

$$p(x|z, y) = p(x|z). \qquad (5.29)$$

Applying (5.28) and (5.29) we have

$$p(x|z,y,w) = p(x|z) = p(x|z,y),$$

which implies that $I(X,W|Z\cup Y)$. The CIS $I(X,Y|Z\cup W)$ can be obtained in a similar way. ∎

5.7.4 Proof of the Contraction Property

Given that $I(X,W|Z\cup Y)$ in (5.4) holds in the JPD $p(x,y,w,z)$, then

$$p(x|z,y,w) = p(x|z,y). \tag{5.30}$$

Similarly, if $I(X,Y|Z)$ holds, then we have

$$p(x|z,y) = p(x|z). \tag{5.31}$$

From (5.30) and (5.31) we have

$$p(x|z,y,w) = p(x|z,y) = p(x|z).$$

Therefore, the CIS $I(X, Y\cup W|Z)$ also holds. ∎

5.7.5 Proof of the Intersection Property

Given that $I(X,W|Z\cup Y)$ holds in a non-extreme JPD $p(x,y,w,z)$, then

$$p(x|z,y,w) = p(x|z,y). \tag{5.32}$$

Similarly, if $I(X,Y|Z\cup W)$ holds, then

$$p(x|z,y,w) = p(x|z,w). \tag{5.33}$$

Equations (5.32) and (5.33) imply

$$p(x|z,y,w) = p(x|z,y) = p(x|z,w),$$

which (since the JPD is non-extreme) implies $p(x|z,y,w) = p(x|z)$. Thus, $I(X, Y\cup W|Z)$ also holds. ∎

Exercises

5.1 Referring to the undirected graph in Figure 5.10, determine whether or not each of the following CISs is true using the U-separation criterion:

(a) $I(F, H|\phi)$.
(b) $I(F, H|D)$.
(c) $I(A, G|\{D, E\})$.
(d) $I(C, \{B, G\}|D)$.
(e) $I(\{A, B\}, \{F, G\}|\{C, D\})$.
(f) $I(\{C, F\}, \{G, E\}|\{A, D\})$.

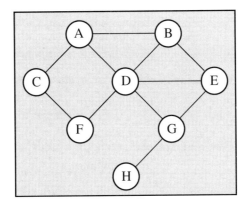

FIGURE 5.10. An undirected graph.

5.2 Referring to the directed graph in Figure 5.11, determine whether or not each of the following CISs is true using the D-separation criterion in Definition 5.3:

(a) $I(E, G|\phi)$.
(b) $I(C, D|\phi)$.
(c) $I(C, D|G)$.
(d) $I(B, C|A)$.
(e) $I(\{C, D\}, E|\phi)$.
(f) $I(F, \{E, H\}|A)$.
(g) $I(\{A, C\}, \{H, E\}|D)$.

5.3 Repeat the previous exercise using the D-separation criterion in Definition 5.4.

5.4 Consider a set of four variables $\{X, Y, Z, W\}$, which are related by

$$I(X, Y|\phi) \text{ and } I(X, Z|\{Y, W\}).$$

Find the minimal list of CISs generated by the above two CISs and satisfying

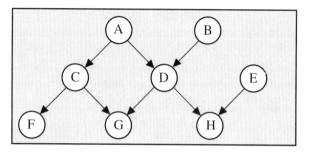

FIGURE 5.11. A directed graph.

(a) The symmetry property.

(b) The symmetry and decomposition properties.

(c) The semigraphoid properties.

(d) The graphoid properties.

5.5 Repeat the previous exercise with the following CISs:

$$I(X, W|\{Y, Z\}) \text{ and } I(Y, Z|\{X, W\}).$$

5.6 Write the set of all possible independence statements for the case of three variables.

5.7 Find the probabilistic input list corresponding to the JPD

$$p(x, y, z) = 0.3^{x+y} 0.7^{2-x-y} \left(\frac{x+y}{2}\right)^z \left(1 - \frac{x+y}{2}\right)^{1-z},$$

where $x, y, z \in \{0, 1\}$.

5.8 Consider the set of variables $\{X, Y, Z, W\}$ and the JPD $p_1(x, y, z, w)$ in Table 5.3,

(a) Prove that the JPD satisfies the CIS $I(X, Y \cup W|Z)$.

(b) Is this the most general family of JPDs satisfying this property?

5.9 Consider the set of four variables $\{X, Y, Z, W\}$ and the parametric family of JPD $p_2(x, y, z, w)$ in Table 5.3,

(a) Prove that the JPD satisfies the CISs $I(X, Y|Z)$ and $I(X, W|Z)$.

(b) Is this the most general family of JPDs satisfying these properties?

(c) Are the constraints $p_6 = p_{14}p_7/p_{15}$ and $p_4 = p_{12}p_5/p_{13}$ sufficient for the above family to satisfy $I(X, Y \cup W|Z)$?

x	y	z	w	$p_1(x,y,z,w)$	$p_2(x,y,z,w)$
0	0	0	0	$p_5 p_8 / p_{13}$	$(-p_{13} p_4 + p_{12} p_5 + p_4 p_8 + p_5 p_8)/a$
0	0	0	1	$p_5 p_9 / p_{13}$	$(p_{13} p_4 - p_{12} p_5 + p_4 p_9 + p_5 p_9)/a$
0	0	1	0	$p_{10} p_7 / p_{15}$	$(p_{10} p_6 - p_{15} p_6 + p_{10} p_7 + p_{14} p_7)/b$
0	0	1	1	$p_{11} p_7 / p_{15}$	$(p_{11} p_6 + p_{15} p_6 + p_{11} p_7 - p_{14} p_7)/b$
0	1	0	0	$p_{12} p_5 / p_{13}$	p_4
0	1	0	1	p_5	p_5
0	1	1	0	$p_{14} p_7 / p_{15}$	p_6
0	1	1	1	p_7	p_7
1	0	0	0	p_8	p_8
1	0	0	1	p_9	p_9
1	0	1	0	p_{10}	p_{10}
1	0	1	1	p_{11}	p_{11}
1	1	0	0	p_{12}	p_{12}
1	1	0	1	p_{13}	p_{13}
1	1	1	0	p_{14}	p_{14}
1	1	1	1	p_{15}	p_{15}

TABLE 5.3. Two parametric families of JPDs, where $a = p_{12} + p_{13}$ and $b = p_{14} + p_{15}$.

5.10 Express the JPD in Example 5.8 in a factorized form when the variables are partitioned as

(a) $Y_1 = \{X_1, X_3\}$, $Y_2 = \{X_2, X_4\}$.

(b) $Y_1 = \{X_4\}$, $Y_2 = \{X_2\}$, $Y_3 = \{X_1, X_3\}$.

(c) $Y_1 = \{X_2\}$, $Y_2 = \{X_1, X_3, X_4\}$.

5.11 Consider the set of four variables in Example 5.9 and suppose that X_1 is ternary and the other three variables are binary.

(a) What is the maximum number of free parameters in the JPD?

(b) How many free parameters are there in the JPD if it satisfies the CISs in (5.17)?

5.12 Repeat the previous exercise when all three variables are ternary.

5.13 Consider again the set of four variables in Example 5.9. In each of the following cases write the JPD in a factored form and indicate how many free parameters there are in

(a) The JPD satisfying $I(X_1, X_4 | \{X_2, X_3\})$.

(b) The JPD satisfying $I(X_2, X_3 | X_1)$, $I(X_3, X_4 | X_1)$, and $I(X_2, X_4 | X_1)$.

5.14 Find the probabilistic input list corresponding to the JPD given in Table 3.2.

5.15 Suppose that the JPD of four variables $\{X, Y, Z, W\}$ can be factorized as
$$p(x, y, z, w) = p(x)p(y|x)p(z|x)p(w|y, z).$$
Determine whether or not each of the following CISs holds:

(a) $I(X, W|Y)$.

(b) $I(X, W|Z)$.

(c) $I(X, W|Y, Z)$.

(d) $I(Y, Z|X, W)$.

Chapter 6
Graphically Specified Models

6.1 Introduction

We have seen in Chapter 3 that the performance of a probabilistic expert system hinges on the correct specification of the probabilistic model, which is represented by a joint probability distribution (JPD) of the set of variables of interest. The JPD is needed for the knowledge base of probabilistic expert systems. We have also seen in Chapter 3 that the most general JPD involves an infeasible large number of parameters. For this reason, simplifications of the most general JPD are needed. In Section 3.5 we presented several models, which are obtained by imposing certain global or special cases of independence assumptions. However, these models are ad hoc because they are suitable only for the diseases-symptoms paradigms. In this chapter we show how more general dependency models are obtained using graphs. The basic idea consists of using undirected or directed graphs to build a dependency model.

It may be useful at this point to clarify some notation and terminology. The term *probabilistic model* refers to a complete specification of the JPD. Therefore, the terms probabilistic model and JPD are used synonymously. The JPD contains qualitative as well as quantitative information about the relationships among the variables. We also use the terms *dependency model* and *independency model* synonymously. These terms refer only to the *qualitative structure* of the relationships among a set of variables from which one can determine whether sets of variables are conditionally or unconditionally dependent or independent. For every probabilistic model, there exists a corresponding dependency model M that can be obtained by generating all possible combinations of three disjoint subsets of variables and checking to see whether they are conditionally dependent or independent. For example, if X, Y, and Z are three disjoint subsets and $p(x|y,z) = p(x|z)$ for all x, y, and z, then we have the conditional independence statement (CIS) $I(X,Y|Z)$ and we conclude that X and Y are conditionally independent given Z. On the other hand, if $p(x|y,z) \neq p(x|z)$ for some x, y, z, then X and Y are conditionally dependent given Z. Thus, a JPD contains a complete specification (qualitative and quantitative) of the relationships among the variables, whereas the associated dependency model M can contain only a qualitative description of these relationships. When we say probabilistic dependency model M, we refer to the dependency model M associated with a JPD.

On the other hand, a dependency model can alternatively be described by a graph (directed or undirected), an input list of CISs, or a set of conditional probability distributions (CPDs). These lead to three approaches for building dependency models:

- Graphically specified models (GSM).
- Models specified by input lists.
- Conditionally specified models.

These approaches are more general than the four models described in Section 3.5 and can be employed in medical as well as in other fields of applications. These approaches, however, require some background material such as graph separation criteria and properties of conditional independence. This material was presented in Sections 5.2 and 5.3. This chapter is devoted entirely to GSM, or more precisely, models specified by a single undirected or directed graph. Models specified by multiple graphs, models specified by input lists, and conditionally specified models are presented in Chapter 7.

As we have seen in Chapter 4, a set of variables $X_1, \ldots, X_n$ can be represented by a graph in which each variable is represented by a node and the relationship between two variables is represented by a link, which is either an undirected line connecting the corresponding nodes (undirected graph) or a directed arrow starting from a node and pointing to another node (directed graphs). Thus, the words *variable* and *node* are used synonymously. Graph representations (directed and undirected) have the advantages that they display the relationships among the variables explicitly and preserve them qualitatively (i.e., for any numerical specification of the parameters). Graphs are also intuitive and easy to explain.

In Chapter 5 we introduced two different graph separation criteria to derive CISs from undirected and directed graphs. Accordingly, graphically specified models can be classified according to the type of graphs used in their constructions as follows:

- Undirected graph dependency models, presented in Section 6.3.

- Directed graph dependency models, presented in Section 6.4.

We should mention here that a third class of graphical models can be represented by a mixed graph (a graph containing some directed and some undirected links). These graphs are known as *chain graphs*. In this book we focus on undirected and directed graphical models. For models specified by chain graphs, the interested reader is referred to Lauritzen and Wermuth (1989) and Frydenberg (1990).

We use the word *dependency* in the above models because graphs can be used only to describe the dependency or qualitative structure of a probabilistic model. Once the dependency (qualitative) structure is specified, the corresponding set of CPDs is identified and the corresponding set of parameters needs to be either specified by the human experts or estimated from the available data; see Section 5.6. The set of CPDs, together with the corresponding set of parameters, is known as the *quantitative structure*.

In this chapter, we investigate the capabilities and incapabilities of undirected and directed graphs in capturing various types of dependency structures found in probabilistic models. We start with Section 6.2, where we state some definitions and problems to be investigated. Sections 6.3 and 6.4 discuss undirected and directed graph dependency models, respectively. Section 6.5 defines and characterizes equivalent graphical models. Section 6.6 discusses the expressiveness of graphs in representing probabilistic models.

6.2 Some Definitions and Questions

Our goal here is to represent probabilistic dependency models by graphs. Thus, a question that arises naturally is whether graphs provide a complete representation of all dependency models. To address this question, we start with the following definition.

Definition 6.1 Perfect map. *A graph G is said to be a perfect map of a dependency model M if every CIS derived from G can also be derived from M and vice versa, that is,*

$$I(X,Y|Z)_M \Leftrightarrow I(X,Y|Z)_G \Leftrightarrow Z \; separates \; X \; from \; Y.$$

Depending on the directed or undirected character of G, perfect maps are called directed or undirected perfect maps, respectively.

Since we deal with two types of graphs, the question stated above can be restated as follows:

- **Question 6.1:** Can any dependency model be represented by a directed or undirected perfect map?

Unfortunately, not every dependency model can be represented by a directed or undirected perfect map. The following are two examples of dependency models that do not have perfect maps. Other examples are given in Sections 6.3 and 6.4.

Example 6.1 Dependency model with no undirected perfect map.
Consider a set of three variables $\{X, Y, Z\}$ that are related by a dependency model
$$M = \{I(X, Y|\phi), I(Y, X|\phi)\}, \tag{6.1}$$
with only one CIS and its symmetric counterpart. Let us now try to represent this dependency model by an undirected graph. There are $2^{n(n-1)/2}$ different undirected graphs with n nodes (see Whittaker (1990)). Figure 6.1 shows all eight possible undirected graphs for three variables. The graphs are arranged in rows such that graphs in the same row have the same number of links. Thus, graph (a) is the empty graph (a graph with no links), each of the three graphs in (b)–(d) contains one link, each of the graphs in (e)–(g) contains two links, and the last graph is the complete graph (a graph in which there is a link between every pair of nodes). The second column in Table 6.1 shows some of the CISs implied by each graph but not found in M. The reader can verify these relationships using the U-separation criterion introduced in Section 5.2. The last column shows whether the only CIS in M is also found in G. As can be seen from Table 6.1, in every graph G one can either find a CIS in G but not in M and/or vice versa. Thus, none of the graphs in Figure 6.1 is a perfect map of M in (6.1). Since this set of graphs is exhaustive, the dependency model M does not have an undirected perfect map. ∎

Although the dependency model M in Example 6.1 has no undirected perfect map, it has a directed perfect map. It is left as an exercise for the reader to demonstrate that the directed graph in Figure 6.2 is a directed perfect map of M. In this case directed graphs are more useful than undirected graphs. However, not every probabilistic model can be represented by a directed perfect map. The following is an example of such models.

Example 6.2 Dependency model with no directed perfect map.
Consider the set of three variables $\{X, Y, Z\}$ and the dependency model
$$M = \{I(X, Y|Z), I(Y, Z|X), I(Y, X|Z), I(Z, Y|X)\}. \tag{6.2}$$
There is no directed acyclic graph (DAG) D that is a perfect map of the dependency model M. ∎

In cases where a perfect map does not exist, we want to be sure that no false independencies are derived from the graphical structure and that the

6.2 Some Definitions and Questions   215

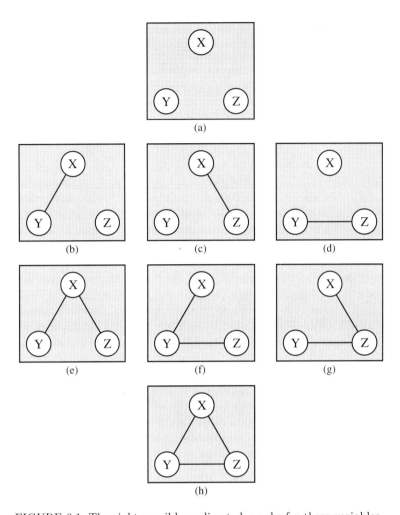

FIGURE 6.1. The eight possible undirected graphs for three variables.

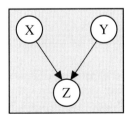

FIGURE 6.2. A directed perfect map of the dependency model M in (6.1).

216   6. Graphically Specified Models

Graph G	CIS in G but not in M	CIS in M but not in G
(a)	$I(X,Z\|\phi)$	ϕ
(b)	$I(X,Z\|\phi)$	$I(X,Y\|\phi)$
(c)	$I(Y,Z\|\phi)$	ϕ
(d)	$I(X,Z\|\phi)$	ϕ
(e)	$I(Y,Z\|X)$	$I(X,Y\|\phi)$
(f)	$I(X,Z\|Y)$	$I(X,Y\|\phi)$
(g)	$I(X,Y\|Z)$	$I(X,Y\|\phi)$
(h)	ϕ	$I(X,Y\|\phi)$

TABLE 6.1. Some CISs found in G in Figure 6.1 but not in the dependency model M in (6.1), and CISs found in M but not in G.

number of nonreproduced independencies and dependencies is kept to a minimum. This motivates the following definitions.

Definition 6.2 Independency map. *A graph G is said to be an independency map (I-map) of a dependency model M if*

$$I(X,Y|Z)_G \Rightarrow I(X,Y|Z)_M,$$

that is, if all CISs derived from G hold in M.

Note that an I-map G of a dependency model M includes some, but not necessarily all, of the independencies in M. Using an I-map G of M instead of M itself means that all derived independencies from G are true in M, but not all independencies are represented in the I-map. Thus, using an independency map G of M leads to

$$I(X,Y|Z)_G \;\Rightarrow\; I(X,Y|Z)_M,$$

which implies

$$D(X,Y|Z)_M \;\Rightarrow\; D(X,Y|Z)_G.$$

Hence, all dependencies in M are represented in G. For example, only the complete graph in Figure 6.1(h) is an I-map of the dependency model given in (6.1). Each of the remaining graphs implies some CISs not found in M (see Table 6.1). In general, a complete graph is always a trivial I-map of any dependency model.

Definition 6.3 Dependency map. *A graph G is said to be a dependency map (D-map) of a dependency model M if*

$$D(X,Y|Z)_G \Rightarrow D(X,Y|Z)_M,$$

that is, all CISs derived from G hold in M.

Using a D-map G of M leads to

$$D(X,Y|Z)_G \Rightarrow D(X,Y|Z)_M,$$

which implies

$$I(X,Y|Z)_M \Rightarrow I(X,Y|Z)_G,$$

that is, all independencies in M are represented in G.

Note that a D-map of a dependency model M is a model that includes some of the dependencies in M. Using a D-map of M instead of M itself means that all derived dependencies are true in M, but not all dependencies are represented in the D-map. For example, the totally disconnected graph in Figure 6.1(a) is a trivial, but useless, D-map of the dependency model given in (6.1). The graphs in Figures 6.1(c) and (d) are also D-maps of the dependency model.

It follows from this definition that every dependency model has a trivial I-map and a trivial D-map. For example, any completely disconnected graph (a graph with no links) is a trivial D-map of every dependency model. Also, any complete graph is a trivial I-map of every dependency model. Therefore, for a graph to be a perfect map it must be both an I-map and a D-map.

Definition 6.4 Minimal I-map. *A graph G is said to be a minimal I-map of a dependency model M if it is an I-map of M, but it is not an I-map of M when removing any link from it.*

Although dependency models and graphical representations have a wide applicability beyond probability, in this book we are interested in the construction of probabilistic models. Thus, it is of interest to know whether or not such graphical representations exists for any given JPD. Therefore, a natural question to ask is whether the intuitive notion of informational relevancy or the formal notions of probabilistic dependencies can be captured by a graphical representation, in the sense that all dependencies and independencies in a given probability model would be deducible from the topological properties of the graph.

In graphical structures, the task of testing connectedness (using graph separation criteria discussed in Chapter 5) is easier than that of testing conditional independence using conditional probability formulas given in Section 3.2. A D-map guarantees that nodes found to be connected are, indeed, dependent; however, it may occasionally display dependent variables as separated nodes. An I-map works the opposite way: it guarantees that nodes found to be separated always correspond to genuinely independent variables, but does not guarantee that all those shown to be connected are, in fact, dependent. As mentioned above, empty graphs are trivial D-maps, while complete graphs are trivial I-maps.

The task of capturing dependencies by graphs, however, is not at all trivial. When we deal with a phenomenon where the notion of neighborhood or connectedness is explicit (e.g., family relations, electronic circuits, communication networks, etc.), we usually have little problem configuring a graph that represents the main features of the phenomenon. However, in modeling conceptual relations such as causation, association, and relevance, it is often hard to distinguish direct neighbors from indirect neighbors; hence, the task of constructing a graphical representation then becomes more difficult. The notion of conditional independence in probability theory provides a perfect example of such a task. Given a JPD and any three variables X, Y, Z, while it is fairly easy to verify whether knowing Z renders X independent of Y, the JPD does not dictate which variable is the cause and which is the effect, or which variables should be regarded as direct neighbors.

In Chapter 4 we introduced some elemental concepts of graph theory and saw that nodes in a graph represent propositional variables and the links represent local dependencies among conceptually-related propositions. Graph representations are perfectly suited for meeting the requirements of explicitness, saliency, and stability, i.e., the links in the graph permit us to qualitatively encode dependence relationships, and the graph topology displays these relationships explicitly and preserves them under any assignment of numerical parameters. We shall discuss in this chapter how some probability models can be represented by either a directed or an undirected graph.

Since not every probabilistic model can be represented by a perfect map, the answers to the following questions are of interest:

- **Question 6.2:** What are the dependency models and in particular the probabilistic dependency models, that can be represented by a perfect map?

- **Question 6.3:** What are the probabilistic models that can be represented by a unique minimal I-map?

- **Question 6.4:** If a probabilistic model has a unique minimal I-map, how can we construct it?

- **Question 6.5:** Given a graph G, is there any probabilistic model P such that G is a minimal I-map of P? If yes, how do we build it?

We shall address these questions for undirected graphs in Section 6.3 and for directed graphs in Section 6.4. Note that Question 5.7 "What is the class of JPDs that can be represented by graphs?" is now split into two parts: Questions 6.2 and 6.3.

6.3 Undirected Graph Dependency Models

In this section we discuss building dependency models using undirected graphs. Our goal is to find a graph that produces as many independencies associated with a given probabilistic model P (or its associated dependency model M) as possible. We start with the problem of mapping probabilistic models using undirected graphs and then discuss an important class of probabilistic models given by undirected graphs known as *Markov network models*.

6.3.1 From Models to Undirected Graphs

In this section we discuss the problem of mapping probabilistic models using undirected graphs, that is, we wish to find an undirected graph corresponding to a given a probabilistic dependency model. As we have seen in Example 6.1, not all probabilistic models can be represented by undirected perfect maps. Pearl and Paz (1987) give the following theorem, which characterizes the dependency models that can be represented by undirected perfect maps. The theorem refers not only to probabilistic models but to dependency models in general.

Theorem 6.1 Models with undirected perfect maps. *A necessary and sufficient condition for a dependency model M to have an undirected perfect map is that M must satisfy the following properties:*

- **Symmetry:**
$$I(X,Y|Z)_M \Leftrightarrow I(Y,X|Z)_M.$$

- **Decomposition:**
$$I(X, Y \cup W|Z)_M \Rightarrow I(X,Y|Z)_M \text{ and } I(X,W|Z)_M.$$

- **Intersection:**
$$I(X,W|Z \cup Y)_M \text{ and } I(X,Y|Z \cup W)_M \Rightarrow I(X, Y \cup W|Z)_M.$$

- **Strong union:**
$$I(X,Y|Z)_M \Rightarrow I(X,Y|Z \cup W)_M.$$

- **Strong transitivity:**
$$I(X,Y|Z)_M \Rightarrow I(X,A|Z)_M \text{ or } I(Y,A|Z)_M,$$

  where A is a single node not in $\{X,Y,Z\}$.

Thus, the answer to Question 6.2 for the case of undirected graphs is that only dependency models that satisfy the above properties can be represented by an undirected perfect map in the sense that all CISs implied by the corresponding dependency model M can be derived from the undirected perfect map and that all CISs derived from the undirected perfect map are in M. We address this question for directed graphs in Section 6.4.1.

Note that in general, graphoids and semigraphoids do not have undirected perfect maps, because semigraphoids satisfy only the symmetry and decomposition properties and graphoids satisfy only the symmetry, decomposition, and intersection properties. For example, the dependency model introduced in Example 6.1 is a graphoid, but it has no perfect map because it violates the strong union and strong transitivity properties.

Unfortunately, probabilistic dependency models can also violate the last two properties, and hence, not all probabilistic dependency models can be represented by an undirected perfect map. We give below examples illustrating this fact.

Example 6.3 Strong union and strong transitivity violation. The strong union property states that if X and Y are independent given Z, then they are also independent given a larger set $Z \cup W$:

$$I(X, Y|Z) \Rightarrow I(X, Y|Z \cup W).$$

For example, for $Z = \phi$, this property implies: $I(X, Y|\phi) \Rightarrow I(X, Y|W)$, which says that if X and Y are unconditionally independent, then they must be conditionally independent given some other subset W. This is not always true. For example, for the family of JPDs given by the factorization

$$p(x, y, z) = p(x)p(y)p(z|x, y),$$

we have $I(X, Y|\phi)$, but $I(X, Y|Z)$ is not true in general. Therefore, $p(x, y, z)$ violates the strong union property, and for this reason, it cannot be represented by an undirected perfect map, as we have seen in Example 6.1. Furthermore, this family of JPDs also violates the strong transitivity property. According to this property, and taking $Z = \phi$, we have

$$I(X, Y|\phi) \Rightarrow I(X, A|\phi) \text{ or } I(Y, A|\phi),$$

where A is a single node other than $\{X, Y\}$. Thus, in this case $A = Z$. However, although $I(X, Y|\phi)$ neither $I(X, Z|\phi)$ nor $I(X, Z|\phi)$ is satisfied in general by the above family of distributions. We have left as an exercise to the reader to specify some numerical values for the parameters associated with CPDs of this family to obtain a JPD $p(x, y, z)$ violating both properties. Note that these parameters cannot be taken arbitrarily because some specification of the parameters may render variables X and Z, or Y and Z, independent, and hence, the above properties will not be violated. ∎

The next example illustrates the strong union and strong transitivity violation using a continuous JPD. This example requires the following property of the multivariate normal distribution (see any multivariate analysis book such as Anderson (1984), Johnson and Wichern (1988), or Rencher (1995)).

6.3 Undirected Graph Dependency Models

Theorem 6.2 Multivariate normal distribution. *Let X and Y be two sets of random variables having a multivariate normal distribution with mean vector and covariance matrix given by*

$$\mu = \begin{pmatrix} \mu_X \\ \mu_Y \end{pmatrix} \quad \text{and} \quad \Sigma = \begin{pmatrix} \Sigma_{XX} & \Sigma_{XY} \\ \Sigma_{YX} & \Sigma_{YY} \end{pmatrix},$$

where μ_X and Σ_{XX} are the mean vector and covariance matrix of X, μ_Y and Σ_{YY} are the mean vector and covariance matrix of Y, and Σ_{XY} is the covariance of X and Y. Then the CPD of X given $Y = y$ is multivariate normal with mean vector $\mu_{X|Y=y}$ and covariance matrix $\Sigma_{X|Y=y}$, which are given by

$$\mu_{X|Y=y} = \mu_X + \Sigma_{XY}\Sigma_{YY}^{-1}(y - \mu_y), \tag{6.3}$$

$$\Sigma_{X|Y=y} = \Sigma_{XX} - \Sigma_{XY}\Sigma_{YY}^{-1}\Sigma_{YX}. \tag{6.4}$$

Note that the conditional mean $\mu_{X|Y=y}$ depends on y but the conditional variance $\Sigma_{X|Y=y}$ does not.

Example 6.4 Strong union and strong transitivity violation. Suppose that (X_1, X_2, X_3) are multivariate normally distributed random variables with

$$\mu = \begin{pmatrix} \mu_1 \\ \mu_2 \\ \mu_3 \end{pmatrix} \quad \text{and} \quad \Sigma = \begin{pmatrix} 1 & 0 & 1/4 \\ 0 & 1 & 1/2 \\ 1/4 & 1/2 & 1 \end{pmatrix}. \tag{6.5}$$

The strong union property implies that if two sets of variables X and Y are unconditionally independent, then they must be conditionally independent given some other subset W. That is $I(X, Y|\phi) \Rightarrow I(X, Y|W)$. In this example, the only two unconditionally independent variables are X_1 and X_2, since $\Sigma_{X_1 X_2} = \Sigma_{X_2 X_1} = 0$. Therefore, we have $I(X_1, X_2|\phi)$. However, X_1 and X_2 are not conditionally independent given X_3, that is, $I(X_1, X_2|X_3)$ does not hold. To see this, we use (6.4) to calculate

$$\Sigma_{X_1|X_3} = \Sigma_{X_1 X_1} - \Sigma_{X_1 X_3}\Sigma_{X_3 X_3}^{-1}\Sigma_{X_3 X_1}$$

$$= 1 - \frac{1}{4} \times 1 \times \frac{1}{4} = \frac{15}{16}, \tag{6.6}$$

$$\Sigma_{X_1|X_2,X_3} = \Sigma_{X_1 X_1} - \Sigma_{X_1(X_2 X_3)}\Sigma_{(X_2 X_3)(X_2 X_3)}^{-1}\Sigma_{(X_2 X_3)X_1}$$

$$= 1 - \begin{pmatrix} 0 & \frac{1}{4} \end{pmatrix} \begin{pmatrix} 1 & \frac{1}{2} \\ \frac{1}{2} & 1 \end{pmatrix}^{-1} \begin{pmatrix} 0 \\ \frac{1}{4} \end{pmatrix} = \frac{11}{12}. \tag{6.7}$$

From (6.6) and (6.7), we obtain that the normal distributions of the variables $(X_1|X_3)$ and $(X_1|X_2, X_3)$ are different; thus, the normal distribution whose covariance matrix is given in (6.5) violates the strong union property. Hence it cannot be represented by an undirected perfect map.

Similarly, for the strong transitivity property, taking $Z = \phi$, we have

$$I(X,Y|\phi) \Rightarrow I(X,A|\phi) \text{ or } I(Y,A|\phi),$$

where A is a single node other than $\{X,Y\}$. This property does not hold in the multivariate normal distribution in (6.5). To see this, take $X = X_1$, $Y = X_2$, and $A = X_3$. We know that X_1 and X_2 are independent, but each depends on X_3. Equations (6.6) and (6.5) show that X_1 and X_3 are not independent, since $\Sigma_{X_1|X_3} \neq \Sigma_{X_1 X_1}$. On the other hand, using (6.4), we have

$$\Sigma_{X_2|X_3} = \Sigma_{X_2 X_2} - \Sigma_{X_2 X_3} \Sigma_{X_3 X_3}^{-1} \Sigma_{X_3 X_2} = 1 - \frac{1}{2}\frac{1}{2} = \frac{3}{4} \neq \Sigma_{X_2 X_2},$$

which shows that X_2 is not independent of X_3. Thus, the multivariate normal distribution given in (6.5) does not satisfy the strong transitivity property. ∎

In cases where it is impossible to construct a perfect map, we try to build an I-map. From Definition 6.2, it follows that every probabilistic model has an I-map. But in order for the I-map to represent as many independencies in M as possible, we should use a minimal I-map. Note, however, that a probabilistic dependency model may not have a unique I-map. The following theorem by Pearl and Paz, (1987), see also Verma and Pearl (1990), specifies the conditions under which a probabilistic model has a unique minimal undirected I-map. The theorem also shows how to construct such an I-map.

Theorem 6.3 Minimal undirected I-map. *Every dependency model M of a set of variables $X = \{X_1, \ldots, X_n\}$ satisfying the symmetry, decomposition, and intersection properties has a unique minimal undirected I-map that is produced by deleting from the complete graph every link (X_i, X_j) for which $I(X_i, X_j | X \setminus \{X_i, X_j\})_M$ holds, where $X \setminus \{X_i, X_j\}$ denotes the set of all variables in X except X_i and X_j.*

Note that all three properties required for having a unique minimal undirected I-map are satisfied for nonextreme JPDs. Thus, every nonextreme JPD has a unique minimal undirected I-map. Theorem 6.3 provides the answer for the case of undirected graphs to Question 6.3: What are the dependency models that can be represented by a unique minimal I-map?. The answer for the case of directed graphs is given in Section 6.4.1.

Note that graphoids satisfy the symmetry, decomposition, and intersection properties. Thus, an important property of graphoids is that they possess unique minimal undirected I-maps, and permit the construction of I-maps from local dependencies. By connecting each variable X_i in X to any subset of the variables that renders X_i conditionally independent of all other variables, we obtain a graph that is an I-map of the graphoid. Such local construction is not guaranteed for semigraphoids.

Example 6.5 Minimal undirected I-map (I). Suppose we have a set of four variables $\{X_1, X_2, X_3, X_4\}$ that are related by the set of CISs:

$$M = \{I(X_1, X_2|X_3), \quad I(X_1, X_4|X_2), \quad I(X_1, X_4|\{X_2, X_3\}), \\ I(X_2, X_1|X_3), \quad I(X_4, X_1|X_2), \quad I(X_4, X_1|\{X_2, X_3\})\}, \quad (6.8)$$

as in Example 5.7. This dependency model satisfies the three properties required in Theorem 6.3. Therefore, we can obtain the minimal undirected I-map by checking which independencies of the form $I(X_i, X_j|X \setminus \{X_i, X_j\})_M$ are satisfied in M. All the possible CISs of this form with four variables are

$$I(X_1, X_2|\{X_3, X_4\}), \quad I(X_1, X_3|\{X_2, X_4\}), \quad I(X_1, X_4|\{X_2, X_3\}), \\ I(X_2, X_3|\{X_1, X_4\}), \quad I(X_2, X_4|\{X_1, X_3\}), \quad I(X_3, X_4|\{X_1, X_2\}),$$

that is, those corresponding to all possible pairs of nodes. From this list, only $I(X_1, X_4|\{X_2, X_3\})$ holds in M. Then, we must remove link (X_1-X_4) from the complete graph in Figure 6.3(a) to obtain the minimal undirected I-map of M in Figure 6.3(b).

Lets now consider the new dependency model M' created by adding the additional CIS $I(X_1, X_2|\{X_3, X_4\})$ to the model M given in (6.8), that is,

$$M' = M \cup \{I(X_1, X_2|\{X_3, X_4\})\}. \quad (6.9)$$

If we apply to M' the same procedure to build an associated undirected graph (see Theorem 6.3) we will obtain the graph given in Figure 6.3(c). However, this graph is not an I-map of M'. For example, $I(X_1, X_4|X_3)_G$, but this independency is not contained in M'. The reason is that M' does not satisfy the properties given in Theorem 6.3. For example, applying the intersection property to $I(X_1, X_2|\{X_3, X_4\})$ and $I(X_1, X_4|\{X_2, X_3\})$ we get the CIS $I(X_1, \{X_2, X_4\}|X_3)$; then, applying the decomposition property to $I(X_1, \{X_2, X_4\}|X_3)$ we get $I(X_1, X_2|X_3)$ and $I(X_1, X_4|X_3)$, which are not in M. Similarly, applying the intersection property to $I(X_1, X_4|X_2)$ and $I(X_1, X_2|\{X_3, X_4\})$ we get the new CIS $I(X_1, \{X_2, X_4\}|\phi)$ and applying the decomposition to this CIS we get $I(X_1, X_2|\phi)$ and $I(X_1, X_4|\phi)$. Therefore, the dependency model

$$M \cup C, \quad (6.10)$$

has the graph in Figure 6.3(c) as the unique minimal I-map, where C is the set containing the following CISs and their symmetric counterparts:

$$I(X_1, X_2|\{X_3, X_4\}), \quad I(X_1, \{X_2, X_4\}|X_3), \quad I(X_1, X_4|X_3), \\ I(X_1, \{X_2, X_4\}|\phi) \quad I(X_1, X_2|\phi), \quad I(X_1, X_4|\phi).$$

∎

224   6. Graphically Specified Models

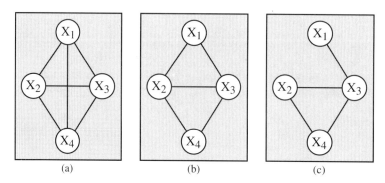

FIGURE 6.3. (a) The complete graph of four nodes, (b) the minimal undirected I-map for the dependency M model defined by the CIS in (6.8), and (c) minimal undirected I-map for the model in (6.10).

Note also that from Theorem 6.3, every nonextreme JPD has an associated minimal undirected I-map produced by deleting from the complete graph every link L_{ij} between X_i and X_j, if $I(X_i, X_j | X \setminus \{X_i, X_j\})_P$. Note also that the CIS $I(X_i, X_j | X \setminus \{X_i, X_j\})_P$ is equivalent to

$$p(x_i | x \setminus \{x_i, x_j\}) = p(x_i | x \setminus x_i), \qquad (6.11)$$

which implies that

$$\frac{p(x \setminus x_j)}{p(x \setminus \{x_i, x_j\})} = \frac{p(x)}{p(x \setminus x_i)}. \qquad (6.12)$$

This suggests the following algorithm, which provides the answer for the case of undirected graphs to Question 6.4: If a probabilistic model has a unique minimal I-map, how can we construct it? The answer for the case of directed graphs is given in Section 6.4.1.

Algorithm 6.1 Minimal Undirected I-Map of a Nonextreme JPD.

- **Input:** A set of variables $X = \{X_1, \ldots, X_n\}$ and their nonextreme JPD $p(x)$.

- **Output:** The minimal undirected I-map corresponding to $p(x)$.

1. Start with a complete graph with n nodes, in which there is a link between every pair of nodes.

2. For every pair of nodes (X_i, X_j) calculate

$$p(x \setminus x_i) = \sum_{x_i} p(x),$$

$$p(x \setminus x_j) = \sum_{x_j} p(x),$$

$$p(x \setminus \{x_i, x_j\}) = \sum_{x_j} p(x \setminus x_i).$$

6.3 Undirected Graph Dependency Models

Then, if
$$p(x)p(x \setminus \{x_i, x_j\}) = p(x \setminus x_i)p(x \setminus x_j),$$
eliminate the link L_{ij} between X_i and X_j. ∎

The following examples illustrate Algorithm 6.1.

Example 6.6 Minimal undirected I-map (I). Suppose that the JPD of a set of seven binary variables $X = \{X_1, \ldots, X_7\}$ is given by the factorization

$$p(x) = p(x_1)p(x_2|x_1)p(x_3|x_1)p(x_4|x_2,x_3)p(x_5|x_3)p(x_6|x_4)p(x_7|x_4), \quad (6.13)$$

which depends on the 15 parameters $\theta_1, \ldots, \theta_{15}$. We wish to construct the minimal undirected I-map corresponding to $p(x)$. A computer program that implements Algorithm 6.1 and achieves this objective is given in Figure 6.4. This program is written in *Mathematica* (see Wolfram (1991)) but of course, it can be written in other symbolic environments as well. The program starts with defining the JPD in (6.13). For this purpose, the functions $PA[i]$, $PB[i,j]$, etc., are defined symbolically using the parameters $p1, \ldots, p15$. Then, the JPD of the seven variables is defined as the product of the above functions. The functions $P1[i]$ and $P2[i,j]$ are introduced to marginalize with respect to the variable X_i, or the variables X_i and X_j, respectively. Finally, the last part of the program checks whether (6.12) is satisfied for all possible combinations of nodes X_i and X_j. Upon executing the program, we find that the following links can be deleted from the complete undirected graph:

$$L_{14}, \; L_{15}, \; L_{16}, \; L_{17}, \; L_{23}, \; L_{25}, \; L_{26},$$
$$L_{27}, \; L_{36}, \; L_{37}, \; L_{45}, \; L_{56}, \; L_{57}, \; L_{67}.$$

Thus, starting with the full graph, in which all pairs of nodes are connected, and after deleting the above links, we obtain the graph in Figure 6.5, which is the minimal undirected I-map corresponding to the probabilistic model in (6.13). ∎

Example 6.7 Minimal undirected I-map (II). Suppose now that the JPD of five variables can be written as

$$p(x) = \psi_1(x_1, x_2, x_3)\psi_2(x_1, x_3, x_4)\psi_3(x_1, x_4, x_5), \quad (6.14)$$

where ψ_1, ψ_2, and ψ_3 are unspecified positive functions. A *Mathematica* program that finds the corresponding minimal undirected I-map is given in Figure 6.6. The I-map resulting from executing the program is shown in Figure 6.7. We see that the minimal undirected I-map associated with $p(x)$ in (6.14) is obtained by deleting three links (L_{24}, L_{25}, L_{35}) from the complete graph. ∎

```
T={p1,1-p1}; n=1;
Do[PA[i1]=T[[n]];n++,{i1,0,1}];
T={p2,p3,1-p2,1-p3}; n=1;
Do[PB[i1,i2]=T[[n]];n++,{i1,0,1},{i2,0,1}];
T={p4,p5,1-p4,1-p5}; n=1;
Do[PC[i1,i2]=T[[n]];n++,{i1,0,1},{i2,0,1}];
T={p6,p7,p8,p9,1-p6,1-p7,1-p8,1-p9}; n=1;
Do[PD[i1,i2,i3]=T[[n]];n++,{i1,0,1},{i2,0,1},{i3,0,1}];
T={p10,p11,1-p10,1-p11}; n=1;
Do[PE[i1,i2]=T[[n]];n++,{i1,0,1},{i2,0,1}];
T={p12,p13,1-p12,1-p13}; n=1;
Do[PF[i1,i2]=T[[n]];n++,{i1,0,1},{i2,0,1}];
T={p14,p15,1-p14,1-p15}; n=1;
Do[PG[i1,i2]=T[[n]];n++,{i1,0,1},{i2,0,1}];
P[x1_,x2_,x3_,x4_,x5_,x6_,x7_]=PA[x1]*PB[x2,x1]*PC[x3,x1]*
    PD[x4,x2,x3]*PE[x5,x3]*PF[x6,x4]*PG[x7,x4];
P1=Sum[P[x[1],x[2],x[3],x[4],x[5],x[6],x[7]],
{x[#1],0,1}]&;
P2=Sum[P[x[1],x[2],x[3],x[4],x[5],x[6],x[7]],
    {x[#1],0,1},{x[#2],0,1}]& ;
Do[
    Do[
        a=Simplify[P[x[1],x[2],x[3],x[4],x[5],x[6],x[7]]*
           P2[i,j]-P1[i]*P1[j]];
        If[a==0,Print["Eliminate link ",i,"--",j]],
    {j,i+1,7}],
{i,1,7}]
```

FIGURE 6.4. A *Mathematica* program that finds the minimal undirected I-map corresponding to the JPD in (6.13).

6.3.2 From Undirected Graphs to JPD

So far we have assumed that either the probabilistic model $p(x)$ or the corresponding dependency model M is known, hence one can always find an undirected I-map that portrays as many independencies in M as possible. In practice, however, we do not usually know $p(x)$ or M. Actually, our objectives are

1. To construct an undirected graph G that describes the dependency structure among the set of variables in X.

2. To find a corresponding probability function $p(x)$ for which G is an I-map.

6.3 Undirected Graph Dependency Models 227

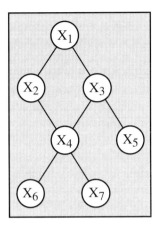

FIGURE 6.5. The minimal undirected I-map associated with the JPD given in (6.13).

```
P[x1_,x2_,x3_,x4_,x5_]=f1[x1,x2,x3]*f2[x1,x3,x4]*
f3[x1,x4,x5]
P1=Sum[P[x[1],x[2],x[3],x[4],x[5]],{x[#1],0,1}]&;
P2=Sum[P[x[1],x[2],x[3],x[4],x[5]],
    {x[#1],0,1},{x[#2],0,1}]&;
Do[
    Do[
        a=Simplify[P[x[1],x[2],x[3],x[4],x[5]]*
            P2[i,j]-P1[i]*P1[j]];
        If[a==0,Print["Eliminate link ",i,"--",j]],
    {j,i+1,5}],
{i,1,5}]
```

FIGURE 6.6. A *Mathematica* program that finds the minimal undirected I-map corresponding to the JPD in (6.14).

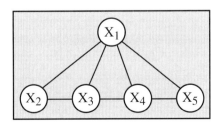

FIGURE 6.7. The minimal undirected I-map associated with the JPD in (6.14).

The dependency structure needed to build the undirected graph is given by the human experts in the particular domain of application. The rest of this section is devoted to the second task. We first need some definitions.

Definition 6.5 JPD factorized according to an undirected graph. *A JPD $p(x)$ is said to factorize according to an undirected graph G if it can be written as*

$$p(x) = \prod_{i=1}^{m} \psi_i(c_i), \qquad (6.15)$$

where $\psi_i(c_i), i = 1, \ldots, m$, are nonnegative functions; C_i, $i = 1, \ldots, m$, are the cliques of G; and c_i is a realization of C_i. The functions ψ_i are called factor potentials of the JPD, and the pair $(\{C_1, \ldots, C_m\}, \{\psi_1, \ldots, \psi_m\})$ is called a potential representation of the JPD.

This definition illustrates the idea of obtaining the probabilistic model associated with a graphical model by factorizing the JPD according to the graph. The resulting factorization encodes the local independencies portrayed in the graph.

The following theorems relate factorization according to undirected graphs and I-maps (see Lauritzen et al. (1990)).

Theorem 6.4 Factorization implications. *Given an arbitrary JPD, $p(x)$, and an undirected graph G, if $p(x)$ factorizes according to G, then G is an I-map of $p(x)$.*

Then, any independency derived from the graph also holds in $p(x)$. The above theorem implies several local properties, such as the *local Markov property*, that give us information about the local independency structure of a probabilistic model. For example, the local Markov property states that for any node $X_i \in X$ we have

$$I(X_i, X \setminus (\{X_i\} \cup Bnd(X_i)) | Bnd(X_i))_G,$$

and therefore

$$p(x_i | x \setminus x_i) = p(x_i | Bnd(X_i)),$$

where $Bnd(X_i)$ represents the boundary of the node X_i in the graph G (see Definition 4.13).

Theorem 6.5 Factorizations of nonextreme JPD. *Given a nonextreme JPD $p(x)$ and a graph G, the following two conditions are equivalent:*

- *$p(x)$ factorizes according to G.*
- *G is an I-map of $p(x)$.*

Since any nonextreme JP, $p(x)$ has a unique minimal undirected I-map, we can always factorize $p(x)$ according to its minimal undirected I-map.

6.3 Undirected Graph Dependency Models

Theorem 6.6 Factorization according to a minimal undirected I-map. *Every nonextreme probability distribution factorizes according to its associated minimal undirected I-map.*

Given an undirected graph G, the following algorithm (see, for example, Isham (1981) or Lauritzen (1982)) suggests a procedure for constructing a factorized JPD from an undirected graph.

Algorithm 6.2 JPD from an Undirected Graph G.

- **Input:** An undirected graph G.
- **Output:** A JPD, $p(x)$, having G as an I-map.

1. Identify all the cliques $\{C_1, \ldots, C_m\}$ of the graph.
2. Assign to each clique C_i a nonnegative function $\psi_i(c_i)$ (the factor potential).
3. Form the product of all the factor potentials.
4. Normalize the above function:

$$p(x_1, \ldots, x_n) = \frac{\prod_{i=1}^{m} \psi_i(c_i)}{\sum_{x_1, \ldots, x_n} \prod_{i=1}^{m} \psi_i(c_i)}. \qquad (6.16)$$

■

The results above guarantee that the undirected graph G is an I-map of the resulting JPD. However, the potential functions do not have clear physical interpretations, and their assessment is not easy. In the following, we show that triangulated undirected graphs allow us to factorize the JPD as a product of CPDs. The resulting JPDs are called *decomposable models*.

Definition 6.6 Decomposable probabilistic model. *A probabilistic model is said to be decomposable if it has a minimal I-map that is chordal.*

The running intersection property[1] allows us to obtain a factorization of the JPD from a triangulated undirected graph. Let $\{C_1, \ldots, C_m\}$ be the cliques given in an order that satisfies the running intersection property.[2] Let

$$S_i = C_i \cap (C_1 \cup \ldots \cup C_{i-1}) \qquad (6.17)$$

---

[1] See Definition 4.34.
[2] The cliques can be ordered as in Algorithm 4.3.

be the separator sets. Since $S_i \subset C_i$, define the residual sets as

$$R_i = C_i \setminus S_i. \qquad (6.18)$$

In this situation, the running intersection property guarantees that the separator sets S_i separate the residuals R_i from the sets $(C_1 \cup \ldots \cup C_{i-1}) \setminus S_i$ in the undirected graph. Since the residual R_i contains all elements in C_i that are not in $C_1 \cup \ldots \cup C_{i-1}$, we also have the CIS $I(R_i, R_1 \cup \ldots \cup R_{i-1}, |S_i)$. This fact allows us to factorize the JPD by applying the chain rule to the partition given by the residuals (see Pearl (1988) and Lauritzen and Spiegelhalter (1988)):

$$\begin{aligned} p(x_1, \ldots, x_n) &= \prod_{i=1}^{m} p(r_i | r_1, \ldots, r_{i-1}) \\ &= \prod_{i=1}^{m} p(r_i | s_i), \end{aligned} \qquad (6.19)$$

where m is the number of cliques. Note that (6.19) provides a factorization of the JPD in terms of CPDs. Thus, we get a practical procedure for obtaining the JPD associated with a triangulated undirected graph. This suggests the following theorem and algorithm.

Theorem 6.7 Decomposable models. *If $p(x)$ is decomposable according to G, then it can be written as the product of the CPDs of the residuals of the cliques of G, given its separator sets.*

Algorithm 6.3 Factorizing a Decomposable Model.

- **Input:** A triangulated undirected graph G.

- **Output:** A factorization of the JPD $p(x)$ having G as an I-map.

1. Identify all the cliques of the graph.

2. Use Algorithm 4.3 to obtain an ordering of the cliques $\{C_1, \ldots, C_m\}$ satisfying the running intersection property.

3. Calculate the separator sets $S_i = C_i \cap (C_1 \cup \ldots \cup C_{i-1})$ and the residual sets $R_i = C_i \setminus S_i$.

4. Return $p(x)$ as

$$p(x) = \prod_{i=1}^{m} p(r_i | s_i).$$

∎

Equation (6.19) indicates that the potential functions in (6.16) can be defined as $\psi_i(c_i) = p(r_i|s_i)$, $i = 1, \ldots, m$. Note also that this is an example of the factorization of the JPD we referred to in Section 5.5. This provides the answer for the case of undirected graphs to

- **Question 6.5:** Given a graph G, is there any probabilistic model P such that G is a minimal I-map of P? If yes, how do we build it?,

The answer for the case of directed graphs is given in Section 6.4.2.

An important practical implication of the above factorization is that calculation of conditional probabilities associated with a JPD $p(x_1, \ldots, x_n)$ using a factorization such as (6.19) implies an important saving in computational time. The more the factors and the fewer the variables involved in each term, the larger the savings.

Another advantage of decomposable models is that they yield easily interpretable factor potentials because the potential functions can be interpreted as CPDs.

Example 6.8 Factorizing a JPD from a triangulated graph. Given the undirected triangulated graph in Figure 6.8(a), let us apply Algorithm 6.3 to obtain a factorization of the associated decomposable JPD. The cliques associated with this graph are shown in Figure 6.8(a). For example, the ordering of the cliques $C_1 = \{X_1, X_2, X_3\}$, $C_2 = \{X_2, X_3, X_4\}$, $C_3 = \{X_3, X_5\}$, $C_4 = \{X_4, X_6\}$, $C_5 = \{X_4, X_7\}$ satisfies the running intersection property. To see this, note that $C_1 \cap C_2 = \{X_2, X_3\} \subset C_1$, $C_3 \cap (C_1 \cup C_2) = \{X_3\}$, which is contained either in C_1 or in C_2, and so on. A graphical interpretation of this property is given by a join tree. For example, Algorithm 4.4 uses the running intersection property to build a join tree associated with the triangulated graph by joining each clique with another clique containing its separator set. For example, the graph in Figure 6.8(b) shows one of the join trees associated with the undirected graph in Figure 6.8(a). Figure 6.9 shows the separator sets S_2, S_3, S_4, and S_5 for the cliques in the join tree in Figure 6.8(b).

Now, using the separators S_i and the residuals R_i, we can obtain a factorization of $p(x)$. Table 6.2 shows the corresponding sets S_i and R_i associated with clique C_i. From this table we have

$$p(x) = \prod_{i=1}^{5} p(r_i|s_i)$$

$$= p(x_1, x_2, x_3)p(x_4|x_2, x_3)p(x_5|x_3)p(x_6|x_4)p(x_7|x_4), \quad (6.20)$$

which is the JPD $p(x)$ having the undirected graph in Figure 6.8(a) as a minimal I-map. ∎

Note that if a minimal I-map G of a probabilistic model $p(x)$ is not triangulated, we can still factorize $p(x)$ as in (6.19) according to some triangulated

232   6. Graphically Specified Models

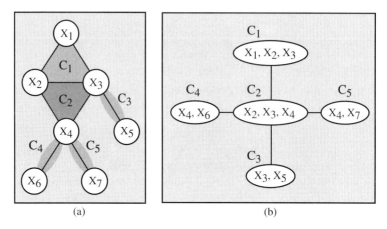

FIGURE 6.8. (a) A triangulated graph and its associated cliques, and (b) one of its associated join tree.

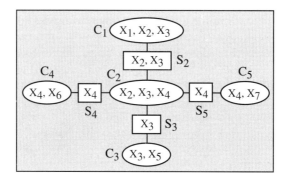

FIGURE 6.9. The separator sets corresponding to the cliques in Figure 6.8(a).

i	Clique C_i	Separator S_i	Residual R_i
1	X_1, X_2, X_3	ϕ	X_1, X_2, X_3
2	X_2, X_3, X_4	X_2, X_3	X_4
3	X_3, X_5	X_3	X_5
4	X_4, X_6	X_4	X_6
5	X_4, X_7	X_4	X_7

TABLE 6.2. Separator and residual sets associated with the cliques in Figure 6.9.

graph by losing some independency. If G is nonchordal, we can add the corresponding links to make it chordal, keeping the I-mapness character of the modified graph. Thus, to obtain a decomposable JPD from G, we first need to triangulate G if it is nontriangulated. Then we can factorize $p(x)$ accordingly. In this case, some of the conditional independencies contained

i	Clique C_i	Separator S_i	Residual R_i
1	X_1, X_2, X_3	ϕ	X_1, X_2, X_3
2	X_1, X_3, X_4	X_1, X_3	X_4
3	X_1, X_4, X_5	X_1, X_4	X_5

TABLE 6.3. Separator and residual sets associated with the cliques in (6.21).

in the original nontriangulated graph will not be included in the resulting JPD, unless we create them by a careful numerical assessment of the CPDs involved.

Example 6.9 Factorizing a JPD from a triangulated graph. Suppose that the graph in Figure 6.7 was constructed by a human expert to describe the relationships among the five variables $X = \{X_1, \ldots, X_5\}$. This graph is triangulated and contains three cliques:

$$C_1 = \{X_1, X_2, X_3\}, \quad C_2 = \{X_1, X_3, X_4\}, \quad C_3 = \{X_1, X_4, X_5\}. \quad (6.21)$$

The ordering of the cliques (C_1, C_2, C_3) satisfies the running intersection property. The separators of these cliques are $S_1 = \phi$, $S_2 = \{X_1, X_3\}$, and $S_3 = \{X_1, X_4\}$. Thus the JPD can be expressed as

$$p(x) = p(x_1, x_2, x_3) p(x_4|x_1, x_3) p(x_5|x_1, x_4), \quad (6.22)$$

which is the JPD for which the graph in Figure 6.7 is a minimal undirected I-map. Since the graph is triangulated, then the corresponding JPD is a decomposable model. Note that (6.22) has the same structure as the JPD in (6.14) that was used to construct the graph. Here we have

$$\psi_1(x_1, x_2, x_3) = p(x_1, x_2, x_3),$$
$$\psi_2(x_1, x_3, x_4) = p(x_4|x_1, x_3),$$
$$\psi_3(x_1, x_4, x_5) = p(x_5|x_1, x_4).$$

Thus, in this case we were able to recover the JPD structure from its associated undirected I-map. ∎

Because the graphs in Examples 6.8 and 6.9 are triangulated, they lead to decomposable models. The following example uses a nontriangulated graph.

Example 6.10 Finding a JPD from a nontriangulated graph. Consider the graph in Figure 6.5, whose triangulated graph is shown in Figure 6.8(a). In Example 6.8 we have shown that the JPD, which is decomposable according to this graph, can be factorized as

$$p(x_1, \ldots, x_7) = p(x_1, x_2, x_3) p(x_4|x_2, x_3) p(x_5|x_3) p(x_6|x_4) p(x_7|x_4). \quad (6.23)$$

Note that the JPD in (6.23) is derived from the triangulated graph in Figure 6.8(a), hence the original graph in Figure 6.5 is not an I-map of the JPD in (6.23) unless some constraints are imposed on the parameters to ensure that the lost independencies due to the graph triangulation are still valid in the JPD. By factorizing $p(x_1, x_2, x_3) = p(x_1)p(x_2|x_1)p(x_3|x_1, x_2)$ in (6.23), it can be shown that this model contains some extra dependencies when compared with (6.13). Thus, the required constraint for (6.23) to have the nontriangulated graph in Figure 6.5 as a minimal undirected I-map is given by $p(x_3|x_1, x_2) = p(x_3|x_1)$. ∎

6.3.3 Markov Network Models

In the previous sections we studied the relationship between undirected graphs and dependency models. In this section we present an important class of dependency models associated with undirected I-maps. This class is known as *Markov network models*.

Definition 6.7 Markov network model. *A Markov network model, or simply a Markov network, is a pair (G, Ψ) where G is an undirected graph and $\Psi = \{\psi_1(c_1), \ldots, \psi_m(c_m)\}$ is a set of positive potential functions defined on the cliques $C_1, \ldots, C_m$ of G (see Definition 6.5) that defines the JPD $p(x)$ as*

$$p(x) = \prod_{i=1}^{n} \psi_i(c_i). \tag{6.24}$$

If the undirected graph G is triangulated, then $p(x)$ can also be factorized, using probability functions $P = \{p(r_1|s_1), \ldots, p(r_m|s_m)\}$, as

$$p(x_1, \ldots, x_n) = \prod_{i=1}^{m} p(r_i|s_i), \tag{6.25}$$

where R_i and S_i are the separator and residual of the cliques as defined in (6.17) and (6.18). In this case, the Markov network model is defined by (G, P). The graph G is an undirected I-map of $p(x)$.

Thus, a Markov network can be used to define the qualitative structure of a probabilistic model through a factorization of the corresponding JPD in terms of potential functions or probability functions. The quantitative structure is then obtained by numerically specifying the functions appearing in the factorization.

Example 6.11 Markov network model. In this example, we shall build a Markov network model using the undirected triangulated graph G given in Figure 6.10(a). Figure 6.10(b) shows the cliques of this graph:

$$\begin{array}{ll} C_1 = \{A, B, C\}, & C_2 = \{B, C, E\}, \\ C_3 = \{B, D\}, & C_4 = \{C, F\}. \end{array} \tag{6.26}$$

6.3 Undirected Graph Dependency Models 235

Using (6.24), we obtain the following factorization associated with the graph:

$$\begin{aligned}p(a,b,c,d,e,f) &= \psi_1(c_1)\psi_2(c_2)\psi_3(c_3)\psi_4(c_4)\\ &= \psi_1(a,b,c)\psi_2(b,c,e)\psi_3(b,d)\psi_4(c,f). \quad (6.27)\end{aligned}$$

Therefore, a Markov network model is defined by the graph G and the set of potential functions $\Psi = \{\psi_1(a,b,c), \psi_2(b,c,e), \psi_3(b,d), \psi_4(c,f)\}$.

On the other hand, since the graph given in Figure 6.10(a) is triangulated, another factorization of the JPD in terms of probability functions can be obtained using (6.25). To obtain this factorization, we need an ordering of the cliques satisfying the running intersection property. The reader can verify that the cliques (C_1, C_2, C_3, C_4) in (6.26) satisfy the running intersection property. Table 6.4 shows the separator and residual sets obtained from the cliques (C_1, C_2, C_3, C_4) (see Figure 6.11). From this table and (6.25), we have

$$\begin{aligned}p(a,b,c,d,e,f) &= \prod_{i=1}^{4} p(r_i|s_i)\\ &= p(a,b,c)p(e|b,c)p(d|b)p(f|c). \quad (6.28)\end{aligned}$$

Thus, another way of obtaining a Markov network model associated with the graph given in Figure 6.10(a) is by specifying the probability functions $P = \{p(a,b,c), p(e|b,c), p(d|b), p(f|c)\}$. Table 6.5 shows one example of the numerical values for these probability functions. Note that each of the potential functions in (6.27) can be taken to be equal to the corresponding CPD in (6.28). Therefore, (G, Ψ) and (G, P) are two representations of the Markov network model. ∎

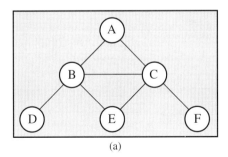
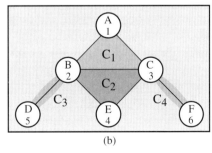

FIGURE 6.10. (a) Undirected triangulated graph and (b) its cliques.

i	Clique C_i	Separator S_i	Residual R_i
1	A, B, C	ϕ	A, B, C
2	B, C, E	B, C	E
3	B, D	B	D
4	C, F	C	F

TABLE 6.4. Separator and residual sets associated with the cliques of the graph in Figure 6.10(a).

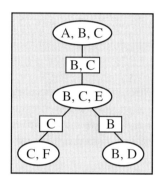

FIGURE 6.11. Join tree with separator sets.

a	b	c	$p(a,b,c)$
0	0	0	0.024
0	0	1	0.096
0	1	0	0.036
0	1	1	0.144
1	0	0	0.035
1	0	1	0.035
1	1	0	0.315
1	1	1	0.315

e	b	c	$p(e|b,c)$
0	0	0	0.4
0	0	1	0.6
0	1	0	0.5
0	1	1	0.5
1	0	0	0.7
1	0	1	0.3
1	1	0	0.2
1	1	1	0.8

f	c	$p(f|c)$
0	0	0.1
0	1	0.9
1	0	0.4
1	1	0.6

b	d	$p(d|b)$
0	0	0.3
0	1	0.7
1	0	0.2
1	1	0.8

TABLE 6.5. An example of probability functions required to define the JPD in (6.28).

6.4 Directed Graph Dependency Models

The main weakness of undirected graphs stems from their inability to represent nontransitive dependencies; two independent variables will end up being connected if there exists some other variable that depends on both. As a result, many useful independencies cannot be represented by undirected graphs. For example, in Example 6.1 we showed a simple dependency model, $M = \{I(X, Y | \phi)\}$, that cannot be represented by an undirected graph because it does not satisfy the transitivity property: we have $I(X, Y | \phi)$, but $D(X, Z | \phi)$ and $D(Y, Z | \phi)$. To overcome this deficiency, one can employ directed graphs and use the arrow directionality to distinguish between dependencies in various contexts. Thus, for example, the directed graph given in Figure 6.2 represents $I(X, Y | \phi)$, $D(X, Z | \phi)$, and $D(Y, Z | \phi)$. Hence, this graph is a perfect representation of the above nontransitive dependencies.

Another example is given in the graph shown in Figure 6.12, where happiness is thought to be determined by *health*, *wealth* and *love*. The pattern of converging arrows in the graph is interpreted as stating that health, wealth and love are unconditionally independent but may become dependent upon knowing the outcome of happiness.

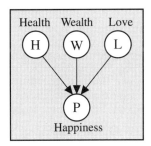

FIGURE 6.12. An example illustrating nontransitive dependencies.

Having observed that undirected graphs are insufficient to reproduce some common independencies, we move now to directed graphs with the hope of solving the problem, or at least covering a wider spectrum. Directed graphs are treated in this section in parallel to our treatment of undirected graphs in Section 6.3. We start with the problem of mapping probabilistic models using directed graphs, and then we discuss an important class of probabilistic models given by directed graphs known as *Bayesian network models*.

6.4.1 From Models to Directed Graphs

In this section we discuss the problem of mapping probabilistic models using directed graphs, that is, we wish to find a directed graph corresponding to a given probabilistic model. Theorem 6.1 gives a complete characterization for a dependency model to have an undirected perfect map. The following theorem gives only necessary conditions for a given dependency model to have a directed perfect map (see, for example, Pearl (1988)).

Theorem 6.8 Necessary conditions for directed perfect maps. *A necessary condition for a dependency model M to have a directed perfect map is that M must satisfy the following properties:*

- **Symmetry:**
$$I(X, Y|Z)_M \Leftrightarrow I(Y, X|Z)_M.$$

- **Composition-Decomposition:**
$$I(X, Y \cup W|Z)_M \Leftrightarrow I(X, Y|Z)_M \text{ and } I(X, W|Z)_M.$$

- **Intersection:**
$$I(X, W|Z \cup Y)_M \text{ and } I(X, Y|Z \cup W)_M \Rightarrow I(X, Y \cup W|Z)_M.$$

- **Weak union:**
$$I(X, Y \cup Z|W)_M \Rightarrow I(X, Y|W \cup Z)_M.$$

- **Weak transitivity:**
$$I(X, Y|Z)_M \text{ and } I(X, Y|Z \cup A)_M \Rightarrow I(X, A|Z)_M \text{ or } I(Y, A|Z)_M,$$
where A is a single node not in $\{X, Y, Z\}$.

- **Contraction:**
$$I(X, Y|Z \cup W)_M \text{ and } I(X, W|Z)_M \Rightarrow I(X, Y \cup W|Z)_M.$$

- **Chordality:**
$$I(A, B|C \cup D)_M \text{ and } I(C, D|A \cup B)_M \Rightarrow I(A, B|C)_M \text{ or } I(A, B|D)_M,$$
where A, B, C, D are single nodes.

Example 6.2 shows a dependency model M that has no directed perfect map. It can be seen, for example, that the model is not closed under the intersection property. Thus, it does not satisfy the necessary conditions to have a directed perfect map.

As we mentioned above, Theorem 6.8 provides only a necessary condition for a model to have a directed perfect map, but it is not a complete characterization, that is, there exist dependency models that satisfy all seven properties in the theorem, yet they do not have directed perfect maps. The following example, provided by Milam Studený in a personal communication, gives a dependency model with no directed perfect map.

6.4 Directed Graph Dependency Models

Example 6.12 A dependency model satisfying Theorem 6.8 but having no directed perfect map. The dependency model defined by

$$M = \{I(X,Y|Z), I(Y,X|Z), I(X,Y|W), I(Y,X|W)\} \qquad (6.29)$$

satisfies the seven properties in Theorem 6.8, but it does not have a directed perfect map. ∎

Unfortunately, probabilistic models can violate the weak transitivity, composition (see Example 5.5), and chordality properties. Hence, not all probabilistic models can be represented by directed perfect maps. However, as we shall see in Section 6.6, the violation of the chordality property is not a problem because auxiliary nodes can be added to the graph.

Thus, Theorem 6.8 provides a partial answer for the case of directed graphs to Question 6.2: What are the dependency models, and in particular, the probabilistic models that can be represented by a perfect map? Whether or not models that admit a perfect direct map can be completely characterized by a finite set of properties (like those for the undirected graphs) is an open question (see Geiger (1987)).

In cases where it is impossible to build a directed perfect map, we try to build a minimal directed I-map. Recall that a directed graph D is said to be a directed I-map of a dependency model M if $I(X,Y|Z)_D \Rightarrow I(X,Y|Z)_M$, that is, all CISs derived from D are in M. Thus, minimal directed I-maps are graphs that display every dependency in M but fail to do so if a single link is removed from them.

The following theorem, the equivalent of Theorem 6.3 for undirected graphs, gives the necessary and sufficient conditions for a dependency model M to have a minimal directed I-map (see Verma and Pearl (1990) and Lauritzen et al. (1990)).

Theorem 6.9 Minimal directed I-map of a dependency model. *Every dependency model M of a set of variables $X = \{X_1, \ldots, X_n\}$ that is a semigraphoid, i.e., satisfying the symmetry, decomposition, weak union, and contraction properties, has a minimal directed I-map. The minimal directed I-map is created by considering an arbitrary ordering of the variables $(Y_1, \ldots, Y_n)$ and designating as parents of each node Y_i any minimal set of predecessors Π_i satisfying*

$$I(Y_i, B_i \setminus \Pi_i | \Pi_i)_M, \qquad (6.30)$$

where $\Pi_i \subseteq B_i = \{Y_1, \ldots, Y_{i-1}\}$.

Example 6.13 Minimal directed I-map of a dependency model. Suppose that we have the dependency model

$$M = \{I(A,C|B), I(C,A|B)\}$$

defined on a set of three binary variables $\{A, B, C\}$. This model satisfies the four properties given in Theorem 6.9. Therefore, by considering an ordering of the variables we can build an I-map for M by calculating the sets of parents Π_i satisfying (6.30). Figure 6.13 shows all possible I-maps associated with the different orderings of the nodes. For example, given the ordering (A, B, C), we obtain the following sets of parents:

- For node A, $\Pi_A = \phi$, because it has no predecessors.
- For node B, $\Pi_B = \{A\}$, since the only predecessor is A, and $I(B, A|\phi)$ does not hold in M.
- For node C, $\Pi_C = \{B\}$, since $I(C, A|B)$ holds in M.

The resulting I-map is shown in Figure 6.13(a). Note that two different orderings of the variables can yield the same graph. For example, the graph in Figure 6.13(c) is an I-map associated with the orderings (B, A, C) and (B, C, A).

Note that the graphs in Figures 6.13(a), (c), and (e) are directed perfect maps of M, whereas the graphs in Figure 6.13(b) and (d) are not directed perfect maps, but they are minimal I-maps of M. ■

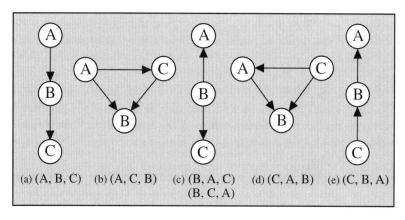

FIGURE 6.13. Minimal directed I-maps associated with the dependency model M defined in Example 6.13.

Note that any JPD satisfies the four properties given in Theorem 6.9 (any probabilistic dependency model is a semigraphoid). The following theorem gives a procedure for finding a minimal I-map for any given JPD.

Theorem 6.10 Minimal directed I-map of a JPD. *Given a permutation (ordering)* $Y = \{Y_1, \ldots, Y_n\}$ *of a set of variables* $X = \{X_1, \ldots, X_n\}$. *Let* $p(x)$ *be a JPD of* X. *The DAG created by designating as parents of each node* Y_i *any minimal set of predecessors,* Π_i *satisfying*

$$p(y_i|b_i) = p(y_i|\pi_i), \tag{6.31}$$

6.4 Directed Graph Dependency Models

for all values π_i of the variables $\Pi_i \subseteq B_i = \{Y_1, \ldots, Y_{i-1}\}$, is a directed minimal I-map of $p(x)$.

In general, the sets of minimal predecessors in the above definition do not have to be unique. The following theorem gives the necessary conditions for the uniqueness of these sets and hence, for a dependency model M to have a unique minimal directed I-map.

Theorem 6.11 Minimal I-map of a nonextreme JPD. *If the JPD $p(x)$ is nonextreme, then the parent sets $\{\Pi_1, \ldots, \Pi_n\}$ satisfying (6.31) are unique, and the minimal I-map is unique.*

Thus, Theorems 6.10 and 6.11 provide the answer for the case of directed graphs to Question 6.3: What are the dependency models that can be represented by a unique minimal I-map? The theorems also suggest the following algorithm for building a minimal I-map from a JPD, which provides the answer for the case of directed graphs to Question 6.4: If a probabilistic model P has a unique minimal I-map, how can we construct it?

Algorithm 6.4 Building a Minimal I-Map from a JPD.

- **Input:** A set of variables X and a JPD $p(x)$.
- **Output:** A minimal I-map D corresponding to the JPD $p(x)$.

1. Arrange the variables in X in any given order $(X_1, \ldots, X_n)$.

2. For each variable X_i, identify a minimal set of predecessors Π_i that renders X_i independent of $\{X_1, \ldots, X_{i-1}\} \setminus \Pi_i$.

3. Construct the directed graph D by assigning a direct link from every variable in Π_i to X_i. ∎

The resulting DAG is a directed minimal I-map of $p(x)$ in the sense that no edge can be deleted without destroying its I-map character. The following example illustrates Algorithm 6.4.

Example 6.14 Minimal I-map for a normal distribution. Consider a multivariate normal distribution of (X_1, X_2, X_3, X_4) with mean vector and covariance matrix given by

$$\mu = \begin{pmatrix} \mu_1 \\ \mu_2 \\ \mu_3 \\ \mu_4 \end{pmatrix} \quad \text{and} \quad \Sigma = \begin{pmatrix} 1 & 1/2 & 1/8 & 1/4 \\ 1/2 & 1 & 1/4 & 1/2 \\ 1/8 & 1/4 & 1 & 0 \\ 1/4 & 1/2 & 0 & 1 \end{pmatrix}.$$

From the covariance matrix it can be seen that the only pairwise independent variables are X_3 and X_4 ($\sigma_{34} = \sigma_{43} = 0$). We use Algorithm 6.4 to obtain minimal I-maps for two different orderings of the variables:

$$(X_1, X_2, X_3, X_4) \quad \text{and} \quad (X_4, X_3, X_2, X_1).$$

242   6. Graphically Specified Models

To proceed as indicated in Step 2 of the algorithm we need to know whether or not several CISs hold. To this aim we use Theorem 6.2. Table 6.6 shows the means and variances of the normal variables $X_i|\pi_i$ that appear in the I-map building process. Figure 6.14 shows a *Mathematica* program that gives the means and variances of the CPDs of a normal random variable (see Theorem 6.2).

Assuming that the mean is zero, we can calculate any of the CPDs shown in Table 6.6. For example, the conditional mean and variance of $(X_4|X_1, X_2, X_3)$, which is the first node in the table, can be obtained by

```
In:=M={0,0,0,0};
    V={{1,1/2,1/8,1/4},{1/2,1,1/4,1/2},
       {1/8,1/4,1,0},{1/4,1/2,0,1}};
    CondMedVar[4,{3,2,1},M,V]

Out:=Mean = 2 (4 x2 - x3)/15
     Variance = 11/15
```

For the ordering (X_1, X_2, X_3, X_4), we find that

$$p(x_2|x_1) \neq p(x_2); \quad p(x_3|x_2) \neq p(x_3); \quad p(x_4|x_3, x_2, x_1) = p(x_4|x_3, x_2),$$

which explains the first two columns in Table 6.7. Similarly, for the ordering (X_4, X_3, X_2, X_1), we have

$$p(x_3|x_4) = p(x_3);$$

$$p(x_2|x_3, x_4) = p(x_2|x_3) \text{ or } p(x_2|x_3, x_4) = p(x_2|x_4);$$

$$p(x_1|x_2, x_3, x_4) = p(x_1|x_2),$$

which explains the last two columns in Table 6.7. The corresponding I-maps are shown in Figure 6.15.   ∎

6.4.2   From Directed Graphs to JPD

In this section we discuss how to build a JPD from a given directed graph. When the probabilistic model or its associated dependency model M is known, one can always find a directed I-map that portrays as many independencies in M as possible. In practice, since we do not usually know M, our goals are

1. To construct a directed graph D that describes the dependency structure among the set of variables X.

2. To find the corresponding probability function $p(x)$ for which D is an I-map.

6.4 Directed Graph Dependency Models   243

```
CondMedVar[i_,CondVar_,M_,V_]:=
  Module[{
    Listvar={x1,x2,x3,x4,x5,x6,x7,x8,x9,x10},
    dim=Length[M],n=Length[CondVar]},
    w11=Array[v11,1];
    w21=Array[v21,{n,1}];
    w12=Array[v12,{1,n}];
    w22=Array[v22,{n,n}];
    wchi=Array[chi,n];
    wz=Array[variab,n];
    v11[1]=V[[i]][[i]];
    weta={M[[i]]};
    Do[
        v21[k1,1]=V[[i]][[CondVar[[k1]]]];
        chi[k1]=M[[CondVar[[k1]]]];
        variab[k1]=Listvar[[CondVar[[k1]]]],
    {k1,1,n}];
    Do[
        v22[k1,k2]=V[[CondVar[[k1]]]][[CondVar[[k2]]]],
    {k1,1,n},{k2,1,n}];
    w12=Transpose[w21];
    waux=w12.Inverse[w22];
    Mean=Simplify[weta+waux.(wz-wchi)];
    wVar=Simplify[w11-waux.w21];
    Print["Mean = ",Mean];
    Print["Variance = ",wVar]
  ]
```

FIGURE 6.14. A *Mathematica* program for obtaining the mean and variance of the CPD of the multivariate normal distribution in Example 6.14.

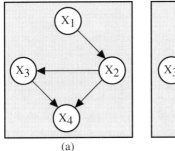
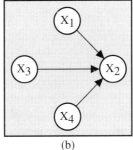

FIGURE 6.15. Minimal *I*-maps associated with the orderings (X_1, X_2, X_3, X_4) and (X_4, X_3, X_2, X_1).

244   6. Graphically Specified Models

X_i	π_i	Mean of $X_i\|\pi_i$	Variance of $X_i\|\pi_i$
X_4	$\{x_3, x_2, x_1\}$	$2(4x_2 - x_3)/15$	$11/15$
X_4	$\{x_2, x_1\}$	$x_2/2$	$3/4$
X_4	$\{x_3, x_1\}$	$2(8x_1 - x_3)/63$	$59/63$
X_4	$\{x_3, x_2\}$	$2(4x_2 - x_3)/15$	$11/15$
X_4	$\{x_1\}$	$x_1/4$	$15/16$
X_4	$\{x_2\}$	$x_2/2$	$3/4$
X_4	$\{x_3\}$	0	1
X_3	$\{x_2, x_1\}$	$x_2/4$	$15/16$
X_3	$\{x_1\}$	$x_1/8$	$63/64$
X_3	$\{x_2\}$	$x_2/4$	$15/16$
X_2	$\{x_1\}$	$x_1/2$	$3/4$
X_1	$\{x_2, x_3, x_4\}$	$x_2/2$	$3/4$
X_1	$\{x_3, x_4\}$	$x_3/8 + x_4/4$	$59/64$
X_1	$\{x_2, x_4\}$	$x_2/2$	$3/4$
X_1	$\{x_2, x_3\}$	$x_2/2$	$3/4$
X_1	$\{x_4\}$	$x_4/4$	$15/16$
X_1	$\{x_3\}$	$x_3/8$	$63/64$
X_1	$\{x_2\}$	$x_2/2$	$3/4$
X_2	$\{x_3, x_4\}$	$x_3/4 + x_4/2$	$11/16$
X_2	$\{x_4\}$	$x_4/2$	$3/4$
X_2	$\{x_3\}$	$x_3/4$	$15/16$
X_3	$\{x_4\}$	0	1

TABLE 6.6. Conditional means and variances of the normal variables $(X_i|\pi_i)$ in Example 6.14.

Ordering (X_1, X_2, X_3, X_4)		Ordering (X_4, X_3, X_2, X_1)	
X_i	Π_i	X_i	Π_i
X_1	ϕ	X_4	ϕ
X_2	$\{X_1\}$	X_3	ϕ
X_3	$\{X_2\}$	X_2	$\{X_3, X_4\}$
X_4	$\{X_3, X_2\}$	X_1	$\{X_2\}$

TABLE 6.7. Minimal sets of predecessors rendering X_i independent of all its other predecessors for two different orderings.

The dependency structure needed to build the directed graph is given by the human experts in the particular domain of application. The second goal is discussed below.

Definition 6.8 Recursive factorization of a JPD according to a DAG. *A JPD is said to admit a recursive factorization according to a*

DAG D if the JPD can be expressed as

$$p(x_1,\ldots,x_n) = \prod_{i=1}^{n} p(x_i|\pi_i), \quad (6.32)$$

where $p(x_i|\pi_i)$ is the conditional probability distribution of X_i given its parents Π_i.

Theorem 6.12 Recursive factorization. *Let D be a DAG and $p(x)$ a probabilistic model on D. The following conditions are equivalent:*

1. *$p(x)$ admits a recursive factorization according to D.*

2. *D is an I-map of $p(x)$.*

Thus, given a directed graph D, one can construct a JPD that is the product of CPDs as in (6.32). Then, D is an I-map of the resultant JPD P as given by Theorem 6.12 (see Pearl, (1988)). This is illustrated by the following example.

Example 6.15 Factorization of a JPD according to a DAG. Suppose that we have the two DAGs shown in Figure 6.15. We use the above definition to obtain the recursive factorizations of the JPD associated with these graphs. From the graph in Figure 6.15(a), the JPD can be factorized as

$$\begin{aligned} p(x_1,x_2,x_3,x_4) &= p(x_1|\pi_1)p(x_2|\pi_2)p(x_3|\pi_3)p(x_4|\pi_4) \\ &= p(x_1)p(x_2|x_1)p(x_3|x_2)p(x_4|x_2,x_3), \end{aligned}$$

whereas the DAG in Figure 6.15(b) suggests the following factorization:

$$\begin{aligned} p(x_1,x_2,x_3,x_4) &= p(x_1|\pi_1)p(x_2|\pi_2)p(x_3|\pi_3)p(x_4|\pi_4) \\ &= p(x_1)p(x_2|x_1,x_3,x_4)p(x_3)p(x_4). \end{aligned}$$

■

6.4.3 Causal Models

Although not every probabilistic model can be represented by a perfect map, many probability models have perfect maps. In this section we characterize the type of models that can be represented by perfect maps. We have the following definitions.

Definition 6.9 Causal input lists. *Let $Y = \{Y_1,\ldots,Y_n\}$ be a permutation of $X = \{X_1,\ldots,X_n\}$. A causal input list is a list that consists of n CISs, one for each variable, each of the form*

$$I(Y_i, B_i \setminus \Pi_i | \Pi_i), \quad (6.33)$$

where $B_i = \{Y_1, \ldots, Y_{i-1}\}$ is the set of predecessors of Y_i, and Π_i is a subset of B_i that renders Y_i conditionally independent of its other predecessors, $B_i \setminus \Pi_i$.

When the variables are ordered such that a cause always precedes its effect (i.e., parents come before their children), then the minimal subset of the predecessors of a variable X_i that renders it conditionally independent from all its other predecessors is thought of as the direct causes of X_i, hence the name *causal list*. A graphical representation of a causal list can be obtained by linking every direct cause X_j with the corresponding variable X_i by a link $X_j \to X_i$. All the CISs derived from the resulting graph will be satisfied by the dependency obtained by completing the causal list. If we complete the initial causal list by considering the semigraphoid properties, then the resulting dependency model will have the graph as a minimal directed I-map (see Theorem 6.9).

Definition 6.10 Causal model.[3] *A causal model is a probabilistic dependency model generated by a causal input list.*

Causal models satisfy the semigraphoid properties. Thus, any causal model has associated a minimal I-map produced by considering the DAG in which sets of parents Π_i are given by $I(Y_i, B_i \setminus \Pi_i | \Pi_i)$, where B_i is the set of predecessors of node X_i given the ordering associated with the causal list. The causal list also allows one to factorize the JPD of the probabilistic model by considering the CPDs. Any of the CISs above imply a CPD

$$p(y_i|b_i) = p(y_i|\pi_i), \quad i = 1, \ldots, n. \tag{6.34}$$

Any of these CPDs is a standard canonical component (see Section 5.5). Therefore, each CIS in a causal input list gives rise to a standard canonical component. Thus, causal input lists allow the specification of the JPD in an easy way by establishing a correspondence between a graphical representation and a standard canonical factorization of the JPD. These graphical representations are the graphical components of the probabilistic network models known as *Bayesian network models*.

Example 6.16 Generating causal lists. Suppose that a causal input list describing the dependency structure of four variables is as given in Table 6.8. This causal input list can be represented by the DAG in Figure 6.16. Both the causal input list in Table 6.8 and the DAG in Figure 6.16 suggest the following standard canonical form:

$$\begin{aligned} p(x) &= p(x_1)p(x_2|x_1)p(x_3|x_1,x_2)p(x_4|x_1,x_2,x_3) \\ &= p(x_1)p(x_2)p(x_3|x_1,x_2)p(x_4|x_1,x_3). \end{aligned} \tag{6.35}$$

---

[3]Some authors use the term *causal model* to refer to a dependency model M that has a perfect directed acyclic graph (DAG).

6.4 Directed Graph Dependency Models

Node	B_i	Π_i	$I(X_i, B_i \setminus \Pi_i | \Pi_i)$
X_1	ϕ	ϕ	$I(X_1, \phi | \phi)$
X_2	X_1	ϕ	$I(X_2, X_1 | \phi)$
X_3	X_1, X_2	X_1, X_2	$I(X_3, \phi | X_1, X_2)$
X_4	X_1, X_2, X_3	X_1, X_3	$I(X_4, X_2 | X_1, X_3)$

TABLE 6.8. An example of a causal input list.

The second equality in (6.35) follows from the first because

$$I(X_2, X_1|\phi) \Leftrightarrow p(x_2|x_1) = p(x_2)$$

and

$$I(X_4, X_2|X_1, X_3) \Leftrightarrow p(x_4|x_1, x_2, x_3) = p(x_4|x_1, x_3).$$

Consequently, to quantify the JPD of all four variables, we need only to quantify four smaller CPDs:

$$p(x_1), \quad p(x_2),$$
$$p(x_3|x_1, x_2), \quad p(x_4|x_1, x_3).$$

∎

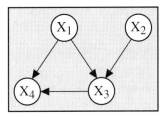

FIGURE 6.16. A DAG representing the causal input list in Table 6.8.

The most important properties of causal lists are illustrated in the following three theorems (see Verma and Pearl (1990) and Geiger and Pearl (1990)).

Theorem 6.13 Soundness of causal lists. *Let D be a DAG defined by a causal input list M. Then every CIS derived from D (using the D-separation criterion) is also satisfied by the minimal semigraphoid generated by M.*

Theorem 6.14 Closure of causal lists. *Let D be a DAG defined by a causal input list M. Then the set of graphically-verified statements is exactly the closure of M under the symmetry, decomposition, weak union, and contraction (semigraphoid) properties.*

Theorem 6.15 Completeness of causal lists. *Let D be a DAG defined by a causal input list M. Then every CIS in the minimal semigraphoid generated by M is also satisfied by D (using the D-separation criterion).*

Theorems 6.13 and 6.14 guarantee that all the CISs displayed in a DAG are also derivable from M via the semigraphoid properties. On the other hand, Theorem 6.15 assures that the DAG displays all CISs that can be derived from M using the semigraphoid properties. Thus, the symmetry, decomposition, weak union, and contraction axioms are complete, capable of deriving all valid consequences of a causal input list. Thus, the semigraphoid generated by an input list has the DAG associated with the causal list as a perfect map.

DAGs are viewed as an economical scheme for representing CISs. The nodes of a DAG represent variables in some domain of knowledge and its topology is specified by a list of CISs elicited from the human expert in this domain. The input list designates parents to each variable X by asserting that X is conditionally independent of all its predecessors given its parents (in some total order of the variables). This input list, which recursively specifies the relation of each variable to its predecessors in some order, is called a *causal list*. Dependency models generated by causal lists are called *causal models*.

6.4.4 Bayesian Network Models

Having discussed the relationship between directed graphs and dependency models, we now introduce an important class of dependency models associated with directed minimal I-maps. This class is known as *Bayesian network models*.

Definition 6.11 Bayesian network model. *A Bayesian network model, or simply a Bayesian network, is a pair (D, P), where D is a DAG, $P = \{p(x_1|\pi_1), \ldots, p(x_n|\pi_n)\}$ is a set of n CPDs, one for each variable, and Π_i is the set of parents of node X_i in D. The set P defines the associated JPD as*

$$p(x) = \prod_{i=1}^{n} p(x_i|\pi_i). \qquad (6.36)$$

The DAG D is a minimal directed I-map of $p(x)$.

Note that Theorem 6.12 shows that any CIS derived from a DAG using the D-separation criterion is also satisfied by the corresponding probabilistic model.

Note also that in the case of Bayesian networks, the factorization of the JPD is easily obtained from the associated DAG by considering a set of CPDs each of which involves only a node and its parents. On the other hand, the factorization of a JPD associated with Markov networks involves several steps such as finding the cliques of the undirected graph, ordering them according to the running intersection property, finding the separator and residual sets, etc. Therefore, it is easier to construct a probabilistic model using a Bayesian network than using a Markov network.

There are several types of Bayesian networks depending on whether the variables are discrete, continuous, or mixed, and on the assumed type of distribution for each variable. Two important types of Bayesian networks, the *multinomial* and the *normal*, or *Gaussian*, Bayesian networks are described below.

Multinomial Bayesian networks

In a multinomial Bayesian network we assume that all variables in X are discrete, that is, each variable has a finite set of possible values. We also assume that the CPD of each variable is multinomial. These CPDs can be specified either parametrically or numerically using probability tables that specify the numerical values associated with different combinations of values of the variables involved. An example of a multinomial Bayesian network is given below.

Example 6.17 Multinomial Bayesian network. Consider the DAG given in Figure 6.17 and suppose that all variables, $\{A, B, C, D, E, F, G\}$, are binary, that is, each can take only one of two possible values (e.g., 0 or 1). Given this DAG, a Bayesian network is defined by specifying the set of CPDs appearing in the factorization (6.36), which gives

$$p(a,b,c,d,e,f,g) = p(a)p(b)p(c|a)p(d|a,b)p(e)p(f|d)p(g|d,e). \quad (6.37)$$

In this case, the CPDs are tables of probabilities for the different combinations of values for the variables. An example of the numerical values needed to define the set of CPDs in (6.37) is shown in Table 6.9. Thus, the DAG in Figure 6.17 and the CPDs in Table 6.9 define a multinomial Bayesian network. ∎

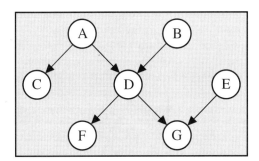

FIGURE 6.17. An example of a Bayesian network used in Example 6.17.

250   6. Graphically Specified Models

a	p(a)
0	0.3
1	0.7

b	p(b)
0	0.6
1	0.4

e	p(e)
0	0.1
1	0.9

c	a	p(c|a)
0	0	0.25
0	1	0.50
1	0	0.75
1	1	0.50

f	d	p(f|d)
0	0	0.80
0	1	0.30
1	0	0.20
1	1	0.70

d	a	b	p(d|a,b)
0	0	0	0.40
0	0	1	0.45
0	1	0	0.60
0	1	1	0.30
1	0	0	0.60
1	0	1	0.55
1	1	0	0.40
1	1	1	0.70

g	d	e	p(g|d,e)
0	0	0	0.90
0	0	1	0.70
0	1	0	0.25
0	1	1	0.15
1	0	0	0.10
1	0	1	0.30
1	1	0	0.75
1	1	1	0.85

TABLE 6.9. CPDs associated with the multinomial Bayesian network in Figure 6.17.

Gaussian Bayesian networks

In a normal Bayesian network, the variables in X are assumed to have a multivariate normal distribution, $N(\mu, \Sigma)$, whose joint probability density function is given by $N(\mu, \Sigma)$, that is,

$$f(x) = (2\pi)^{-n/2} |\Sigma|^{-1/2} \exp\left\{-1/2(x-\mu)^T \Sigma^{-1}(x-\mu)\right\}, \quad (6.38)$$

where μ is the n-dimensional mean vector, Σ is the $n \times n$ covariance matrix, $|\Sigma|$ is the determinant of Σ, and μ^T denotes the transpose of μ.

The JPD of the variables in a Gaussian Bayesian network is specified as in (6.36) by the product of a set of CPDs whose joint probability density function is given by

$$f(x_i|\pi_i) \sim N\left(\mu_i + \sum_{j=1}^{i-1} \beta_{ij}(x_j - \mu_j), v_i\right), \quad (6.39)$$

where β_{ij} is the regression coefficient of X_j in the regression of X_i on the parents of X_i, Π_i, and

$$v_i = \Sigma_i - \Sigma_{i\Pi_i} \Sigma_{\Pi_i}^{-1} \Sigma_{i\Pi_i}^T$$

is the conditional variance of X_i, given $\Pi_i = \pi_i$, where Σ_i is the unconditional variance of X_i, $\Sigma_{i\Pi_i}$ is the covariances between X_i and the variables in Π_i, and Σ_{Π_i} is the covariance matrix of Π_i. Note that β_{ij} measures the strength of the relationship between X_i and X_j. If $\beta_{ij} = 0$, then X_j is not a parent of X_i.

Note that while the conditional mean $\mu_{x_i|\pi_i}$ depends on the values of the parents π_i, the conditional variance does not depend on these values. Thus, the set of CPDs defining a normal Bayesian network is given by a collection of parameters $\{\mu_1, \ldots, \mu_n\}$, $\{v_1, \ldots, v_n\}$, and $\{\beta_{ij} \,|\, j < i\}$, as shown in (6.39).

Alternatively, we can define a normal JPD function by giving its mean μ vector and its precision matrix $W = \Sigma^{-1}$. Shachter and Kenley (1989) describe the general transformation from $\{v_1, \ldots, v_n\}$ and $\{\beta_{ij} : j < i\}$ to W. They use the following recursive formula, in which $W(i)$ denotes the $i \times i$ upper left submatrix of W and β_i denotes the column vector $\{\beta_{ij} : j < i\}$:

$$W(i+1) = \begin{pmatrix} W(i) + \dfrac{\beta_{i+1}\beta_{i+1}^T}{v_{i+1}} & \dfrac{-\beta_{i+1}}{v_{i+1}} \\ \dfrac{-\beta_{i+1}^T}{v_{i+1}} & \dfrac{1}{v_{i+1}} \end{pmatrix}, \quad (6.40)$$

with $W(1) = 1/v_1$.

Thus, we have two alternative representations of the JPD of a normal Bayesian network. The following is an illustrative example of a normal Bayesian network.

Example 6.18 Normal Bayesian network. Consider the DAG given in Figure 6.18. Suppose that the four variables, $\{A, B, C, D\}$, are normally distributed, that is, $f(a, b, c, d) \sim N(\mu, \Sigma)$. A normal Bayesian network is defined by specifying the set of CPDs appearing in the factorization (6.36), which gives

$$f(a, b, c, d) = f(a)f(b)f(c|a)f(d|a, b), \quad (6.41)$$

where

$$\begin{aligned} f(a) &\sim N(\mu_A, v_A), \\ f(b) &\sim N(\mu_B, v_B), \\ f(c) &\sim N(\mu_C + \beta_{CA}(a - \mu_A), v_C), \\ f(d) &\sim N(\mu_D + \beta_{DA}(a - \mu_A) + \beta_{DB}(b - \mu_B), v_D). \end{aligned} \quad (6.42)$$

This set of CPDs constitutes one of two equivalent representations of the normal Bayesian network. The parameters involved in this representation are $\{\mu_A, \mu_B, \mu_C, \mu_D\}$, $\{v_A, v_B, v_C, v_D\}$, and $\{\beta_{CA}, \beta_{DA}, \beta_{DB}\}$.

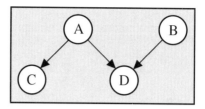

FIGURE 6.18. A Bayesian network used in Example 6.18.

An alternative representation can be obtained using (6.40). In this case, after four iterations, we finally obtain the matrix

$$W = \begin{pmatrix} \dfrac{1}{v_A} + \dfrac{\beta_{CA}^2}{v_C} + \dfrac{\beta_{DA}^2}{v_D} & \dfrac{\beta_{DA}\beta_{DB}}{v_D} & -\dfrac{\beta_{CA}}{v_C} & -\dfrac{\beta_{DA}}{v_D} \\ \dfrac{\beta_{DA}\beta_{DB}}{v_D} & \dfrac{1}{v_B} + \dfrac{\beta_{DB}^2}{v_D} & 0 & -\dfrac{\beta_{DB}}{v_D} \\ -\dfrac{\beta_{CA}}{v_C} & 0 & \dfrac{1}{v_C} & 0 \\ -\dfrac{\beta_{DA}}{v_D} & -\dfrac{\beta_{DB}}{v_D} & 0 & \dfrac{1}{v_D} \end{pmatrix}.$$

Inverting this matrix, we obtain the final covariance matrix for the JPD:

$$\Sigma = \begin{pmatrix} v_A & 0 & \beta_{CA} v_A & \beta_{DA} v_A \\ 0 & v_B & 0 & \beta_{DB} v_B \\ \beta_{CA} v_A & 0 & \beta_{CA}^2 v_A + v_C & \beta_{CA}\beta_{DA} v_A \\ \beta_{DA} v_A & \beta_{DB} v_B & \beta_{CA}\beta_{DA} v_A & \beta_{DA}^2 v_A + \beta_{DB}^2 v_B + v_D \end{pmatrix}.$$

Note that so far, all parameters have been considered in symbolic form. Thus, we can specify a Bayesian model by assigning numerical values to the parameters above. For example, setting the means equal to zero, the variances equal to one, and the link parameters $\beta_{CA} = 1$, $\beta_{DA} = 0.2$, $\beta_{DB} = 0.8$, we get the covariance matrix

$$\Sigma = \begin{pmatrix} 1.0 & 0.0 & 1.0 & 0.20 \\ 0.0 & 1.0 & 0.0 & 0.80 \\ 1.0 & 0.0 & 2.0 & 0.20 \\ 0.2 & 0.8 & 0.2 & 1.68 \end{pmatrix}. \qquad (6.43)$$

This matrix and the vector of means form a normal Bayesian network associated with the DAG in Figure 6.18. ∎

6.5 Independence Equivalent Graphical Models

Different undirected graphs lead to different dependency models. However, different DAGs can lead to the same dependency model. In this section,

6.5 Independence Equivalent Graphical Models

we characterize networks that lead to the same dependency model. We start with a definition of *independence equivalent graphs* (Verma and Pearl (1991)).

Definition 6.12 Independence equivalent of two graphs. *Two graphs are said to be independence equivalent if and only if they lead to the same dependency model, i.e., if they lead to the same set of CISs.*

Clearly, two Markov networks (undirected graphs) are independence equivalent if and only if they are identical. However, in directed graphs reversing the directionality of a link can leave the dependency structure unchanged. Accordingly, two different Bayesian networks can be independence equivalent. Thus, it is of interest to characterize Bayesian networks that lead to the same dependency model. This raises the following question:

- **Question 6.6:** How can we determine whether or not two given Bayesian networks are independence equivalent?

To answer this question, we need the following definition.

Definition 6.13 V-structure. *An ordered triplet of nodes (X, Z, Y) in a Bayesian network is said to be a v-structure iff Z has converging arrows from X and Y and there is no link between X and Y. The node Z in a v-structure is sometimes referred to as an uncoupled head-to-head node in the undirected path $X - Z - Y$.*

Note that v-structures represent the nontransitive CISs in a Bayesian network. Given a v-structure (X, Z, Y), then X and Y may be unconditionally independent, but conditionally dependent given Z. For example, the nodes (X, Z, Y) in the DAG in Figure 6.2 represent a v-structure that allows the nontransitive dependency model in Example 6.1 to be represented by a directed graph. Also, the DAG in Figure 6.12 includes three v-structures: (H, P, W), (H, P, L), and (W, P, L).

The answer to Question 6.6 is given by the following theorem (Verma and Pearl (1991)).

Theorem 6.16 Independence equivalent Bayesian networks. *Two Bayesian networks are independence equivalent iff they have: (a) the same associated undirected graphs and (b) the same v-structures.*

The following examples illustrate this theorem.

Example 6.19 Independence equivalent Bayesian networks. Consider the six different directed graphs in Figure 6.19. Graphs (a)–(c) lead to the same dependency model, since their associated undirected graphs are the same and there are no v-structures in any of them. Graphs (a)–(d) have the same associated undirected graphs, but there is a v-structure, (X, Z, Y),

254   6. Graphically Specified Models

in graph (d). Therefore, graph (d) is not independence equivalent to any of the the graphs (a)–(c).

Graphs (e) and (f), are independence equivalent since they have the same associated undirected graph and there are no v-structures. Note that (X, Z, Y) in graph (e) is not a v-structure, since nodes X and Y are connected by a link.

Thus, the six graphs given in Figure 6.19 define only three different dependency models associated with three equivalence classes of equivalent graphs: $\{(a), (b), (c)\}$, $\{(d)\}$, and $\{(e), (f)\}$, where the letters are associated with the graphs in Figure 6.19. ∎

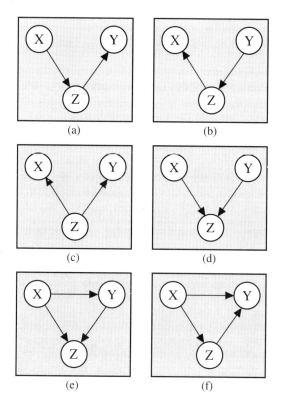

FIGURE 6.19. Six different DAGs of three nodes.

Example 6.20 The set of all complete DAGs is independence equivalent. Recall that a complete graph is a graph in which there is a link between every pair of nodes. Since all complete DAGs have the same associated undirected graph and there is no V-structure in complete DAGs, it follows that the set of all complete DAGs is an independence equivalent set. To illustrate, for a set of n variables, there are 2^n complete directed graphs. For example, for $n = 3$, the $2^3 = 8$ graphs are shown in Figure

6.20. Since we are considering only acyclic graphs, the last two graphs are cyclic, hence they are excluded. The number of complete DAGs for a set of n nodes is $n!$; one graph for each possible ordering of the variables. Thus, for $n = 3$ nodes, there are $3! = 6$ different directed graphs. The first six DAGs in Figure 6.20 correspond to the following orderings of the variables.

$$(X,Y,Z), \quad (Y,X,Z),$$
$$(Z,Y,X), \quad (Z,X,Y),$$
$$(X,Z,Y), \quad (Y,Z,X).$$

Graphs (a)–(f) lead to the same dependency model, since their associated undirected graphs coincide and there are no V-structures in any of them. This dependency model contains no CIS, that is, any two variables are conditionally dependent given the third. ∎

An important consequence of independence equivalence is that in a Bayesian network, some directed links can be reversed without changing the corresponding dependency structure. This leads to the definition of *reversible* and *irreversible* links.

Definition 6.14 Reversible and Irreversible Links. Let $D = (X, L)$ be a DAG. A link $L_{ij} = (X_i \to X_j) \in L$ is said to be an irreversible link iff $L_{ij} \in L'$ for any DAG $D' = (X, L')$ that is independence equivalent to D. Any link that is not irreversible is said to be reversible.

Note, however, that if the links in a Bayesian network have causal interpretations, then reversing a link between two variables will necessarily reverse their cause-effect relationship.

The following algorithms, due to Chickering (1995b), first order the links of a directed graph and then classify the links as either irreversible or reversible.

Algorithm 6.5 Orders the Links in a DAG.

- **Input:** A directed acyclic graph $D = (X, L)$.

- **Output:** An ordering of the links in L.

1. Perform an ancestral ordering of the nodes in D (Algorithm 4.6).

2. Set $i \leftarrow 1$.

3. While there are unordered links in L do:

    (a) Let Y be the lowest ordered node that has an unordered link incident into it.

    (b) Let X be the highest ordered node for which the link L_{XY} is not ordered.

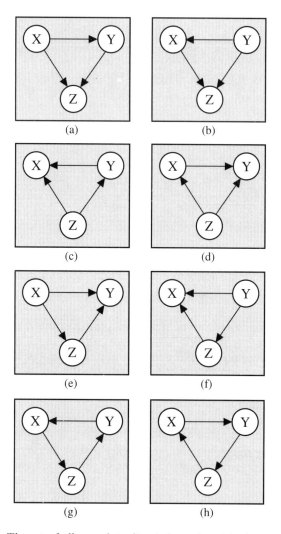

FIGURE 6.20. The set of all complete directed graphs with three nodes. Graphs (a)–(f) are independence equivalent DAGs. Graphs (g) and (h) are cyclic.

    (c) Label link L_{XY} with order i.
    (d) $i \leftarrow i + 1$. ∎

Algorithm 6.6 Find the Irreversible Links of a DAG.

- **Input:** A directed acyclic graph $D = (X, L)$.
- **Output:** Links in L classified as either *irreversible* or *reversible*.

1. Order the links in L using Algorithm 6.5.

6.5 Independence Equivalent Graphical Models

2. Label every link in L as *unknown*.

3. While there are links labeled *unknown* in L do:

    (a) Let L_{XY} be the lowest ordered link that is labeled *unknown*.

    (b) For every link L_{WX} labeled *irreversible*, if W is not a parent of Y, then label L_{XY} and every link incident into Y as *irreversible* and go to 3; otherwise, label L_{WY} as *irreversible*.

    (c) If there exists a link L_{ZY} such that $Z \neq X$ and Z is not a parent of X, then label L_{XY} and all *unknown* links incident into Y as *irreversible*; otherwise, label L_{XY} and all *unknown* links incident into Y as *reversible*. ∎

Example 6.21 Reversible and irreversible links. In Figure 6.21 we show one example of application of Algorithms 6.5 and 6.6.

First, we illustrate the use of Algorithm 6.5.

- **Step 1**: We perform an ancestral ordering of the nodes, which is shown in Figure 6.21.

- **Step 2**: We set $i = 1$.

- **Step 3**:

    - (a): The lowest ordered node that has an unordered link incident to it is $Y = C$.
    - (b): The highest ordered node for which the link L_{XC} is not ordered is $X = A$.
    - (c): We label the link L_{AC} with number 1.
    - (d): We make $i = 2$.
    - (a): The lowest ordered node that has an unordered link incident to it is $Y = D$.
    - (b): The highest ordered node for which the link L_{XD} is not ordered is $X = B$.
    - (c): We label the link L_{BD} with number 2.
    - (d): We make $i = 3$ and so on.

In Table 6.10 we summarize the substeps corresponding to Step 3. Next, we illustrate the use of Algorithm 6.6.

- **Step 1**: The first step consists of the link ordering above.

- **Step 2**: We label every link as *unknown*.

(a)	(b)	(c)	(d)
Y	X	L_{XY}	i
C	A	$Order(L_{AC}) = 1$	2
D	B	$Order(L_{BD}) = 2$	3
D	A	$Order(L_{AD}) = 3$	4
F	D	$Order(L_{DF}) = 4$	5
G	E	$Order(L_{EG}) = 5$	6
G	D	$Order(L_{DG}) = 6$	7

TABLE 6.10. Details of Step 3 of Algorithm 6.5 for the DAG in Example 6.21.

- **Step 3**:
    - (a): The lowest ordered link that is labeled *unknown* is L_{AC}.
    - (b): There is no link L_{WA} labeled *irreversible*.
    - (c): There is no link L_{ZC} such that $Z \neq A$, thus, we label L_{AC} as *reversible*.
    - (a): The lowest ordered link that is labeled *unknown* is L_{BD}.
    - (b): There is no link L_{WB} labeled *irreversible*.
    - (c): There is a link L_{AD} such that $A \neq B$ and A is not a parent of B, thus, we label L_{BD} and L_{AD} as *irreversible*.
    - (a): The lowest ordered link that is labeled *unknown* is L_{DF}.
    - (b): We have two *irreversible* links (L_{AD} and L_{BD}) of the form L_{WD}. Since A and B are not parents of F, then we label L_{DF} as *irreversible*. There are no more links incident into F.
    - (a): The lowest ordered link that is labeled *unknown* is L_{EG}.
    - (b): There is no link L_{WE} labeled *irreversible*.
    - (c): There is a link L_{DG} such that $D \neq E$ and D is not a parent of E, thus, we label L_{EG} and L_{DG} as *irreversible*.

Table 6.11 summarizes the steps corresponding to Step 3 of Algorithm 6.6. ∎

Definition 6.15 Distribution equivalent Bayesian networks. *Two Bayesian networks (D_1, P_1) and (D_2, P_2) are said to be distribution equivalent iff D_1 and D_2 are independence equivalent and $P_1 = P_2$.*

Note that distribution equivalence implies independence equivalence but the converse is not always true.

Using the notion of independence or distribution equivalence, one can divide the set of all possible Bayesian networks over a set of n variables or nodes into a number of equivalence classes. This is particularly useful in learning Bayesian networks, a topic that is covered in detail in Chapter 11.

(a) L_{XY}	(b)	(c)
$A \to C$	—	$L_{AC} \leftarrow Reversible$
$B \to D$	—	$L_{AD} \leftarrow Irreversible$
		$L_{BD} \leftarrow Irreversible$
$D \to F$	$L_{DF} \leftarrow Irreversible$	—
$E \to G$		$L_{EG} \leftarrow Irreversible$
		$L_{DG} \leftarrow Irreversible$

TABLE 6.11. Details of Step 3 of Algorithm 6.5 for the DAG in Example 6.21.

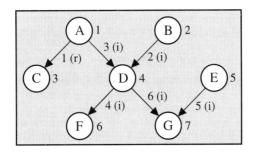

FIGURE 6.21. DAG that links have been ordered and classified as reversible (r) and irreversible (i).

6.6 Expressiveness of Graphical Models

In the previous sections we studied in detail building probabilistic models using undirected and directed graphs. We have seen that not all probabilistic models can be represented by a directed or an undirected perfect map. In cases where a perfect map does not exist, we use an I-map. Two important classes of models that can be represented by I-maps have emerged from this discussion: the Markov network models (undirected graphs) and the Bayesian network models (directed graphs). The following question then arises:

- **Question 6.7:** Can all dependency models that are representable by one type of graph also be represented by the other type?

The answer is generally no. For instance, Example 6.1 gives a case of a dependency model that can be represented by a directed perfect map but cannot be represented by an undirected perfect map. Also, the following example gives a case of a dependency model that can be represented by an undirected perfect map but cannot be represented by a directed perfect map graph.

Example 6.22 Models obtained from nonchordal graphs. The undirected graph given in Figure 6.22(a) defines an undirected dependency model formed by the set of CISs $M = \{I(X,Y|\{W,Z\}), I(W,Z|\{X,Y\})\}$. Thus, this undirected graph is a perfect map of M. However, it is not possible to find a directed perfect map reproducing the two CISs. Note that every DAG associated with this undirected graph will contain at least one node with converging arrows. This leads to a v-structure that induces some dependencies in the directed model that are not contained in the undirected model. For example, from the DAG in Figure 6.22(b), where the dotted line corresponds to the link required for the associated moral graph (see Definition 5.4), we can derive the first CIS but not the second. Note that in this case, the graph contains the v-structure (W,Y,Z), which renders W and Z conditionally dependent given Y. Similarly, from the DAG in Figure 6.22(c), we can derive the second CIS but not the first. ■

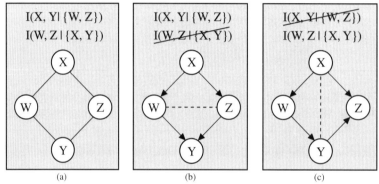

FIGURE 6.22. An undirected dependency model and two directed graphs that partially represent the model.

In general, neither of the two types of graphs is more powerful than the other (Ur and Paz (1994)). However, as we shall see shortly, by adding auxiliary nodes, directed graphs can represent models representable by undirected graphs.

From Theorems 6.1 and 6.8 we can see that some of the properties characterizing directed perfect maps are implied by those characterizing undirected perfect maps. In fact, *strong union* implies *weak union*, *strong transitivity* implies *weak transitivity*, and *strong union* and *intersection* imply *composition* and *contraction* (see Section 5.3). Consequently, every undirected dependency model satisfies the first six properties in Theorem 6.8. However, an undirected dependency model does not have necessarily to be chordal. In this case, the model cannot be represented by a directed perfect map.

This implies that for a model to be perfectly represented by both undirected and directed graphs it has to be a chordal or decomposable model.

Furthermore, if the graph is chordal, then all dependencies that can be represented by undirected graphs can also be represented by directed graphs, as given in the following theorems (see Pearl and Verma (1987) and Pearl (1988)).

Theorem 6.17 Intersection of directed and undirected graphical models. *Undirected graph dependency models and directed graph dependency models intersect in a class of dependency models representable by chordal graphs (decomposable models).*

Example 6.23 Decomposable models. Figure 6.23(a) is an undirected chordal graph. The corresponding dependency model contains one CIS, $I(X,Y|\{W,Z\})$. Then, the corresponding Markov network model can be factorized as

$$\begin{aligned} p(x,y,z,w) &= \psi_1(x,w,z)\psi_2(y,w,z) \\ &= p(x,w,z)p(y|w,z). \end{aligned} \quad (6.44)$$

On the other hand, the DAG in Figure 6.23(b) defines the following Bayesian network model factorization of the JPD:

$$\begin{aligned} p(x,y,z,w) &= p(x)p(w|x)p(z|x,w)p(y|w,z) \\ &= p(x,w,z)p(y|w,z). \end{aligned} \quad (6.45)$$

Therefore, the undirected and directed graphs in Figures 6.23(a) and (b), respectively, define the same probabilistic network model as can be seen from (6.44) and (6.45). ∎

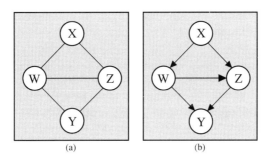

FIGURE 6.23. (a) Undirected chordal graph and (b) an associated DAG.

When an undirected dependency model is given by a nonchordal undirected graph, we can obtain an equivalent directed map with the help of some auxiliary nodes (see Pearl (1988)). Therefore, any dependency model representable by an undirected graph can also be represented by a directed graph.

Theorem 6.18 Auxiliary nodes. *Every dependency model expressible by an undirected graph is also expressible by a DAG, with the addition of some auxiliary nodes.*

Example 6.24 Auxiliary nodes. Consider the nonchordal undirected graph given in Figure 6.22(a). Example 6.22 shows that there is no DAG that generates the same dependency model. However, when we add an auxiliary node A (see Figure 6.24) we obtain a DAG that contains the same CISs as those implied by the undirected graph. Therefore, by instantiating the auxiliary nodes to arbitrary values, we obtain the same probabilistic model as the one represented by the undirected graph. ∎

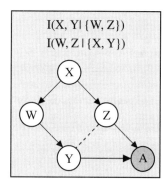

FIGURE 6.24. An illustration of how directed graphs can be augmented by auxiliary nodes and represent the same dependency models that are representable by undirected graphs.

From this discussion, one can conclude that Bayesian networks provide an intuitive, easy, and general framework for probabilistic expert systems.

Exercises

6.1 Show that the dependency model M over $\{X, Y, Z\}$ given in Example 6.2 has no directed perfect map. Consider the following steps:

- Build a list with the 18 different CISs with three variables.
- Build all possible DAGs with three variables as shown in Example 6.1 for the case of undirected graphs.
- Show that for any of those DAGs there is either a CIS in the graph that is not in M or vice versa.

Using all the possible DAGs with three variables obtained above, calculate all the dependency models with three variables that have a directed perfect map.

6.2 Consider a set of four variables $\{X_1, X_2, X_3, X_4\}$ and the following graphoid:

$I(X_3, X_1|\phi), I(X_2, X_3|X_1), I(X_2, X_3|\phi), I(X_2, X_3|X_1X_4),$
$I(X_2, X_4|X_1), I(X_2, X_4|X_1X_3), I(X_3, X_2|X_1), I(X_3, X_2|X_1X_4),$
$I(X_4, X_2|X_1), I(X_4, X_2|X_1X_3), I(X_3, X_1X_2|\phi), I(X_3, X_1|X_2),$
$I(X_3, X_2|\phi), I(X_1X_2, X_3|\phi), I(X_1, X_3|X_2), I(X_3X_4, X_2|X_1).$

Obtain the minimal undirected I-map using Theorem 6.3.

6.3 Show that the JPD in Example 6.4 does not satisfy the strong union property.

6.4 Write a computer program, similar to that in Figure 6.4, to construct a minimal undirected I-map for the JPD

$$p(x) = p(x_1)p(x_2|x_1)p(x_3|x_2, x_1)p(x_4|x_1)p(x_5|x_3, x_4). \qquad (6.46)$$

6.5 Consider the JPD in (6.46) and the graph in Figure 6.25. Then

(a) Obtain all the cliques.

(b) Factorize the joint probability of the variables in the nodes.

(c) Is the graph a minimal undirected I-map of (6.46)?

(d) Is the probability induced by (6.46) decomposable?

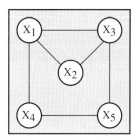

FIGURE 6.25. An undirected graph with five nodes.

6.6 Write the factorization of the JPD suggested by each of the undirected graphs in Figure 6.1.

6.7 Write the factorization of the JPD suggested by each of the undirected graphs in Figure 6.3.

6.8 Show that the directed graph in Figure 6.2 is a directed perfect map of M in Example 6.1. In this case directed graphs are more useful than undirected graphs.

6.9 Is the graph in Figure 6.26 a perfect map of the JPD

$$p(x, y, z, u) = p_1(x)p_2(y)p_3(z|x, y)p_4(w|y, z)?$$

Consider the cases

(a) $p_1(x)$ is Binomial(1, 0.3) and $p_2(y)$ is Binomial(1, 0.7).
(b) $p_3(z|x, y)$ is Binomial(1, $(x+y)/2$) and $p_4(w|z, y)$ is Binomial(1, $(y+z)/2$).

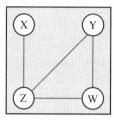

FIGURE 6.26. An undirected graph with four nodes.

6.10 Given the variables $(X_1, X_2, X_3, X_4, X_5)$ that are multivariate normally distributed with mean vector and covariance matrix given by

$$\mu = \begin{pmatrix} \mu_1 \\ \mu_2 \\ \mu_3 \\ \mu_4 \end{pmatrix} \quad \text{and} \quad \Sigma = \begin{pmatrix} 1 & 0.3 & 0 & 0.4 & 0 \\ 0.3 & 1 & 0 & 0.2 & 0 \\ 0 & 0 & 1 & 0 & 0.1 \\ 0.4 & 0.2 & 0 & 1 & 0 \\ 0 & 0 & 0.1 & 0 & 1 \end{pmatrix},$$

use Theorem 6.10, to find the DAG associated with

(a) the ordering $(X_1, X_2, X_3, X_4, X_5)$.
(b) the ordering $(X_5, X_4, X_1, X_3, X_2)$.

6.11 Consider the multivariate normal random variables (X_1, X_2, X_3, X_4) with mean vector and covariance matrix given by

$$\mu = \begin{pmatrix} \mu_1 \\ \mu_2 \\ \mu_3 \\ \mu_4 \end{pmatrix} \quad \text{and} \quad \Sigma = \begin{pmatrix} 1 & 1/2 & 1/8 & 1/4 \\ 1/2 & 1 & 1/4 & 1/2 \\ 1/8 & 1/4 & 1 & 1/4 \\ 1/4 & 1/2 & 1/4 & 1 \end{pmatrix}.$$

Use Algorithm 6.4 to obtain an I-map using the two different orderings, (X_1, X_2, X_3, X_4), and (X_4, X_3, X_2, X_1), and check whether or not the graphs in Figure 6.27 are correct.

Note: In Table 6.12 we show the conditional mean and variances of the normal variables $X_i|\pi_i$ that appear in the independency map building process.

6.6 Expressiveness of Graphical Models 265

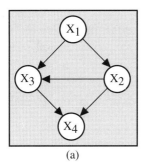
(a)

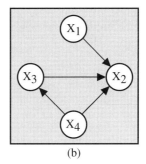
(b)

FIGURE 6.27. Graphs associated with two different orderings of the nodes (a) (X_1, X_2, X_3, X_4) and (b) (X_4, X_3, X_2, X_1).

X_i	π_i	Mean of $X_i|\pi_i$	Variance of $X_i|\pi_i$.
X_1	$\{x_2\}$	$x_2/2$	$3/4$
X_1	$\{x_3\}$	$x_3/8$	$63/64$
X_1	$\{x_4\}$	$x_4/4$	$15/16$
X_1	$\{x_2, x_3\}$	$x_2/2$	$3/4$
X_1	$\{x_2, x_4\}$	$x_2/2$	$3/4$
X_1	$\{x_3, x_4\}$	$(2x_3 + 7x_4)/30$	$14/15$
X_1	$\{x_2, x_3, x_4\}$	$x_2/2$	$3/4$
X_2	$\{x_3\}$	$x_3/4$	$15/16$
X_2	$\{x_4\}$	$x_4/2$	$3/4$
X_2	$\{x_3, x_4\}$	$(2x_3 + 7x_4)/15$	$11/15$
X_3	$\{x_4\}$	$x_4/4$	$15/16$
X_4	$\{x_3\}$	$x_3/4$	$15/16$
X_4	$\{x_2\}$	$x_2/2$	$3/4$
X_4	$\{x_1\}$	$x_1/4$	$15/16$
X_4	$\{x_3, x_2\}$	$(7x_2 + 2x_3)/15$	$11/15$
X_4	$\{x_3, x_1\}$	$2(x_1 + x_3)/9$	$3/4$
X_4	$\{x_2, x_1\}$	$x_2/2$	$3/4$
X_4	$\{x_3, x_2, x_1\}$	$(7x_2 + 2x_3)/15$	$11/15$
X_3	$\{x_2\}$	$x_2/4$	$15/16$
X_3	$\{x_1\}$	$x_1/8$	$63/64$
X_3	$\{x_2, x_1\}$	$x_2/2$	$3/4$
X_2	$\{x_1\}$	$x_1/2$	$3/4$

TABLE 6.12. Conditional means and variances of the normal variables $X_i|\pi_i$ in the previous exercise.

6.12 In Example 6.12, show the following:

(a) The model M satisfies the seven properties in Theorem 6.8.

(b) The CISs in M can be obtained from the directed graph D in Figure 6.28 using the D-separation criterion.

(c) The set of CISs $I(\{X, Z\}, Y|W)$ and $I(X, \{W, Y\}|Z)$ are found in D but not in M; hence D is not a perfect map of M.

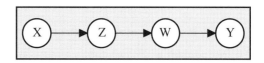

FIGURE 6.28. A directed graph that includes the CISs in M in (6.29) but has some independencies not included in M.

6.13 Write the factorization of the JPD suggested by each of the directed graphs in Figure 6.13.

6.14 Given the set of four variables $\{X, Y, Z, W\}$ and the list

$$M = \{I(Y, X|\phi), I(Z, Y|X), I(W, X|\{Y, Z\})\},$$

choose an ordering of the variables and obtain the causal list generated by M.

6.15 Given a set with five nodes $\{X_1, X_2, X_3, X_4, X_5\}$ and the causal list

$I(I(X_2, X_1|\phi),\qquad I(X_3, X_1|X_2),$
$I(X_4, X_1|\{X_2, X_3\}),\quad I(X_5, X_2|\{X_1, X_3, X_4\}),$

(a) Calculate the minimum set of independencies required for a semigraphoid.

(b) Is the obtained list also a graphoid?

6.16 Generate the factorizations associated with the undirected and the directed graphs given in Figure 6.23 and check that they are the same as the ones used in Example 6.23.

Chapter 7
Extending Graphically Specified Models

7.1 Introduction

In Chapter 6 we introduced graphically specified models using undirected or directed graphs to describe the dependency structures of probabilistic models. We have seen that not every probabilistic model can be specified by a perfect map. Thus, in general, graphs can only be thought of as minimal independence maps (I-maps), from which every conditional independence statement (CIS) derived from the graph holds in the associated probability model, though some CISs in the probability model may not be represented by the graph. Consequently, the main limitation of graphical models is that they can only represent certain types of independence structures. The following example illustrates this limitation.

Example 7.1 A model with no directed perfect map. Consider the set of variables $\{X, Y, Z\}$ that are related by the following conditional independence statements (CISs):

$$M = \{I(X, Y|Z), I(Y, X|Z), I(Y, Z|X), I(Z, Y|X)\}. \qquad (7.1)$$

Note that M includes two CISs and their symmetric counterparts. Although this model is very simple, it is not possible to find a directed acyclic graph (DAG) that is a perfect map of M. For example, using the D-separation criterion (see Section 5.2.2), the DAG in Figure 7.1(a), implies only the first two CISs. Similarly, from the DAG in Figure 7.1(b) we can only derive the last two CISs. Thus, none of the graphs is a perfect representation of the model, and thus, we can only think of them as I-maps of the dependency model M. Due to this limitation, a single graph cannot be used to specify a joint probability distribution (JPD) whose dependency structure is given by M. ∎

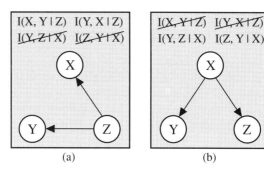

FIGURE 7.1. An example illustrating the fact that not all CISs can be simultaneously represented by a single DAG. Only two of the four CISs in (7.1) can be derived from each of the two graphs.

In this chapter we introduce some methods for extending the capabilities of graphs to represent a wider class of dependency and probabilistic models. These methods include:

1. Models specified by multiple graphs.

2. Models specified by a set of CISs (an input list).

3. Models specified by a combination of graphs and CISs.

4. Models specified by a set of conditional probability distributions.

Although these methods provide us with four different alternatives to extend graphical representations, they present some similarities. For example, using the graph separation criteria discussed in Section 5.2, one can derive the set of CISs implied by a given graph. Thus, one can convert any graph to a set of equivalent CISs, thus reducing the models in 1 and 3 above to the models in 2. On the other hand, we shall see that both models specified by multiple graphs and models specified by an input list of CISs lead to a set of factorizations of the corresponding JPD. Since these factorizations consist of a set of conditional probability distributions (CPDs), the models in 1–3 lead to the models in 4. Thus, the models in 4 provide us with the basic notions underlying all four types of model.

These models and their relationships are presented and discussed in this chapter. Sections 7.2 and 7.3 deal with models specified by multiple graphs and input lists, respectively. Section 7.4 presents multifactorized models. Two specific examples (one discrete and one continuous) of multifactorized models are given in Sections 7.5 and 7.6, respectively. Conditionally specified models are presented in Section 7.7. Finally, the issues of uniqueness and compatibility that may arise in any of these models are discussed in Sections 7.7.1 and 7.7.2, respectively.

7.2 Models Specified by Multiple Graphs

7.2.1 Definition and Example

Since not every dependency model can be represented by a single graph, graphically specified models can be extended by using multiple graphs and obtained by what we refer to as *models specified by multiple graphs*, or *multigraph models*. For example, Geiger (1987) introduced multi-DAG as a graphical representation of dependency models based on a set of DAGs. Similarly, Paz (1987) and Shachter (1990b) use sets of undirected graphs to represent general dependency models. However, Verma (1987) shows that a dependency model may require an exponential number of graphs to be perfectly represented. Therefore, from a practical point of view, multigraph models can only be used with the aim of extending models specified by single graphs. Thus, even when using multiple graphs, some independence relations in the model can escape from this representation. Hence, a multigraph can be thought of as improved I-map of a given dependency model. Note that by *multigraph* we mean the set (the union) of CISs that are implied by a set of graphs. Thus, models specified by multigraphs are, in effect, equivalent to models specified by input lists of CISs. These latter models are discussed in Section 7.3.

This simple idea of combining several graphs offers an important extension to the graphically specified models that are based on single graphs.

Definition 7.1 Multinetwork models. *Let* $X = \{X_1, \ldots, X_n\}$ *be a set of variables. A multinetwork model (MNM) over* X *is a set of compatible network models over* X

$$\{(G^\ell, P^\ell), \ell = 1, \ldots, m\}, \qquad (7.2)$$

where G^ℓ *is a Markov or Bayesian network and* P^ℓ *is the corresponding set of CPDs defining a factorization of the JPD. Compatibility requires the JPD* $p(x)$ *defined by all networks in (7.2) to be identical, that is,*

$$p(x) = \prod_{i=1}^{n} p^\ell(x_i^\ell | s_i^\ell), \quad \ell = 1, \ldots, m. \qquad (7.3)$$

The set of network models in (7.2) defines both the conditional dependency structure (given by the multigraph defined by $\{G^1, \ldots, G^m\}$) and the resulting probabilistic model (given by the different factorizations).

The resulting probabilistic models can contain a more general independence structure than models created using a single network.

Example 7.2 Multi-Bayesian network model. Let D^1 and D^2 be the DAGs given in Figures 7.1(a) and (b), respectively. Each of these graphs is a directed I-map of the dependency model M given in (7.1). The multigraph $\{D^1, D^2\}$ implies the following set of CISs:

$$M = \{I(X,Y|Z), I(Y,X|Z), I(Y,Z|X), I(Z,Y|X)\}, \qquad (7.4)$$

which is the same as M in (7.1). Note that the first two CISs in M are implied by D^1 and the last two are implied by D^2. A multi-Bayesian network model can then be defined in terms of these two graphs. Here $m = 2$, and (7.2) becomes

$$\{(D^1, P^1), (D^2, P^2)\},$$

where

$$P^1 = \{p^1(x|z), p^1(y|z), p^1(z)\},$$
$$P^2 = \{p^2(x), p^2(y|x), p^2(z|x)\}.$$

From (7.3), the two JPDs obtained from P^1 and P^2 must be identical, that is,

$$p(x,y,z) = p^1(x|z)p^1(y|z)p^1(z) = p^2(x)p^2(y|x)p^2(z|x), \qquad (7.5)$$

for the model to be consistent. The consistency problem, that is, the conditions under which (7.3) is satisfied, is addressed later in this chapter. ■

When dealing with MNM, the following questions are of interest:

- **Question 7.1:** How do we interpret conditional independence graphically?

- **Question 7.2:** Can the set of graphs $\{G^\ell, \ell = 1, \ldots, m\}$ be reduced to an equivalent smaller set?

- **Question 7.3:** How can we obtain a probabilistic model compatible with this dependency structure?

We address these questions in the next three subsections.

7.2.2 Interpreting Conditional Independence in a MNM

Question 7.1 is easy to answer. We already know that Markov and Bayesian networks are I-maps of dependency models. Thus, every conditional independence in the graph is a conditional independence in the probability model. But this is true for each of the graphs. Thus, to answer queries such as "Does the conditional independence statement $I(X,Y|Z)$ hold in

7.2 Models Specified by Multiple Graphs

a multigraph model?" we just need to use all graphs, and if any of them leads to a positive answer, the final answer will be positive; otherwise we answer negatively. Thus, a CIS holds in a multigraph model if and only if it holds in at least one of the graphs.

7.2.3 Reducing a Set of Multiple Graphs

Question 7.2 raises the problem of redundancy. In some situations, the CISs implied by one graph can also be derived from another. For example, Shachter (1990b) introduces some graphical transformations that allow simplifying the topologic structure of the graphs by eliminating redundant independences. In some cases, the set of graphs can be reduced, yielding a simpler and more efficient representation of the model. To this end, we start with the following definition.

Definition 7.2 Independence redundant graphs. *Two graphs are said to be independence redundant if the set of CISs implied by one graph is included in the set of CISs implied by the other.*

As can be seen in the next theorem, the redundancy problem in undirected graphs is easy to solve.

Theorem 7.1 Redundancy in multiple undirected graphs. *Given two undirected graphs $G^1 = (X, L^1)$ and $G^2 = (X, L^2)$ over the same set of variables in X, where L^1 and L^2 are the set of links in G^1 and G^2, respectively, if $L^1 \subset L^2$, then G^2 is redundant, that is, G^1 subsumes G^2.*

Example 7.3 Redundancy in multiple undirected graphs. Let G^1 and G^2 be the undirected graphs in Figure 7.2(a) and (b), respectively. It can be seen that G^1 subsumes G^2 because $L^1 = \{L_{12}, L_{13}, L_{34}, L_{35}\}$ is a subset of $L^2 = \{L_{12}, L_{13}, L_{34}, L_{35}, L_{24}\}$. Consequently, the set of CISs implied by both graphs is the same as the set implied by the first graph alone. ∎

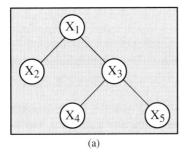 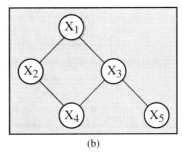

(a)   (b)

FIGURE 7.2. Two undirected graphs where graph (a) subsumes graph (b).

272   7. Extending Graphically Specified Models

The problem of redundancy in directed graphs is not trivial. The next example illustrates this point.

Example 7.4 Reducing a set of DAGs. Consider the three DAGs D^1, D^2, and D^3 given in Figures 7.3(a)–(c), respectively. All the CISs derived from D^2 can also be derived from D^1. To see this, note that if we add the link L_{24} to D^1, any CIS derived from the resulting graph can be also derived form the original graph (adding links does not introduce new CISs). Note also that the links L_{13} and L_{35} can be reversed without changing the dependency model associated with the graph. Thus, in D^1, if we add the link L_{24} and reverse the links L_{13} and L_{35}, we obtain D^2. Therefore, all CISs in the graph D^2 are contained in D^1. Consequently, the dependency model generated by the two graphs $\{D^1, D^2\}$ is equivalent to the one generated by D^1 alone.

On the other hand, D^1 and D^3 are not redundant because D^1 contains the CIS $I(X_2, X_4|X_1)$, which cannot be derived from D^3, and D^3 implies the CIS $I(X_1, X_2|X_3)$, which cannot be derived from D^1. ∎

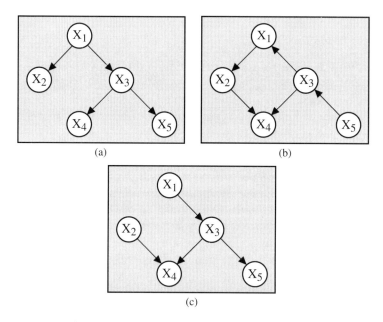

FIGURE 7.3. Three DAGs defining a multigraph.

To determine whether two directed graphs are independence redundant, we have the following theorem.

Theorem 7.2 Redundancy in multiple directed graphs. *Given two directed graphs D^1 and D^2 over the same set of variables in X, let G^1 and*

7.2 Models Specified by Multiple Graphs   273

G^2 be the corresponding underlying undirected graphs. Then D^2 is redundant, that is, D^1 subsumes D^2 if (a) G^1 subsumes G^2, (b) every v-structure in D^1 is also contained in D^2, and (c) every v-structure (X_i, X_j, X_k) in D^2 is also in D^1 whenever G^1 contains the path $X_i - X_j - X_k$.

This theorem is illustrated by the following example.

Example 7.5 Redundancy in multiple directed graphs. In Example 7.4, we have seen that D^1 subsumes D^2. We now use Theorem 7.2 to reach the same conclusion. It can be seen from Figure 7.3 that every link in G^1 (the underlying undirected graph associated with D^1) is found in G^2 (the underlying undirected graph associated with D^2), as shown in Example 7.3. Therefore, G^1 subsumes G^2 and the first condition of Theorem 7.2 holds. The second condition also holds because D^1 has no v-structure. The third condition is satisfied because D^2 has one v-structure (X_2, X_4, X_3), but G^1 does not contain the path $X_2 - X_4 - X_3$. Therefore, D^1 subsumes D^2. ■

7.2.4 Compatibility of Multiple Graphs

Question 7.3 refers to the existence of a JPD, $p(x)$, satisfying (7.3). Since every graph leads to a factorization, for this type of model, the JPD is obtained from a set of factorizations. Thus, we need to solve the compatibility problem of finding a JPD satisfying all these factorizations.

Example 7.6 Compatibility of multiple graphs. Consider again the problem introduced in Example 7.1 with the two graphs D^1 and D^2 given in Figures 7.1(a) and (b), respectively. The Bayesian network model defined by D^1 implies the following factorization:

$$p(x, y, z) = p^1(z)p^1(x|z)p^1(y|z), \qquad (7.6)$$

whereas the one defined by D^2 implies

$$p(x, y, z) = p^2(x)p^2(y|x)p^2(z|x), \qquad (7.7)$$

where the superindices indicate the different factorizations. The set of Bayesian network models $\{(D^1, P^1), (D^2, P^2)\}$ defines a multi-Bayesian network model. By combining the CISs implied by both factorizations in a probabilistic model, we may also add some extra independences derived by the properties of conditional independence (see Chapter 5). Thus, a multigraph may not be, in general, a perfect map of the resulting probabilistic models. For example, note that the multigraph defined by D^1 and D^2 consists of

$$M = \{I(X, Y|Z), I(Y, X|Z), I(Y, Z|X), I(Z, Y|X)\}. \qquad (7.8)$$

Applying the intersection property to these CISs, we obtain $I(Y, \{X, Z\}|\phi)$, which from the decomposition property leads to $I(X, Y|\phi)$ and $I(Y, Z|\phi)$.

274   7. Extending Graphically Specified Models

Therefore, the family of JPDs compatible with the multigraph defined by the two graphs in Figure 7.1 will contain at least the following CISs:

$$M_1 = \{I(X,Y|Z), I(Y,Z|X), I(Y,\{X,Z\}|\phi), I(X,Y|\phi), I(Y,Z|\phi)\}, \quad (7.9)$$

and their symmetric counterparts. By comparing M in (7.8) and M_1 in (7.9), one can see that the original multigraph M is only an I-map of M_1.

The new CISs in M_1 allow us to rewrite the factorizations given in (7.6) and (7.7) as

$$p(x,y,z) = p^1(z)p^1(x|z)p^1(y|z) = p^1(z)p^1(x|z)p^1(y) \quad (7.10)$$

and

$$p(x,y,z) = p^2(x)p^2(y|x)p^2(z|x) = p^2(x)p^2(y)p^2(z|x), \quad (7.11)$$

which are two equivalent factorizations of the same family of JPDs. These factorizations are associated with the graphs given in Figure 7.4, which are two equivalent perfect maps of the dependency model M_1 in (7.9), but not perfect maps of the original multigraph model in (7.8). Therefore, the family of JPDs compatible with both factorizations is given by (7.10) or (7.11). Note that the two graphs in Figure 7.4 have been obtained by removing the links $Z \rightarrow Y$ and $X \rightarrow Y$ from the graphs in Figure 7.1. In this case, there is a graph containing all the independences in both dependency models, from which we can obtain a factorization of a JPD compatible with both models.

The problem of compatibility here consists of finding the family of JPDs that can be factorized according to both (7.6) and (7.7). In this example, solving the compatibility problem was easy. In general, however, the compatibility problem is complicated and requires general techniques. These techniques are discussed later in this chapter. ∎

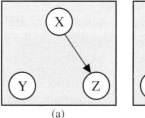
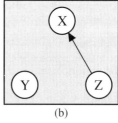

FIGURE 7.4. Two directed perfect maps of the dependency model in (7.9).

Example 7.7 Compatibility of multiple graphs. Consider the MNM given by the graphs D^1 and D^2 in Figures 7.5(a) and (b), respectively. The Bayesian network model defined by D^1 implies the following factorization:

$$p(x_1,x_2,x_3,x_4) = p^1(x_1)p^1(x_2|x_1)p^1(x_3|x_1)p^1(x_4|x_2,x_3), \quad (7.12)$$

whereas the one defined by D^2 implies

$$p(x_1, x_2, x_3, x_4) = p^2(x_1)p^2(x_2|x_1)p^2(x_4|x_2)p^2(x_3|x_1, x_4). \qquad (7.13)$$

Note that the CPDs in (7.12) and (7.13) are given according to the ancestral orderings of the variables that are implied by the corresponding graphs in Figure 7.5. Thus, the set of Bayesian network models $\{(D^1, P^1), (D^2, P^2)\}$ is a MNM (in particular a multi-Bayesian network model). Unlike Example 7.6, the problem of compatibility associated with this example is not trivial and requires general techniques that will be introduced later in this chapter. ∎

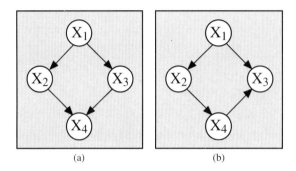

FIGURE 7.5. Example of two DAGs defining a multigraph.

Models specified by multigraphs are a special case of a more general class of models known as *multifactorized probabilistic models*. This class of model is dealt with in Section 7.4.

7.3 Models Specified by Input Lists

As we mentioned in Chapter 5, an alternative to graphs for building probabilistic models consists of using an input list of CISs. Recall that an input list is simply a list of CISs usually given by the human expert to define the relationships among a set of variables. In this section we discuss the relationship between CISs and factorizations of the JPD. This relationship can be summarized as follows:

- Given any CIS, one can always obtain a factorization that implies the given CIS.
- A factorization can imply zero, one, or more CISs.

Example 7.8 From a CIS to a factorization. Consider a set of four variables $\{X_1, X_2, X_3, X_4\}$ whose relationships satisfy $I(X_1, X_2|X_3)$. Then we express their JPD as

$$p(x_1, x_2, x_3, x_4) = p(x_2, x_3)p(x_1|x_2, x_3)p(x_4|x_1, x_2, x_3)$$
$$= p(x_2, x_3)p(x_1|x_3)p(x_4|x_1, x_2, x_3). \quad (7.14)$$

We get the first equality by applying the chain rule to $\{\{X_2, X_3\}, X_1, X_4\}$. The second equality is obtained by using the CIS $I(X_1, X_2|X_3)$, which implies that $p(x_1|x_2, x_3) = p(x_1|x_3)$. Thus, any JPD factorized according to (7.14) contains at least the CIS $I(X_1, X_2|X_3)$. Note that a JPD may also contain other CISs derived from the probability axioms (for example, the symmetric CIS $I(X_2, X_1|X_3)$). Therefore, the input list formed by this single CIS is an I-map of the resulting probability model. ∎

Note that there are some special types of input lists that define a unique factorization collectively. For example, given a set of variables $\{X_1, \ldots, X_n\}$, a causal list $\{I(Y_1, B_1 \setminus S_1|S_1), \ldots, I(Y_n, B_n \setminus S_n|S_n)\}$, where $(Y_1, \ldots, Y_n)$ is a permutation of $\{X_1, \ldots, X_n\}$ and $S_i \subset B_i = \{Y_1, \ldots, Y_{i-1}\}$, defines the factorization of the JPD

$$p(y_1, \ldots, y_n) = \prod_{i=1}^{n} p(y_i|s_i), \quad (7.15)$$

which includes all the CISs in the causal list.

Example 7.9 From a factorization to a set of CISs. Consider a set of four variables $\{X_1, X_2, X_3, X_4\}$. By the chain rule factorization, any JPD of the variables can be expressed as

$$p(x_1, x_2, x_3, x_4) = p(x_1)p(x_2|x_1)p(x_3|x_1, x_2)p(x_4|x_1, x_2, x_3). \quad (7.16)$$

This factorization implies no CISs because it is a standard canonical factorization (see Section 5.5), and therefore, it contains no independences among the variables.

On the other hand, suppose that the JPD is specified by

$$p(x_1, x_2, x_3, x_4) = p(x_1)p(x_2|x_1)p(x_3|x_1)p(x_4|x_2, x_3). \quad (7.17)$$

The factorizations in (7.16) and (7.17) imply the same ancestral ordering of the variables (X_1, X_2, X_3, X_4). Thus, we can derive the corresponding CISs by comparison as follows. No CISs are implied by the first two CPDs, $p(x_1)$ and $p(x_2|x_1)$, because they are found in both factorizations. For the third CPD, we have $p(x_3|x_1, x_2) = p(x_3|x_1)$, which implies $I(X_2, X_3|X_1)$. Finally, $p(x_4|x_1, x_2, x_3) = p(x_4|x_2, x_3)$, which implies $I(X_1, X_4|X_2, X_3)$. Therefore, the factorization in (7.17) implies the following list of CISs:

$$M_1 = \{I(X_2, X_3|X_1), I(X_1, X_4|X_2, X_3)\}. \quad (7.18)$$

7.3 Models Specified by Input Lists 277

Note that other CISs can be derived from this list using the properties of probabilistic conditional independence (using, for example, semigraphoid properties).

As a final example, suppose that the JPD is specified by the factorization

$$p(x_1, x_2, x_3, x_4) = p(x_1)p(x_2|x_1)p(x_4|x_2)p(x_3|x_1, x_4), \quad (7.19)$$

which implies the ancestral ordering of the variables (X_1, X_2, X_4, X_3). The chain rule factorization corresponding to this ordering is

$$p(x_1, x_2, x_3, x_4) = p(x_1)p(x_2|x_1)p(x_4|x_1, x_2)p(x_3|x_1, x_2, x_4). \quad (7.20)$$

By comparing (7.19) and (7.20), we obtain the following CISs:

$$\begin{aligned} p(x_1) &= p(x_1) &\Leftrightarrow &\quad \text{no CISs},\\ p(x_2|x_1) &= p(x_2|x_1) &\Leftrightarrow &\quad \text{no CISs},\\ p(x_4|x_1, x_2) &= p(x_4|x_2) &\Leftrightarrow &\quad I(X_1, X_4|X_2),\\ p(x_3|x_1, x_2, x_4) &= p(x_3|x_1, x_4) &\Leftrightarrow &\quad I(X_2, X_3|\{X_1, X_4\}). \end{aligned}$$

Thus, the factorization in (7.19) implies the following list of CISs:

$$M_2 = \{I(X_1, X_4|X_2), I(X_2, X_3|\{X_1, X_4\})\}. \quad (7.21)$$

Note that this list can be completed using the properties of conditional independence. ∎

In general, given a set of variables $\{X_1, \ldots, X_n\}$, a canonical chain rule factorization (see Definition 5.13)

$$p(y_1, \ldots, y_n) = \prod_{i=1}^{n} p(y_i|s_i), \quad (7.22)$$

where $S_i \subset B_i = \{Y_1, \ldots, Y_{i-1}\}$ and $(Y_1, \ldots, Y_n)$ is a $\{X_1, \ldots, X_n\}$ permutation, defines the following causal list:

$$\{I(Y_1, B_1 \setminus S_1|S_1), \ldots, I(Y_n, B_n \setminus S_n|S_n)\}. \quad (7.23)$$

From the above examples, we see that every CIS implies a factorization of the JPD. Thus, given an input list (a set of CISs), we can obtain a set of equivalent factorizations. This set may or may not be reduced to a single equivalent factorization. The following examples illustrate this point.

Example 7.10 Reducible set of factorizations. The input list

$$M_1 = \{I(X_2, X_3|X_1), I(X_1, X_4|\{X_2, X_3\})\} \quad (7.24)$$

is equivalent to the set of two factorizations

$$p(x_1, x_2, x_3, x_4) = p^1(x_1, x_2)p^1(x_3|x_1)p^1(x_4|x_1, x_2, x_3)$$

and
$$p(x_1, x_2, x_3, x_4) = p^2(x_1, x_2, x_3)p^2(x_4|x_2, x_3),$$
one factorization for each of the CISs in M_1, respectively, where the superindices represent the number of CISs implying the factorization. However, this set is equivalent to the single factorization
$$p(x_1, x_2, x_3, x_4) = p(x_1, x_2)p(x_3|x_1)p(x_4|x_2, x_3). \qquad (7.25)$$
This is because
$$\begin{aligned} p(x_1, x_2, x_3, x_4) &= p(x_1, x_2)p(x_3|x_1, x_2)p(x_4|x_1, x_2, x_3) \\ &= p(x_1, x_2)p(x_3|x_1)p(x_4|x_2, x_3). \end{aligned}$$
The first of the above equalities is obtained by the chain rule, and the second follows because of the two CISs in M_1. ∎

Example 7.11 Irreducible set of factorizations. Consider two input lists M_1 and M_2, where M_1 is given in (7.24) and
$$M_2 = \{I(X_1, X_4|X_2), I(X_2, X_3|\{X_1, X_4\})\}. \qquad (7.26)$$
We have seen in Example 7.10 that M_1 gives rise to the single factorization in (7.25). Similarly, M_2, leads to the single factorization
$$p(x_1, x_2, x_3, x_4) = p(x_1)p(x_2|x_1)p(x_4|x_2)p(x_3|x_1, x_4). \qquad (7.27)$$
First, note that the factorizations in (7.25) and (7.27), which are obtained from M_1 and M_2, respectively, are the same as the factorizations in (7.12) and (7.13), which are obtained from D^1 and D^2 in Figures 7.5(a) and (b), respectively. This illustrates the fact that dependency models can equivalently be specified by graphs or by input lists.

Now, suppose we wish to construct a model that satisfies both M_1 and M_2, or equivalently, both factorizations in (7.25) and (7.27). These factorizations cannot be reduced to a single factorization unless some constraints are imposed on their parameters. This leads to the compatibility problem, which requires the JPD $p(x)$ defined by the two factorizations to be identical. ∎

When a set of CISs is equivalent to a single factorization, the parameters associated with the CPDs in the factorization can be given independently, that is, without constraints. This is the case, for example, of causal lists that always lead to a single factorization of the JPD. However, when the factorizations are irreducible to a single one without imposing constraints on the parameters, we have to solve the same compatibility problem arising in models specified by multigraphs. That is, we have to obtain the constraints that the parameters of one of the factorizations have to satisfy in order for the JPD also to be factorized according to the remaining factorizations. The compatibility problem is discussed in detail later in this chapter.

7.4 Multifactorized Probabilistic Models

In the last two sections we have seen that the specification of the JPD in both multigraph models and models defined by input lists reduces to finding a JPD compatible with a set of factorizations. These two models are special cases of a more general class of models known as multifactorized probabilistic models (MFPM).

Definition 7.3 Multifactorized probabilistic models. *A multifactorized probabilistic model (MFPM), over a set of variables $X = \{X_1, \ldots, X_n\}$, is a set of compatible chain rule factorizations*

$$P = \{P^\ell, \ell = 1, \ldots, m\}, \tag{7.28}$$

where $P^\ell = \{p^\ell(y_1^\ell|s_1^\ell), \ldots, p^\ell(y_n^\ell|s_n^\ell)\}$ with $S_i^\ell \subset B_i^\ell = \{Y_1^\ell, \ldots, Y_{i-1}^\ell\}$, and $(Y_1^\ell, \ldots, Y_n^\ell)$ is a permutation of $(X_1, \ldots, X_n)$ leading to a JPD, $p(x)$, over X, that is,

$$p(x) = \prod_{i=1}^{n} p^\ell(y_i^\ell|s_i^\ell), \quad \ell = 1, \ldots, m. \tag{7.29}$$

For example, the factorizations (7.12) and (7.13) define an MFPM. The above definition raises the following question:

- **Question 7.4:** What are the conditions under which the sets of CPDs P^ℓ determine the same JPD?

This problem is known as the *consistency* or *compatibility* problem. We address this question in the case of multinomial (discrete) random variables in Section 7.5 and in the case of multivariate normal (continuous) random variables in Section 7.6.

7.5 Multifactorized Multinomial Models

To answer Question 7.4 for the case of discrete random variables, we first need to analyze the algebraic structure of probabilities implied by a factorization.

7.5.1 Parametric Structure of JPDs

Consider a set of discrete variables $\{X_1, \ldots, X_n\}$, where variable X_i can take a value in $\{0, \ldots, r_i\}$. Since the CPDs $p^\ell(y_i^\ell|s_i^\ell)$ in the factorizations of the JPD can be considered as parametric families, a convenient representation of the parameters for the ℓth CPD is given by

$$\theta_{ijs}^\ell = p(Y_i^\ell = j|S_i^\ell = s), \quad j \in \{0, \ldots, r_i^\ell\}, \tag{7.30}$$

where s is any possible instantiation of S_i^ℓ. Thus, the first subscript in θ_{ijs}^ℓ refers to the node number, the second subscript refers to the state of the node, and the remaining subscripts refer to the possible instantiations of S_i^ℓ. Since the parameters are associated with probabilities, they have to satisfy the following relationships:

$$\sum_{j=0}^{r_i^\ell} \theta_{ijs}^\ell = 1, \quad \ell = 1, \ldots, m,$$

for all i and s. Then any one of the parameters can be written as one minus the sum of all others. For example, $\theta_{ir_i s}^\ell$ is

$$\theta_{ir_i s}^\ell = 1 - \sum_{j=0}^{r_i^\ell - 1} \theta_{ijs}^\ell, \quad \ell = 1, \ldots, m. \tag{7.31}$$

We denote the set of parameters θ_{ijs}^ℓ by Θ^ℓ.

Example 7.12 Parametric structure of a JPD. Consider the multi-factorized probabilistic model defined by the two factorizations (7.12) and (7.13) obtained by the two DAGs in Figure 7.5 (see Example 7.7). These factorizations are the same as those in (7.25) and (7.27) obtained by the two input lists M_1 and M_2 in Examples 7.10 and 7.11. We can use two different sets of parameters for the probabilistic model associated with these two factorizations. For example, if all variables are binary, then each of these two factorizations has nine free parameters, which are given in Table 7.1, where $\bar{x}_i$ and x_i denote $X_i = 0$ and $X_i = 1$, respectively. The two sets of free parameters are

$$\Theta^1 = \{\theta_{10}^1, \theta_{200}^1, \theta_{201}^1, \theta_{300}^1, \theta_{301}^1, \theta_{4000}^1, \theta_{4001}^1, \theta_{4010}^1, \theta_{4011}^1\},$$
$$\Theta^2 = \{\theta_{10}^2, \theta_{200}^2, \theta_{201}^2, \theta_{3000}^2, \theta_{3001}^2, \theta_{3010}^2, \theta_{3011}^2, \theta_{400}^2, \theta_{401}^2\}.$$

Note that each factorization contains 18 parameters, but 9 of the 18 parameters are related to the 9 parameters in Table 7.1 by $\theta_{i0s}^\ell + \theta_{i1s}^\ell = 1$, for $\ell = 1, 2$ and $i = 1, \ldots, 4$. ∎

The algebraic structure of marginal and conditional probabilities as functions of the parameters is valuable information for several problems (see Castillo, Gutiérrez, and Hadi (1995c, 1996c)). We start with the case of marginal probabilities, and later we analyze the case of conditional probabilities. For the sake of simplicity we shall consider a general factorization

$$p(x_1, \ldots, x_n) = \prod_{i=1}^n p(x_i | s_i) \tag{7.32}$$

and its associated parameters throughout this section.

7.5 Multifactorized Multinomial Models

Variable	Θ^1	Θ^2
X_1	$\theta^1_{10} = p^1(\bar{x}_1)$	$\theta^2_{10} = p^2(\bar{x}_1)$
X_2	$\theta^1_{200} = p^1(\bar{x}_2\|\bar{x}_1)$	$\theta^2_{200} = p^2(\bar{x}_2\|\bar{x}_1)$
	$\theta^1_{201} = p^1(\bar{x}_2\|x_1)$	$\theta^2_{201} = p^2(\bar{x}_2\|x_1)$
X_3	$\theta^1_{300} = p^1(\bar{x}_3\|\bar{x}_1)$	$\theta^2_{3000} = p^2(\bar{x}_3\|\bar{x}_1, \bar{x}_4)$
	$\theta^1_{301} = p^1(\bar{x}_3\|x_1)$	$\theta^2_{3001} = p^2(\bar{x}_3\|\bar{x}_1, x_4)$
		$\theta^2_{3010} = p^2(\bar{x}_3\|x_1, \bar{x}_4)$
		$\theta^2_{3011} = p^2(\bar{x}_3\|x_1, x_4)$
X_4	$\theta^1_{4000} = p^1(\bar{x}_4\|\bar{x}_2, \bar{x}_3)$	$\theta^2_{400} = p^2(\bar{x}_4\|\bar{x}_2)$
	$\theta^1_{4001} = p^1(\bar{x}_4\|\bar{x}_2, x_3)$	$\theta^2_{401} = p^2(\bar{x}_4\|x_2)$
	$\theta^1_{4010} = p^1(\bar{x}_4\|x_2, \bar{x}_3)$	
	$\theta^1_{4011} = p^1(\bar{x}_4\|x_2, x_3)$	

TABLE 7.1. The set of parameters, Θ^1 and Θ^2, associated with the two factorizations in (7.12) and (7.13), respectively.

Theorem 7.3 *The probability of any instantiation $\{x_1, \ldots, x_n\}$ of the variables is a polynomial in the parameters of degree less than or equal to the number of variables. However, it is a first-degree polynomial in each parameter.*

Proof: According to (7.32) the probability of an instantiation $(x_1, \ldots, x_n)$, that is, any combination of values for all variables, is

$$p(x_1, \ldots, x_n) = \prod_{i=1}^n p(x_i|s_i) = \prod_{i=1}^n \theta_{ix_is_i}.$$

Note that all the parameters appearing in the above product are associated with different variables. Thus, $p(x_1, \ldots, x_n)$ is a monomial of degree less than or equal to the number of variables. Note also that $p(x_1, \ldots, x_n)$ may become a polynomial when only the set of free parameters is considered (see (7.31)). This simply requires replacing the parameters $\theta_{ir_is_i}$ by

$$\theta_{ir_is_i} = 1 - \prod_{j=0}^{r_i-1} \theta_{ijs_i}.$$

This substitution creates as many different monomials as the cardinality of X_i, but each of the resulting monomials is still of first degree in each parameter. ∎

The following corollary determines the algebraic structure of marginal probabilities.

Corollary 7.1 *The marginal probability of any set of nodes $Y \subset X$ is a polynomial in the parameters of degree less than or equal to the number of variables. However, it is a first-degree polynomial in each parameter.*

Proof: For simplicity, assume $Y = \{X_1, \ldots, X_r\}$. Then $p(y)$ is the sum of the probabilities of a subset of instantiations:

$$\begin{aligned}
p(y) &= p(x_1, \ldots, x_r) \\
&= \sum_{x_{r+1}, \ldots, x_n} p(x_1, \ldots, x_r, x_{r+1}, \ldots, x_n) \\
&= \sum_{x_{r+1}, \ldots, x_n} \prod_{i=1}^{n} \theta_{ix_i s_i}.
\end{aligned}$$

Therefore, the marginal probabilities of any node are also polynomials on the variables of first-degree in each parameter. ∎

Example 7.13 Structure of marginal probabilities. Given the factorization in (7.12) with the associated parameters Θ^1 shown in Table 7.1, we can compute the marginal probability of a set of nodes using the definition of marginal probability formula given in (3.4). For instance, the marginal probabilities for the single node X_2 are

$$p(X_2 = 0) = \sum_{x_1, x_3, x_4} p(x_1, 0, x_3, x_4)$$
$$= \theta_{10}^1 \theta_{200}^1 + \theta_{201}^1 - \theta_{10}^1 \theta_{201}^1.$$

and

$$p(X_2 = 1) = \sum_{x_1, x_3, x_4} p(x_1, 1, x_3, x_4)$$
$$= 1 - \theta_{10}^1 \theta_{200}^1 - \theta_{201}^1 + \theta_{10}^1 \theta_{201}^1.$$

These are polynomial expressions in the parameters of degree 2 (which is less than the number of variables). ∎

Corollary 7.2 *The posterior marginal probability of any set of nodes Y, i.e., the conditional of the set Y given some evidence $E = e$, is a ratio of two polynomial functions of the parameters. Furthermore, the denominator polynomial depends only on the evidence.*

7.5 Multifactorized Multinomial Models

Proof: We have
$$p(y|e) = \frac{p(y,e)}{p(e)}. \qquad (7.33)$$
Using Corollary 1, both the numerator and the denominator are first-degree polynomials in the parameters, since they are marginal probabilities of a subset of variables. ∎

Note that in (7.33), the denominator polynomial is the same for any conditional probability $p(y|e)$, for a given evidence set $E = e$. Then, in practical situations and for implementation purposes, it is more convenient to calculate and store only the numerator polynomials for each node and calculate the common denominator polynomial by normalization.

Example 7.14 Structure of conditional probabilities. Consider again the factorization in (7.12). The posterior probabilities of X_2 given the evidence $X_3 = 0$ can be obtained as

$$p(X_2 = 0 | X_3 = 0) = \frac{\sum_{x_1, x_4} p(x_1, 0, 0, x_4)}{\sum_{x_1, x_2, x_4} p(x_1, x_2, 0, x_4)}$$

$$= \frac{\theta_{10}\theta_{200}\theta_{300} + \theta_{201}\theta_{301} - \theta_{10}\theta_{201}\theta_{301}}{\theta_{10}\theta_{300} + \theta_{301} - \theta_{10}\theta_{301}}, \qquad (7.34)$$

and

$$p(X_2 = 1 | X_3 = 0) = \frac{\sum_{x_1, x_4} p(x_1, 1, 0, x_4)}{\sum_{x_1, x_2, x_4} p(x_1, x_2, 0, x_4)}$$

$$= \frac{\theta_{10}\theta_{300} - \theta_{10}\theta_{200}\theta_{300} + \theta_{301} - \theta_{10}\theta_{301} - \theta_{201}\theta_{301} + \theta_{10}\theta_{201}\theta_{301}}{\theta_{10}\theta_{300} + \theta_{301} - \theta_{10}\theta_{301}},$$

which are ratios of polynomial expressions in the parameters. Note that the above expressions have been obtained by directly applying the corresponding probability formulas (a brute force method). In Chapter 10 we present some methods for calculating these symbolic expressions in an efficient way (symbolic propagation). ∎

7.5.2 Solving the Compatibility Problem

The analysis of the parametric structure of probabilities introduced in the previous section can be used to solve the compatibility problem arising in MFPM, that is, to obtain the family of JPDs compatible with all factorizations in (7.29). It is worthwhile mentioning that there is at least one solution for the compatibility problem, since the complete independence case satisfies all possible CISs. Thus, we are interested in obtaining a JPD satisfying all the desired CISs, but including as few extra CISs as possible.

The idea of the method proposed by Castillo, Gutiérrez, and Hadi (1996b) is to choose one of the factorizations, say P^1, and designate it as the *reference factorization* for the JPD. The associated parameters Θ^1 are also designated as the *reference parameters*. Once the parametric probabilistic structure is specified with the reference factorization (the corresponding set of CPDs), the compatibility problem can be solved by calculating the constraints among the parameters for the JPD to factorize according to the remaining factorizations. Then, we sequentially impose the following constraints according to the remaining factorizations. Thus, for any P^ℓ, with $\ell = 2, \ldots, m$, we have

$$p^1(y_i^\ell | s_i^\ell) = p^1(y_i^\ell | b_i^\ell), \; i = 1, \ldots, n, \qquad (7.35)$$

where $B_i = \{Y_1^\ell, \ldots, Y_{i-1}^\ell\}$ and $S_i^\ell \subset B_i^\ell$. Note that the equations in (7.35) determine the constraints necessary for the parameters to satisfy the CISs $I(Y_i^\ell, B_i^\ell \setminus S_i^\ell | S_i^\ell)$, $i = 1, \ldots, n$, that is, the causal list associated with the factorization P^ℓ.

The equalities in (7.35) determine a system of equations. Each of the CPDs appearing in (7.35) is a ratio of polynomials (see Corollary 7.2). Thus, the system (7.35) is a system of polynomial equations that can be solved either simultaneously or sequentially. In the simultaneous approach we find the set of solutions by directly solving the system of equations. In the sequential approach we initially find the constraints among the parameters given by the first equation; then we impose the additional constraints on the solution set of the second equation, and so on. Note that this iterative approach can be simplified by saving in the kth step the equations associated with CISs that already hold in the dependency model defined by the first k CISs. This point is illustrated later by an example.

The above procedure is described in the following algorithm.

Algorithm 7.1 Compatibility of MFPM.

- **Input:** An MFPM $\{\{p^\ell(y_1^\ell | s_1^\ell), \ldots, p^\ell(y_n^\ell | s_n^\ell)\}, \ell = 1, \ldots, m\}$, where the parameters of the first factorization, Θ^1, are designated as the reference parameters.

- **Output:** A set of constraints to be satisfied by the reference parameters Θ^1 in order for P^1 to define a JPD of the MFPM.

1. Let $\ell \leftarrow 2$ and $Eqns = \phi$.

2. For $i \leftarrow 1, \ldots, n$ **do**:

    For every value j of Y_i^ℓ and every instantiation s of S_i^ℓ **do**:

    - Generate all possible instantiations of $B_i^\ell \setminus S_i^\ell$: $\{z_1, \ldots, z_k\}$.

7.5 Multifactorized Multinomial Models 285

- Add equations $\theta^\ell_{ijs} = p(y^\ell_i | z_1 \cup s) = \ldots = p(y^\ell_i | z_k \cup s)$ to the *Eqns* list.

3. If $\ell = m$ go to Step 4. Otherwise, set $\ell \leftarrow \ell + 1$ and go to Step 2.

4. Calculate the CPDs appearing in *Eqns* symbolically, using the reference parameters Θ^1. Solve the resulting system of polynomial equations, finding a simplified and logically equivalent set of equations providing the constraint among the parameters.

5. Return the resulting equations. ∎

Note that in Step 2, a total of $card(S^\ell_i)$ equations, each containing $|B^\ell_i \setminus S^\ell_i|$ terms, are added to the system. Each of these equations contains one non-reference parameter and several reference parameters resulting from the symbolic expression of the CPDs $p^1(y^\ell_i | z_1 \cup s)$, which are ratios of polynomials. Then, the resulting systems of equations determine the constraints among the reference parameters and the relationships among reference and nonreference parameters.

The systems of equations arising from this algorithm can be automatically reduced by using any program with symbolic computation capabilities such as *Mathematica* (Wolfram (1991), Castillo et al. (1993)) or Maple (see Char et al. (1991) and Abell and Braselton (1994)).

Algorithm 7.1 solves the compatibility problems arising in models specified by multiple graphs (Section 7.2) and models specified by input lists (Section 7.3). The following examples illustrate its application.

Example 7.15 Solving compatibility by imposing constraints. The JPD of the multi-Bayesian network model defined by the two Bayesian networks given in Figure 7.5 can be factorized as in (7.12) and (7.13), which are repeated here for convenience:

$$p(x_1, x_2, x_3, x_4) = p^1(x_1)p^1(x_2|x_1)p^1(x_3|x_1)p^1(x_4|x_2, x_3), \quad (7.36)$$
$$p(x_1, x_2, x_3, x_4) = p^2(x_1)p^2(x_2|x_1)p^2(x_4|x_2)p^2(x_3|x_1, x_4). \quad (7.37)$$

The parameters associated with both factorizations are given in Table 7.1. For the model to be consistent, the two JPDs in (7.36) and (7.37) must be identical. Thus, the question of interest is, What are the conditions under which the two JPDs in (7.36) and (7.37) are identical? In other words, what are the constraints that the parameter sets Θ^1 and Θ^2 in Table 7.1 must satisfy so that the two JPDs are the same?

To answer this question, we select one of the Bayesian networks in Figure 7.1 as the reference network, and then derive the conditions under which the other network leads to the same JPD. Let us apply Algorithm 7.1 to solve the compatibility problem in this case. Here $m = 2$. Suppose we select

(7.36) as the reference factorization. Note that the ancestral orderings[1] implied by the two factorizations are (X_1, X_2, X_3, X_4) and (X_1, X_2, X_4, X_3), respectively. Then Algorithm 7.1 proceeds as follows:

Step 1: Set $\ell = 2$ and $Eqns = \phi$. The ancestral ordering of the second factorization implies the following permutation:

$$(Y_1, Y_2, Y_3, Y_4) = (X_1, X_2, X_4, X_3).$$

Step 2: For $i = 1$, we consider $Y_1 = X_1$ and $p^2(x_1)$. Here we see that $B_1^2 = S_1^2 = \phi$. Therefore, no equations are generated. For $i = 2$, we consider $Y_2 = X_2$ and $p^2(x_2|x_1)$. We have $B_2^2 = S_2^2 = \{X_1\}$ and again, no equations are generated.

For $i = 3$, we consider $Y_3 = X_4$ and $p^2(x_4|x_2)$. We have $B_3^2 = \{X_1, X_2\}$ but $S_3^2 = \{X_2\}$. Then, for every instantiation x_1 of X_1, we have

$$\theta_{40x_1}^2 = p(X_4 = 0 | x_1, X_2 = 0) = p(X_4 = 0 | x_1, X_2 = 1), \quad x_1 = 0, 1,$$

from which we have

$$\begin{aligned}\theta_{400}^2 &= p(X_4 = 0|0,0) = p(X_4 = 0|0,1), \\ \theta_{401}^2 &= p(X_4 = 0|1,0) = p(X_4 = 0|1,1).\end{aligned} \quad (7.38)$$

We add these two equations to $Eqns$.

For $i = 4$, we consider $Y_4 = X_3$ and $p^2(x_3|x_1, x_4)$. In this case, we have $B_4^2 = \{X_1, X_2, X_4\}$ and $S_4^2 = \{X_1, X_4\}$. Thus, for every instantiation x_2 of X_2 we have

$$\theta_{3000}^2 = p(X_3 = 0 | X_1 = 0, x_2, X_4 = 0),$$

$$\theta_{3001}^2 = p(X_3 = 0 | X_1 = 0, x_2, X_4 = 1),$$

$$\theta_{3010}^2 = p(X_3 = 0 | X_1 = 1, x_2, X_4 = 0),$$

$$\theta_{3011}^2 = p(X_3 = 0 | X_1 = 1, x_2, X_4 = 1).$$

This gives the following four equations:

$$\begin{aligned}\theta_{3000}^2 &= p(X_3 = 0|0,0,0) = p(X_3 = 0|0,1,0), \\ \theta_{3001}^2 &= p(X_3 = 0|0,0,1) = p(X_3 = 0|0,1,1), \\ \theta_{3010}^2 &= p(X_3 = 0|1,0,0) = p(X_3 = 0|1,1,0), \\ \theta_{3011}^2 &= p(X_3 = 0|1,0,1) = p(X_3 = 0|1,1,1).\end{aligned} \quad (7.39)$$

---

[1] Algorithm 4.6 provides an automatic procedure to generate an ancestral ordering for a DAG.

We add these four equations to $Eqns$, which now contains six equations. Since $i = 4 = n$, Step 2 is complete.

Step 3: Since $\ell = 2 = m$, we go to Step 4.

Step 4: Calculating the CPDs in (7.38) and (7.39) symbolically, we obtain the following system of equations:

$$\begin{aligned}
\theta^2_{400} &= \theta^1_{300}\theta^1_{4000} + \theta^1_{4001} - \theta^1_{300}\theta^1_{4001} \\
&= \theta^1_{301}\theta^1_{4000} + \theta^1_{4001}(1-\theta^1_{301}), \\
\theta^2_{401} &= \theta^1_{300}\theta^1_{4010} + \theta^1_{4011}(1-\theta^1_{300}) \\
&= \theta^1_{301}\theta^1_{4010} + \theta^1_{4011}(1-\theta^1_{301}), \\
\theta^2_{3000} &= \frac{\theta^1_{300}\theta^1_{4000}}{\theta^1_{300}\theta^1_{4000}+\theta^1_{4001}(1-\theta^1_{300})} = \frac{\theta^1_{300}\theta^1_{4010}}{\theta^1_{300}\theta^1_{4010}+\theta^1_{4011}(1-\theta^1_{300})}, \\
\theta^2_{3001} &= \frac{\theta^1_{300}(1-\theta^1_{4000})}{1-\theta^1_{300}\theta^1_{4000}+\theta^1_{4001}(\theta^1_{300}-1)} = \frac{\theta^1_{300}(1-\theta^1_{4010})}{1-\theta^1_{300}\theta^1_{4010}+\theta^1_{4011}(\theta^1_{300}-1)}, \\
\theta^2_{3010} &= \frac{\theta^1_{301}\theta^1_{4000}}{\theta^1_{301}\theta^1_{4000}+\theta^1_{4001}(1-\theta^1_{301})} = \frac{\theta^1_{301}\theta^1_{4010}}{\theta^1_{301}\theta^1_{4010}+\theta^1_{4011}(1-\theta^1_{301})}, \\
\theta^2_{3011} &= \frac{\theta^1_{301}(1-\theta^1_{4000})}{1-\theta^1_{301}\theta^1_{4000}+\theta^1_{4001}(\theta^1_{301}-1)} = \frac{\theta^1_{301}(1-\theta^1_{4010})}{1-\theta^1_{301}\theta^1_{4010}+(\theta^1_{301}-1)}.
\end{aligned} \quad (7.40)$$

Note that the first two of the above equations follow from (7.38) and the last four follow from (7.39). Solving the system of equations (7.40) in the parameters Θ^1, we get the following three sets of solutions:

$$\begin{aligned}
&\text{Solution 1}: \{\theta^1_{300} = 0, \ \theta^1_{301} = 0\}, \\
&\text{Solution 2}: \{\theta^1_{300} = 1, \ \theta^1_{301} = 1\}, \\
&\text{Solution 3}: \{\theta^1_{4000} = \theta^1_{4001}, \ \theta^1_{4010} = \theta^1_{4011}\}.
\end{aligned} \quad (7.41)$$

Therefore, the class of JPDs satisfying the sets of conditional independences implied by the two graphs in Figure 7.5 is given by the parameters in Table 7.1 (Factorization (7.12)) with one of the three sets of constraints in (7.41). Note, however, that the first two solutions imply that the JPD is extreme.

Figure 7.6(a) gives a numerical example of a JPD, expressed as in factorization (7.36), which satisfies the last set of constraints in (7.41). Then, this JPD contains all the independences given by the two graphs in Figure 7.5. Once we find the reference parameters, we can obtain the associated nonreference parameters by utilizing the relationships among them given by (7.40). The numerical values of these parameters associated with the values of the reference parameters given in Figure 7.6(a) are shown in Figure 7.6(b). Thus, both Bayesian networks define the same JPD including all the independences given by the two graphs.

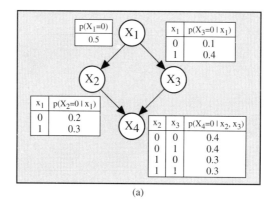

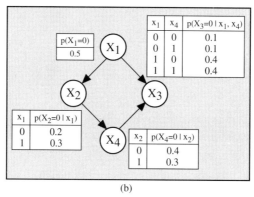

FIGURE 7.6. Two different factorizations of the same JPD, portraying all the independences given by the two Bayesian networks of Example 7.7.

In the above example, we have chosen (7.36) to be the reference factorization and obtained the constraints on its parameters so that the JPD in (7.37) is identical to that in (7.36). In a similar way, one can apply the algorithm using (7.37) as the reference factorization and find the constraints on its parameters so that the JPD in (7.36) is identical to that in (7.37). The reader can verify that in this case, the following constraints are obtained:

$$\text{Solution 1}: \quad \{\theta^2_{3010} = \theta^2_{3011}, \theta^2_{3000} = \theta^2_{3001}\},$$

$$\text{Solution 2}: \quad \{\theta^2_{400} = \theta^2_{401}, \theta^2_{3010} = \theta^2_{3000}, \theta^2_{3011} = \theta^2_{3001}\}.$$

(7.42)

The solution of the compatibility problem is given by Equations (7.41) and (7.42), which provide the constraints for the parameters Θ^1 and Θ^2, respectively. ∎

In cases where the various factorizations have many conditional independences in common, Algorithm 7.1 can be improved by taking advantage

7.5 Multifactorized Multinomial Models   289

of these independences, which results in the reduction of the number of equations to be solved. This idea is illustrated by an example.

Example 7.16 Improving the compatibility method. Consider the MFPM associated with the two Bayesian networks (D^1, P^1) and (D^2, P^2) in Figures 7.7(a) and (b), respectively. This means that the JPD of $X = \{X_1, \ldots, X_7\}$ can be factorized as

$$p(x) = p^1(x_1)p^1(x_2|x_1)p^1(x_3|x_1)p^1(x_4|x_2,x_3)p^1(x_5|x_3)p^1(x_6|x_4)p^1(x_7|x_4), \quad (7.43)$$

and

$$p(x) = p^2(x_2)p^2(x_1|x_2)p^2(x_3)p^2(x_4|x_2,x_3)p^2(x_7|x_4)p^2(x_5|x_7)p^2(x_6|x_4). \quad (7.44)$$

Note that the CPDs in (7.43) and (7.44) are given according to the ancestral orderings of the variables that are implied by the corresponding graphs in Figure 7.7. Table 7.2 shows the parameters associated with these factorizations. We wish to calculate the JPD that satisfies both factorizations.

We select (7.43) as the reference factorization. The factorization (7.44) implies the following relationships among the parameters:

$$p^2(x_2) = p(x_2) = p^1(x_2),$$
$$p^2(x_1|x_2) = p(x_1|x_2) = p^1(x_1|x_2),$$
$$p^2(x_3) = p(x_3|x_1,x_2) = p^1(x_3|x_1),$$
$$p^2(x_4|x_2,x_3) = p(x_4|x_1,x_2,x_3) = p^1(x_4|x_2,x_3),$$
$$p^2(x_7|x_4) = p(x_7|x_1,x_2,x_3,x_4) = p^1(x_7|x_4),$$
$$p^2(x_5|x_7) = p(x_5|x_1,x_2,x_3,x_4,x_7) = p^1(x_5|x_3),$$
$$p^2(x_6|x_4) = p(x_6|x_1,x_2,x_3,x_4,x_7,x_5) = p^1(x_6|x_4).$$

Thus, we have the system of equations

$$\theta^2_{20} = \theta^1_{10}(\theta^1_{200} - \theta^1_{201}) + \theta^1_{201},$$

$$\theta^2_{100} = \frac{\theta^1_{10}\theta^1_{200}}{\theta^1_{10}\theta^1_{200} + \theta^1_{201} - \theta^1_{10}\theta^1_{201}}, \quad \theta^2_{101} = \frac{\theta^1_{10}(1 - \theta^1_{200})}{1 - \theta^1_{10}\theta^1_{200} - \theta^1_{201} + \theta^1_{10}\theta^1_{201}},$$

$$\theta^2_{30} = \theta^1_{300} = \theta^1_{301},$$

$$\theta^2_{4000} = \theta^1_{4000}, \; \theta^2_{4001} = \theta^1_{4001}, \; \theta^2_{4010} = \theta^1_{4010}, \; \theta^2_{4011} = \theta^1_{4011}, \quad (7.45)$$
$$\theta^2_{700} = \theta^1_{700}, \; \theta^2_{701} = \theta^1_{701}.$$

$$\theta^2_{500} = \theta^1_{500} = \theta^1_{501}, \; \theta^2_{501} = \theta^1_{500} = \theta^1_{501},$$

$$\theta^2_{600} = \theta^1_{600}, \; \theta^2_{601} = \theta^1_{601}.$$

290   7. Extending Graphically Specified Models

If we eliminate Θ^2 in (7.45) we get the following solution:

$$\theta^1_{300} = \theta^1_{301} \quad \text{and} \quad \theta^1_{500} = \theta^1_{501}. \tag{7.46}$$

Thus, the class of JPDs satisfying the sets of conditional independences implied by the two graphs in Figure 7.7 are given by the parameters in Table 7.2 (Θ^1) with the constraints in (7.46).

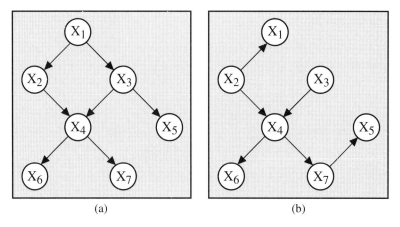

FIGURE 7.7. Two DAGs defining a multi-Bayesian network model used in Example 7.16.

On the other hand, if now we eliminate Θ^1 in (7.45) we get

$$\theta^2_{500} = \theta^2_{501}. \tag{7.47}$$

Thus, the only class of JPDs satisfying the sets of conditional independences implied by the two graphs in Figure 7.7 can also be given by the parameters in Table 7.2 (Θ^2) with the constraint in (7.47).

The previous method can be substantially improved by considering only the equations associated with CISs that cannot be derived from the previous information. Then, we can check whether the CISs associated with the factorization (7.44) already hold in the JPDs defined by the factorization (7.43), that is, we can check whether these CISs hold in the graph given in Figure 7.7(a).

The factorization in (7.44) implies the following ancestral ordering of the nodes $\{X_2, X_1, X_3, X_4, X_7, X_5, X_6\}$ and the following CISs:

7.5 Multifactorized Multinomial Models 291

Variable	Θ^1	Θ^2
X_1	$\theta^1_{10} = p^1(\bar{x}_1)$	$\theta^2_{100} = p^2(\bar{x}_1|\bar{x}_2)$
		$\theta^2_{101} = p^2(\bar{x}_1|x_2)$
X_2	$\theta^1_{200} = p^1(\bar{x}_2|\bar{x}_1)$	$\theta^2_{20} = p^2(\bar{x}_2)$
	$\theta^1_{201} = p^1(\bar{x}_2|x_1)$	
X_3	$\theta^1_{300} = p^1(\bar{x}_3|\bar{x}_1)$	$\theta^2_{30} = p^2(\bar{x}_3)$
	$\theta^1_{301} = p^1(\bar{x}_3|x_1)$	
X_4	$\theta^1_{4000} = p^1(\bar{x}_4|\bar{x}_2\bar{x}_3)$	$\theta^2_{4000} = p^2(\bar{x}_4|\bar{x}_2\bar{x}_3)$
	$\theta^1_{4001} = p^1(\bar{x}_4|\bar{x}_2 x_3)$	$\theta^2_{4001} = p^2(\bar{x}_4|\bar{x}_2 x_3)$
	$\theta^1_{4010} = p^1(\bar{x}_4|x_2\bar{x}_3)$	$\theta^2_{4010} = p^2(\bar{x}_4|x_2\bar{x}_3)$
	$\theta^1_{4011} = p^1(\bar{x}_4|x_2 x_3)$	$\theta^2_{4011} = p^2(\bar{x}_4|x_2 x_3)$
X_5	$\theta^1_{500} = p^1(\bar{x}_5|\bar{x}_3)$	$\theta^2_{500} = p^2(\bar{x}_5|\bar{x}_7)$
	$\theta^1_{501} = p^1(\bar{x}_5|x_3)$	$\theta^2_{501} = p^2(\bar{x}_5|x_7)$
X_6	$\theta^1_{600} = p^1(\bar{x}_6|\bar{x}_4)$	$\theta^2_{600} = p^2(\bar{x}_6|\bar{x}_4)$
	$\theta^1_{601} = p^1(\bar{x}_6|x_4)$	$\theta^2_{601} = p^2(\bar{x}_6|x_4)$
X_7	$\theta^1_{700} = p^1(\bar{x}_7|\bar{x}_4)$	$\theta^2_{700} = p^2(\bar{x}_7|\bar{x}_4)$
	$\theta^1_{701} = p^1(\bar{x}_7|x_4)$	$\theta^2_{601} = p^2(\bar{x}_6|x_4)$

TABLE 7.2. The set of parameters Θ^1 and Θ^2 associated with the two factorizations in (7.43) and (7.44), respectively.

$$I(X_2, \phi|\phi),$$
$$I(X_1, \phi|X_2),$$
$$I(X_3, \{X_1, X_2\}|\phi),$$
$$I(X_4, X_1|\{X_2, X_3\}),$$
$$I(X_7, \{X_1, X_2, X_3\}|X_4),$$
$$I(X_5, \{X_1, X_2, X_3, X_4\}|X_7),$$
$$I(X_6, \{X_1, X_2, X_3, X_5, X_7\}|X_4).$$

The only CISs that are not graphically verified in the graph in Figure 7.7(a) are $I(X_3, \{X_1, X_2\}|\phi)$ and $I(X_5, \{X_1, X_2, X_3, X_4\}|X_7)$. This gives the following equations:

$$\theta^2_{30} = \theta^1_{300} = \theta^1_{301}, \; \theta^2_{500} = \theta^1_{500} = \theta^1_{501}, \; \text{and} \; \theta^2_{501} = \theta^1_{500} = \theta^1_{501}.$$

This is a substantially simplified system as compared to the system in (7.45). Note that now, the solutions in (7.46) and (7.47) can be trivially obtained from this reduced system. ∎

The above improvement can be easily incorporated into Algorithm 7.1.

7.6 Multifactorized Normal Models

In this section we analyze the case of normal probabilistic models. Now we assume that the JPD of $X = \{X_1, \ldots, X_n\}$ is a multivariate normal distribution, $N(\mu, \Sigma)$, with mean vector μ and covariance matrix Σ. Thus, the set of parameters consists of the n means $\{\mu_i; i = 1, \ldots, n\}$ and the $n(n + 1)/2$ variances and covariances $\{\sigma_{ij}; i, j = 1, \ldots, n\}$. The covariance matrix Σ is independent of the mean vector μ, which is a set of location parameters. Thus, for the purpose of determining independence relationships, the only relevant parameters are the variances and covariances.

The compatibility problem associated with a set of factorizations in normal MFPMs reduces to the problem of finding the covariance matrix of a multidimensional random variable compatible with the set of CISs. Similar to the case of multinomial MFPM, discussed in Section 7.5, we can designate the covariance matrix corresponding to the first factorization as the reference parameters. Alternatively, we can start with the fully parameterized (complete) JPD and consider its variances and covariances as reference parameters. We then force the reference parameters to satisfy the CISs implied by all the factorizations in the model $P^\ell, \ell = 1, \ldots, m$ (see Definition 7.3), that is, we must have

$$I(X_i^\ell, B_i^\ell \setminus S_i^\ell | S_i^\ell) \Leftrightarrow p(x_i^\ell | b_i^\ell) = p(x_i^\ell | s_i^\ell); \; i = 1, \ldots, n, \qquad (7.48)$$

where $p(.)$ denotes the complete normal joint probability density function, and x_i^ℓ, b_i^ℓ, and s_i^ℓ are realizations of X_i^ℓ, B_i^ℓ, and S_i^ℓ, respectively. Note that this notation assumes, without loss of generality, that $(X_1^\ell, \ldots, X_n^\ell)$ are given in an ancestral ordering for every ℓ.

System (7.48) gives the set of constraints on the reference parameters to give a compatible JPD. The following theorem provides an expression for calculating these constraints.

Theorem 7.4 Conditional Independence in normal JPDs. *Let* $X = \{X_1, \ldots, X_n\}$ *be a set of random variables and* $\{V, Y, Z\}$ *a partition of* X.

7.6 Multifactorized Normal Models

Suppose the variables in X are normally distributed,

$$N\left(\begin{bmatrix} \mu_V \\ \mu_Y \\ \mu_Z \end{bmatrix}; \begin{bmatrix} \Sigma_{VV} & \Sigma_{VY} & \Sigma_{VZ} \\ \Sigma_{YV} & \Sigma_{YY} & \Sigma_{YZ} \\ \Sigma_{ZV} & \Sigma_{ZY} & \Sigma_{ZZ} \end{bmatrix}\right), \quad (7.49)$$

where we have used the block-decomposition associated with (V, Y, Z) and we assume that the covariance matrix corresponding to (V, Y) is nonsingular. Then, a necessary and sufficient condition for the conditional independence $I(V, Y|Z)$ to hold, that is, for $p(v|y, z) = p(v|z)$ is $\Sigma_{VY} = \Sigma_{VZ}\Sigma_{ZZ}^{-1}\Sigma_{ZY}$.

For the proof of this theorem and other related material see Whittaker (1990), Johnson and Wichern (1988), and Rencher (1995).

Corollary 7.3 Conditional independence in terms of the inverse of the covariance matrix. Let X be a normally distributed random variable and $\{V, Y, Z\}$ a partition of X as in Theorem 7.4. Let $W = \Sigma^{-1}$ be the precision matrix, which is the inverse of the covariance matrix Σ. Then, the CIS $I(V, Y|Z)$ holds if and only if the block W_{VY} of W is the null matrix.

For normal random variables we know that independent and uncorrelated random variables coincide. Similarly, conditionally independent random variables and partially uncorrelated random variables also coincide, as the following theorem shows.

Theorem 7.5 Conditional independence and partial correlation. Let (V, Y, Z) be normally distributed random variables. Then V and Y are uncorrelated given Z if and only if $I(V, Y|Z)$.

Next, we introduce some illustrative examples of application.

Example 7.17 Normal model specified by an input list. Consider a set of normally distributed random variables $X = \{X_1, X_2, X_3\}$ with the associated nonsingular covariance matrix

$$\Sigma = \begin{pmatrix} \sigma_{11} & \sigma_{12} & \sigma_{13} \\ \sigma_{12} & \sigma_{22} & \sigma_{23} \\ \sigma_{13} & \sigma_{23} & \sigma_{33} \end{pmatrix}. \quad (7.50)$$

Given the input list

$$M = \{I(X_1, X_2|X_3), I(X_1, X_3|X_2), I(X_2, X_3|X_1)\}, \quad (7.51)$$

what are the constraints that must be imposed on the elements of Σ so that the JPD of X satisfies the CISs in M? We start with Σ in (7.50), and find the constraints imposed by the first CIS, $I(X_1, X_2|X_3)$. With $(V, Y, Z) = (X_1, X_2, X_3)$, Theorem 7.4 gives the constraint

$$\sigma_{12} = \frac{\sigma_{13}\sigma_{32}}{\sigma_{33}}. \quad (7.52)$$

Note that because Σ is assumed to be nonsingular, $\sigma_{ii} > 0$, for $i = 1, 2, 3$. Thus, the covariance matrix that satisfies the first CIS $I(X_1, X_2 | X_3)$ must be of the form

$$\Sigma = \begin{pmatrix} \sigma_{11} & \dfrac{\sigma_{13}\sigma_{23}}{\sigma_{33}} & \sigma_{13} \\ \dfrac{\sigma_{13}\sigma_{23}}{\sigma_{33}} & \sigma_{22} & \sigma_{23} \\ \sigma_{13} & \sigma_{23} & \sigma_{33} \end{pmatrix}.$$

Similarly, the other two CISs in M imply two constraints:

$$\sigma_{13} = \dfrac{\sigma_{12}\sigma_{23}}{\sigma_{22}}, \tag{7.53}$$

$$\sigma_{23} = \dfrac{\sigma_{21}\sigma_{13}}{\sigma_{11}}. \tag{7.54}$$

Note that because of the symmetry of Σ, we have $\sigma_{ij} = \sigma_{ji}$. Thus, we have six distinct parameters subject to three constraints in (7.52), (7.53), and (7.54). Solving for the covariances in terms of the variances, we obtain the following five sets of solutions for Σ:

$$\begin{pmatrix} \sigma_{11} & 0 & 0 \\ 0 & \sigma_{22} & 0 \\ 0 & 0 & \sigma_{33} \end{pmatrix}, \begin{pmatrix} \sigma_{11} & -\delta_{12} & -\delta_{13} \\ -\delta_{12} & \sigma_{22} & \delta_{23} \\ -\delta_{13} & \delta_{23} & \sigma_{33} \end{pmatrix},$$

$$\begin{pmatrix} \sigma_{11} & \delta_{12} & \delta_{13} \\ \delta_{12} & \sigma_{22} & \delta_{23} \\ \delta_{13} & \delta_{23} & \sigma_{33} \end{pmatrix}, \begin{pmatrix} \sigma_{11} & -\delta_{12} & \delta_{13} \\ -\delta_{12} & \sigma_{22} & -\delta_{23} \\ \delta_{13} & -\delta_{23} & \sigma_{33} \end{pmatrix}, \tag{7.55}$$

$$\begin{pmatrix} \sigma_{11} & \delta_{12} & -\delta_{13} \\ \delta_{12} & \sigma_{22} & -\delta_{23} \\ -\delta_{13} & -\delta_{23} & \sigma_{33} \end{pmatrix},$$

where $\delta_{ij} = \sqrt{\sigma_{ii}\sigma_{jj}}$.

On the other hand, solving for the variances in terms of the covariances, we obtain the following two sets of solutions for Σ:

$$\begin{pmatrix} \sigma_{11} & 0 & 0 \\ 0 & \sigma_{22} & 0 \\ 0 & 0 & \sigma_{33} \end{pmatrix}, \begin{pmatrix} \dfrac{\sigma_{12}\sigma_{13}}{\sigma_{23}} & \sigma_{12} & \sigma_{13} \\ \sigma_{12} & \dfrac{\sigma_{12}\sigma_{23}}{\sigma_{13}} & \sigma_{23} \\ \sigma_{13} & \sigma_{23} & \dfrac{\sigma_{13}\sigma_{23}}{\sigma_{12}} \end{pmatrix}. \tag{7.56}$$

Note that the first solutions in (7.55) and (7.56) are the same and that every one of the last four solutions in (7.55) satisfies the second solution in (7.56). Note also that substituting any two of the three equations (7.52), (7.53), and (7.54) in the third equation, we obtain

$$\sigma_{11}\sigma_{22}\sigma_{33} = \sigma_{12}\sigma_{13}\sigma_{23}, \tag{7.57}$$

7.6 Multifactorized Normal Models

provided that the covariances are different from zero. Thus, the elements of the covariance matrix are such that the product of the variances is equal to the product of the covariances. As can be seen, this property is satisfied by the solutions in (7.55) and (7.56). We therefore conclude that the normal probabilistic model specified by the input list M in (7.51) can be defined by either one of these covariance matrices. ∎

In Example 7.17, we derived the constraints imposed on the parameters of a normal MFPM that is specified by an input list of CISs. In the following example, we derive the constraints imposed on the parameters of a normal MFPM that is specified by a set of factorizations.

Example 7.18 Multifactorized normal model. Consider a set of four normally distributed random variables $\{X_1, X_2, X_3, X_4\}$ whose JPD satisfies the following two factorizations:

$$p(x_1, x_2, x_3, x_4) = p(x_1)p(x_2|x_1)p(x_3|x_1)p(x_4|x_2, x_3), \quad (7.58)$$
$$p(x_1, x_2, x_3, x_4) = p(x_1)p(x_2|x_1)p(x_4|x_2)p(x_3|x_1, x_4). \quad (7.59)$$

Note that these are the same factorizations as those in (7.17) and (7.19). We have seen in Example 7.9 that these two factorizations lead to the two lists

$$\begin{aligned} M_1 &= \{I(X_2, X_3|X_1), I(X_1, X_4|X_2, X_3)\}, \\ M_2 &= \{I(X_1, X_4|X_2), I(X_2, X_3|X_1, X_4)\}, \end{aligned} \quad (7.60)$$

in (7.24) and (7.26), respectively. Combining M_1 and M_2, we obtain

$$M = \{I(X_2, X_3|X_1), I(X_1, X_4|X_2, X_3), I(X_1, X_4|X_2), I(X_2, X_3|X_1, X_4)\}.$$

Thus, the MFPM specified by both in (7.58) and (7.59) can also be specified by the input list M. We can then follow the same procedure as in Example 7.17 and obtain the constraints on the covariance matrix Σ that are imposed by the CISs in M.

Using Theorem 7.4, the following constraints are implied by the four CISs in M, respectively:

$$\begin{aligned} \sigma_{23} &= \frac{\sigma_{21}\sigma_{13}}{\sigma_{11}}, \\ \sigma_{14} &= \begin{pmatrix} \sigma_{12} & \sigma_{13} \end{pmatrix} \begin{pmatrix} \sigma_{22} & \sigma_{23} \\ \sigma_{32} & \sigma_{33} \end{pmatrix}^{-1} \begin{pmatrix} \sigma_{24} \\ \sigma_{34} \end{pmatrix}, \\ \sigma_{14} &= \frac{\sigma_{12}\sigma_{24}}{\sigma_{22}}, \\ \sigma_{23} &= \begin{pmatrix} \sigma_{21} & \sigma_{24} \end{pmatrix} \begin{pmatrix} \sigma_{11} & \sigma_{14} \\ \sigma_{41} & \sigma_{44} \end{pmatrix}^{-1} \begin{pmatrix} \sigma_{13} \\ \sigma_{43} \end{pmatrix}. \end{aligned} \quad (7.61)$$

Using a symbolic program such as *Mathematica* to solve the system of equations in (7.61), we obtain the following solution for the covariance matrix:

$$\begin{pmatrix} \sigma_{11} & \sigma_{12} & \sigma_{13} & \dfrac{\sigma_{11}\sigma_{34}}{\sigma_{13}} \\ \sigma_{12} & \sigma_{22} & \dfrac{\sigma_{12}\sigma_{13}}{\sigma_{11}} & \dfrac{\sigma_{12}\sigma_{13}\sigma_{34}}{\sigma_{11}\sigma_{22}} \\ \sigma_{13} & \dfrac{\sigma_{12}\sigma_{13}}{\sigma_{11}} & \sigma_{33} & \sigma_{34} \\ \dfrac{\sigma_{11}\sigma_{34}}{\sigma_{13}} & \dfrac{\sigma_{12}\sigma_{13}\sigma_{34}}{\sigma_{11}\sigma_{22}} & \sigma_{34} & \sigma_{44} \end{pmatrix}. \quad (7.62)$$

Thus, the normal probabilistic model associated with the MFPM specified by the two factorizations in (7.58) and (7.59) is given by the covariance matrix in (7.62). ■

In Examples 7.17 and 7.18, we derived the constraints imposed on the parameters of normal MFPMs that are specified by either an input list of CISs or by a set of factorizations. In the following and final example, we derive the constraints imposed on the parameters of a normal multigraph model.

Example 7.19 Multigraph normal model. Consider the MFPM given in Example 7.16 with the two Bayesian networks (D^1, P^1) and (D^2, P^2) shown in Figures 7.7(a) and (b), respectively. Suppose that the variables are normally distributed. In this example we calculate the covariance matrix of the probabilistic normal model specified by the two Bayesian networks. Using the D-separation criterion (see Definition 5.4), one can obtain the following sets of CISs:

$$M_1 = \left\{ \begin{array}{l} I(X_7, \{X_1, X_2, X_3\}|X_4), I(X_4, X_1|\{X_2, X_3\}), \\ I(X_3, X_2|X_1), I(X_5, \{X_1, X_2, X_3, X_4, X_7\}|X_3) \\ I(X_6, \{X_1, X_2, X_3, X_5, X_7\}|X_4) \end{array} \right\}, \quad (7.63)$$

$$M_2 = \left\{ \begin{array}{l} I(X_3, \{X_1, X_2\}|\phi), I(X_7, \{X_1, X_2, X_3\}|X_4), \\ I(X_4, X_1|\{X_2, X_3\}), I(X_5, \{X_1, X_2, X_3, X_4\}|X_7) \\ I(X_6, \{X_1, X_2, X_3, X_5, X_7\}|X_4) \end{array} \right\}, \quad (7.64)$$

which are derived from D^1 and D^2, respectively. Note that one can also derive M_1 and M_2 from the factorizations given in (7.43) and (7.44), respectively.

Let us first calculate the covariance matrix of the probabilistic normal model specified only by the second Bayesian network D^2:

- The CIS $I(X_3, \{X_1, X_2\}|\phi)$ implies:

$$\begin{pmatrix} \sigma_{31} \\ \sigma_{32} \end{pmatrix} = \begin{pmatrix} 0 \\ 0 \end{pmatrix}, \quad (7.65)$$

- The CIS $I(X_1, X_4|\{X_2, X_3\})$ implies

$$\sigma_{14} = \begin{pmatrix} \sigma_{42} & \sigma_{43} \end{pmatrix} \begin{pmatrix} \sigma_{22} & \sigma_{23} \\ \sigma_{32} & \sigma_{33} \end{pmatrix}^{-1} \begin{pmatrix} \sigma_{24} \\ \sigma_{34} \end{pmatrix}, \quad (7.66)$$

7.6 Multifactorized Normal Models 297

- The CIS $I(X_7, \{X_1, X_2, X_3\}|X_4)$ implies

$$\begin{pmatrix} \sigma_{17} \\ \sigma_{27} \\ \sigma_{37} \end{pmatrix} = \frac{\sigma_{47}}{\sigma_{44}} \begin{pmatrix} \sigma_{14} \\ \sigma_{24} \\ \sigma_{34} \end{pmatrix}, \quad (7.67)$$

- The CIS $I(X_5, \{X_1, X_2, X_3, X_4\}|X_7)$ implies

$$\begin{pmatrix} \sigma_{51} \\ \sigma_{52} \\ \sigma_{53} \\ \sigma_{54} \end{pmatrix} = \frac{\sigma_{57}}{\sigma_{77}} \begin{pmatrix} \sigma_{71} \\ \sigma_{72} \\ \sigma_{73} \\ \sigma_{74} \end{pmatrix}, \quad (7.68)$$

- The CIS $I(X_6, \{X_1, X_2, X_3, X_5, X_7\}|X_4)$ implies

$$\begin{pmatrix} \sigma_{61} \\ \sigma_{62} \\ \sigma_{63} \\ \sigma_{65} \\ \sigma_{67} \end{pmatrix} = \frac{\sigma_{64}}{\sigma_{44}} \begin{pmatrix} \sigma_{14} \\ \sigma_{24} \\ \sigma_{34} \\ \sigma_{54} \\ \sigma_{74} \end{pmatrix}. \quad (7.69)$$

Solving the system of equations in (7.65)–(7.69), and considering the symmetry of the covariance matrix, $\sigma_{ij} = \sigma_{ji}$, we obtain

$$\begin{pmatrix}
\sigma_{11} & \sigma_{12} & 0 & \alpha & \sigma_{15} & \delta & \beta \\
\sigma_{12} & \sigma_{22} & 0 & \frac{\sigma_{25}\sigma_{44}}{\sigma_{45}} & \sigma_{25} & \frac{\sigma_{25}\sigma_{46}}{\sigma_{45}} & \frac{\sigma_{25}\sigma_{47}}{\sigma_{45}} \\
0 & 0 & \sigma_{33} & \frac{\sigma_{35}\sigma_{44}}{\sigma_{45}} & \sigma_{35} & \frac{\sigma_{35}\sigma_{46}}{\sigma_{45}} & \frac{\sigma_{35}\sigma_{47}}{\sigma_{45}} \\
\alpha & \frac{\sigma_{25}\sigma_{44}}{\sigma_{45}} & \frac{\sigma_{35}\sigma_{44}}{\sigma_{45}} & \sigma_{44} & \sigma_{45} & \sigma_{46} & \sigma_{47} \\
\sigma_{15} & \sigma_{25} & \sigma_{35} & \sigma_{45} & \sigma_{55} & \frac{\sigma_{45}\sigma_{46}}{\sigma_{44}} & \sigma_{57} \\
\delta & \frac{\sigma_{25}\sigma_{46}}{\sigma_{45}} & \frac{\sigma_{35}\sigma_{46}}{\sigma_{45}} & \sigma_{46} & \frac{\sigma_{45}\sigma_{46}}{\sigma_{44}} & \sigma_{66} & \frac{\sigma_{46}\sigma_{47}}{\sigma_{44}} \\
\beta & \frac{\sigma_{25}\sigma_{47}}{\sigma_{45}} & \frac{\sigma_{35}\sigma_{47}}{\sigma_{45}} & \sigma_{47} & \sigma_{57} & \frac{\sigma_{46}\sigma_{47}}{\sigma_{44}} & \frac{\sigma_{47}\sigma_{57}}{\sigma_{45}}
\end{pmatrix}, \quad (7.70)$$

where

$$\alpha = \frac{\sigma_{12}\sigma_{25}\sigma_{44}}{\sigma_{22}\sigma_{45}}, \quad \beta = \frac{\sigma_{12}\sigma_{25}\sigma_{47}}{\sigma_{22}\sigma_{45}}, \quad \delta = \frac{\sigma_{12}\sigma_{25}\sigma_{46}}{\sigma_{22}\sigma_{45}}.$$

Thus, the covariance matrix of a JPD having the DAG in Figure 7.7(b) as an I-map must satisfy (7.70).

Now let us find the covariance matrix of the MFPM specified by the two DAGs in Figure 7.7. The union of M_1 and M_2 adds only one CIS to M_2, which is $I(X_5, \{X_1, X_2, X_4, X_7\}|X_3)$. Therefore, we only need to consider this CIS. This CIS adds the following constraints:

$$\begin{pmatrix} \sigma_{51} \\ \sigma_{52} \\ \sigma_{54} \\ \sigma_{57} \end{pmatrix} = \frac{\sigma_{53}}{\sigma_{33}} \begin{pmatrix} \sigma_{31} \\ \sigma_{32} \\ \sigma_{34} \\ \sigma_{37} \end{pmatrix}. \quad (7.71)$$

Imposing this additional constraint and solving, we get

$$\begin{pmatrix} \sigma_{11} & \sigma_{12} & 0 & 0 & 0 & 0 & 0 \\ \sigma_{12} & \sigma_{22} & 0 & 0 & 0 & 0 & 0 \\ 0 & 0 & \sigma_{33} & \rho & \sigma_{35} & -\dfrac{\sqrt{\sigma_{33}}\,\sigma_{46}}{\sqrt{\sigma_{44}}} & -\dfrac{\sqrt{\sigma_{33}}\,\sigma_{47}}{\sqrt{\sigma_{44}}} \\ 0 & 0 & \rho & \sigma_{44} & -\dfrac{\sigma_{35}\sqrt{\sigma_{44}}}{\sqrt{\sigma_{33}}} & \sigma_{46} & \sigma_{47} \\ 0 & 0 & \sigma_{35} & \tau & \sigma_{55} & -\dfrac{\sigma_{35}\,\sigma_{46}}{\sqrt{\sigma_{33}\sigma_{44}}} & -\dfrac{\sigma_{35}\,\sigma_{47}}{\sqrt{\sigma_{33}\sigma_{44}}} \\ 0 & 0 & -\dfrac{\sqrt{\sigma_{33}}\,\sigma_{46}}{\sqrt{\sigma_{44}}} & \sigma_{46} & -\dfrac{\sigma_{35}\,\sigma_{46}}{\sqrt{\sigma_{33}\sigma_{44}}} & \sigma_{66} & \dfrac{\sigma_{46}\,\sigma_{47}}{\sigma_{44}} \\ 0 & 0 & -\dfrac{\sqrt{\sigma_{33}}\,\sigma_{47}}{\sqrt{\sigma_{44}}} & \sigma_{47} & -\dfrac{\sigma_{35}\,\sigma_{47}}{\sqrt{\sigma_{33}\sigma_{44}}} & \dfrac{\sigma_{46}\,\sigma_{47}}{\sigma_{44}} & \dfrac{\sigma_{47}^{2}}{\sigma_{44}} \end{pmatrix},$$

where

$$\tau = -\dfrac{\sigma_{35}\sqrt{\sigma_{44}}}{\sqrt{\sigma_{33}}}, \quad \rho = -\sqrt{\sigma_{33}\sigma_{44}}.$$

Thus, the covariance of the multi-Bayesian normal model formed by the two graphs in Figure 7.7 must be of this form. ∎

7.7 Conditionally Specified Probabilistic Models

In the previous sections we have seen that a multifactorized probabilistic model (MFPM) can be specified by either a set of input lists and/or a multigraph. In this section we deal with the problem of specifying a JPD directly by a set of CPDs involving any set of variables. Thus, based on a set of CPDs provided by the human experts, we need to find the corresponding JPD of the variables. Models specified in this way are known as *conditionally specified probabilistic models* (CSPM).

Definition 7.4 CSPM. *Consider a set of variables $X = \{X_1, \ldots, X_n\}$. A conditionally specified probabilistic model is a set of conditional and/or marginal probability distributions on X of the form*

$$P = \{p(u_i|v_i);\ i = 1, \ldots, m\}, \qquad (7.72)$$

which uniquely defines the JPD of X, where U_i and V_i are disjoint subsets of X and $U_i \neq \phi$.

Thus, the probabilistic models associated with decomposable Markov network models, Bayesian network models (discussed in Chapter 6) and any probabilistic model defined by a factorization are special cases of CSPM. We have also seen in the previous sections that both input lists and graphs imply certain factorizations of the corresponding JPD of the variables in terms of a product of CPDs. Hence, CSPM can be seen as a generalization

7.7 Conditionally Specified Probabilistic Models

of the MFPM. Therefore, CSPM can be used as a general framework to analyze the problems underlying all of the above probabilistic models.

In the previous sections, we have also discussed how to solve the compatibility problem arising in models specified by multigraphs and/or by input lists. In this section, we discuss the compatibility and other problems associated with CSPM. In particular, we address the following questions:

- **Question 7.5. Uniqueness:**
  Does the given set of CPDs define a unique JPD? In other words, does the set of CPDs imply enough constraints for the existence of *at most* one JPD?

- **Question 7.6. Consistency or compatibility:**
  Is the given set of CPDs compatible with a JPD of the variables?

- **Question 7.7. Parsimony:**
  If the answer to Question 7.5 is yes, can any of the given CPDs be ignored or deleted without loss of information?

- **Question 7.8. Reduction:**
  If the answer to Question 7.6 is yes, can any set of CPDs be reduced to a minimum (e.g., by removing some of the conditioning variables)?

The above questions suggest that general probabilistic models should be handled within a coherent knowledge-base system in order to eliminate inconsistencies, minimize redundancies, and increase the chances of obtaining sensible and accurate results. In the rest of this section, we address the above questions and discuss ways by which CSPM can be structured. In particular, we show, for example, that

1. Any well-behaved (just to ensure existence of conditional probabilities) JPD can be represented by a CSPM.

2. Given any set of CPDs defining a JPD, we can always obtain an equivalent subset of CPDs that form a chain rule factorization for the JPD.

3. Models specified by a Bayesian network are always consistent.

Since any set of CPDs can always be equivalently written in canonical form (see Section 5.5), we shall assume, without loss of generality, that a set of CPDs provided by the human expert to represent the JPD of the variables is given in a canonical form. Each of the CPDs can be specified numerically or can be given as a parametric family. It is possible, however, that the given set defines zero (contains contradictory CPDs), one, or more than one JPD. Thus, the set must be checked for uniqueness (Section 7.7.1) and compatibility (Section 7.7.2) with a JPD.

7.7.1 Checking Uniqueness

We begin with addressing the uniqueness problem raised in Question 7.5, that is, we need to determine whether a given set of CPDs implies enough constraints to define at most one JPD. The following theorem (Gelman and Speed (1993)) states the conditions under which a given canonical representation defines at most one JPD.

Theorem 7.6 Uniqueness. *Let $Y = \{Y_1, \ldots, Y_n\}$ be a permutation of $X = \{X_1, \ldots, X_n\}$, $B_i = \{Y_1, \ldots, Y_{i-1}\}$, and $A_i = \{Y_{i+1}, \ldots, Y_n\}$. A set of CPDs in canonical form determines at most one JPD of X if and only if it contains a nested sequence of probability functions of the form*

$$p(y_i|s_i, a_i), \quad i = 1, \ldots, n, \tag{7.73}$$

or equivalently (by reversing the order of the permutation), of the form

$$p(y_i|b_i, u_i), \quad i = 1, \ldots, n, \tag{7.74}$$

where $S_i \subseteq B_i$ and $U_i \subseteq A_i$. If all sets $S_i = \phi$ or $U_i = \phi, i = 1, \ldots, n$, the JPD exists and is unique.

Example 7.20 Unique but possibly inconsistent JPD. Consider a set of four variables $X = \{X_1, X_2, X_3, X_4\}$ and the set of CPDs

$$\{p(x_4|x_1, x_2, x_3), p(x_3|x_1, x_2, x_4), p(x_2|x_1), p(x_1|x_4)\}, \tag{7.75}$$

that is associated with the permutation $(Y_1, Y_2, Y_3, Y_4) = (X_4, X_3, X_2, X_1)$. In this case, we have

$$\begin{array}{ll} S_1 = \phi, & A_1 = \{X_3, X_2, X_1\}, \\ S_2 = \{X_4\}, & A_2 = \{X_2, X_1\}, \\ S_3 = \phi, & A_3 = \{X_1\}, \\ S_4 = \{X_4\}, & A_4 = \phi. \end{array}$$

Hence, the set in (7.75) satisfies the conditions of Theorem 7.6. Thus, the set is either compatible with exactly one JPD or is not compatible with any JPD.

However, since $S_2 \neq \phi$ and $S_4 \neq \phi$, the CPDs $p(x_3|x_1, x_2, x_4)$ and $p(x_1|x_4)$ must be checked for consistency. In other words, only for some selected forms of $p(x_3|x_1, x_2, x_4)$ and $p(x_1|x_4)$ does the JPD exist. ■

Theorem 7.6 has very important practical implications in modeling the JPD of the variables because

1. There exists a minimal set of CPDs that are required for the JPD to be uniquely defined. This occurs when P is in standard canonical form. This implies, for example, that the number of links in a corresponding DAG can be reduced to a minimum, but below this minimum the JPD is not defined uniquely.

7.7 Conditionally Specified Probabilistic Models

2. To obtain a well-defined JPD of a set of variables one must carry out the following steps:
   - **Step 1.** Order the variables.
   - **Step 2.** Give a set of CPDs that contains the nested sequence in (7.73) or (7.74).
   - **Step 3.** Check the set of CPDs for consistency, as discussed below.

7.7.2 Checking Consistency

Since we are looking for a uniquely defined JPD, in this section we deal only with sets of JPDs that satisfy the uniqueness conditions. Note that if a set of CPDs satisfies the uniqueness conditions, then either there is no JPD compatible with the set, or the set defines exactly one JPD. Then given a set of CPDs that satisfies the uniqueness conditions, we need to check this set to see whether it satisfies compatibility, that is, whether it defines exactly one JPD. The following theorem (Arnold, Castillo, and Sarabia (1992, 1996)) can be used to determine whether a given set of CPDs, which satisfies the uniqueness conditions (7.73) or (7.74), is consistent with a JPD of X.

Theorem 7.7 Consistency. *Let* $Y = \{Y_1, \ldots, Y_n\}$ *be a permutation of a set of random variables* $X = \{X_1, \ldots, X_n\}$. *Given a set of canonical CPDs*

$$\{p(y_1|s_1, a_1), \ldots, p(y_n|s_n, a_n)\}, \qquad (7.76)$$

or equivalently (by reversing the order of the permutation),

$$\{p(y_1|b_1, u_1), \ldots, p(y_n|b_n, u_n)\}, \qquad (7.77)$$

where $U_i \subseteq A_i = \{Y_{i+1}, \ldots, Y_n\}$ *and* $S_i \subseteq B_i = \{Y_1, \ldots, Y_{i-1}\}$ *for all* $i = 1, \ldots, n$, *then a necessary and sufficient condition for the set in (7.76) to be compatible with exactly one JPD of* X *is that either* $S_i = \phi$ *or*

$$R_i = p(y_i|a_i) = \frac{p(y_i|s_i, a_i)/p(s_i|y_i, a_i)}{\sum_{y_i} p(y_i|s_i, a_i)/p(s_i|y_i, a_i)} \qquad (7.78)$$

is independent of S_i, *for* $i = 1, \ldots, n$. *Equivalently, a necessary and sufficient condition for this set in (7.77) to be compatible with exactly one JPD of* X *is that either* $U_i = \phi$ *or*

$$T_i = p(y_i|b_i) = \frac{p(y_i|b_i, u_i)/p(u_i|y_i, b_i)}{\sum_{y_i} [p(y_i|b_i, u_i)/p(u_i|y_i, b_i)]} \qquad (7.79)$$

is independent of U_i, *for* $i = 1, \ldots, n$. *Note that the sum in the denominators of (7.78) and (7.79) is replaced by the integral in the case of continuous random variables.*

Corollary 7.4 *The set of CPDs in (7.76) is compatible with exactly one JPD of Y if each $p(y_i|s_i, a_i)$, $i = 1, \ldots, n$, is of the form*

$$p(y_i|s_i, a_i) = \frac{p(y_i|a_i) \sum_{b_i \setminus s_i} \prod_{j=1}^{i-1} p(y_j|a_j)}{\sum_{b_i \cup \{y_i\} \setminus s_i} \prod_{j=1}^{i} p(y_j|a_j)}, \qquad (7.80)$$

where $p(y_k|a_k)$, $k = 1, \ldots, i-1$, is given by (7.78), and $p(y_i|a_i)$ is an arbitrary probability function.

A similar corollary can be written for the set of CPDs in (7.77). Note that once the set $\{p(y_k|s_k, a_k); k = 1, \ldots, i-1\}$ is specified, the only freedom for specifying $p(y_i|s_i, a_i)$ consists of arbitrarily selecting the probability function $p(y_i|a_i)$. In other words, at this stage, we are completely free to choose $p(y_i|a_i)$ but not $p(y_i|s_i, a_i)$. Note also that if we replace the CPD $p(y_i|s_i, a_i)$ by the obtained $p(y_i|a_i)$ using (7.78), we obtain the same JPD. This implies that once we have defined an ordering $(Y_1, \ldots, Y_n)$ for the variables, we can always replace the CPD $p(y_i|s_i, a_i)$ by $p(y_i|a_i)$ without changing the JPD. We can obtain $p(y_i|a_i)$ by rewriting (7.80) as

$$p(y_i|a_i) = \frac{p(y_i|s_i, a_i) \sum_{b_i \cup \{y_i\} \setminus s_i} \prod_{j=1}^{i} p(y_j|a_j)}{\sum_{b_i \setminus s_i} \prod_{j=1}^{i-1} p(y_j|a_j)}, \qquad (7.81)$$

and then, since the first $p(y_1|a_1)$ must belong to the given set of CPDs for it to satisfy the uniqueness condition, from each $p(y_i|s_i, a_i)$ and the previous standard canonical components $p(y_j|a_j), j = 1, \ldots, i-1$, we can calculate $p(y_i|a_i)$. Thus, we conclude that (7.81) allows calculating the standard canonical components from the given set of CPDs.

From Theorems 7.6 and 7.7, the following important corollary follows:

Corollary 7.5 *The sets of CPDs*

$$\{p(x_i|b_i); i = 1, \ldots, n\} \text{ and } \{p(x_i|a_i); i = 1, \ldots, n\}, \qquad (7.82)$$

where $b_i = \{y_1, \ldots, y_{i-1}\}$ and $a_i = \{y_{i+1}, \ldots, y_n\}$, are consistent and cannot be reduced without affecting uniqueness.

Theorem 7.7 shows that all factorizations implied by Bayesian networks are consistent because these factorizations satisfy the first condition of the theorem, that is, $S_i = \phi$, $i = 1, \ldots, n$.

Theorem 7.7 also suggests an algorithm for checking the consistency of a given set of CPDs, one CPD at a time, and for constructing a canonical form with $S_i = \phi$, $i = 1, \ldots, n$. A version of the algorithm that applies to the set in (7.76) is illustrated in the flow chart in Figure 7.8 and is given below. A similar algorithm, which applies to the set in (7.77), can be obtained.

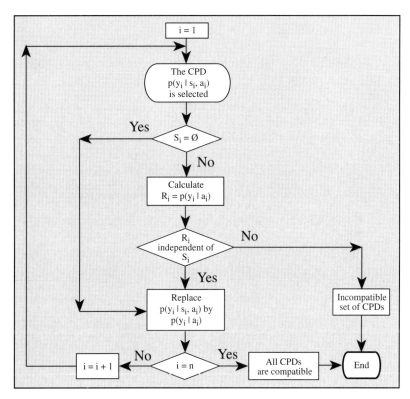

FIGURE 7.8. A flow chart for checking consistency of a canonical set of CPDs that satisfies the uniqueness conditions.

Algorithm 7.2 Checking Compatibility of a Set of CPDs.

- **Input:** A set P of CPDs in standard canonical form that satisfies the uniqueness condition.

- **Output:** True or False, depending on whether or not the set of CPDs in P is consistent.

1. The first conditional probability $p(y_1|a_1)$ must be given in any case. Otherwise, the given set of CPDs does not satisfy the uniqueness condition. Thus, at the initial step, we start with $p(y_1|a_1)$.

2. At the ith step we are given either $p(y_i|a_i)$ or $p(y_i|s_i, a_i)$, where the set $S_i \subset B_i$. If $S_i = \phi$ proceed to Step 5; otherwise, we calculate

$p(s_i|y_i, a_i)$ by marginalizing $p(b_i|y_i, a_i)$ over all variables in B_i other than those in S_i, that is, using

$$p(s_i|y_i, a_i) = \sum_{b_i \setminus s_i} p(b_i|y_i, a_i). \tag{7.83}$$

3. Calculate the standard canonical component $R_i = p(y_i|a_i)$ based on the previous and the new information using (7.78).

4. If R_i is independent of the variables in the set S_i, then go to Step 5; otherwise the given $p(y_i|s_i, a_i)$ is not compatible with the previous conditional probabilities.

5. Calculate $p(b_i, y_i|a_i) = p(b_i|y_i, a_i)p(y_i|a_i)$.

6. Repeat Steps 2 to 5 until all CPDs have been analyzed. ∎

Therefore, given the set P of CPDs we can determine whether or not it is consistent using Algorithm 7.2 or Theorem 7.7. This provides the answer to Question 7.6.

Example 7.21 Consistent set of CPDs. The following set of CPDs is given by a human expert to describe the JPD of four variables, $X = \{X_1, X_2, X_3, X_4\}$:

$$P_1 = \{p(x_4|x_3, x_2, x_1), p(x_3|x_2, x_1), p(x_2|x_1), p(x_1)\}. \tag{7.84}$$

Note that P_1 is given in a form suitable for a chain rule factorization, that is, the JPD can be factorized as

$$p(x_1, \ldots, x_n) = p(x_4|x_3, x_2, x_1)p(x_3|x_2, x_1)p(x_2|x_1)p(x_1),$$

and hence the set is consistent. The corresponding parameters of this JPD are shown in Table 7.3, where $\bar{x}_i$ and x_i denote $X_i = 0$ and $X_i = 1$, respectively. These parameters can be given arbitrarily in the interval $[0, 1]$.

Choosing the permutation $(Y_1, Y_2, Y_3, Y_4) = \{X_4, X_3, X_2, X_1\}$, the reader can verify that P_1 satisfies the uniqueness conditions. Let us now use Algorithm 7.2 to show that it is indeed compatible with the above JPD:

- For $i = 1$, the first set $p(x_4|x_3, x_2, x_1)$ is always compatible because $A_1 = X_3, X_2, X_1$ and $S_1 = \phi$. Therefore, no further calculations are needed.

- For $i = 2$, the second set $p(x_3|x_2, x_1)$ is compatible because $A_2 = \{X_2, X_1\}$ and $S_2 = \phi$. Here, we have $B_2 = \{Y_1\} = \{X_4\}$; hence

$$p(b_2, y_2|a_2) = p(x_4, x_3|x_2, x_1) = p(x_4|x_3, x_2, x_1)p(x_3|x_2, x_1).$$

7.7 Conditionally Specified Probabilistic Models

Variable	Free Parameters
X_1	$\theta_{10} = p(\bar{x}_1)$
X_2	$\theta_{200} = p(\bar{x}_2\|\bar{x}_1)$ $\theta_{201} = p(\bar{x}_2\|x_1)$
X_3	$\theta_{3000} = p(\bar{x}_3\|\bar{x}_1, \bar{x}_2)$ $\theta_{3001} = p(\bar{x}_3\|\bar{x}_1, x_2)$ $\theta_{3010} = p(\bar{x}_3\|x_1, \bar{x}_2)$ $\theta_{3011} = p(\bar{x}_3\|x_1, x_2)$
X_4	$\theta_{40000} = p(\bar{x}_4\|\bar{x}_1, \bar{x}_2, \bar{x}_3)$ $\theta_{40001} = p(\bar{x}_4\|\bar{x}_1, \bar{x}_2, x_3)$ $\theta_{40010} = p(\bar{x}_4\|\bar{x}_1, x_2, \bar{x}_3)$ $\theta_{40011} = p(\bar{x}_4\|\bar{x}_1, x_2, x_3)$ $\theta_{40100} = p(\bar{x}_4\|x_1, \bar{x}_2, \bar{x}_3)$ $\theta_{40101} = p(\bar{x}_4\|x_1, \bar{x}_2, x_3)$ $\theta_{40110} = p(\bar{x}_4\|x_1, x_2, \bar{x}_3)$ $\theta_{40111} = p(\bar{x}_4\|x_1, x_2, x_3)$

TABLE 7.3. The set of parameters of the CPDs given in Example 7.21.

- For $i = 3$, the third set $p(x_2|x_1)$ is compatible because $A_3 = \{X_1\}$ and $S_3 = \phi$. Here, we have $B_3 = \{Y_1, Y_2\} = \{X_4, X_3\}$; hence

$$p(b_3, y_3|a_3) = p(x_4, x_3, x_2|x_1) = p(x_4, x_3|x_2, x_1)p(x_2|x_1).$$

- For $i = 4$, the fourth set $p(x_1)$ is compatible because $S_4 - A_4 = \phi$. Here, we have $B_4 = \{Y_1, Y_2, Y_3\} = \{X_4, X_3, X_2\}$; hence

$$p(b_4, y_4|a_4) = p(x_4, x_3, x_2, x_1) = p(x_4, x_3, x_2|x_1)p(x_1).$$

Therefore, the set P_1 is consistent. ■

Example 7.22 Possibly inconsistent set of CPDs. The following set of CPDs was given by a human expert to describe the JPD of four variables, $X = \{X_1, X_2, X_3, X_4\}$:

$$P_2 = \{p(x_4|x_3, x_2, x_1), p(x_3|x_4, x_2, x_1), p(x_2|x_1), p(x_1)\}.$$

Let us use Algorithm 7.2 to check the consistency of this set:

- The first CPD, $p(x_4|x_3, x_2, x_1)$, can be given without any restrictions.

- The second CPD, $p(x_3|x_4, x_2, x_1)$, must be checked for consistency because $S_2 = \{X_4\} \neq \phi$. From (7.80), this CPD must be of the form

$$p(x_3|x_4, x_2, x_1) = \frac{p(x_3|x_2, x_1)}{\sum_{x_3} p(x_3|x_2, x_1)p(x_4|x_3, x_2, x_1)}, \quad (7.85)$$

306   7. Extending Graphically Specified Models

to be consistent with the first CPD. Using the parameters in Table 7.3, (7.85) gives

$$p(\bar{x}_3|\bar{x}_4, \bar{x}_2, \bar{x}_1) = \frac{\theta_{3000}}{\theta_{3000}\theta_{40000} + \theta_{40001} - \theta_{3000}\theta_{40001}},$$

$$p(\bar{x}_3|x_4, \bar{x}_2, x_1) = \frac{\theta_{3010}}{1 - \theta_{3010}\theta_{40100} - \theta_{40101} + \theta_{3010}\theta_{40101}},$$

$$p(\bar{x}_3|\bar{x}_4, x_2, \bar{x}_1) = \frac{\theta_{3001}}{\theta_{3001}\theta_{40010} + \theta_{40011} - \theta_{3001}\theta_{40011}},$$

$$p(\bar{x}_3|x_4, x_2, x_1) = \frac{\theta_{3011}}{1 - \theta_{3011}\theta_{40110} - \theta_{40111} + \theta_{3011}\theta_{40111}}, \quad (7.86)$$

$$p(\bar{x}_3|x_4, \bar{x}_2, \bar{x}_1) = \frac{\theta_{3000}}{1 - \theta_{3000}\theta_{40000} - \theta_{40001} + \theta_{3000}\theta_{40001}},$$

$$p(\bar{x}_3|\bar{x}_4, \bar{x}_2, x_1) = \frac{\theta_{3010}}{\theta_{3010}\theta_{40100} + \theta_{40101} - \theta_{3010}\theta_{40101}},$$

$$p(\bar{x}_3|x_4, x_2, \bar{x}_1) = \frac{\theta_{3001}}{1 - \theta_{3001}\theta_{40010} - \theta_{40011} + \theta_{3001}\theta_{40011}},$$

$$p(\bar{x}_3|\bar{x}_4, x_2, x_1) = \frac{\theta_{3011}}{\theta_{3011}\theta_{40110} + \theta_{40111} - \theta_{3011}\theta_{40111}}.$$

- The third and fourth CPDs, $p(x_2|x_1)$ and $p(x_1)$, are consistent because $S_3 = S_4 = \phi$.

Therefore, the set of CPDs P_2 must satisfy the constraints in (7.86) to be consistent. ∎

Example 7.23 Possibly inconsistent set of CPDs. Consider the following set P_3 of CPDs, which was given by a human expert to describe the JPD of five variables in X:

$$\{p(x_4|x_2, x_1, x_3, x_5), p(x_2|x_1, x_3, x_5), p(x_1|x_4, x_3, x_5), p(x_3|x_2, x_5), p(x_5)\}.$$

Choosing the permutation $\{Y_1, Y_2, Y_3, Y_4, Y_5\} = \{X_4, X_2, X_1, X_3, X_5\}$, the set P_3 can be shown to satisfy the uniqueness conditions. Thus, we can use Algorithm 7.2 to determine whether this set is compatible with a JPD:

- For $i = 1$, the first set, $p(x_4|x_2, x_1, x_3, x_5)$, is always compatible because $A_1 = \{X_2, X_1, X_3, X_5\}$ and $S_1 = \phi$.

- For $i = 2$, the second set, $p(x_2|x_1, x_3, x_5)$, is compatible because $A_2 = \{X_1, X_3, X_5\}$ and $S_2 = \phi$. Here, we have $B_2 = \{Y_1\} = \{X_4\}$; hence

$$p(b_2, y_2|a_2) = p(x_4, x_2|x_1, x_3, x_5) = p(x_4|x_2, x_1, x_3, x_5)p(x_2|x_1, x_3, x_5).$$

- For $i = 3$, the third set, $p(x_1|x_4, x_3, x_5)$, must be checked for compatibility because we have $A_3 = \{X_3, X_5\}$ and $S_3 = \{X_4\} \neq \phi$. Here, we have $B_3 = \{Y_1, Y_2\} = \{X_4, X_2\}$, and we need to compute $p(s_3|y_3, a_3) = p(x_4|x_1, x_3, x_5)$ using (7.83). We obtain

$$p(x_4|x_1, x_3, x_5) = \sum_{x_2} p(x_4, x_2|x_1, x_3, x_5).$$

We also need to compute R_3 using (7.78). We obtain

$$R_3 = p(y_3|a_3) = p(x_1|x_3, x_5) = \frac{p(x_1|x_4, x_3, x_5)/p(x_4|x_1, x_3, x_5)}{\sum_{x_1}(p(x_1|x_4, x_3, x_5)/p(x_4|x_1, x_3, x_5))}.$$

Then, if R_3 does not depend on X_4, $p(x_1|x_4, x_3, x_5)$ is compatible with the previous two CPDs. Otherwise the set is incompatible. For the sake of illustration, suppose that it is compatible. In this case, we replace $p(x_1|x_4, x_3, x_5)$ by R_3. We also calculate

$$p(b_3, y_3|a_3) = p(x_4, x_2, x_1|x_3, x_5) = p(x_4, x_2|x_1, x_3, x_5)p(x_1|x_3, x_5).$$

- For $i = 4$, the fourth set, $p(x_3|x_2, x_5)$, must be checked for compatibility because we have $A_4 = \{X_5\}$ and $S_4 = \{X_2\} \neq \phi$. Here, we have $B_4 = \{Y_1, Y_2, Y_3\} = \{X_4, X_2, X_1\}$, and we need to compute $p(s_4|y_4, a_4) = p(x_2|x_3, x_5)$ using (7.83). We obtain

$$p(x_2|x_3, x_5) = \sum_{x_1, x_4} p(x_4, x_2, x_1|x_3, x_5).$$

We also need to compute R_4 using (7.78). We obtain

$$R_4 = p(y_4|a_4) = p(x_3|x_5) = \frac{p(x_3|x_2, x_5)/p(x_2|x_3, x_5)}{\sum_{x_3}(p(x_3|x_2, x_5)/p(x_2|x_3, x_5))}.$$

Then, if R_4 does not depend on X_2, $p(x_3|x_2, x_5)$ is compatible with the previous three CPDs. Otherwise the set is incompatible. Again, for the sake of illustration, suppose that it is compatible. In this case, we replace $p(x_3|x_2, x_5)$ by R_4. We also calculate

$$p(b_4, y_4|a_4) = p(x_4, x_2, x_1, x_3|x_5) = p(x_4, x_2, x_1|x_3, x_5)p(x_3|x_5).$$

- For $i = 5$, the fifth set $p(x_5)$ is compatible because $A_5 = \phi$ and $S_5 = \phi$.

Therefore, if R_3 does not depend on X_4 and R_4 does not depend on X_2, then P_3 is consistent; otherwise P_3 is inconsistent. ■

308   7. Extending Graphically Specified Models

From the above examples, we can see that Theorem 7.7 has the following important practical implications:

1. Every well-behaved JPD for the variables can be represented by a CSPM.

2. Increasing the standard canonical form by superfluous information (extra CPD) leads to the need for checking its compatibility.

3. If the set of CPDs is given in standard canonical form, the CPDs can be completely arbitrary, that is, they are not restricted by conditions other than those implied by the probability axioms.

4. Any CPD of the form $p(x_i|s_i, a_i)$, with $S_i \neq \phi$, can be replaced by the conditional probability in standard canonical form $p(x_i|a_i)$ without affecting the JPD of the variables. The standard canonical form can be obtained using (7.78).

5. A minimal I-map of the CSPM can then be constructed, if desired, using Theorems 7.6 and 7.7 and Algorithms 7.1 and 7.3 (below).

We see from the above examples that Theorems 7.6 and 7.7 answer Questions 7.5 and 7.6. Thus, when defining a CSPM, it is preferable to specify only the minimal set of conditional probabilities needed to define the JPD uniquely. Any extra information will require unnecessary additional computational effort not only for its assessment but also for checking that the extra information is indeed consistent with the previously given CPDs. This answers Question 7.7.

Furthermore, given a set P of CPDs that is consistent and leads to a unique JPD of X, we can replace P by another P' in standard canonical form leading to the same JPD. Further reduction of P' violates the uniqueness conditions. Also, redundant information hurts because it requires consistency checking. This answers Question 7.8.

7.7.3 Assessment of CSPM

Note that Theorem 7.7 assumes a set of CPDs that already satisfies uniqueness. Therefore, uniqueness has to be checked first before compatibility. Thus initially, we convert the specified set of CPD to a canonical form, then we check uniqueness using Theorem 7.6, and then we check compatibility by means of Theorem 7.7.

When $S_i = \phi$ or $U_i = \phi$ for all i, the consistency is guaranteed, and we say that the canonical form is a *standard canonical form* and we call the term $p(y_i|a_i)$, or $p(y_i|b_i)$, a *standard canonical component*. Otherwise, the set of CPDs must be checked for consistency.

The following algorithm allows determining whether a set P of CPDs satisfies the uniqueness conditions and which subset must be checked for compatibility.

7.7 Conditionally Specified Probabilistic Models

Algorithm 7.3 Assessment of CSPM.

- **Input:** A set X of n variables and a set of CPDs in canonical form $P = \{p(x_i|s_i); i = 1, \ldots, m\}$.

- **Output:** The collection Q_1 of all subsets of P leading to uniqueness solutions and the collection C_1 of all subsets of P that must be checked for compatibility.

The algorithm consists of a calling procedure, *Compatible*, which initializes the set of *Solutions* to empty, calls the recursive procedure *CompatibleAux* (with X, P, and two empty lists C_2 and Q_2, as its arguments), and prints the *Solutions*. *CompatibleAux* proceeds as follows:

1. Set $i \leftarrow 1$, and $m \leftarrow$ number of CPDs in P.

2. Let $P_1 \leftarrow P$, $V \leftarrow X$, $C_1 \leftarrow C_2$ and $Q_1 \leftarrow Q_2$.

3. If $p(x_i|s_i) \in P_1$ is such that $V \cup S_i \supset V$, do the following:

    - Remove $p(x_i|s_i)$ from P_1 and add it to Q_1.
    - If $V \cup S_i \neq V$ add $p(x_i|s_i)$ to C_1.
    - Remove any remaining $p(x_r|s_r) \in P_1$ such that $X_r = X_i$ from P_1 and add it to C_1.
    - If $P \neq \phi$ call Algorithm 7.3, recursively, with arguments $V \setminus X_i$ and P_1, and add to C_1 its C output and to Q_1 its Q output; otherwise add its output to Solutions.
    - Go to Step 4.

    Otherwise, go to Step 4.

4. If $i < m$, make $i = i + 1$ and repeat Step 3; otherwise return C_1 and Q_1. ∎

Algorithm 7.3 deals with a set of CPDs given in the form in (7.73). The algorithm can be easily modified to deal with CDPs in the form in (7.74).

A recursive program in *Mathematica* that implements Algorithm 7.3 is given in Figure 7.9. The function *Compatible*$[X, P]$ takes two arguments, X and P, where X is the list of variables and P is the list of CPDs. For example, to run the program when $X = \{A, B, C\}$ and

$$P = \{p(a), p(b|a), p(c|a, b)\} \qquad (7.87)$$

we need the execute the following *Mathematica* statements:

```
X={A,B,C};
P=List[{{A},{}},{{B},{A}},{{C},{A,B}}];
Compatible[X,P];
```

```
Remov[CM_,j_]:=Join[Take[CM,j-1],
   Take[CM,{j+1,Length[CM]}]]

Compatible[X_,P_]:=Module[{},
   Solutions={};CompatibleAux[X,P,{},{}];
   Solutions
]
CompatibleAux[X_,P_,C2_,Q2_]:=
Module[{Xi,V,i,Q1,C1},
Do[
   P1=P;V=X;C1=C2;Q1=Q2;
   Uni=Union[P1[[i,1]],P1[[i,2]]];
   If[Uni==Union[V,Uni],AppendTo[Q1,P1[[i]]];
      Xi=P1[[i,1]];If[Uni !=V, AppendTo[C1,P1[[i]]]];
      P1=Remov[P1,i];
      Do[
         If[Xi==P1[[k,1]],
            AppendTo[C1,P1[[k]]];P1=Remov[P1,k],
            True
         ],
      {k,Length[P1],1,-1}];
      If[P1!={},
         Res=CompatibleAux[Complement[V,Xi],P1,C1,Q1];
         C1=Union[C1,Res[[1]]];Q1=Union[Q1,Res[[2]]],
         AppendTo[Solutions,{Q1,C1}]
      ]
   ],
{i,1,Length[P]}];
Return[{C1,Q1}]]
```

FIGURE 7.9. A *Mathematica* program for determining whether a set P of CPDs satisfies the uniqueness and compatibility conditions.

The first of the above statements defines the list X, the second defines the list P, and the third calls the function $Compatible[X, P]$, which takes the two arguments X and P.

The output of the function $Compatible[X, P]$ is two lists, Q and C. The list Q consists of all possible sets of CPDs that can be constructed from P that lead to a unique JPD. The number of sets in the list C is equal to the number of sets in the list Q. For each set in Q, the corresponding set in C is the set of CPDs that need to be checked for consistency before the corresponding set in Q can define a unique JPD.

When the set P defines a unique JPD, the set Q will contain only one set, which is the same as the set P, and the set C will be the empty set. When the set P is consistent but does not define a unique JPD, both Q and C will be empty sets. This program is illustrated by the following examples.

Example 7.24 Unique and compatible set of CPDs. Consider a set of three variables, $\{A, B, C\}$, and the set of CPDs in (7.87). The set P can be seen to satisfy the conditions of Corollary 7.5, and hence it defines a unique JPD, which is given by

$$p(a,b,c) = p(a)p(b|a)p(c|a,b). \tag{7.88}$$

This result can be obtained using the program in Figure 7.9. To run the program, we first define the set X and P as above, and then execute the function $Compatible[X, P]$. The output of this function, in this case, is two lists: $Q = P$ and $C = \phi$. ∎

Example 7.25 Unique but possibly inconsistent JPD. Consider a set of variables, $\{A, B, C\}$, and the set of CPDs

$$P = \{p(a|b,c), p(b|c), p(b|c,a), p(c|a,b)\}. \tag{7.89}$$

We use the *Mathematica* program in Figure 7.9 to determine which of these CPDs uniquely define a JPD of the three variables. We need to execute the following statements:

```
X={A,B,C};
P=List[{{A},{B,C}},{{B},{C}},{{B},{C,A}},{{C},{A,B}}];
Compatible[X,P];
```

Table 7.4 shows the output of the program. We get nine different possible sets in the lists Q (CPDs defining a unique JPD) and C (the set of CPDs to be checked for consistency). For example, the first set in the list C, $p(b|c,a)$ and $p(c|a,b)$ must be checked for consistency with the corresponding set in Q. If this set is consistent then the set $\{p(a|b,c),(b|c),(c|a,b)\}$ defines a unique JPD. This JPD can be expressed as

$$p(a,b,c) = p(a|b,c)p(b|c)p(c),$$

where $p(c)$ must be calculated from Expression (7.81). ∎

Exercises

7.1 Use Algorithm 6.6 to show that the links L_{13} and L_{35} in Figure 7.3(a) are reversible.

Set	The List Q	The List C
1	$p(a\|b,c), p(b\|c), p(c\|a,b)$	$p(b\|c,a), p(c\|a,b)$
2	$p(a\|b,c), p(b\|c,a), p(c\|a,b)$	$p(b\|c,a), p(b\|c), p(c\|a,b)$
3	$p(a\|b,c), p(c\|a,b), p(b\|c)$	$p(c\|a,b), p(b\|c), p(b\|c,a)$
4	$p(a\|b,c), p(c\|a,b), p(b\|c,a)$	$p(c\|a,b), p(b\|c,a), p(b\|c)$
5	$p(b\|c,a), p(a\|b,c), p(c\|a,b)$	$p(b\|c), p(a\|b,c), p(c\|a,b)$
6	$p(b\|c,a), p(c\|a,b), p(a\|b,c)$	$p(b\|c), p(c\|a,b), p(a\|b,c)$
7	$p(c\|a,b), p(a\|b,c), p(b\|c)$	$p(a\|b,c), p(b\|c), p(b\|c,a)$
8	$p(c\|a,b), p(a\|b,c), p(b\|c,a)$	$p(a\|b,c), p(b\|c,a), p(b\|c)$
9	$p(c\|a,b), p(b\|c,a), p(a\|b,c)$	$p(b\|c,a), p(b\|c), p(a\|b,c)$

TABLE 7.4. Results of the *Mathematica* program in Figure 7.9 for determining whether the set P in (7.89) satisfies the uniqueness and consistency conditions. The second column gives all possible sets of CPDs that satisfy the uniqueness conditions. The third column gives the corresponding sets of CPDs that must be checked for consistency.

7.2 Derive the condition under which (7.5) holds.

7.3 Verify that using (7.37) as the reference factorization, Algorithm 7.1 yields the solutions in (7.42).

7.4 Using the D-separation criterion in Definition 5.4

   (a) Verify that each of the CISs in (7.63) is implied by the DAG in Figure 7.7(a).

   (b) Verify that each of the CISs in (7.64) is implied by the DAG in Figure 7.7(b).

7.5 (a) Show that the factorization of the JPD in (7.43) implies the list of CISs in (7.63).

   (b) Show that the factorization of the JPD in (7.44) implies the list of CISs in (7.64).

7.6 Suppose that $\{X_1, X_2, X_3, X_4\}$ are normally distributed random variables whose JPD satisfies the CISs $I(X_2, X_3|X_1)$ and $I(X_1, X_4|X_2, X_3)$. Show that the corresponding form of the covariance matrix is given by

$$\begin{pmatrix} \sigma_{11} & \sigma_{12} & \sigma_{13} & \alpha \\ \sigma_{12} & \sigma_{22} & \dfrac{\sigma_{12}\sigma_{13}}{\sigma_{11}} & \sigma_{24} \\ \sigma_{13} & \dfrac{\sigma_{12}\sigma_{13}}{\sigma_{11}} & \sigma_{33} & \sigma_{34} \\ \alpha & \sigma_{24} & \sigma_{34} & \sigma_{44} \end{pmatrix},$$

7.7 Conditionally Specified Probabilistic Models

where

$$\alpha = \frac{\sigma_{11}(\sigma_{12}\sigma_{24}(\sigma_{11}\sigma_{33} - \sigma_{13}^2) + \sigma_{13}\sigma_{34}(\sigma_{11}\sigma_{22} - \sigma_{12}^2))}{\sigma_{11}^2\sigma_{22}\sigma_{33} - \sigma_{12}^2\sigma_{13}^2}.$$

7.7 Suppose that $\{X_1, \ldots, X_7\}$ are normally distributed random variables whose JPD satisfies the CISs $I(X_3, \{X_1, X_2\}|\phi)$, $I(X_1, X_4|\{X_2, X_3\})$, and $I(X_7, \{X_1, X_2, X_3\}|X_4)$. Show that the corresponding form of the covariance matrix is given by

$$\begin{pmatrix} \sigma_{11} & \sigma_{12} & 0 & \frac{\sigma_{12}\sigma_{24}}{\sigma_{22}} & \sigma_{15} & \sigma_{16} & \frac{\sigma_{12}\sigma_{24}\sigma_{47}}{\sigma_{22}\sigma_{44}} \\ \sigma_{12} & \sigma_{22} & 0 & \sigma_{24} & \sigma_{25} & \sigma_{26} & \frac{\sigma_{24}\sigma_{47}}{\sigma_{44}} \\ 0 & 0 & \sigma_{33} & \sigma_{34} & \sigma_{35} & \sigma_{36} & \frac{\sigma_{34}\sigma_{47}}{\sigma_{44}} \\ \frac{\sigma_{12}\sigma_{24}}{\sigma_{22}} & \sigma_{24} & \sigma_{34} & \sigma_{44} & \sigma_{45} & \sigma_{46} & \sigma_{47} \\ \sigma_{15} & \sigma_{25} & \sigma_{35} & \sigma_{45} & \sigma_{55} & \sigma_{56} & \sigma_{57} \\ \sigma_{16} & \sigma_{26} & \sigma_{36} & \sigma_{46} & \sigma_{56} & \sigma_{66} & \sigma_{67} \\ \frac{\sigma_{12}\sigma_{24}\sigma_{47}}{\sigma_{22}\sigma_{44}} & \frac{\sigma_{24}\sigma_{47}}{\sigma_{44}} & \frac{\sigma_{34}\sigma_{47}}{\sigma_{44}} & \sigma_{47} & \sigma_{57} & \sigma_{67} & \sigma_{77} \end{pmatrix}$$

7.8 Consider the set of variables $X = \{X_1, X_2, X_3, X_4, X_5\}$ and the following two sets of CPDs:

- $p(x_1, x_2, x_3|x_4, x_5), p(x_4, x_5|x_1, x_2, x_3)$ and
- $p(x_1, x_3|x_2), p(x_2, x_4, x_5|x_1)$.

Using the methods described in this chapter

(a) Write the above sets of CPDs in canonical form.
(b) Does each set satisfy the uniqueness condition?
(c) Does each set satisfy the compatibility condition?
(d) Give a simple set of CPDs, in standard canonical form, leading to existence and uniqueness and draw its associated Bayesian network.

7.9 Given the set of variables $X = \{X_1, X_2, X_3, X_4\}$, find conditions for the following CPDs to be compatible:

$$p(x_1|x_2), \quad p(x_2|x_3),$$
$$p(x_3|x_2, x_1) = p(x_3|x_2), \quad p(x_4|x_3, x_2, x_1) = p(x_4|x_3).$$

7.10 For two discrete random variables X_1 and X_2, what are the conditions under which $p(x_1|x_2)$ and $p(x_2|x_1)$ are compatible.

7.11 Find the most general family $X = \{X_1, X_2, X_3, X_4\}$ of normal random variables such that the following CISs hold:

$$\{I(x_1, x_2|\{x_3, x_4\}), I(x_3, x_1|x_2), I(x_4, x_2|x_3)\}.$$

314   7. Extending Graphically Specified Models

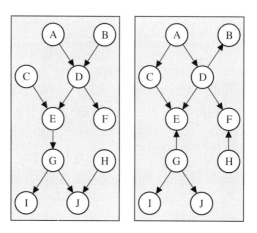

FIGURE 7.10. Two graphs defining an MNM.

7.12 Find the most general family of MNM associated with the graphs in Figure 7.10.

7.13 Given the JPD $p(a,b,c,d) = p(a)p(b|a)p(c|a)p(d|b)$, find

(a) The most general compatible function $p(c|a,b)$.

(b) The most general compatible function $p(c|a,b)$ when we replace $p(d|b)$ by $p(d|a,b,c)$.

7.14 Suppose that a set of CPDs is given in the canonical form in (7.74). Modify Algorithm 7.3 to deal with this case.

7.15 Run the program in Figure 7.9 using the set of CPDs in (7.75) to determine whether it defines a unique JPD.

7.16 Using Algorithm 7.2, find the set of constraints under which the set of CPDs in (7.75) is consistent, and hence, unique.

7.17 Write a corollary, similar to Corollary 7.4, for the set of CPDs in (7.77).

7.18 Draw a flow chart, similar to that in Figure 7.8, that applies to the set of CPDs in (7.77) and write a corresponding algorithm, similar to Algorithm 7.2.

7.19 Two human experts independently give the following sets of CPDs to specify the JPD of three binary variables $X = \{X_1, X_2, X_3\}$:

$$\{p(x_1),\ p(x_2|x_1,x_3),\ p(x_3|x_1,x_2)\}.$$

The first expert gives the CPDs in Table 7.5, whereas the second expert gives the CPDs in Table 7.6. Determine whether each of these two sets specifies a unique and consistent JPD of the variables.

x_1	$p(x_1)$
0	0.3
1	0.7

x_1	x_3	x_2	$p(x_2\|x_1,x_3)$
0	0	0	0.40
0	0	1	0.60
0	1	0	0.40
0	1	1	0.60
1	0	0	0.20
1	0	1	0.80
1	1	0	0.20
1	1	1	0.80

x_1	x_2	x_3	$p(x_3\|x_1,x_2)$
0	0	0	0.90
0	0	1	0.10
0	1	0	0.70
0	1	1	0.30
1	0	0	0.50
1	0	1	0.50
1	1	0	0.60
1	1	1	0.40

TABLE 7.5. CPDs given by the first human expert.

x_1	$p(x_1)$
0	0.3
1	0.7

x_1	x_3	x_2	$p(x_2\|x_1,x_3)$
0	0	0	0.30
0	0	1	0.70
0	1	0	0.40
0	1	1	0.60
1	0	0	0.10
1	0	1	0.90
1	1	0	0.50
1	1	1	0.50

x_1	x_2	x_3	$p(x_3\|x_1,x_2)$
0	0	0	0.90
0	0	1	0.10
0	1	0	0.70
0	1	1	0.30
1	0	0	0.50
1	0	1	0.50
1	1	0	0.60
1	1	1	0.40

TABLE 7.6. CPDs given by the second human expert.

Chapter 8
Exact Propagation in Probabilistic Network Models

8.1 Introduction

In the previous chapters we presented and discussed different methodologies for building a coherent and consistent knowledge base for a probabilistic expert system. The knowledge base of a probabilistic expert system includes the joint probability distribution (JPD) for the variables involved in the model. Once the knowledge base has been defined, one of the most important tasks of an expert system is to draw conclusions when new information, or *evidence*, is observed. For example, in the field of medical diagnosis, the main task of an expert system consists of obtaining a diagnosis for a patient who presents some symptoms (evidence). The mechanism of drawing conclusions in probabilistic expert systems is called *propagation of evidence*,[1] or simply *propagation*. Propagation of evidence consists of updating the probability distributions of the variables according to the newly available evidence. For example, we need to calculate the conditional distribution of each element of a set of variables of interest (e.g., diseases) given the evidence (e.g., symptoms).

There are three types of algorithms for propagating evidence: exact, approximate, and symbolic. By an *exact propagation algorithm* we mean a method that, apart from precision or round-off errors, computes the proba-

---

[1]Some authors refer to propagation of evidence using other terminologies such as *uncertainty propagation*, *probabilistic inference*, etc.

bility distribution of the nodes exactly. In this chapter we discuss in detail some of the most commonly used exact propagation methods.

Approximate propagation algorithms compute the probabilities approximately. Approximate propagation methods are used in cases where no exact algorithm exists for the probabilistic model of interest or where exact algorithms are computationally intensive. These methods are presented in Chapter 9. Finally, a *symbolic propagation algorithm* can deal not only with numerical values, but also with symbolic parameters. As a result, these algorithms compute the probabilities in symbolic form (i.e., as a function of the symbolic parameters). Chapter 10 describes some symbolic propagation methods.

Some of the exact propagation methods described in this chapter are applicable to both Markov and Bayesian network models. Others are only applicable to Bayesian network models because they exploit their particular representation of the JPD. Also, some of these methods are applicable only to discrete network models, while others are applicable to both discrete and continuous network models. Section 8.2 introduces the problem of evidence propagation and analyzes some computational issues associated with it. We then start in Section 8.3 with the simplest propagation algorithm, which applies only to networks with polytree structures. This algorithm illustrates the basic ideas underlying all other propagation algorithms. Exact propagation of evidence methods for the case of networks with more general structures are presented in Sections 8.4–8.7, where *conditioning* (Section 8.5), *clustering* (Section 8.6), and *join tree* (Section 8.7) algorithms are introduced. In Section 8.8 we analyze the problem of goal-oriented propagation. Section 8.9 analyzes the case of continuous network models, showing how one can propagate evidence in Gaussian Bayesian network models.

8.2 Propagation of Evidence

One of the most important tasks of expert systems is the propagation of evidence, which allows queries to be answered when new evidence is observed. Suppose we have a set of discrete variables $X = \{X_1, \ldots, X_n\}$ and a JPD, $p(x)$, over X. Before any evidence is available, the propagation process consists of calculating the marginal probability distribution (MPD) $p(X_i = x_i)$, or simply $p(x_i)$, for each $X_i \in X$, which gives the information about the variables in the model before evidence has been observed.

Now, suppose that some evidence has become available, that is, a set of variables $E \subset X$ are known to take the values $X_i = e_i$, for $X_i \in E$. We refer to the variables in E as evidential variables.

Definition 8.1 Evidential variable. *The set of variables $E \subset X$ that are known to have certain values $E = e$ in a given situation are called evidential* variables. We refer to the rest of the variables, $X \setminus E$, as nonevidential variables, where $X \setminus E$ denotes the variables in X excluding the variables in E.

In this situation, propagation of evidence consists of calculating the conditional probabilities $p(x_i|e)$ for each nonevidential variable X_i given the evidence $E = e$. This conditional probability distribution measures the effect of the new evidence on the given variable. When no evidence is available ($E = \phi$), the conditional probabilities $p(x_i|e)$ are simply the MPD $p(x_i)$.

One way to compute $p(x_i|e)$ is to use the conditional probability formula (3.5) and obtain

$$p(x_i|e) = \frac{p(x_i, e)}{p(e)} \propto p(x_i, e), \tag{8.1}$$

where the proportionality constant is $1/p(e)$. Thus, to compute $p(x_i|e)$, we have only to compute and normalize $p(x_i, e)$. Now, $p(x_i, e)$ can be computed by

$$p(x_i, e) = \sum_{x \setminus \{x_i, e\}} p_e(x_1, \ldots, x_n), \tag{8.2}$$

where $p_e(x_1, \ldots, x_n)$ is the same as $p(x_1, \ldots, x_n)$ with the values of the evidential variables E fixed at the observed evidence e. Thus to compute $p(x_i, e)$, we need to sum $p_e(x_1, \ldots, x_n)$ over all nonevidential variables except X_i. For the case of no evidence, (8.2) reduces to

$$p(x_i) = \sum_{x \setminus x_i} p(x_1, \ldots, x_n). \tag{8.3}$$

Clearly, this brute force method becomes increasingly inefficient when dealing with a large number of variables. Note that for binary variables, (8.3) requires the summation of 2^{n-1} different probabilities. As shown in Figure 8.1, the computation time required for a personal computer to calculate $p(x_i)$ increases exponentially with the number of variables n, which makes this method computationally intensive and at times infeasible.

Equations (8.2) and (8.3) do not take into account the independency structure of the variables as given by the JPD, $p(x)$. When the JPD exhibits some independency structure, the number of calculations can be reduced significantly by exploiting the independency structure of the probabilistic model. The following example illustrates this point.

Example 8.1 Exploiting independency structure. Consider a set of variables $X = \{A, \ldots, G\}$ whose JPD, $p(x)$, can be factorized according to the directed acyclic graph (DAG) in Figure 8.2 as (see Section 6.4.4)

$$p(x) = \prod_{i=1}^{n} p(x_i|\pi_i) = p(a)p(b)p(c|a)p(d|a,b)p(e)p(f|d)p(g|d,e), \tag{8.4}$$

320   8. Exact Propagation in Probabilistic Network Models

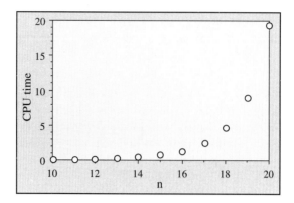

FIGURE 8.1. A scatter plot of the CPU time (minutes) needed to compute $p(x_i)$ in (8.3) versus the number of variables n.

where π_i is a realization of Π_i, the set of parents of X_i. Suppose that we wish to calculate the initial probabilities of the nodes, that is, when no evidence is available. We can obtain $p(x_i)$ by marginalizing the JPD using (8.3). For example, for the variable D, we have

$$p(d) = \sum_{x \setminus d} p(x) = \sum_{a,b,c,e,f,g} p(a,b,c,d,e,f,g). \qquad (8.5)$$

Considering the simplest case, that is, assuming that the variables in X are binary, the above summation includes $2^6 = 64$ different terms.

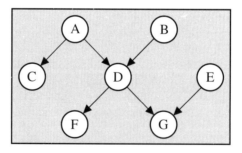

FIGURE 8.2. A Bayesian network.

A more efficient way of obtaining these probabilities is by taking advantage of the independence structure contained in the JPD, $p(x)$, which gives rise to the factorization in (8.4). This factorization allows us to simplify both the assessment of the JPD and the calculations required for the propagation of evidence. The number of operations can be substantially reduced by taking advantage of the local structure of the factorization by arranging the terms in the summation as follows:

$$p(d) = \sum_{a,b,c,e,f,g} p(a)p(b)p(c|a)p(d|a,b)p(e)p(f|d)p(g|d,e)$$

$$= \left(\sum_{a,b,c} p(a)p(b)p(c|a)p(d|a,b)\right)\left(\sum_{e,f,g} p(e)p(g|d,e)p(f|d)\right), \quad (8.6)$$

where each of the two summations can be calculated independently of the other. Thus, the original problem of marginalizing a JPD over six variables is reduced to marginalizing two functions depending only on three variables. Since the computation time is exponential in the number of variables, the computations are then greatly simplified by considering the local terms within the factorizations given by the independences contained in the graph. In this example, not only has the number of terms appearing in the summations been reduced from 64 to $2^3 + 2^3 = 16$, but the number of factors in each term reduces from 7 to 4 and from 7 to 3, respectively. A further saving can be obtained by rewriting the right-hand-side term in (8.6) as

$$\sum_a \left[p(a)\sum_c \left[p(c|a)\sum_b p(b)p(d|a,b)\right]\right]\sum_e \left[p(e)\sum_f \left[p(f|d)\sum_g p(g|d,e)\right]\right],$$

which gives an extra reduction in the number of terms appearing in each summation. ∎

The above example illustrates the savings that can be obtained by exploiting the independency structure of probabilistic models. In this chapter we describe several propagation algorithms that exploit such independency structure. We start in Section 8.3 with the most simple algorithm, which applies only to polytrees. This algorithm is subsequently modified in the remaining sections to suit other types of probabilistic models.

8.3 Propagation in Polytrees

One of the simplest probabilistic models is a Bayesian network model whose associated graph has a polytree structure. In this section we present an algorithm for propagating evidence in a polytree Bayesian network model (see Kim and Pearl (1983) and Pearl (1986b)). The main feature of this algorithm is that it runs in linear time with respect to the number of nodes in the network, as opposed to the brute force method, which requires an exponential number of computations (see Figure 8.1).

As we have seen in Chapter 4, in a polytree every two nodes are joined by a unique path, which implies that every node X_i divides the polytree

322   8. Exact Propagation in Probabilistic Network Models

into two disconnected polytrees: one includes the nodes that can be accessed through the parents of X_i and the other includes the nodes that can be accessed through the children of X_i. For example, node D divides the polytree in Figure 8.2 into two disconnected polytrees, the first of which, $\{A, B, C\}$, includes its parents and the nodes connected to D through its parents and the second, $\{E, F, G\}$, includes its children and the nodes connected to D through its children. This fact is illustrated in Figure 8.3, from which it can also be seen that the conditional independence statement (CIS) $I(\{A, B, C\}, \{E, F, G\}|D)$ holds in the directed graph.

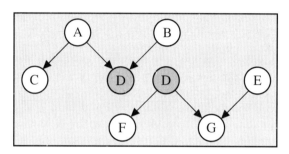

FIGURE 8.3. A typical node D divides a polytree in two disconnected polytrees.

Propagation of evidence can be done in an efficient way by combining the information coming from different subgraphs through passing messages (local computations) from one subgraph to another.

Suppose a set of evidential nodes E, with evidential values e, is given, and we wish to calculate the conditional probabilities $p(x_i|e)$ for all values x_i of a typical nonevidential node X_i. To facilitate the computations, the evidence E can be decomposed into two disjoint subsets, each corresponding to one of the two subgraphs separated by the node X_i in the polytree, as follows:

- E_i^+, the subset of E that can be accessed from X_i through its parents.

- E_i^-, the subset of E that can be accessed from X_i through its children.

Thus, $E = E_i^+ \cup E_i^-$. In some cases the notation $E_{X_i}^+$ is used in place of E_i^+ for clarity of exposition. Using (8.1) we can write

$$p(x_i|e) = p(x_i|e_i^-, e_i^+) = \frac{1}{p(e_i^-, e_i^+)} p(e_i^-, e_i^+|x_i) p(x_i).$$

Since X_i separates E_i^- from E_i^+ in the polytree, then the CIS $I(E_i^-, E_i^+|X_i)$ holds; hence we have

8.3 Propagation in Polytrees

$$p(x_i|e) = \frac{1}{p(e_i^-, e_i^+)} p(e_i^-|x_i) p(e_i^+|x_i) p(x_i)$$

$$= \frac{1}{p(e_i^-, e_i^+)} p(e_i^-|x_i) p(x_i, e_i^+)$$

$$= k\, p(e_i^-|x_i)\, p(x_i, e_i^+)$$

$$= k\, \lambda_i(x_i)\, \rho_i(x_i),$$

where $k = 1/p(e_i^-, e_i^+)$ is a normalizing constant;

$$\lambda_i(x_i) = p(e_i^-|x_i), \tag{8.7}$$

which accounts for the evidence coming through the children of X_i; and

$$\rho_i(x_i) = p(x_i, e_i^+), \tag{8.8}$$

which accounts for the evidence coming through the parents of X_i. Let

$$\beta_i(x_i) = \lambda_i(x_i)\rho_i(x_i) \tag{8.9}$$

be the unnormalized conditional probability $p(x_i|e)$, then

$$p(x_i|e) = k\, \beta_i(x_i) \propto \beta_i(x_i). \tag{8.10}$$

To compute the functions $\lambda_i(x_i)$, $\rho_i(x_i)$, and $\beta_i(x_i)$, suppose that a typical node X_i has p parents and c children. For notational simplicity, we denote the parents by $U = \{U_1, \ldots, U_p\}$ and the children by $Y = \{Y_1, \ldots, Y_c\}$, as illustrated in Figure 8.4.

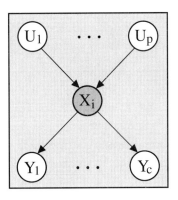

FIGURE 8.4. The parents and children of a typical node X_i.

Taking advantage of the polytree structure, the evidence E_i^+ can be partitioned into p disjoint components, one for each parent of X_i:

$$E_i^+ = \{E_{U_1 X_i}^+, \ldots, E_{U_p X_i}^+\}, \tag{8.11}$$

where the evidence $E^+_{U_j X_i}$ is the subset of E^+_i contained in the U_j-side of the link $U_j \to X_i$. Similarly, the evidence E^-_i can be partitioned into c disjoint components, one for each child of X_i:

$$E^-_i = \{E^-_{X_i Y_1}, \ldots, E^-_{X_i Y_c}\}, \qquad (8.12)$$

where $E^-_{X_i Y_j}$ is the subset of E^-_i contained in the Y_j-side of the link $X_i \to Y_j$. Figure 8.5 illustrates the different evidential sets associated with node X_i.

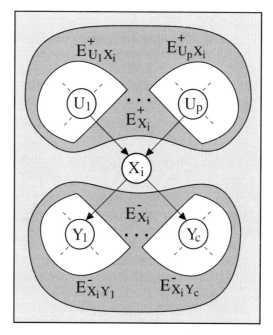

FIGURE 8.5. Partitioning the set E into subsets associated with the parents and children of a typical node X_i.

Let $u = \{u_1, \ldots, u_p\}$ be a realization (an instantiation) of the parents of X_i, then the function $\rho_i(x_i)$ can be computed as follows:

$$\begin{aligned}
\rho_i(x_i) &= p(x_i, e^+_i) = \sum_u p(x_i, u \cup e^+_i) \\
&= \sum_u p(x_i | u \cup e^+_i) p(u \cup e^+_i) \\
&= \sum_u p(x_i | u \cup e^+_i) p(u \cup e^+_{U_1 X_i} \cup \ldots \cup e^+_{U_p X_i}).
\end{aligned}$$

Due to the fact that $\{U_j, E^+_{U_j X_i}\}$ is independent of $\{U_k, E^+_{U_k X_i}\}$, for $j \ne k$, we have

$$\rho_i(x_i) = \sum_u p(x_i | u \cup e^+_i) \prod_{j=1}^p p(u_j \cup e^+_{U_j X_i})$$

$$= \sum_u p(x_i | u \cup e^+_i) \prod_{j=1}^p \rho_{U_j X_i}(u_j), \qquad (8.13)$$

where

$$\rho_{U_j X_i}(u_j) = p(u_j \cup e^+_{U_j X_i}) \qquad (8.14)$$

is the ρ-message that node U_j sends to its child X_i. This message depends only on the information contained in the U_j-side of the link $U_j \to X_i$. Note that if U_j is an evidential variable, then the corresponding message $\rho_{U_j X_i}(u_j)$ is the trivial function

$$\rho_{U_j X_i}(u_j) = \begin{cases} 1, & \text{if } u_j = e_j, \\ 0, & \text{if } u_j \ne e_j. \end{cases} \qquad (8.15)$$

The function $\lambda_i(x_i)$ can be computed in a similar way:

$$\lambda_i(x_i) = p(e^-_i | x_i) = p(e^-_{X_i Y_1}, \ldots, e^-_{X_i Y_s} | x_i).$$

Since X_i D-separates $E^-_{X_i Y_j}$ from $E^-_{X_i Y_k}$, for $j \ne k$, we have[2]

$$\lambda_i(x_i) = \prod_{j=1}^c \lambda_{Y_j X_i}(x_i), \qquad (8.16)$$

where

$$\lambda_{Y_j X_i}(x_i) = p(e^-_{X_i Y_j} | x_i) \qquad (8.17)$$

is the λ-message that node Y_j sends to its parent X_i.

From (8.13) we see that a node X_i can calculate its ρ-function $\rho_i(x_i)$ only after it receives all the ρ-messages from its parents. Similarly, from (8.16), the λ-function $\lambda_i(x_i)$ can be calculated only after X_i receives the λ-messages from all of its children. Figure 8.6 shows the different messages associated with a typical node X_i.

Substituting (8.13) and (8.16) in (8.10), we obtain

$$p(x_i | e) \propto \beta_i(x_i) = \left(\sum_u p(x_i | u \cup e^+_i) \prod_{j=1}^p \rho_{U_j X_i}(u_j) \right) \left(\prod_{j=1}^c \lambda_{Y_j X_i}(x_i) \right).$$

---

[2] Recall from Definition 5.4 that Z is said to D-separate X and Y, iff Z separates X and Y in the moral graph of the smallest ancestral set containing X, Y, and Z. Recall also from Definition 4.20 that an ancestral set is a set that contains the ancestors of all its nodes.

326   8. Exact Propagation in Probabilistic Network Models

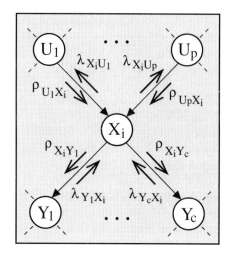

FIGURE 8.6. Typical ρ and λ messages sent to and from X_i.

Let us now calculate the different messages appearing in the above formula. Considering again the node X_i and a typical child Y_j and using the equality

$$E^+_{X_i Y_j} = E^+_i \bigcup_{k \neq j} E^-_{X_i Y_k},$$

we get

$$\begin{aligned}
\rho_{X_i Y_j}(x_i) &= p(x_i \cup e^+_{X_i Y_j}) \\
&= p(x_i \cup e^+_i \bigcup_{k \neq j} e^-_{X_i Y_k}) \\
&= p(e^+_i | x_i \bigcup_{k \neq j} e^-_{X_i Y_k}) p(x_i \bigcup_{k \neq j} e^-_{X_i Y_k}) \\
&= p(e^+_i | x_i) p(\bigcup_{k \neq j} e^-_{X_i Y_k} | x_i) p(x_i) \\
&\propto p(x_i | e^+_i) \prod_{k \neq j} p(e^-_{X_i Y_k} | x_i) \\
&\propto \rho_i(x_i) \prod_{k \neq j} \lambda_{Y_k X_i}(x_i). \quad (8.18)
\end{aligned}$$

Note that (8.18) is valid when X_i is an evidential node by assigning to node X_i the ρ-function $\rho_i(x_i) = 1$ if $x_i = e_i$, and $\rho_i(x_i) = 0$ if $x_i \neq e_i$. In this situation, the value of $\rho_{X_i Y_j}(x_i)$ resulting from (8.18) is the same given in (8.15). This will simplify the implementation of this propagation method.

On the other hand, in order to calculate $\lambda_{Y_j X_i}(x_i)$, let $V = \{V_1, \ldots, V_q\}$ be the set of parents of Y_j other than X_i. Thus the child Y_j has $q + 1$

8.3 Propagation in Polytrees

parents, as illustrated in Figure 8.7. Then,

$$e^-_{X_i Y_j} = e^-_{Y_j} \cup e^+_{V Y_j},$$

where $e^+_{V Y_j}$ denotes the evidence coming through the parents of Y_j, other than X_i. Thus, we have

$$\begin{aligned}
\lambda_{Y_j X_i}(x_i) &= p(e^-_{X_i Y_j}|x_i) = \sum_{y_j, v} p(y_j, v, e^-_{X_i Y_j}|x_i) \\
&= \sum_{y_j, v} p(y_j, v, e^-_{Y_j}, e^+_{V Y_j}|x_i) \\
&= \sum_{y_j, v} p(e^-_{Y_j}|y_j, v, e^+_{V Y_j}, x_i) p(y_j|v, e^+_{V Y_j}, x_i) p(v, e^+_{V Y_j}|x_i) \\
&= \sum_{y_j} p(e^-_{Y_j}|y_j) \sum_v p(y_j|v, x_i) p(v, e^+_{V Y_j}),
\end{aligned} \qquad (8.19)$$

where the last equality is obtained by considering different independences among the evidential subsets. Equation (8.19) can be written as

$$\lambda_{Y_j X_i}(x_i) = \sum_{y_j} \lambda_{Y_j}(y_j) \sum_{v_1, \ldots, v_q} p(y_j|\pi_{Y_i}) \prod_{k=1}^q \rho_{V_k Y_j}(v_k). \qquad (8.20)$$

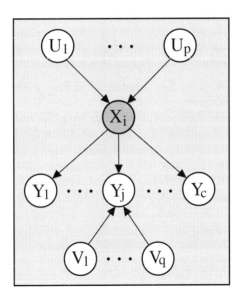

FIGURE 8.7. The parents of one child Y_j of node X_i.

From Equations (8.13), (8.16), (8.18), and (8.20) we can conclude the following:

328   8. Exact Propagation in Probabilistic Network Models

- Equation (8.13) shows that $\rho_i(x_i)$ can be computed as soon as node X_i receives the ρ-messages from all of its parents.

- Equation (8.16) shows that $\lambda_i(x_i)$ can be computed as soon as node X_i receives the λ-messages from all of its children.

- Equation (8.18) shows that node X_i can send the message $\rho_{X_i Y_j}(x_i)$ to its child Y_j as soon as it knows its own $\rho_i(x_i)$ and receives the λ-messages from all of its other children.

- Equation (8.20) shows that node X_i can send the message $\lambda_{X_i U_j}(u_j)$ to its parent U_j as soon as it knows its own $\lambda_i(x_i)$ and receives the ρ-messages from all of its other parents.

The above discussion suggests an iterative algorithm for computing $p(x_i|e)$ for all nonevidential nodes X_i.

Algorithm 8.1 Propagation in Polytrees.

- **Input:** A Bayesian network model (D, P) over a set of variables X and a set of evidential nodes E with evidential values $E = e$, where D is a polytree.

- **Output:** The CPD $p(x_i|e)$ for every nonevidential node X_i.

Initialization Steps:

1. For all evidential nodes $X_i \in E$ set

    - $\rho_i(x_i) = 1$ if $x_i = e_i$, or $\rho_i(x_i) = 0$ if $x_i \neq e_i$.
    - $\lambda_i(x_i) = 1$ if $x_i = e_i$, or $\lambda_i(x_i) = 0$ if $x_i \neq e_i$.

    (In effect, this reduces the feasible values of X_i to those matching the evidence and facilitates the message-passing process for evidential nodes.)

2. For all nonevidential nodes X_i with no parents, set $\rho_i(x_i) = p(x_i)$.

3. For all nonevidential nodes X_i with no children, set $\lambda_i(x_i) = 1$ for all x_i.

Iteration Steps:

4. For every nonevidential node X_i do:

    (a) If X_i has received the ρ-messages from all of its parents, then calculate $\rho_i(x_i)$ using (8.13).

    (b) If X_i has received the λ-messages from all of its children, then calculate $\lambda_i(x_i)$ using (8.16).

(c) If $\rho_i(x_i)$ has been calculated, then for every child Y_j of X_i such that X_i has received the λ-messages from all of its other children, calculate and send the message $\rho_{X_i Y_j}(x_i)$ using (8.18). This implies that if X_i has received the λ-messages from all of its children, then it can send its ρ-message to all of them.

(d) If $\lambda_i(x_i)$ has been calculated, then for every parent U_j of X_i such that X_i has received the ρ-messages from all of its other parents, calculate and send the message $\lambda_{X_i U_j}(u_i)$ using (8.20). This implies that if X_i has received the ρ-messages from all of its parents, then it can send its λ-message to all of them.

5. Repeat Step 4 as many times as needed to calculate the ρ- and λ-functions for all nonevidential nodes, that is, until no new message is calculated in one iteration step.

6. For all nonevidential nodes X_i, compute $\beta_i(x_i)$ using (8.9). This is the unnormalized probability $p(x_i|e)$.

7. For all nonevidential nodes X_i compute $p(x_i|e)$ by normalizing $\beta_i(x_i)$, that is, $p(x_i|e) = \beta_i(x_i)/k$, where $k = \sum_{x_i} \beta_i(x_i)$. ∎

Note that during the propagation process, the functions ρ and λ are calculated at different times for each node. Thus, if we are interested in only one goal variable X_i, the algorithm can be stopped immediately after the functions $\rho_i(x_i)$ and $\lambda_i(x_i)$ have been calculated. Further simplifications of the propagation process for a goal-oriented implementation are presented in Section 8.8.

The structure of message-passing used in Algorithm 8.1 makes it suitable for a parallel implementation. Assume that each node has its own processor. The processor of a typical node X_i needs the following information to compute the conditional probability $p(x_i|e)$:

- Two lists: one contains the parents of X_i and the other contains the children of X_i. This information is independent of the evidence E.

- The conditional probability distribution (CPD), $p(x_i|\pi_{X_i})$. This distribution is also independent of the evidence E. If X_i has no parents, then $p(x_i|\pi_{X_i}) = p(x_i)$, which is the MPD of X_i.

- The function $\rho_i(x_i)$, which is computed at X_i using (8.13).

- The function $\lambda_i(x_i)$, which is computed at X_i using (8.16).

- The message $\rho_{U_j X_i}(u_j)$, received from every parent U_j of X_i. It is computed at U_j, using (8.18).

- The message $\lambda_{Y_j X_i}(x_i)$, received from every child Y_j of X_i. It is computed at Y_j, using (8.20).

Once all the above information has been received in the processor associated with node X_i, it can calculate the unnormalized probabilities $\beta_i(x_i)$, using (8.9), and then normalize to obtain $p(x_i|e)$.

On the other hand, every processor has to perform the following computations to send the necessary messages to the corresponding neighbors:

- The message $\rho_{X_i Y_j}(x_i)$, for any children Y_j of X_i. It is computed at X_i, using (8.18), and is sent, as a message, to its child Y_j.

- The message $\lambda_{X_i U_j}(u_j)$, for any parent U_j of X_i. It is computed at X_i, using (8.20), and is sent, as a message, to its parent U_j.

The computations performed at a typical node X_i and the received and sent messages are shown schematically in Figure 8.8, where hardware and software implementations of the model are suggested. This figure illustrates the basic operations needed for a parallel implementation of the algorithm (for further discussion see Díez and Mira (1994)).

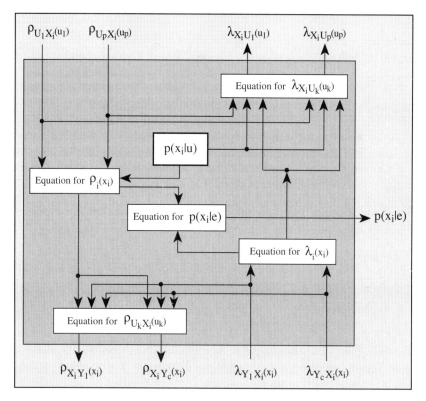

FIGURE 8.8. The computations performed at a typical node X_i and the messages that are received from and sent to the parents and children.

8.3 Propagation in Polytrees

The complexity of Algorithm 8.1 is linear in the number of nodes of the network (the size of the network). Thus, propagation of evidence in polytrees can be carried out using this efficient algorithm. In more general structures, propagation of evidence has been proven to be NP-hard[3] (see Cooper (1990)). This means that there is no linear-time propagation algorithm for Bayesian network models with general topology. Sections 8.5–8.7 present several exact propagation methods for network models with general structure.

Some modifications of the algorithm for exact propagation in polytrees, which improve its efficiency, are presented in Peot and Shachter (1991) and in Delcher et al. (1995).

Example 8.2 Propagation in polytrees (I). Consider the Bayesian network introduced in Example 8.1 whose DAG is given in Figure 8.2. The joint probability distribution of the model can be factorized as

$$p(a,b,c,d,e,f,g) = p(a)p(b)p(c|a)p(d|a,b)p(e)p(f|d)p(g|d,e).$$

For the sake of simplicity, suppose that the variables are binary. The associated sets of CPDs are given in Table 8.1.

We first consider the case of no evidence and use the polytree propagation Algorithm 8.1 to calculate the MPD, $p(x_i)$, of the nodes in the graph. Then, we introduce an evidential node $E = \{D\}$ with evidential value $e = \{D = 0\}$ and use the same algorithm to calculate the conditional probabilities $p(x_i|e)$ of all nonevidential nodes.

For illustrative purpose, we carry out the steps of Algorithm 8.1 sequentially, following the alphabetical ordering of the nodes in every iteration step. In the case of no evidence, Algorithm 8.1 proceeds as follows:

Initialization Steps:

- The first step of the initialization algorithm does not apply because there is no evidence available.

- The ρ functions for the nodes without parents, $A, B,$ and E, are

$$\rho_A(a) = p(a), \quad \rho_B(b) = p(b), \quad \rho_E(e) = p(e).$$

Thus, from Table 8.1, we have

$$\rho_A(0) = 0.3, \quad \rho_B(0) = 0.6, \quad \rho_E(0) = 0.1,$$
$$\rho_A(1) = 0.7, \quad \rho_B(1) = 0.4, \quad \rho_E(1) = 0.9.$$

---

[3] An introduction to algorithm complexity and NP-hard problems is found in the book by Garey and Johnson (1979).

332   8. Exact Propagation in Probabilistic Network Models

a	p(a)
0	0.3
1	0.7

b	p(b)
0	0.6
1	0.4

e	p(e)
0	0.1
1	0.9

a	c	p(c|a)
0	0	0.25
0	1	0.75
1	0	0.50
1	1	0.50

d	f	p(f|d)
0	0	0.80
0	1	0.20
1	0	0.30
1	1	0.70

a	b	d	p(d|a,b)
0	0	0	0.40
0	0	1	0.60
0	1	0	0.45
0	1	1	0.55
1	0	0	0.60
1	0	1	0.40
1	1	0	0.30
1	1	1	0.70

d	e	g	p(g|d,e)
0	0	0	0.90
0	0	1	0.10
0	1	0	0.70
0	1	1	0.30
1	0	0	0.25
1	0	1	0.75
1	1	0	0.15
1	1	1	0.85

TABLE 8.1. Conditional probability distributions for the Bayesian network in Example 8.2.

- The λ-functions for the nodes without children, C, F, and G, are

$$\lambda_C(0) = 1.0, \quad \lambda_F(0) = 1.0, \quad \lambda_G(0) = 1.0,$$
$$\lambda_C(1) = 1.0, \quad \lambda_F(1) = 1.0, \quad \lambda_G(1) = 1.0.$$

Figure 8.9 illustrates the ρ- and λ-functions calculated in the initialization step. The numbers in parentheses indicate the order in which the functions are calculated.

Iteration Step 1:

- *Node A:* Applying the rules in the iteration step of Algorithm 8.1 to node A, we obtain

    (a) $\rho_A(a)$ has been calculated in the initialization step.

    (b) $\lambda_A(a)$ cannot be calculated, because A did not receive the λ-messages from its children C and D.

    (c) $\rho_A(a)$ has been calculated, but A cannot send the messages $\rho_{AC}(a)$ and $\rho_{AD}(a)$ to its children, because it has not received the λ-messages from D and C, respectively.

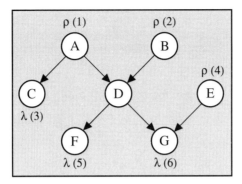

FIGURE 8.9. Initialization step of the polytree propagation algorithm.

(d) Because node A has no parents, no λ-message has to be sent from this node.

Thus, no calculation associated with node A can be performed at this stage.

- Node B: Since D is the only child of node B, the message ρ_{BD} can be calculated and sent to node D using (8.18):

$$\rho_{BD}(b) = \rho_B(b) \prod_{y_j \backslash d} \lambda_{Y_j B}(b),$$

where Y_j is the jth child of node B. Since B has only one child, this equation reduces to the simple expression $\rho_{BD}(b) = \rho(b)$. Thus, we have $(\rho_{BD}(0), \rho_{BD}(1)) = (0.6, 0.4)$.

- Node C: Since A is the only parent of node C, the message λ_{CA} can be computed using (8.20), which gives

$$\lambda_{CA}(a) = \sum_c \lambda_C(c) p(c|a),$$

from which we have

$$\begin{aligned}
\lambda_{CA}(0) &= \lambda_C(0)\, p(C=0|A=0) + \lambda_C(1)\, p(C=1|A=0) \\
&= 1 \times 0.25 + 1 \times 0.75 = 1.00,
\end{aligned}$$

$$\begin{aligned}
\lambda_{CA}(1) &= \lambda_C(0)\, p(C=0|A=1) + \lambda_C(1)\, p(C=1|A=1) \\
&= 1 \times 0.5 + 1 \times 0.5 = 1.00.
\end{aligned}$$

- Node D: No calculations can be made, because node D did not receive any message from its parents or children.

- **Node E:** Since node E has only one child, G, the message $\rho_{EG}(e)$ can be calculated and sent to node G using (8.18). Proceeding in the same way as in the case of node B, we obtain $\rho_{EG}(e) = \rho(e)$. Thus, we have $(\rho_{EG}(0), \rho_{EG}(1)) = (0.1, 0.9)$.

- **Node F:** The situation for node F is similar to that of node C. Thus, we obtain
$$\lambda_{FD}(d) = \sum_f \lambda(f) p(f|d),$$
from which we have $(\lambda_{FD}(0), \lambda_{FD}(1)) = (1.0, 1.0)$.

- **Node G:** Node G has two parents, D and E; $\lambda_G(g)$ was calculated in the initialization step, and G received the ρ-message from node E. Thus, node G can calculate and send the λ-message to its other parent, D. Using (8.20) we have
$$\lambda_{GD}(d) = \sum_g \lambda_G(g) \sum_e p(g|d, e) \rho_{EG}(e),$$
from which we obtain
$$\begin{aligned}
\lambda_{GD}(0) &= \lambda_G(0) \sum_e p(G=0|D=0, e) \rho_{EG}(e) \\
&\quad + \lambda_G(1) \sum_e p(G=1|D=0, e) \rho_{EG}(e) \\
&= 1.0 \times (0.9 \times 0.1 + 0.7 \times 0.9) \\
&\quad + 1.0 \times (0.1 \times 0.1 + 0.3 \times 0.9) = 1.0,\\
\lambda_{GD}(1) &= \lambda_G(0) \sum_e p(G=0|D=1, e) \rho_{EG}(e) \\
&\quad + \lambda_G(1) \sum_e p(G=1|D=1, e) \rho_{EG}(e) \\
&= 1.0 \times (0.25 \times 0.1 + 0.15 \times 0.9) \\
&\quad + 1.0 \times (0.75 \times 0.1 + 0.85 \times 0.9) = 1.0.
\end{aligned}$$

Thus, we have the message $(\lambda_{GD}(0), \lambda_{GD}(1)) = (1.0, 1.0)$.

Figure 8.10 illustrates the order of calculating the ρ- and λ-functions and the messages calculated in the first iteration step. All the functions corresponding to the previous steps are dimmed.

Iteration Step 2:

- **Node A:** Node A has two children, C and D. The function $\rho_A(a)$ was computed, and A received the λ-message from node C. Thus, it can

8.3 Propagation in Polytrees   335

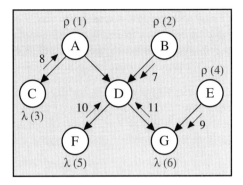

FIGURE 8.10. First iteration step of the polytree propagation algorithm.

calculate and send the ρ-message to D using (8.18):

$$\rho_{AD}(a) = \rho_A(a)\lambda_{CA}(a).$$

Thus, we have $(\rho_{AD}(0), \rho_{AD}(1)) = (0.3, 0.7)$.

- Nodes B and C have not received the messages from nodes D and A, respectively. Thus, no calculation can be made at nodes B and C in this iteration step.

- *Node D:* Node D received the ρ-messages from its two parents A and B. Therefore, $\rho_D(d)$ can be calculated using (8.13):

$$\rho_D(d) = \sum_{a,b} p(d|a,b)\rho_{AD}(a)\rho_{BD}(b),$$

$$\begin{aligned}
\rho_D(0) &= p(D=0|A=0, B=0)\,\rho_{AD}(0)\rho_{BD}(0) \\
&+ p(D=0|A=0, B=1)\,\rho_{AD}(0)\rho_{BD}(1) \\
&+ p(D=0|A=1, B=0)\,\rho_{AD}(1)\rho_{BD}(0) \\
&+ p(D=0|A=1, B=1)\,\rho_{AD}(1)\rho_{BD}(1) \\
&= 0.4 \times 0.3 \times 0.6 + 0.45 \times 0.3 \times 0.4 \\
&+ 0.6 \times 0.7 \times 0.6 + 0.3 \times 0.7 \times 0.4 = 0.462.
\end{aligned}$$

Similarly, for $D = 1$ we get $\rho_D(1) = 0.538$. Additionally, node D received the λ-messages from its two children, F and G. This implies that $\lambda_D(d)$ can be calculated using (8.16):

$$\lambda_D(d) = \lambda_{FD}(d)\lambda_{GD}(d),$$

which gives $(\lambda_D(0), \lambda_D(1)) = (1.0, 1.0)$.

Now, node D has received messages from all of its parents and children. Thus, it can send the ρ- and λ- messages to its children and parents, respectively. Thus, for example, using (8.18) we can compute the messages $\rho_{DF}(d)$ and $\rho_{DG}(d)$ as follows:

$$\rho_{DF}(d) = \rho_D(d)\lambda_{GD}(d),$$
$$\rho_{DG}(d) = \rho_D(d)\lambda_{FD}(d).$$

Similarly, using (8.20) we can compute the messages $\lambda_{DA}(a)$ and $\lambda_{DB}(b)$ as follows:

$$\lambda_{DA}(a) = \sum_d \lambda_D(d) \sum_b p(d|a,b)\rho_{BD}(b),$$
$$\lambda_{DB}(b) = \sum_d \lambda_D(d) \sum_a p(d|a,b)\rho_{AD}(a).$$

The numerical values corresponding to these messages are given in Figure 8.13.

- Node E has not received the message from its child, G. Thus, no calculation can be made at node E.

- Node F: Node F has received the message $\rho_{DF}(d)$ from its only parent, D. Thus, it can compute the function $\rho_F(f)$:

$$\rho_F(f) = \sum_d p(f|d)\rho_{DF}(d),$$

which gives $(\rho_F(0), \rho_F(1)) = (0.531, 0.469)$.

- Node G: Node G has received the two ρ-messages coming from its two parents, D and E. Thus, the function $\rho_G(g)$ can be calculated using (8.13):

$$\rho_G(g) = \sum_{d,e} p(g|d,e)\rho_{DG}(d)\rho_{EG}(e).$$

Also, $\lambda_G(g)$ has already been computed. Thus, we can calculate and send the λ-message to E. Using (8.20) we have:

$$\lambda_{GE}(e) = \sum_g \lambda_G(g) \sum_d p(g|d,e)\rho_{DG}(d).$$

Figure 8.11 illustrates the order of the messages and functions calculated in the above iteration step.

Proceeding as in the two previous iteration steps, in the last iteration step we calculate the messages and functions $\lambda_A(a)$, $\rho_{AC}(a)$, $\lambda_B(b)$, $\rho_C(c)$, and $\lambda_E(e)$. Figure 8.12 illustrates the last step of the algorithm. From this

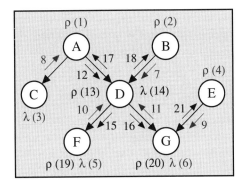

FIGURE 8.11. The second iteration step of the polytree propagation algorithm.

figure we can see that all the messages have been sent, and the ρ- and λ-functions for all nodes have been calculated. Thus, Steps 4 and 5 are now complete.

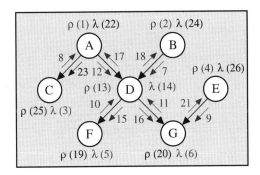

FIGURE 8.12. The last iteration step of the polytree propagation algorithm.

- In Step 6, we calculate the β-functions. Since there is no evidence, we have constant λ functions and messages for all nodes. Thus, in this case, $\beta_i(x_i) = \rho_i(x_i)$ for all nodes X_i.

- In Step 7, the MPD $p(x_i)$ is obtained by normalizing the corresponding β-functions. In this case, however, no normalization is needed, as shown in Figure 8.13, which gives complete information on all messages and functions calculated in the propagation process. ∎

Note that when no evidence is available, all of the λ-functions and messages (see Equations (8.16) and (8.20)) are equal to one. Thus, in the case of no evidence, they need not be calculated. However, when some evidence is available, the λ-messages need to be calculated because they propagate the information in the evidence from children to parents. This fact is illustrated in the following example.

338   8. Exact Propagation in Probabilistic Network Models

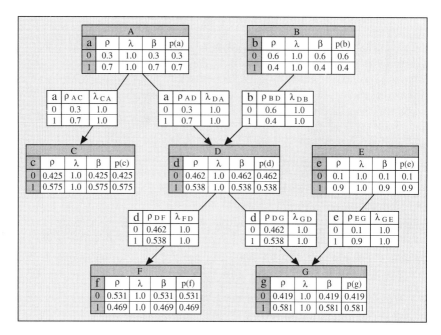

FIGURE 8.13. Numerical values of the messages and functions calculated by the polytree propagation algorithm when no evidence is available.

Example 8.3 Propagation in polytrees (II). Continuing with the Bayesian network of Example 8.2, but now suppose that we are given the evidence $D = 0$ and that we wish to update the probabilities of the nodes accordingly. Algorithm 8.1 proceeds as follows:

Initialization Steps:

- We have an evidential node $E = \{D\}$ with $e = \{D = 0\}$. Thus, we initialize the ρ and λ functions as follows:

$$\rho_D(0) = 1.0, \quad \lambda_D(0) = 1.0,$$
$$\rho_D(1) = 0.0, \quad \lambda_D(1) = 0.0.$$

- The ρ functions for the nodes without parents are calculated directly from the MPDs:

$$\rho_A(0) = 0.3, \quad \rho_B(0) = 0.6, \quad \rho_E(0) = 0.1,$$
$$\rho_A(1) = 0.7, \quad \rho_B(1) = 0.4, \quad \rho_E(1) = 0.9.$$

8.3 Propagation in Polytrees 339

- The λ functions for the nodes without children, E, F, and G, are assigned a constant value:

$$\lambda_C(0) = 1.0, \quad \lambda_F(0) = 1.0, \quad \lambda_G(0) = 1.0,$$
$$\lambda_C(1) = 1.0, \quad \lambda_F(1) = 1.0, \quad \lambda_G(1) = 1.0.$$

Iteration Step 1:

- The calculations corresponding to the first iteration step are the same as in the case of no evidence. Therefore, the messages $\rho_{BD}(b), \lambda_{CA}(a), \rho_{EG}(e), \lambda_{FD}(d), \lambda_{GD}(d)$ are computed as in the case of no evidence.

Iteration Step 2:

- *Node A:* The message $\rho_{AD}(a)$ is computed as in the case of no evidence, using (8.18). Thus, we have the message $(\rho_{AD}(0), \rho_{AD}(1)) = (0.3, 0.7)$.

- As in the case of no evidence, no calculation can be made with nodes B and C in this iteration step.

- *Node D:* The ρ and λ functions of node D have been calculated in the initialization step. Node D has received all messages from its parents and children. Thus, it can send the ρ and λ messages to its children and parents, respectively. Using (8.18) we compute the messages $\rho_{DF}(d)$ and $\rho_{DG}(d)$ as follows:

$$\rho_{DF}(d) = \rho_D(d)\lambda_{GD}(d),$$
$$\rho_{DG}(d) = \rho_D(d)\lambda_{FD}(d).$$

For example, the numerical values associated with $\rho_{DF}(d)$ are

$$\rho_{DF}(0) = \rho_D(0)\lambda_{GD}(0) = 1.0 \times 1.0 = 1.0,$$
$$\rho_{DF}(1) = \rho_D(1)\lambda_{GD}(1) = 0.0 \times 1.0 = 0.0.$$

Similarly, for the message $\rho_{DG}(d)$ we have $\rho_{DG}(0) = 1.0$ and $\rho_{DG}(1) = 0.0$. Using (8.18) we can compute the messages $\lambda_{DA}(a)$ and $\lambda_{DB}(b)$ as follows:

$$\lambda_{DA}(a) = \sum_d \lambda_D(d) \sum_b p(d|a,b)\rho_{BD}(b),$$
$$\lambda_{DB}(b) = \sum_d \lambda_D(d) \sum_a p(d|a,b)\rho_{AD}(a).$$

For example, for $\lambda_{DA}(a)$ we have

340   8. Exact Propagation in Probabilistic Network Models

$$\lambda_{DA}(0) = \lambda_D(0) \sum_b p(D=0|A=0,b)\rho_{BD}(b)$$
$$+ \lambda_D(1) \sum_b p(D=1|A=0,b)\rho_{BD}(b)$$
$$= 1.0 \times (0.4 \times 0.6 + 0.45 \times 0.4)$$
$$+ 0.0 \times (0.6 \times 0.6 + 0.55 \times 0.4) = 0.42,$$

$$\lambda_{DA}(0) = \lambda_D(0) \sum_b p(D=0|A=1,b)\rho_{BD}(b)$$
$$+ \lambda_D(1) \sum_b p(D=1|A=1,b)\rho_{BD}(b)$$
$$= 1.0 \times (0.6 \times 0.6 + 0.3 \times 0.4)$$
$$+ 0.0 \times (0.4 \times 0.6 + 0.7 \times 0.4) = 0.48.$$

- Just as the case of no evidence, no calculation can be made with node E. Node F computes the function $\rho_F(f)$, and node G calculates the function $\rho_G(g)$ and sends the λ-message $\lambda_{GE}(e)$ to E.

Figure 8.14 shows the numerical values associated with the remaining messages.

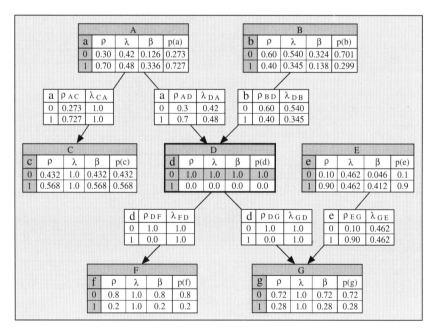

FIGURE 8.14. Numerical values of the messages and functions calculated by the polytree propagation algorithm with evidence $D = 0$.

Figure 8.15(a) shows the MPDs of the nodes when no evidence is observed, and Figure 8.15(b) shows the conditional probabilities obtained when the evidence $D = 0$ is considered.[4] From these figures, it can be shown that node E is not affected by the evidence (the marginal and conditional probabilities are the same). However, the probability of some nodes such as F and G are substantially modified. The dependency structure of the graph allows us to determine which variables will be influenced by the evidence, but not the degree to which the probabilities are modified. Chapter 10 introduces symbolic propagation methods that allow us to determine the degree as well. ∎

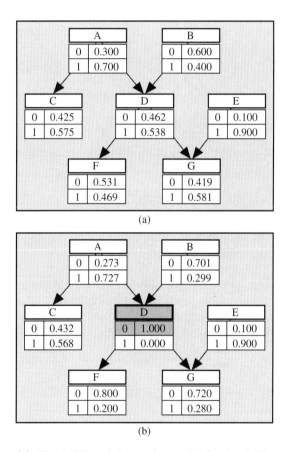

FIGURE 8.15. (a) The MPDs of the nodes and (b) The CPDs with evidence $D = 0$.

---

[4]The probabilistic network expert system *X-pert Nets* and the files needed for solving this example can be obtained from the World Wide Web site http://ccaix3.unican.es/˜AIGroup.

8.4 Propagation in Multiply-Connected Networks

The propagation method for polytrees developed in Section 8.3 applies only to networks with polytree structures. However, this type of structure lacks generality and in practice one often encounters cases in which the probabilistic model cannot be represented by a polytree. In these cases, we may have to work with multiply-connected networks (graphs that contain loops). Two propagation methods for multiply-connected networks are the *conditioning* and *clustering* methods. Conditioning involves breaking the communication pathways along the loops by instantiating a selected group of variables (see Pearl (1986a) and Suermondt and Cooper (1991b)). This results in a singly-connected network in which the polytree propagation algorithm can be applied. Clustering consists of building associated graphs in which each node corresponds to a set of variables rather than just to a single variable (e.g., a join tree). This also results in a network with polytree structure (see Lauritzen and Spiegelhalter (1988), Jensen, Olesen, and Andersen (1990), and Shachter, Andersen, and Szolovits (1994)).

Although the computational complexity of the polytree algorithm is linear in the number of nodes, propagation of evidence in multiply-connected networks is NP-hard (see Cooper (1990)). In general, both conditioning and clustering algorithms suffer from this complexity problem. However, the special characteristics of these methods render each efficient for different types of network structures and neither of them is more efficient in general (Suermondt and Cooper (1991a)). This has motivated the appearance of some mixed algorithms that combine the advantages of both methods (see, for example, Suermondt and Cooper (1991a) and Suermondt, Cooper, and Heckerman (1991)).

Section 8.5 introduces the conditioning algorithm. Since conditioning uses the polytree algorithm, it is only applicable to Bayesian network models. However, clustering algorithms (Section 8.6) can be applied to both Markov and Bayesian network models. Thus, the clustering algorithm is more general than the conditioning algorithm.

8.5 Conditioning Method

When dealing with multiply-connected graphs, the property that a typical node separates the graph into two disconnected parts does not hold. Consequently, some of the assumptions used in the polytrees algorithm cannot be applied in this situation. For example, suppose we have the graph given in Figure 8.16, resulting from the addition of the link $C \to F$ to the polytree given in Figure 8.2. The addition of this link introduces a loop involving the variables A, C, D, and F, which renders the graph multiply-connected. In this situation, none of the nodes contained in the loop D-separates the graph into two disconnected subgraphs. For example, we can associate the subgraphs $\{A, B, C\}$ and $\{E, F, G\}$ with node D, one containing its parents and the other containing its children, respectively. However, these two subgraphs are not D-separated by node D because the link $C \to F$ provided by the loop is a communication pathway between them.

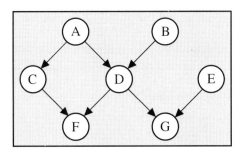

FIGURE 8.16. Multiply-connected graph.

The idea behind the conditioning algorithm is to break the communication pathways along the loops by instantiating a selected subset of nodes. This set of nodes is referred to as a *loop-cutset*. For example, node D does not D-separate the graph in Figure 8.16 into two subgraphs, but if we consider the loop-cutset formed by node C, then $\{C, D\}$ D-separates $\{A, B\}$ from $\{E, F, G\}$, two subgraphs containing the parents and the children of D, respectively. Thus, we can *break* the loop by considering C as an evidential node, that is, by instantiating it to an arbitrary value.

This idea of breaking loops can be practically implemented by using a method called *evidence absorption* (see Shachter (1988, 1990a)). This method shows that evidence can be *absorbed* by the graph, thereby changing its topology. More specifically, if X_i is an evidential node, we can remove any link outgoing from X_i, say $X_i \to X_j$, by substituting the CPD associated with node X_j, $p(x_j|\pi_j)$, by a CPD one dimension smaller:

$$p_1(x_j|\pi_j \setminus x_i) = p(x_j|\pi_j \setminus x_i, X_i = e_i).$$

This operation will keep the CPDs unchanged. Note that the set $\Pi_j \setminus X_i$ is the new set of parents of node X_j in the modified graph. For example, if we instantiate node C to an arbitrary value $C = c$ in the graph in Figure 8.16, we can absorb the evidence $C = c$ in the link $C \to F$, obtaining a new graph with polytree structure (see Figure 8.17). To keep the CPD of the nonevidential variables $p(y|C = c)$ unchanged, we replace the CPD $p(f|c, d)$ by $p_1(f|d) = p(f|C = c, d)$, which eliminates the dependency of node F on the evidential variable C.

Using the evidence absorption method, we can reduce a multiply-connected graph to a polytree structure by instantiating the nodes in a loop-cutset $C = \{C_1, \ldots, C_m\}$. Then, the polytrees propagation algorithm introduced

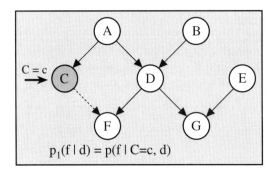

FIGURE 8.17. Absorbing evidence $C = c$ through the link $C \to F$.

in Section 8.3 can be applied to compute the probabilities $p(x_i|e, c_1, \ldots, c_m)$ for each of the instantiations of the loop-cutset $(c_1, \ldots, c_m)$, given some evidence $E = e$. The conditional probability of any node can then be obtained by combining the different probabilities obtained for the different loop-cutset instantiations:

$$p(x_i|e) = \sum_{c_1,\ldots,c_m} p(x_i|e, c_1, \ldots, c_m) p(c_1, \ldots, c_m|e). \qquad (8.21)$$

The complexity of this algorithm results from the number of different instantiations that must be considered. This implies that the complexity grows exponentially with the size of the loop-cutset. Thus, it is desirable to obtain a loop-cutset with the minimum number of nodes.

The problem of finding a minimal loop-cutset has proven to be NP-hard (Suermondt and Cooper (1990)), but there are several efficient heuristic methods to obtain loop-cutsets (see, for example, Stillman (1991) and Suermondt and Cooper (1990)). Some of these methods provide a bound for the size of the set. For example, Becker and Geiger (1994) introduce an algorithm for obtaining a loop-cutset that contains fewer than 2 times the number of variables contained in a minimal loop-cutset.

Note that in (8.21) the weights $p(c_1, \ldots, c_m|e)$ cannot be directly calculated because the evidence e does not render the graph a polytree. Pearl (1986a), Peot and Shachter (1991), and Suermondt and Cooper (1991b) present several algorithms for the calculation of these weights. For example, the algorithm given by Suermondt and Cooper (1991b) uses the decomposition of the weights

$$p(c_1, \ldots, c_m|e) = \frac{p(e|c_1, \ldots, c_m) p(c_1, \ldots, c_m)}{p(e)},$$

which when substituted in (8.21), gives

$$p(x_i|e) \propto \sum_{c_1,\ldots,c_m} p(x_i|e, c_1, \ldots, c_m) p(e|c_1, \ldots, c_m) p(c_1, \ldots, c_m). \qquad (8.22)$$

Thus, the desired probabilities can be obtained by computing three different functions for every possible combination of values of the variables in C. As we have already mentioned, $p(x_i|c,e)$ can be calculated using the polytree algorithm. Similarly, $p(e|c)$ can be calculated using the polytrees algorithm with evidence c. The MPD $p(c)$ is calculated by sequentially instantiating the nodes in the loop-cutset in such a way that only one part of the multiply-connected graph with polytree structure is needed in the propagation process. For details see Suermondt and Cooper (1991b).

The conditioning algorithm is illustrated by the following example.

Example 8.4 Conditioning algorithm. Consider the multiply-connected DAG given in Figure 8.18, which implies the following factorization of the joint probability distribution of the six variables:

$$p(a,b,c,d,e,f) = p(a)p(b|a)p(c|a)p(d|b)p(e|b,c)p(f|c). \qquad (8.23)$$

The associated numerical values of the set of CPDs are given in Table 8.2.

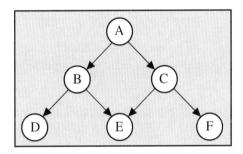

FIGURE 8.18. A multiply-connected graph.

One way to break the loop $A-B-E-C-A$ is to condition on the variable A. Figure 8.19 shows two different options for rendering the graph singly-connected by absorbing the evidence $A = a$. In both cases, the resulting graph is a polytree, and hence, we can apply the polytrees algorithm. Note that in each case only one of the two possible links is absorbed, thus leaving the graph connected.

Let us consider the situation of the graph in Figure 8.19(a). For any instantiation of the loop-cutset $\{A\}$, the new JPD associated with the resulting polytree is obtained from (8.23) by replacing $p(b|a)$ by $p_1(b) = p(b|A = a)$. We have

$$p(a,b,c,d,e,f|A=a) \propto p(a)p_1(b)p(c|a)p(d|b)p(e|b,c)p(f|c). \qquad (8.24)$$

We illustrate the conditioning algorithm using two cases: First we analyze the case of no evidence; then, we consider the evidence $\{C = 1, D = 1\}$.

For the case of no evidence, (8.21) reduces to

$$p(x_i) = \sum_{c_1,\ldots,c_m} p(x_i|c_1,\ldots,c_m)p(c_1,\ldots,c_m).$$

a	p(a)
0	0.3
1	0.7

a	b	p(b|a)
0	0	0.4
0	1	0.6
1	0	0.1
1	1	0.9

a	c	p(c|a)
0	0	0.2
0	1	0.8
1	0	0.5
1	1	0.5

b	d	p(d|b)
0	0	0.3
0	1	0.7
1	0	0.2
1	1	0.8

c	f	p(f|c)
0	0	0.1
0	1	0.9
1	0	0.4
1	1	0.6

b	c	e	p(e|b,c)
0	0	0	0.4
0	0	1	0.6
0	1	0	0.5
0	1	1	0.5
1	0	0	0.7
1	0	1	0.3
1	1	0	0.2
1	1	1	0.8

TABLE 8.2. Numerical values for the CPDs in (8.23).

Therefore, in this example we have

$$p(x_i) = \sum_a p(x_i|a)p(a),$$

for all nonconditioning nodes X_i. Note that $p(a)$ is simply the MPD of node A, which can be directly obtained from Table 8.2:

$$(p(A=0), p(A=1)) = (0.3, 0.7).$$

Thus, we only need to compute $p(x_i|a)$ for the two possible values of the conditioning node A. These probabilities can be obtained by applying the polytrees algorithm to the graph in Figure 8.19(a) with the corresponding JPD given in (8.24). Figures 8.20(a) and (b) show the resulting probabilities associated with $A = 0$ and $A = 1$, respectively. Using the numerical values

8.5 Conditioning Method   347

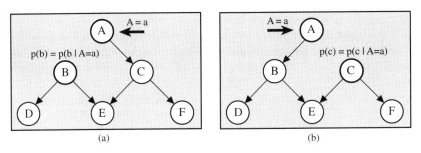

FIGURE 8.19. Two different possibilities to absorb evidence $A = a$.

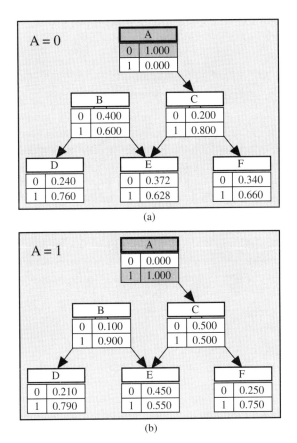

FIGURE 8.20. Probabilities obtained using the polytrees algorithm for the two possible values of the conditioning variable A.

shown in these figures, we can calculate the MPD of the nodes in the Bayesian network.

For example, for node B we have

$$p(B = 0) = \sum_{a=0}^{1} p(B = 0|a)p(a),$$
$$= 0.4 \times 0.3 + 0.1 \times 0.7 = 0.19,$$

$$p(B = 1) = \sum_{a=0}^{1} p(B = 1|a)p(a),$$
$$= 0.6 \times 0.3 + 0.9 \times 0.7 = 0.81.$$

The probabilities for the rest of the nodes can be calculated in a similar way. Figure 8.21 shows the resulting numerical values corresponding to the MPDs of all the nodes in the network.

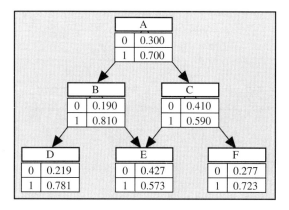

FIGURE 8.21. The MPDs of the nodes.

Suppose now that we are given the evidence $\{C = 1, D = 1\}$. Then, using (8.22) we have

$$p(x_i|C = 1, D = 1) \propto \sum_{a=0}^{1} p(x_i|a, C = 1, D = 1)p(C = 1, D = 1|a)p(a). \tag{8.25}$$

As in the previous case, $p(a)$ is given in Table 8.2. Thus, we only need to calculate $p(C = 1, D = 1|a)$ and $p(x_i|a, C = 1, D = 1)$ for the two possible values of node A. On the one hand, the probabilities $p(x_i|a, C = 1, D = 1)$ can be obtained by applying the polytrees algorithm with evidence $\{A = a, C = 1, D = 1\}$. Figure 8.22 shows the numerical values associated with the two possible values of A.

On the other hand, the probability $p(C = 1, D = 1|a)$ cannot be directly obtained using the polytrees algorithm because it does not depend on a single node. However, applying the chain rule, we can decompose this

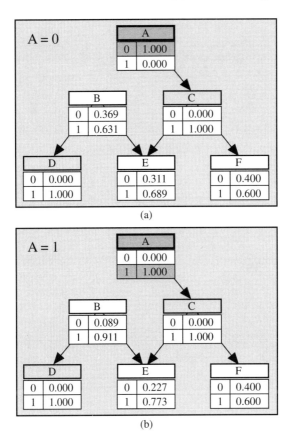

FIGURE 8.22. Probabilities obtained using the polytrees algorithm for the two possible values of the conditioning variable A, given the evidence $\{C = 1, D = 1\}$.

probability in a more suitable form for computational purposes:

$$\begin{aligned} p(C=1, D=1|a) &= \frac{p(C=1, D=1, a)}{p(a)} \\ &= \frac{p(C=1|D=1, a)p(D=1|a)p(a)}{p(a)} \\ &= p(C=1|D=1, a)p(D=1|a). \end{aligned}$$

The probabilities $p(D = 1|a)$ are given in Table 8.20 and $p(C = 1|D = 1, a)$ can be computed simultaneously with $p(x_i|a, C = 1, D = 1)$ by a sequential instantiation of evidence. The numerical values for these probabilities are

$$(p(C = 1|D = 1, A = 0), p(C = 1|D = 1, A = 1)) = (0.8, 0.5).$$

Thus, the numerical values of the probabilities $p(C = 1, D = 1|a)$ are

$$p(C = 1, D = 1|A = 0) = 0.8 \times 0.760 = 0.608,$$

350   8. Exact Propagation in Probabilistic Network Models

$$p(C = 1, D = 1|A = 1) = 0.5 \times 0.790 = 0.395.$$

Now we can compute the conditional probability of the nodes by replacing the obtained numerical values in (8.25). For example, for node B we have

$$p(B = 0|C = 1, D = 1)$$
$$\propto \sum_{a=0}^{1} p(B = 0|a, C = 1, D = 1)p(C = 1, D = 1|a)p(a),$$
$$= 0.369 \times 0.608 \times 0.3 + 0.089 \times 0.395 \times 0.7 = 0.092,$$

$$p(B = 1|C = 1, D = 1)$$
$$\propto \sum_{a=0}^{1} p(B = 1|a, C = 1, D = 1)p(C = 1, D = 1|a)p(a),$$
$$= 0.631 \times 0.608 \times 0.3 + 0.911 \times 0.395 \times 0.7 = 0.367.$$

The final probabilities are obtained by normalizing (dividing by $0.092 + 0.367 = 0.459$):

$$p(B = 0|C = 1, D = 1) \quad = \quad 0.092/0.459 = 0.200,$$

$$p(B = 1|C = 1, D = 1) \quad = \quad 0.367/0.459 = 0.800.$$

Figure 8.23 shows the resulting numerical values for the conditional probabilities of all the nodes in the network. ∎

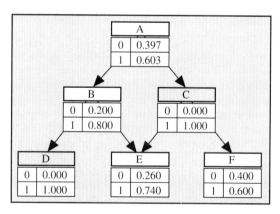

FIGURE 8.23. Conditional probabilities of the nodes, given the evidence $\{C = 1, D = 1\}$.

We have seen that the complexity of the conditioning algorithm resides in the multiple application of the polytrees algorithm. However, these multiple applications involve many repetitive and redundant computations corresponding to different combinations of values of the nodes in the loop-cutset.

Some improvements on this method have been proposed in the literature. For example, the *local conditioning* algorithm (Díez (1996)) and the *dynamic conditioning* algorithm (Darwiche (1995)) take advantage of the local structure of the graph to avoid redundant computations. This leads to substantial savings in computation.

8.6 Clustering Methods

The polytrees and conditioning algorithms introduced above use the particular structure of directed graphs for propagating evidence. Therefore, they can only be applied to Bayesian networks. As we shall see, clustering methods transform the local structure given by the graph into alternative local representations. Thus, these methods do not depend on the type of graph being considered and hence, are applicable to both Markov and Bayesian networks.

More specifically, clustering methods aggregate nodes into sets of nodes called *clusters* to capture the local structure of the JPD given by the associated graph. This method, which we refer to as the *clustering method*, was introduced by Lauritzen and Spiegelhalter (1988) and provides an efficient propagation algorithm. It is based on the recursive calculation of conditional probabilities associated with the cliques of the graph.

The clustering method takes advantage of the structure given by the graph to propagate evidence by calculating local probabilities (involving a small number of variables) and avoiding global expressions (involving a large number of variables). First, cliques are obtained from the original graph; then the JPD of each clique is obtained by recursively calculating several local CPDs. The conditional probability of a given node can then be obtained by marginalizing the JPD of any clique containing the node. In this section we give two versions of the clustering method; one applies to Markov network models and the other applies to Bayesian network models.

8.6.1 Clustering Algorithm in Markov Networks

In Chapter 6 we introduced two alternative representations of the JPD of a Markov network. The basic representation of a Markov network is given by the *clique potentials*, a set of positive functions $\Psi = \{\psi(c_1), \ldots, \psi(c_m)\}$ defined on the cliques of the graph $C = \{C_1, \ldots, C_m\}$. These functions factorize the JPD as

$$p(x) = \frac{1}{k} \prod_{i=1}^{m} \psi_i(c_i), \qquad (8.26)$$

where $k = \sum_x \prod_{i=1}^{m} \psi_i(c_i)$ is a normalization constant.

If the Markov network is decomposable, that is, if the associated graph is triangulated, then an alternative local representation of the JPD can be

352   8. Exact Propagation in Probabilistic Network Models

obtained by considering an ordering of the cliques $(C_1, \ldots, C_m)$ satisfying the running intersection property (see Section 4.5 for more details). This *chain of cliques* provides the following factorization of the JPD in terms of conditional probabilities:

$$p(x) = \prod_{i=1}^{m} p(r_i|s_i), \tag{8.27}$$

where $S_i = C_i \cap (C_1, \ldots, C_{i-1})$ is the *separator* set of clique C_i and $R_i = C_i \setminus S_i$ is the *residual* set. The running intersection property guarantees that S_i is contained in at least one of the previous cliques, $C_1, \ldots, C_{i-1}$. The cliques containing S_i are referred to as the neighbors of C_i and are denoted by B_i. As we shall see below, the representation of the JPD given by a chain of cliques provides a simple procedure used in the clustering algorithm for calculating the JPDs of the cliques and hence, the MPDs of the nodes.

If a Markov network is not decomposable, we first need to triangulate the corresponding graph, obtaining an associated undirected graph. As we shall see later, the original potential representation will provide a potential representation for the new decomposable Markov network. Therefore, without loss of generality, we assume that we have a decomposable Markov network and that we are given a potential representation (C, Ψ).

The clustering algorithm is based on the following steps:

1. Obtain a factorization of the JPD as in (8.27).

2. Compute the JPDs of the cliques using the above CPDs.

3. Compute the MPDs of the nodes.

Step 1: In this step we need to obtain a chain of cliques satisfying the running intersection property. We can use Algorithm 4.3 for this purpose. Then, the CPD $p(r_i|s_i)$ associated with clique C_i can be recursively obtained by marginalizing the JPD $p(x)$ over C_m, then over C_{m-1}, and so on. These marginal probabilities are obtained from the potential representation in the following way: For the last clique in the chain, C_m, we have

$$\begin{aligned} p(c_1, \ldots, c_{m-1}) &= \sum_{c_m \setminus \{c_1, \ldots, c_{m-1}\}} p(x) \\ &= \sum_{r_m} k^{-1} \prod_{i=1}^{m} \psi_i(c_i) \\ &= k^{-1} \prod_{i=1}^{m-1} \psi_i(c_i) \sum_{r_m} \psi_m(c_m). \end{aligned} \tag{8.28}$$

Thus, marginalizing the JPD $p(c_1, \ldots, c_m)$ over C_m is basically the same as marginalizing the associated potential $\psi_m(c_m)$. Once $p(c_1, \ldots, c_{m-1})$ is obtained, we can apply the same idea to obtain $p(c_1, \ldots, c_{m-2})$, and so on, until we obtain $p(c_1)$. Note that the term $\sum_{r_m} \psi_m(c_m)$ in (8.28) depends only on the variables in S_m. Thus, we can include this term in the potential representation of any clique C_j containing S_m, that is, in any neighboring clique of C_m. Thus we obtain the new potential function

$$\psi_j^*(c_j) = \psi_j(c_j) \sum_{r_m} \psi_m(c_m). \qquad (8.29)$$

For the rest of cliques we set $\psi_k^*(c_k) = \psi_k(c_k)$, $k \neq j$. Thus, considering these new potential functions and (8.28), we have

$$p(c_1, \ldots, c_{m-1}) = k^{-1} \prod_{i=1}^{m-1} \psi_i^*(c_i). \qquad (8.30)$$

Note that $\{\psi_1^*(c_1), \ldots, \psi_{m-1}^*(c_{m-1})\}$ is a potential representation of the JPD $p(c_1, \ldots, c_{m-1})$.

After computing $p(c_1, \ldots, c_{m-1})$, we can then calculate $p(c_1, \ldots, c_{m-2})$, again using (8.28) in a similar way. This sequential marginalization of the JPD allows us to obtain the CPDs needed in (8.27) by proceeding recursively starting with $p(r_m|s_m)$, the last element in (8.27). Taking into account that S_m separates R_m from $\{C_1, C_2, \ldots, C_{m-1}\}$, we can write

$$p(r_m|s_m) = p(r_m|c_1, c_2, \ldots, c_{m-1})$$

$$= \frac{p(c_1, c_2, \ldots, c_{m-1}, r_m)}{p(c_1, c_2, \ldots, c_{m-1})}.$$

Now, since $\{C_1, \ldots, C_{m-1}, R_m\} = \{C_1, \ldots, C_{m-1}, C_m\} = X$, and using (8.28), we have

$$p(r_m|s_m) = \frac{k^{-1} \prod_{i=1}^{m} \psi_i(c_i)}{k^{-1} \prod_{i=1}^{m-1} \psi_i(c_i) \sum_{r_q} \psi_m(c_m)}$$

$$= \frac{\psi_m(C_m)}{\sum_{r_m} \psi_m(c_m)}. \qquad (8.31)$$

Thus, we obtain the last element, $p(r_m|s_m)$, in the chain representation. Considering the new JPD $p(c_1, \ldots, c_{m-1})$ given in (8.30), we can apply (8.28) again to obtain $p(c_1, \ldots, c_{m-2})$ by marginalizing now in C_{m-1}. Then, using (8.31) gives $p(r_{m-1}|s_{m-1})$. The same process is repeated until $p(r_1|s_1)$ is obtained.

Step 2: Once a representation given by a chain of cliques has been obtained, the JPDs of the cliques can be easily obtained by calculating the probabilities of the separator sets. Note that, since $S_1 = \phi$, the last CPD is the joint probability of the first clique: $p(r_1|s_1) = p(c_1)$. The separator set of clique C_2 is contained in C_1. Thus, $p(s_2)$ can be obtained by marginalizing $p(c_1)$, which is already known. Finally, the probability $p(c_2)$ can be obtained by

$$p(c_2) = p(r_2, s_2) = p(r_2|s_2)p(s_2).$$

Once $p(c_1)$ and $p(c_2)$ have been obtained, the same iterative process can be repeated to obtain the JPDs of the rest of the cliques.

Step 3: Finally, once the JPDs of the cliques have been obtained, we can calculate the MPD of a node X_i by marginalizing the JPD of any clique containing the node. If C_j contains node X_i, then we can obtain the MPD of the node by

$$p(x_i) = \sum_{c_j \setminus x_i} p(c_j). \tag{8.32}$$

If X_i belongs to more than one clique, one should choose the clique with the smallest size to minimize the number of computations performed in the marginalization process. Note that the size of a clique is the product of the cardinalities of all the variables in the clique.

In the above discussion, we assumed we have no evidence. When evidence is available, the same algorithm can be applied with minor modifications as follows. Suppose that a set of evidential variables E is known to take the values $E = e$. The CPD $p(x \setminus e|e) = p(x \setminus e, e)/p(e)$ is proportional to $p(x \setminus e, e)$, which can be obtained by modifying the original potential functions by substituting the observed values of the evidential variables. This process is called *absorption of evidence*. Two alternative ways for the absorption of evidence are available:

1. Maintain the same sets of nodes X and cliques C. In this case, we just modify the potential functions containing evidential nodes in the following way: For every clique C_i containing some evidential node the new potential function $\psi_i^*(c_i)$ is defined as

$$\psi_i^*(c_i) = \begin{cases} 0, & \text{if some value in } c_i \text{ is inconsistent with } e, \\ \psi_i(c_i), & \text{otherwise.} \end{cases} \tag{8.33}$$

For the rest of the cliques, no change is needed. Then, we have

$$p(x|e) \propto \prod_{i=1}^{m} \psi_i^*(c_i).$$

2. Remove the evidential nodes from X. This implies also changing the set of cliques and the potential representation. The new potential representation, (C^*, Ψ^*), is defined on X^*, where $X^* = X \setminus E$, C^* is the new list of cliques and Ψ^* the new evidence potentials that are obtained in the following way: For each clique C_i in C such that $C_i \cap E \neq \phi$, we include $C_i \setminus E$ in C^* and define

$$\psi_i^*(c_i^*) = \psi_i(c_i \setminus e, \ E = e). \tag{8.34}$$

For the cliques that do not contain evidential nodes, no changes are needed in the cliques or the potential functions. In this case we have

$$p(x^*|e) \propto \prod_{i=1}^{m} \psi_i^*(c_i).$$

Thus, in both cases, the previous method can be applied to calculate the conditional probabilities of the nodes when some evidence is observed. In the first case, we continue with the same structure but we use unnecessary memory resources. In the latter case, we get a reduction in computer memory resources but we need a change of the data and storage structure. Thus, a balance is required in order to make an optimal selection.

Algorithm 8.2 Clustering Propagation Algorithm for Decomposable Markov Networks.

- **Input:** A decomposable Markov network model (C, Ψ) over a set of variables X and a set of evidential nodes E with evidential values $E = e$.

- **Output:** The conditional probabilities $p(x_i|e)$ for every nonevidential node X_i.

Initialization Steps:

1. Absorb the evidence $E = e$ in the potential functions Ψ by using (8.33) or (8.34).

2. Use Algorithm 4.3 to obtain a chain of cliques $(C_1, \ldots, C_m)$ satisfying the running intersection property.

3. For each clique C_i, choose as neighbor any clique C_j, with $j < i$, such that $S_i \subset C_j$.

Iteration Steps:

4. For $i = m$ to 1 (backwards) do

   (a) Compute $m_i(s_i) = \sum_{r_i} \psi_i(c_i)$.

   (b) Let $p(r_i|s_i) = \psi_i(c_i)/m_i(s_i)$.

   (c) Replace the potential function $\psi_j(c_j)$ of the neighboring clique C_j of clique C_i by $\psi_j(c_j) \leftarrow \psi_j(c_j) m_i(s_i)$.

5. Let $p(c_1) = p(r_1|s_1) = p(r_1)$.

6. For $i = 2$ to m (forwards) do

   (a) Compute $p(s_i)$ by marginalizing the JPD $p(c_j)$ of the neighboring clique of C_i, C_j.

   (b) Let $p(c_i) = p(r_i|s_i) p(s_i)$.

7. For $i = 1$ to n do

   (a) Choose the smallest size clique C_j containing X_i.

   (b) Let $p(x_i|e) \propto \sum_{c_j \setminus x_i} p(c_j)$.

   (c) Normalize the obtained values. ∎

Note that Algorithm 8.2 can be used to compute the CPDs of sets of nodes contained in some of the cliques of the graph. Xu (1995) adapts this algorithm to compute the CPD of a set of nodes that is not a subset of any of the cliques. The proposed method modifies the potential representations given by the chain of cliques to include the new sets of variables. Then, the marginal probabilities of these sets are obtained together with the marginal probabilities of the cliques in the propagation process. For details see Xu (1995).

Example 8.5 Clustering algorithm in decomposable Markov network model. Consider the triangulated Markov network given in Figure 8.24(a). Figure 8.24(b) shows the associated cliques

$$C_1 = \{A, B, C\}, \quad C_2 = \{B, C, E\}, \quad C_3 = \{B, D\}, \quad C_4 = \{C, F\}.$$

A potential representation of the Markov network is given by

$$p(a,b,c,d,e,f) = \psi_1(a,b,c)\psi_2(b,c,e)\psi_3(b,d)\psi_4(c,f). \tag{8.35}$$

The associated numerical values are given in Table 8.3.

Suppose that no evidence is given and that we want to obtain the MPDs of the nodes. In the first step of the clustering algorithm we need to order the cliques according to the running intersection property. To do this, we

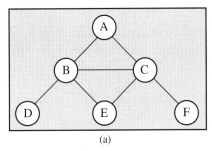 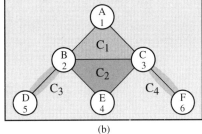

FIGURE 8.24. (a) An undirected triangulated graph and (b) its associated cliques.

a	b	c	$\psi_1(a,b,c)$
0	0	0	0.048
0	0	1	0.192
0	1	0	0.072
0	1	1	0.288
1	0	0	0.070
1	0	1	0.070
1	1	0	0.630
1	1	1	0.630

b	c	e	$\psi_2(b,c,e)$
0	0	0	0.08
0	0	1	0.12
0	1	0	0.10
0	1	1	0.10
1	0	0	0.14
1	0	1	0.06
1	1	0	0.04
1	1	1	0.16

b	d	$\psi_3(b,d)$
0	0	0.06
0	1	0.14
1	0	0.04
1	1	0.16

c	f	$\psi_4(c,f)$
0	0	0.02
0	1	0.18
1	0	0.08
1	1	0.12

TABLE 8.3. Numerical values for the potential functions in (8.35).

apply Algorithm 4.3 (see Chapter 4), which first calculates a perfect ordering for the nodes and then, orders the cliques according to the largest perfect number of the nodes included in each. The perfect numbers associated with the graph in Figure 8.24(a), when applying the maximum cardinality search (Algorithm 4.1) are shown below the node names in Figure 8.24(b). Note that the natural ordering of the cliques (C_1, C_2, C_3, C_4) satisfies the running intersection property. Table 8.4 shows the separators, residuals, and neighbors associated with the cliques. From this table, the set of CPDs in (8.27) is

$$p(a,b,c,d,e,f) = p(r_1|s_1)p(r_2|s_2)p(r_3|s_3)p(r_4|s_4)$$
$$= p(a,b,c)p(e|b,c)p(d|b)p(f|c). \qquad (8.36)$$

358   8. Exact Propagation in Probabilistic Network Models

Clique	Separator S_i	Residual R_i	Neighbor B_i
$C_1 = \{A, B, C\}$	ϕ	A, B, C	–
$C_2 = \{B, C, E\}$	B, C	E	C_1
$C_3 = \{B, D\}$	B	D	$\{C_1, C_2\}$
$C_4 = \{C, F\}$	C	F	$\{C_1, C_2\}$

TABLE 8.4. Residuals, separators, and neighbors of the cliques.

We are now ready to start the iteration steps of Algorithm 8.2: Step 4 considers the cliques in reverse order. Note that the number of cliques here is $m = 4$, thus we have

- For $i = 4$, the corresponding CPD in (8.36) is $p(f|c)$. We first compute the term

$$m_4(s_4) = \sum_{r_4} \psi_4(c_4) = \sum_f \psi_4(c, f)$$

and obtain $(m_4(C = 0), m_4(C = 1)) = (0.2, 0.2)$. Next, using (8.31), the CPD $p(f|c)$ is calculated as

$$p(f|c) = \frac{\psi_4(c, f)}{m_4(c)},$$

from which we have

$$\begin{aligned}
p(F = 0 | C = 0) &= 0.02/0.2 = 0.1, \\
p(F = 0 | C = 1) &= 0.18/0.2 = 0.9, \\
p(F = 1 | C = 0) &= 0.08/0.2 = 0.4, \\
p(F = 1 | C = 1) &= 0.12/0.2 = 0.6.
\end{aligned}$$

This distribution is given in Table 8.6. Finally, we choose a neighboring clique of C_4, say C_2, and multiply the potential function $\psi_2(b, c, e)$ by $m_4(s_4)$, which gives the potential function $\psi_2^*(b, c, e)$ shown in Table 8.5.

- For $i = 3$, the corresponding CPD in (8.36) is $p(d|b)$. We first compute the term

$$m_3(s_3) = \sum_{r_3} \psi_3(c_3) = \sum_d \psi_3(b, d),$$

and obtain $(m_3(B = 0), m_3(B = 1)) = (0.2, 0.2)$. The CPD $p(d|b)$ is then calculated as

$$p(d|b) = \frac{\psi_3(b, d)}{m_3(b)},$$

b	c	e	$\psi_2^*(b,c,e)$
0	0	0	0.016
0	0	1	0.024
0	1	0	0.020
0	1	1	0.020
1	0	0	0.028
1	0	1	0.012
1	1	0	0.008
1	1	1	0.032

a	b	c	$\psi_1^*(a,b,c)$
0	0	0	0.0096
0	0	1	0.0384
0	1	0	0.0144
0	1	1	0.0576
1	0	0	0.0140
1	0	1	0.0140
1	1	0	0.1260
1	1	1	0.1260

TABLE 8.5. New potential function for cliques C_2 and C_1.

c	f	$p(f\|c)$
0	0	0.1
0	1	0.9
1	0	0.4
1	1	0.6

b	d	$p(d\|b)$
0	0	0.3
0	1	0.7
1	0	0.2
1	1	0.8

b	c	e	$p(e\|b,c)$
0	0	0	0.4
0	0	1	0.6
0	1	0	0.5
0	1	1	0.5
1	0	0	0.7
1	0	1	0.3
1	1	0	0.2
1	1	1	0.8

a	b	c	$p(a,b,c)$
0	0	0	0.024
0	0	1	0.096
0	1	0	0.036
0	1	1	0.144
1	0	0	0.035
1	0	1	0.035
1	1	0	0.315
1	1	1	0.315

TABLE 8.6. The CPD of the residual given the separator sets, $p(r_i|s_i)$.

which gives

$$\begin{aligned} p(D=0|B=0) &= 0.06/0.2 = 0.3, \\ p(D=0|B=1) &= 0.14/0.2 = 0.7, \\ p(D=1|B=0) &= 0.04/0.2 = 0.2, \\ p(D=1|B=1) &= 0.16/0.2 = 0.8. \end{aligned}$$

This distribution is given in Table 8.6.

Finally, we choose a neighboring clique of C_3, say C_1, and multiply the potential function $\psi_1(a,b,c)$ of C_1 by $m_3(s_3)$, which gives the potential function $\psi_1^*(a,b,c)$ shown in Table 8.5.

a	b	c	$\psi_1^*(a,b,c)$
0	0	0	0.000384
0	0	1	0.001536
0	1	0	0.000576
0	1	1	0.002304
1	0	0	0.000560
1	0	1	0.000560
1	1	0	0.005040
1	1	1	0.005040

TABLE 8.7. Potential function for clique C_1 in the last marginalization step.

- For $i = 2$, the corresponding CPD in (8.36) is $p(e|b,c)$. We proceed in a similar way and calculate $m_2(b,c) = \sum_e \psi_2^*(b,c,e) = 0.04$, for all values of b and c. The CPD $p(e|b,c)$ is then calculated by

$$p(e|b,c) = \frac{\psi_2^*(b,c,e)}{m_2(b,c)},$$

which gives, for example,

$$p(E=0|B=0,C=0) = 0.016/0.04 = 0.4,$$
$$p(E=1|B=0,C=0) = 0.024/0.04 = 0.6.$$

The entire distribution $p(e|b,c)$ is given in Table 8.6.

Finally, since C_1 is the only neighbor clique of C_2, we multiply the potential function $\psi_1^*(a,b,c)$ of C_1 in Table 8.5 by $m_2(s_2)$ and obtain the new potential function $\psi_1^*(a,b,c)$ shown in Table 8.7. The last MPD $p(c_1) \propto \psi_1^*(a,b,c)$.

- For $i = 1$, the corresponding distribution in (8.36) is $p(a,b,c)$. Here $S_1 = \phi$, $R_1 = \{A,B,C\}$, and $B_1 = \phi$. Hence

$$m_1(\phi) = \sum_{a,b,c} \psi_1^*(a,b,c) = 0.016$$

and

$$p(r_1|s_1) = p(a,b,c) = \psi_1^*(a,b,c)/0.016.$$

This distribution is given in Table 8.6. Iteration step 4 is now complete.

- In iteration step 5, we have $p(c_1) = p(r_1) = p(a,b,c)$, which is given in Table 8.8. We now start iteration step 6 with $i = 2$.

8.6 Clustering Methods

a	b	c	p(a,b,c)
0	0	0	0.024
0	0	1	0.096
0	1	0	0.036
0	1	1	0.144
1	0	0	0.035
1	0	1	0.035
1	1	0	0.315
1	1	1	0.315

b	c	e	p(b,c,e)
0	0	0	0.0236
0	0	1	0.0354
0	1	0	0.0655
0	1	1	0.0655
1	0	0	0.2457
1	0	1	0.1053
1	1	0	0.0918
1	1	1	0.3672

b	d	p(b,d)
0	0	0.057
0	1	0.133
1	0	0.162
1	1	0.648

c	f	p(c,f)
0	0	0.041
0	1	0.369
1	0	0.236
1	1	0.354

TABLE 8.8. Numerical values for the JPDs of the cliques.

- For $i = 2$, $S_2 = \{B, C\}$. To compute $p(b, c)$, we marginalize $p(a, b, c)$ over A and obtain
$$p(b, c) = \sum_a p(a, b, c),$$
which gives

$$p(B = 0, C = 0) = 0.059,$$
$$p(B = 0, C = 1) = 0.131,$$
$$p(B = 1, C = 0) = 0.351,$$
$$p(B = 1, C = 1) = 0.459.$$

Thus,
$$p(b, c, e) = p(e|b, c)p(b, c).$$
The numerical values of this distribution are shown in Table 8.8.

- For $i = 3$, $S_3 = \{B\}$. To compute $p(b)$, we marginalize either $p(a, b, c)$ over A and C, or $p(b, c, e)$ over C and E. We obtain $p(B = 0) = 0.19$, $p(B = 1) = 0.81$. Then we obtain the JPD of clique C_3 as
$$p(b, d) = p(d|b)p(b).$$
The resulting numerical values of this distribution are shown in Table 8.8.

a	b	c	$\psi_1^*(a,b,c)$
0	0	0	0.000
0	0	1	0.192
0	1	0	0.000
0	1	1	0.288
1	0	0	0.000
1	0	1	0.070
1	1	0	0.000
1	1	1	0.630

b	d	$\psi_3^*(b,d)$
0	0	0.00
0	1	0.14
1	0	0.00
1	1	0.16

TABLE 8.9. Absorbing evidence $\{C = 1, D = 1\}$ in the potential functions in (8.35).

- Finally, for $i = 4$, $S_4 = \{C\}$. To compute $p(c)$, we marginalize either $p(a, b, c)$ over A and B, or $p(b, c, e)$ over B and E. We obtain $p(C = 0) = 0.41$, $p(C = 1) = 0.59$. Then we obtain the JPD of clique C_4 as

$$p(c, f) = p(f|c)p(c).$$

The resulting numerical values of this distribution are shown in Table 8.8.

This completes the iteration step 6.

In the final step of the clustering algorithm, we compute the MPD of the nodes using the JPDs shown in Table 8.8. Node A is found only in C_1; thus we compute $p(a)$ by marginalizing $p(a, b, c)$ over B and C. Node B is found in three cliques, the smallest of which is C_3. Thus, we marginalize $p(b, d)$ over D and obtain $p(b)$. Node C is also found in three cliques, the smallest of which is C_4. Thus, we marginalize $p(c, f)$ over F and obtain $p(c)$. Node D is found only in C_3; thus we marginalize $p(b, d)$ over B and obtain $p(d)$. Node E is found only in C_2; thus we marginalize $p(b, c, e)$ over B and C and obtain $p(e)$. Finally, node F is found only in C_4, so we marginalize $p(c, f)$ over C and obtain $p(f)$. All of these MPDs have been previously obtained in Example 8.4 applying the conditioning algorithm and are shown in Figure 8.21. This completes Algorithm 8.2 for the case of no evidence.

We now consider a case where some evidence is available. Suppose that we are given the evidence $\{C = 1, D = 1\}$. We can absorb the evidence in the potential representation given in Table 8.3 using one of the two options described in (8.33) and (8.34). In the first case, we keep the same set of nodes and the same graphical and clique structure, but we change the potential functions by including the evidential values. We absorb evidence $C = 1$ in clique C_1 and evidence $D = 1$ in clique C_3. Thus, we only modify the potential functions $\psi_1(a, b, c)$ and $\psi_3(b, d)$, as shown in Table 8.9.

Now, proceeding in a similar way as we did in the case of no evidence, the conditional probabilities of nodes given the evidence $\{C = 1, D = 1\}$

can be obtained. These probabilities have also been computed using the conditioning algorithm and are shown in Figure 8.23. ■

The clustering algorithm introduced above assumes that the Markov network is decomposable. This property is needed in order to guarantee the existence of a chain of cliques that factorizes the JPD as in (8.27). However, as we have already mentioned, if a Markov network is not decomposable we can triangulate the associated graph, obtain an associated decomposable Markov network, and perform evidence propagation on the decomposable network using the clustering algorithm. Since the triangulation process adds new links, the cliques of the new graph will contain the cliques of the original graph, and hence, the new potential functions can be created by grouping the original potential functions in the new cliques. The next example illustrates this process.

Example 8.6 nondecomposable Markov network model. Consider the undirected graph given in Figure 8.25. This graph is not triangulated because it contains the loop $A - B - E - C - A$ of length four without any chord. Thus, the corresponding Markov network is not decomposable. The graph has six cliques: $C_1 = \{A, B\}$, $C_2 = \{A, C\}$, $C_3 = \{B, E\}$, $C_4 = \{C, E\}$, $C_5 = \{B, D\}$, and $C_6 = \{C, F\}$. Then, a potential representation of this Markov network model is given by

$$p(a,b,c,d,e,f) = \psi_1(a,b)\psi_2(a,c)\psi_3(b,e)\psi_4(c,e)\psi_5(b,d)\psi_6(c,f). \quad (8.37)$$

The clustering algorithm cannot be applied to this nondecomposable Markov network. However, if we triangulate the graph by adding the link $B - C$, we obtain the same graph used in Example 8.5, which is given in Figure 8.24. Note that we can also triangulate the graph by adding the link $A - E$ instead. The cliques associated with this graph are $C_1^* = \{A, B, C\}$, $C_2^* = \{B, C, E\}$, $C_3^* = \{B, D\}$, and $C_4^* = \{C, F\}$. Thus, we can obtain potential representation for the new graph using (8.37) in the following way:

$$\psi_1^*(a,b,c) = \psi_1(a,b)\psi_2(a,c),$$
$$\psi_2^*(b,c,e) = \psi_3(b,e)\psi_4(c,e),$$
$$\psi_3^*(b,d) = \psi_5(b,d),$$
$$\psi_4^*(c,f) = \psi_6(c,f).$$

The graph given in Figure 8.24(a) and the new potential representation

$$p(a,b,c,d,e,f) = \psi_1^*(a,b,c)\psi_2^*(b,c,e)\psi_3^*(b,d)\psi_4^*(c,f), \quad (8.38)$$

provide an associated decomposable Markov network model that can be used for the purpose of evidence propagation. ■

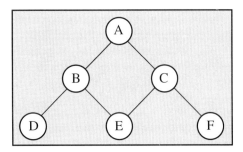

FIGURE 8.25. Undirected nontriangulated graph.

8.6.2 Clustering Algorithm in Bayesian Network Models

In the previous section we introduced the clustering algorithm for evidence propagation in Markov network models. In this section we adapt the clustering algorithm to perform evidence propagation in Bayesian network models. Given a Bayesian network model (D, P) defined over a set of variables $\{X_1, \ldots, X_n\}$, we have seen in Section 6.4.4 that the corresponding JPD can be factorized as

$$p(x_1, \ldots, x_n) = \prod_{i=1}^{n} p(x_i|\pi_i), \tag{8.39}$$

where Π_i is the set of parents of node X_i in D.

The idea here is to transform the directed graph into a triangulated undirected graph and then use the clustering algorithm on the obtained graph. We have seen in Chapter 4 that directed graphs can be transformed into triangulated undirected graphs by first moralizing the directed graph, dropping the directionality of the links, and then triangulating the resultant graph. By doing so, we obtain a triangulated undirected graph in which each family is contained in at least one clique. This property allows us to define a potential representation for the resulting undirected graph and hence, solve the problem of evidence propagation in Bayesian networks by applying Algorithm 8.2. This process is presented in the next algorithm.

Algorithm 8.3 Clustering Algorithm in Bayesian Networks.

- **Input:** A Bayesian network model (D, P) over a set of variables X and a set of evidential nodes E with evidential values $E = e$.

- **Output:** The CPD $p(x_i|e)$ for every nonevidential node X_i.

1. Moralize and triangulate the graph D. Let G be the resulting triangulated undirected graph.

2. Obtain the set of cliques C of G.

3. Assign every node X_i in X to one and only one clique containing the family of X_i. Let A_i be the set of nodes associated with clique C_i.

4. For every clique C_i in C define $\psi_i(c_i) = \prod_{x_i \in A_i} p(x_i|\pi_i)$. If $A_i = \phi$, then define $\psi_i(c_i) = 1$.

5. Use Algorithm 8.2 with the Markov network (C, Ψ) and evidence $E = e$ to obtain the probabilities of the nodes. ∎

The following example illustrates this algorithm.

Example 8.7 Clustering algorithm in Bayesian network models.
Consider the DAG given in Figure 8.26. As we have shown in Example 8.4, this graph implies the following factorization of the JPD:

$$p(a, b, c, d, e, f) = p(a)p(b|a)p(c|a)p(d|b)p(e|b, c)p(f|c), \quad (8.40)$$

with the numerical values of the associated CPDs given in Table 8.2. We apply Algorithm 8.3 to obtain an equivalent Markov network to propagate evidence using the clustering algorithm.

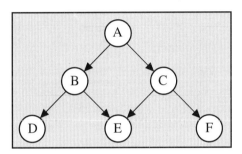

FIGURE 8.26. A multiply-connected DAG.

To moralize the DAG given in Figure 8.26 we need to add the link $B-C$, because nodes B and C have a common child E. The resulting graph G is shown in Figure 8.24(a). The graph is already triangulated, so we do not need to add more links to the graph. The cliques, C, of this graph are $C_1 = \{A, B, C\}$, $C_2 = \{B, C, E\}$, $C_3 = \{B, D\}$, and $C_4 = \{C, F\}$. Now, each of the nodes is assigned to one clique containing its family. We assign nodes A, B, and C to clique C_1, node E to clique C_2, and nodes D and F to cliques C_3 and C_4, respectively. Thus, we can obtain the following potential representation for the Markov network given in Figure 8.24(a):

$$\begin{aligned}\psi_1(a,b,c) &= p(a)p(b|a)p(c|a), \\ \psi_2(b,c,e) &= p(e|b,c), \\ \psi_3(b,d) &= p(d|b), \\ \psi_4(c,f) &= p(f|c).\end{aligned} \quad (8.41)$$

a	b	c	$\psi_1(a,b,c)$
0	0	0	0.024
0	0	1	0.096
0	1	0	0.036
0	1	1	0.144
1	0	0	0.035
1	0	1	0.035
1	1	0	0.315
1	1	1	0.315

b	c	e	$\psi_2(b,c,e)$
0	0	0	0.40
0	0	1	0.60
0	1	0	0.50
0	1	1	0.50
1	0	0	0.70
1	0	1	0.30
1	1	0	0.20
1	1	1	0.80

b	c	$\psi_3(b,d)$
0	0	0.30
0	1	0.70
1	0	0.20
1	1	0.80

c	f	$\psi_4(c,f)$
0	0	0.10
0	1	0.90
1	0	0.40
1	1	0.60

TABLE 8.10. Numerical values for the potential functions in (8.42).

The corresponding numerical values are shown in Table 8.10.

The graph given in Figure 8.24(a) and the new potential representation,

$$p(a,b,c,d,e,f) = \psi_1(a,b,c)\psi_2(b,c,e)\psi_3(b,d)\psi_4(c,f), \qquad (8.42)$$

provide an associated decomposable Markov network that can be used to propagate the evidence. Note that the values for every potential function are proportional to the corresponding values given in Table 8.3, used in Example 8.5. Then, both potential representations define the same Markov network model. Thus, applying the clustering algorithm leads to the same MPDs of the nodes as in Example 8.5. ∎

8.7 Propagation Using Join Trees

The clustering algorithm introduced in Section 8.6 aggregates the nodes of a network into cliques to obtain a suitable local structure to propagate evidence. Some improvements of the clustering algorithm, however, are possible. For example, the *belief universes method* introduced by Jensen, Olesen and Andersen (1990) (see also Jensen, Lauritzen, and Olesen (1990)) transforms the multiply-connected network into a join tree associated with the graph. The method also provides the basic operations of *distributing evidence* and *collecting evidence*, both of which are needed to propagate evidence among the cliques in the tree. Another alternative improvement, introduced by Shachter, Andersen, and Szolovits (1994), provides a unifying framework for the transformations and operations performed by the different propagation algorithms (see Shachter, Andersen, and Szolovits (1994) for more details).

This section presents a propagation algorithm that is based on message passing in join trees. Like the clustering algorithm, this method is suitable for both Markov and Bayesian network models. We refer to this method as the *join tree propagation algorithm*.

We have seen in Section 8.6 that a potential representation of JPDs can be obtained for both Markov and Bayesian network models. For Markov network models, the JPD of all the nodes in the network can be directly written as the product of potential functions defined on the cliques. If the graph is not triangulated, we can triangulate the graph and obtain the new potential representation from the original untriangulated graph. For Bayesian network models, we first build an undirected graph by moralizing and triangulating the original graph and dropping the directions of the links. Then, a potential representation is obtained by assigning the CPD $p(x_i|\pi_i)$ associated with every variable X_i to one and only one clique containing the family of X_i.

Then, without loss of generality, suppose we have a triangulated undirected graph with cliques $\{C_1, \ldots, C_m\}$ and a potential representation $\{\psi_1(c_1), \ldots, \psi_m(c_m)\}$. We first need to build a join tree and then use it to propagate the available evidence. The following example illustrates how to build a join tree (for further details see Section 4.6).

Example 8.8 Building a join tree. Consider the triangulated undirected graph given in Figure 8.24(a). In Example 8.5 we obtained the chain of cliques

$$C_1 = \{A, B, C\}, \quad C_2 = \{B, C, E\}, \quad C_3 = \{B, D\}, \quad C_4 = \{C, F\}$$

that satisfies the running intersection property. This chain of cliques defines a structure of separator sets and determines the possible neighbors of each clique, that is, the previous cliques in the chain containing its separator set. The separator sets and the neighbors of the cliques are shown in Table 8.4. Algorithm 4.4 shows that a chain of cliques can be organized into a tree structure, called a join tree, by linking each clique with some of its neighbors. In this case we have four[5] different options, as shown in Figure 8.27. Note, for example, that the join tree in Figure 8.27(a) was constructed by assigning to C_2 the only possible neighbor, C_1, and by choosing C_2 as a neighbor for both C_3 and C_4.

Thus given the DAG in Figure 8.26, we can apply Algorithm 4.5 (which includes the operations of moralization and triangulation of the DAG) to obtain a family tree, that is, a join tree where the family of each node is

---

[5] We have four options because C_1 has no neighbors, C_2 has only one neighbor, and C_3 and C_4 each has two neighbors. Thus there are two possibilities for C_3 for each of two possibilities for C_4.

contained in at least one clique. This guarantees the existence of a potential representation of the resulting undirected graph defined by the CPDs associated with the DAG. ■

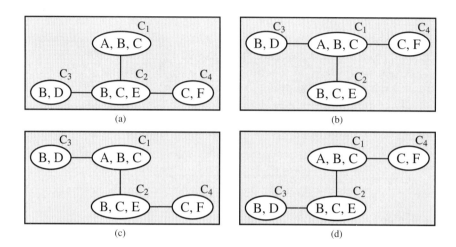

FIGURE 8.27. Four possible join trees for the undirected graph in Figure 8.24(a).

The polytree structure given by a join tree provides a graphical framework for performing the calculations of the clustering algorithm in an intuitive way. For example, the separators given by the chain of cliques can be graphically displayed as the intersection of neighboring cliques in the join tree. If cliques C_i and C_j are neighbors in a join tree, we define their separator set S_{ij}, or S_{ji}, as $S_{ij} = C_i \cap C_j$. For example, given the join tree in Figure 8.27(a), the three possible separator sets are $S_{12} = C_1 \cap C_2 = \{B, C\}$, $S_{23} = \{B\}$, and $S_{24} = \{C\}$, as shown in Figure 8.28. Note that these sets coincide with the separators obtained from the chain of cliques in Example 8.5 (see Table 8.4).

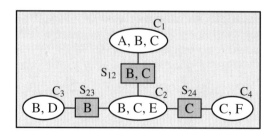

FIGURE 8.28. Separator sets of the join tree.

8.7 Propagation Using Join Trees

Assuming that the evidence $E = e$ has been absorbed in the potential functions, to compute the CPD $p(x_i|e)$ of a nonevidential node X_i given the evidence $E = e$, we need to perform the following calculations:

- Compute the messages between cliques in the join tree.
- Compute the JPD of each clique using these messages.
- Marginalize the JPD of a clique containing X_i over all variables other than X_i.

The join tree propagation algorithm performs basically the same operations as the clustering algorithm. The local calculations are now computed in the form of messages and propagated in the join tree via message passing similar to the algorithm for polytrees. To illustrate this process, suppose that a typical clique C_i has q neighboring cliques $\{B_1, \ldots, B_q\}$. Let C_{ij} and X_{ij} denote the set of cliques and the set of nodes on the subtree containing clique C_i when dropping the link $C_i - B_j$, respectively. Figure 8.29 shows the sets C_{ij} for a typical clique C_i and a typical neighbor B_j. Note that the set X_{ij} is simply the union of the cliques contained in C_{ij}. This figure also shows that the sets C_{ij} and C_{ji} are complementary, that is, $X = C_{ij} \cup C_{ji} = X_{ij} \cup X_{ji}$.

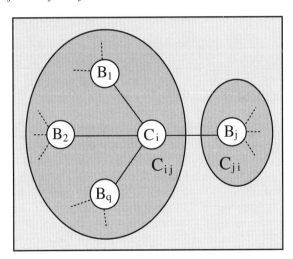

FIGURE 8.29. Decomposition into disjoint sets.

To compute the JPD of a typical separator set S_{ij}, we decompose the set $X \setminus S_{ij}$ as

$$X \setminus S_{ij} = (X_{ij} \cup X_{ji}) \setminus S_{ij} = (X_{ij} \setminus S_{ij}) \cup (X_{ji} \setminus S_{ij}) = R_{ij} \cup R_{ji},$$

where $R_{ij} = X_{ij} \setminus S_{ij}$ is the set of nodes contained in the C_i's side but not in the B_j's side of the link $C_i - B_j$ in the join tree. Since every node contained in two different cliques of the join tree is also contained in the path joining these cliques, the only nodes in common between X_{ij} and X_{ji} must be contained in the separator set S_{ij}. Therefore, R_{ij} and R_{ji} are disjoint subsets. This fact can be used to obtain the JPD of the separator set S_{ij} by

$$\begin{aligned} p(s_{ij}) &= \sum_{x \setminus s_{ij}} \prod_{k=1}^{m} \psi_k(c_k) \\ &= \sum_{r_{ij} \cup r_{ji}} \prod_{k=1}^{m} \psi_k(c_k) \\ &= \left(\sum_{r_{ij}} \prod_{c_k \in C_{ij}} \psi_k(c_k) \right) \left(\sum_{r_{ji}} \prod_{c_k \in C_{ji}} \psi_k(c_k) \right) \\ &= M_{ij}(s_{ij}) \, M_{ji}(s_{ij}), \end{aligned}$$

where

$$M_{ij}(s_{ij}) = \sum_{r_{ij}} \prod_{c_k \in C_{ij}} \psi_k(c_k) \qquad (8.43)$$

is the message sent from clique C_i to the neighbor clique B_j and

$$M_{ji}(s_{ij}) = \sum_{r_{ji}} \prod_{c_k \in C_{ji}} \psi_k(c_k) \qquad (8.44)$$

is the message sent from clique B_j to clique C_i.

Note that the JPD of the separator set S_{ij} is simply the product of these two messages. The information needed to calculate each of these messages is contained in different sides of the link $C_i - B_j$ in the tree. Thus, these messages gather the information coming from one side of the link and propagate it to the other side. These two messages, which contain all relevant information, can be computed asynchronously, which facilitates a parallel implementation.

To compute the JPD of a typical clique C_i, we first decompose the set $X \setminus C_i$ according to the information coming from the different neighboring cliques:

$$X \setminus C_i = \left(\bigcup_{k=1}^{q} X_{ki} \right) \setminus C_i = \bigcup_{k=1}^{q} (X_{ki} \setminus C_i) = \bigcup_{k=1}^{q} R_{ki}.$$

The last equality follows because $X_{ki} \setminus C_i = R_{ki}$, which is given by the property of join trees (every node in X_{ki} that is also in C_i has to be contained in S_{ki}), that is, $X_{ki} \setminus C_i = X_{ki} \setminus S_{ki} = R_{ki}$. This decomposition is illustrated in Figure 8.30.

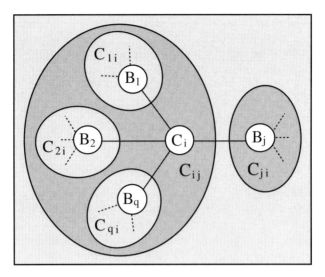

FIGURE 8.30. Decomposition into disjoint sets.

Then the JPD of C_i can be written as

$$
\begin{aligned}
p(c_i) &= \sum_{x \setminus c_i} \prod_{j=1}^{m} \psi_j(c_j) \\
&= \psi_i(c_i) \sum_{x \setminus c_i} \prod_{j \neq i} \psi_j(c_j) \\
&= \psi_i(c_i) \sum_{r_{1i} \cup \ldots \cup r_{qi}} \prod_{j \neq i} \psi_j(c_j) \\
&= \psi_i(c_i) \left(\sum_{r_{1i}} \prod_{c_k \in C_{1i}} \psi_k(c_k) \right) \ldots \left(\sum_{r_{qi}} \prod_{c_k \in C_{qi}} \psi_k(c_k) \right) \\
&= \psi_i(c_i) \prod_{j=1}^{q} M_{ki}(s_{ij}),
\end{aligned}
\qquad (8.45)
$$

where $M_{ji}(s_{ij})$ is the message sent from clique B_j to clique C_i as defined in (8.44).

Once all possible messages between cliques have been calculated, the JPD of the cliques can be obtained by using (8.45). Note that the calculation of message $M_{ij}(s_{ij})$ using (8.43) involves a summation over the set $X_{ij} \setminus S_{ij}$. This computation can be simplified by considering the decomposition

$$ X_{ij} \setminus S_{ij} = (C_i \setminus S_{ij}) \cup \left(\bigcup_{k \neq j} X_{ki} \setminus S_{ki} \right). $$

372   8. Exact Propagation in Probabilistic Network Models

Thus, we obtain the following recurrence relationship between messages:

$$\begin{aligned} M_{ij}(s_{ij}) &= \sum_{x_{ij}\setminus s_{ij}} \prod_{c_s \in C_{ij}} \psi_s(c_s) \\ &= \sum_{c_i \setminus s_{ij}} \sum_{(x_{ki}\setminus s_{ki}), k \neq j} \prod_{c_s \in C_{ij}} \psi_s(c_s) \\ &= \sum_{c_i \setminus s_{ij}} \psi_i(c_i) \prod_{k \neq j} \sum_{x_{ki}\setminus s_{ki}} \prod_{c_s \in C_{ki}} \psi_s(c_s) \\ &= \sum_{c_i \setminus s_{ij}} \psi_i(c_i) \prod_{k \neq j} M_{ki}(s_{ki}). \end{aligned} \qquad (8.46)$$

Finally, the conditional probabilities of any node X_i can be computed by marginalizing the JPD of any clique containing X_i. If the node belongs to more that one clique, the clique with the smallest size is used to minimize the number of computations performed.

From the above discussion, one can conclude the following:

- Equation (8.45) shows that the probability $p(c_i)$ of clique C_i can be computed as soon as C_i receives the messages from all of its adjacent cliques.

- Equation (8.46) shows that the message $M_{ij}(s_{ij})$ that clique C_i sends to its neighbor B_j can be computed as soon as C_i receives the messages $M_{ki}(s_{ki})$ coming from the rest of the neighboring cliques.

This leads to the following algorithm.

Algorithm 8.4 Propagation in Markov Network Models Using a Join Tree.

- **Input:** A Markov network (C, Ψ) over a set of variables X and a set of evidential nodes E with evidential values $E = e$.
- **Output:** The CPD $p(x_i|e)$ for every nonevidential node X_i.

Initialization Steps:

1. Absorb the evidence $E = e$ in the potential functions Ψ by using (8.33) or (8.34).

2. Use Algorithm 4.4 to obtain a join tree for the Markov network.

Iteration Steps:

3. For $i = 1, \ldots, m$ do: For every neighbor B_j of clique C_i, if C_i received the messages from all other neighbors, calculate and send the message $M_{ij}(s_{ij})$ to B_j, where

$$M_{ij}(s_{ij}) = \sum_{c_i \setminus s_{ij}} \psi_i(c_i) \prod_{k \neq j} M_{ki}(s_{ki}). \qquad (8.47)$$

4. Repeat Step 3 until no new message is calculated.

5. Calculate the JPD of each clique C_i using

$$p(c_i) = \psi_i(c_i) \prod_k M_{ki}(s_{ik}). \qquad (8.48)$$

6. For every node X_i in the network calculate $p(x_i|e)$ using

$$p(x_i|e) = \sum_{c_k \setminus x_i} p(c_k), \qquad (8.49)$$

where C_k is the smallest size clique containing X_i. ∎

Note that in Step 3 of the algorithm, there are three possible cases for each clique C_i:

- **Case 1:** C_i received the messages from all of its neighbors. In this case, C_i can compute and send messages to each of its neighbors.

- **Case 2:** C_i received the messages from all of its neighbors except B_j. In this case, C_i can only compute and send a message to B_j.

- **Case 3:** C_i has not received messages from two or more of its neighbors. In this case, no message can be calculated.

This algorithm can be easily modified to propagate evidence in Bayesian networks as follows.

Algorithm 8.5 Propagation in Bayesian Network Models Using a Join Tree.

- **Input:** A Bayesian network (D, P) over a set of variables X and a set of evidential nodes E with evidential values $E = e$.

- **Output:** The CPD $p(x_i|e)$ for every nonevidential node X_i.

1. Use Algorithm 4.5 to obtain a family tree for the DAG D. Let C be the resulting set of cliques.

2. Assign every node X_i in X to one and only one clique containing X_i. Let A_i be the set of nodes assigned to clique C_i.

3. For every clique C_i in C define $\psi_i(c_i) = \prod_{x_i \in A_i} p(x_i|\pi_i)$. If $A_i = \phi$, then define $\psi_i(c_i) = 1$.

4. Use Algorithm 8.4 with the Markov network (C, Ψ) and evidence $E = e$ to obtain the CPDs of the nodes. ∎

374   8. Exact Propagation in Probabilistic Network Models

The structures of message-passing used in Algorithms 8.4 and 8.5 are suitable for a parallel implementation. In this case, each clique in the join tree can be assigned its own processor. The processor of a typical clique C_i needs the following information to compute its conditional joint probability $p(c_i|e)$ (using formula (8.48)) and to send the corresponding messages to its neighbors:

- One list containing the neighbors of C_i. This information is independent of the evidence E.

- The associated potential function $\psi(c_i)$. This function may depend on the evidence E, due to the evidence absorption process.

- The message $M_{ji}(s_{ij})$, received from every neighbor B_j of C_i. This message is computed at B_j, using formula (8.47).

The computations performed at a typical clique C_i and the received and sent messages are shown schematically in Figure 8.31.

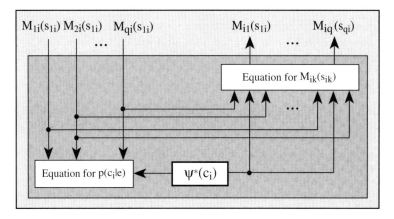

FIGURE 8.31. Computations performed at a typical clique C_i and the messages received from and sent to its neighbors.

Example 8.9 Propagation using join trees. Consider the Bayesian network introduced in Example 8.7, defined by the multiply-connected DAG given in Figure 8.26 and the CPDs in (8.40), whose numerical values are given in Table 8.2. In that example we obtained the associated moralized and triangulated graph (see Figure 8.24(a)) and the corresponding set of cliques (see Figure 8.24(b)): $C_1 = \{A, B, C\}$, $C_2 = \{B, C, E\}$, $C_3 = \{B, D\}$, and $C_4 = \{C, F\}$. The potential functions are defined in terms of the CPDs as shown in (8.41) with the corresponding numerical values given in Table 8.10. In Example 8.8 we obtained the four different join trees associated with this graph (see Figure 8.27). As an example, let

8.7 Propagation Using Join Trees

us use the join tree in Figure 8.27(a) and Algorithm 8.4 to compute the MPDs of the nodes (no evidence is available).

The algorithm proceeds by sequentially passing messages between neighboring cliques in the join tree. Figure 8.32 shows the order in which these messages are computed and sent. The arrows indicate the messages between cliques and the numbers indicate the order in which they have been calculated.

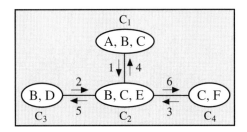

FIGURE 8.32. Order in which messages are sent.

These messages, computed in Step 3 of Algorithm 8.4, are as follows:

- Clique C_1 has only one neighbor, C_2; thus the message $M_{12}(s_{12})$ can be computed and sent to C_2. Using (8.47), we have

$$M_{12}(s_{12}) = \sum_{c_1 \setminus s_{12}} \psi_1(c_1).$$

Thus,

$$M_{12}(b,c) = \sum_a \psi_1(a,b,c),$$

from which we get

$$
\begin{aligned}
M_{12}(0,0) &= \psi_1(0,0,0) + \psi_1(1,0,0) = 0.024 + 0.035 = 0.059, \\
M_{12}(0,1) &= \psi_1(0,0,1) + \psi_1(1,0,1) = 0.096 + 0.035 = 0.131, \\
M_{12}(1,0) &= \psi_1(0,1,0) + \psi_1(1,1,0) = 0.036 + 0.315 = 0.351, \\
M_{12}(1,1) &= \psi_1(0,1,1) + \psi_1(1,1,1) = 0.144 + 0.315 = 0.459.
\end{aligned}
$$

- Clique C_2 has three neighbors, C_1, C_3, and C_4, but it has received a message only from C_1. Therefore, no calculation is possible at this stage (Case 3).

- Since clique C_3 has only one neighbor, C_2, the message $M_{32}(s_{23})$ can be computed and sent to C_2. Using (8.47), we have

$$M_{32}(s_{23}) = \sum_{c_3 \setminus s_{32}} \psi_3(c_3),$$

or
$$M_{32}(b) = \sum_d \psi_3(b,d),$$

from which we obtain $(M_{32}(B=0), M_{32}(B=1)) = (1.0, 1.0)$.

- Clique C_4 is in the same situation as C_3. Thus, the message $M_{42}(s_{24})$ can be computed and sent to C_2. Thus, we have

$$M_{42}(s_{24}) = \sum_{c_4 \setminus s_{42}} \psi_4(c_4),$$

or

$$M_{42}(c) = \sum_f \psi_4(c,f),$$

from which we obtain $(M_{42}(C=0), M_{42}(C=1)) = (1.0, 1.0)$. We now repeat Step 3.

- C_1 already sent a message to C_2. Clique C_2 has received messages from all of its neighbors. Therefore, we calculate and send the messages $M_{21}(b,c)$, $M_{23}(b)$, and $M_{24}(c)$ to C_1, C_3, and C_4, respectively. For example,

$$M_{23}(s_{23}) = \sum_{c_2 \setminus s_{23}} \psi_2(c_2) M_{12}(s_{12}) M_{42}(s_{24}),$$

or

$$M_{23}(b) = \sum_{c,e} \psi_2(b,c,e) M_{12}(b,c) M_{42}(c),$$

from which we have

$$\begin{aligned} M_{23}(0) &= \sum_{c,e} \psi_2(0,c,e) M_{12}(0,c) M_{42}(c) \\ &= 0.4 \times 0.059 \times 1 + 0.6 \times 0.059 \times 1 \\ &\quad + 0.5 \times 0.131 \times 1 + 0.5 \times 0.131 \times 1 = 0.19, \end{aligned}$$

$$\begin{aligned} M_{23}(1) &= \sum_{c,e} \psi_2(1,c,e) M_{12}(1,c) M_{42}(c) \\ &= 0.7 \times 0.351 \times 1 + 0.3 \times 0.351 \times 1 + \\ &\quad + 0.2 \times 0.459 \times 1 + 0.8 \times 0.459 \times 1 = 0.81. \end{aligned}$$

The numerical values corresponding to these and other functions are shown in Figure 8.33. Now all messages have been calculated and sent, and Steps 3 and 4 are complete.

- In Step 5, the JPDs of the cliques are calculated using (8.48). We have

$$p(c_1) = p(a,b,c) = \psi_1(a,b,c)M_{21}(b,c),$$
$$p(c_2) = p(b,c,e) = \psi_2(b,c,e)M_{12}(b,c)M_{32}(b)M_{42}(c),$$
$$p(c_3) = p(b,d) = \psi_3(b,d)M_{23}(b),$$
$$p(c_4) = p(c,f) = \psi_4(c,f)M_{24}(c).$$

The numerical values of this function are the same as those in Table 8.8, which have been calculated by the clustering algorithm.

- In Step 6, The MPDs of the nodes are calculated as in Example 8.5. This completes the calculations in the case of no evidence.

Suppose now that we have the evidence $E = \{C = 1, D = 1\}$. We can absorb the evidence in the potential representation using (8.33). This step has been performed in in Example 8.5 (see Table 8.9), where evidence $C = 1$ is absorbed in clique C_1 and evidence $D = 1$ is absorbed in clique C_3. The reader can execute Algorithm 8.4 and obtain the messages shown in Figure 8.34. ∎

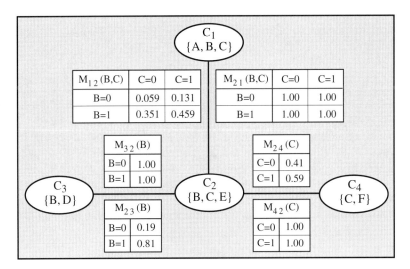

FIGURE 8.33. Numerical values of the messages and functions calculated by Algorithm 8.4 when no evidence is available.

8.8 Goal-Oriented Propagation

In the previous sections we introduced several algorithms for evidence propagation in Markov and Bayesian networks. The objective of these algorithms is to obtain the CPDs of all variables in the network when some

378   8. Exact Propagation in Probabilistic Network Models

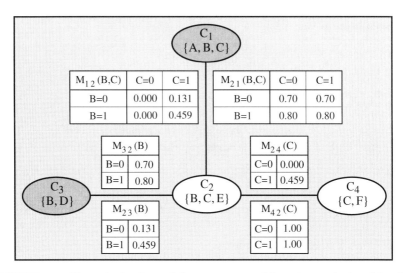

FIGURE 8.34. Numerical values of the messages and functions calculated by the join tree algorithm after observing evidence $e = \{C = 1, D = 1\}$. Dark ovals show the cliques where evidence has been absorbed.

evidence $E = e$ is observed. However, in some cases, we are interested in only a certain set of variables, and our goal is to obtain the CPD of these variables given the evidence. In this situation, some variables in the network may not be relevant to the computations of the required CPDs. Thus, we can avoid unnecessary computations by finding the set of irrelevant variables. These variables can then be removed from the graph, leaving a smaller graph containing only the relevant nodes.

We illustrate this idea using Bayesian network models. In a Bayesian network model (D, P) defined on the set $X = \{X_1, \ldots, X_n\}$, every node X_i has a set of parameters assigned to it that is associated with the conditional distribution $p(x_i|\pi_i)$. Suppose that we are interested in a set of variables $Y \subset X$ and we wish to calculate the CPD $p(x_i|e)$, for $X_i \in Y$, given some evidence $E = e$. The variables in Y are called the *goal variables*.

The parameters irrelevant to the computations of the required probabilities are associated with nodes whose CPDs can be assigned arbitrary values without affecting the CPDs of the goal variables. The problem of identifying relevant parameters was addressed by Shachter (1988, 1990b) and Geiger, Verma, and Pearl (1990a, 1990b). The latter authors give a simple algorithm for finding the relevant parameters in polynomial time in the number of variables. The parameters associated with the CPD $p(x_i|\pi_i)$ of node X_i are represented by a dummy parent Θ_i of node X_i. The relevant set of nodes can then be identified using the D-separation criterion criterion in the new graph (see Section 5.2.2).

8.8 Goal-Oriented Propagation

Algorithm 8.6 Identifying Relevant Nodes.

- **Input:** A Bayesian network (D, P) and two sets of nodes: a goal set Y and an evidential set E (possibly empty).
- **Output:** The set of relevant nodes R needed to compute $p(y|e)$.

1. Construct a DAG, D', by augmenting D with a dummy node Θ_i and adding a link $\Theta_i \to X_i$ for every node X_i in D.

2. Identify the set Θ of dummy nodes in D' not D-separated from Y by E.

3. Let R be the set of nodes X_i whose associated dummy nodes Θ_i are in Θ. ■

The dummy node Θ_i represents the parameters associated with node X_i. Step 2 of Algorithm 8.6 can be carried out in linear time using an algorithm provided by Geiger, Verma, and Pearl (1990a, 1990b). Thus, if we consider the reduced graph obtained by eliminating the irrelevant nodes given by this algorithm, we can significantly reduce the set of parameters to be considered in the analysis. This algorithm is illustrated by the following example.

Example 8.10 Goal-oriented propagation. Consider the Bayesian network given by the directed graph in Figure 8.35, which implies the following factorization of the JPD:

$$p(a,b,c,d,e,f) = p(a)p(b|a)p(c|a)p(d|b)p(e|b,c)p(f|c). \qquad (8.50)$$

Suppose we are only interested in the probability of node C after knowing the evidence $D = 0$. Node C is highlighted in Figure 8.35 to indicate that C is a goal variable.

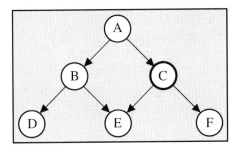

FIGURE 8.35. Example of a DAG used for goal-oriented propagation, where C is the goal node.

We can use one of the exact propagation algorithms described in this chapter and compute $p(x_i|e)$ for all nonevidential nodes including node

380   8. Exact Propagation in Probabilistic Network Models

C. This clearly involves unneeded calculations. Alternatively, we can first reduce the graph by considering the set of relevant nodes and perform the propagation on the reduced graph. This set can be obtained using Algorithm 8.6 as follows:

- First we construct a new DAG by adding a dummy node for each node in the graph. Figure 8.36 shows the new graph obtained by adding the nodes $\{\Theta_A, \ldots, \Theta_F\}$ to the graph in Figure 8.35.

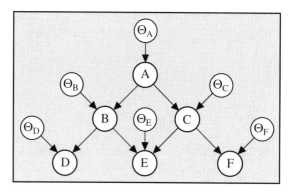

FIGURE 8.36. Augmented graph obtained by adding a dummy node Θ_i and a link $\Theta_i \to X_i$, for every node X_i.

- Using the D-separation criterion we can obtain the set of dummy nodes not D-separated from the goal node C by the evidential node D in the graph in Figure 8.36. For example, the dummy node Θ_A is not D-separated from C by D. To see this, we construct the moralized ancestral graph associated with nodes C, D, and Θ_A, as shown in Figure 8.37 (see Definition 5.4). Since there is a path between Θ_A and C that does not include D, then Θ_A is not D-separated from C by D. From the same graph, we can see that the dummy nodes Θ_B, Θ_C, and Θ_D are not D-separated from C by D. However, nodes Θ_E and Θ_F are D-separated from C by D. (The reader can construct the moralized ancestral graphs for these two cases.) Thus, the set of relevant parameters to calculate $p(c|d)$ is $\Theta = \{\Theta_A, \Theta_B, \Theta_C, \Theta_D\}$.

- The set of relevant nodes is $R = \{A, B, C, D\}$.

Once the set of relevant nodes has been obtained, we can simplify the graph by eliminating the irrelevant nodes. The resulting graph is given in Figure 8.38. Note that the JPD associated with the reduced graph is given by

$$p(a,b,c,d) = p(a)p(b|a)p(c|a)p(d|b). \tag{8.51}$$

Note also that while the original graph is multiply connected, the reduced graph is a polytree, and hence, evidence can be efficiently propagated.

8.8 Goal-Oriented Propagation   381

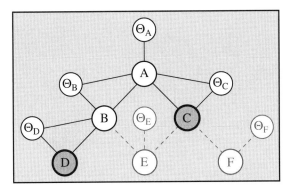

FIGURE 8.37. Moralized ancestral graph used to determine whether each of the dummy nodes Θ_A, Θ_B, Θ_C, and Θ_D is D-separated from C by D.

Thus, identification of relevant parameters can lead to a simplification of the topology of the graph.

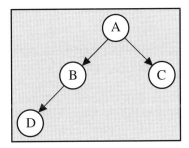

FIGURE 8.38. Reduced graph to perform the computation $p(c|d)$.

Figures 8.39(a) and (b) show the probabilities resulting from the propagation of evidence $D = 0$ in the original graph (using the clustering or join tree algorithms) and in the reduced graph (using the polytree algorithm), respectively. As expected, the two methods provide the same probabilities, $(p(C = 0|D = 0), p(C = 1|D = 0)) = (0.402, 0.598)$. ∎

The above example illustrates the computational savings that can be obtained by using Algorithm 8.6. An improvement of this algorithm is given in Baker and Boult (1991), which introduces the concept of *computationally equivalent networks* to formalize the idea described in this section. They give an algorithm for finding the minimum computationally equivalent network associated with a given task to be performed in an original graph.

In light of the above discussion, the following conclusions can be made:

- Given the evidence $E = e$, the CPD of any nonevidential node X_i in a Bayesian network depends only on the ancestors of the set $\{X_i\} \cup E$.

382   8. Exact Propagation in Probabilistic Network Models

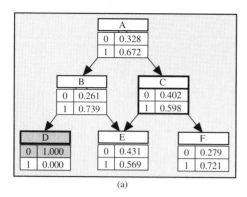
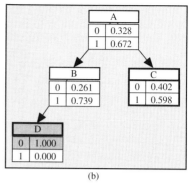

FIGURE 8.39. Probabilities resulting from the propagation of evidence $D = 0$ in the graphs in Figures 8.35 and 8.38.

As we have shown in Example 8.10, the probability of node C given $D = 0$ depends on the set $\{A, B, C, D\}$, which is the ancestral set of nodes C and D.

- In the case of no evidence, the MPD of any node X_i in a Bayesian network depends only on its ancestral set. For example, the MPD of node C in Example 8.10 only depends on nodes A and C.

- The MPD of a root node X_i (a node without parents) can be directly calculated from its associated CPD $p(x_i|\pi_i) = p(x_i)$. For example, the probability of node A in Example 8.10 is given in Table 8.2: $(p(A = 0), p(A = 1)) = (0.3, 0.7)$.

8.9 Exact Propagation in Gaussian Networks

In the previous sections we have introduced several algorithms for evidence propagation in discrete multinomial probabilistic networks. However, although these models are widely used in practical applications, several other types of probabilistic network models have been introduced to deal with continuous variables. Some examples are the Gaussian models (see Kenley (1986) and Shachter and Kenley (1989)), the Dirichlet models (Geiger and Heckerman (1995) and Castillo, Hadi, and Solares (1996)), and even mixed models of both discrete and continuous variables (see, for example, Lauritzen and Wermuth (1989), Olesen (1993), and Castillo, Gutiérrez, and Hadi (1995b)).

Several algorithms have been proposed in the literature to solve the problems of evidence propagation in these models. Some of them have originated from the methods for discrete models introduced in the previous sections.

For example, Normand and Tritchler (1992) introduce an algorithm for evidence propagation in Gaussian network models using the same idea of the polytrees algorithm (see Section 8.3). Lauritzen (1992) suggests a modification of the join tree algorithm (Section 8.7) to propagate evidence in mixed models.

Gaussian Bayesian network models have been introduced in Section 6.4.4 in two different ways: using the covariance matrix of the Gaussian JPD, or using a factorization of the JPD as a product of Gaussian CPDs. Several algorithms use the structure provided by the second representation for evidence propagation (see Xu and Pearl (1989), and Chang and Fung (1991)). In this section we present a conceptually simple and efficient algorithm that uses the covariance matrix representation. This algorithm illustrates the basic concepts underlying exact propagation in Gaussian network models. An incremental implementation of the algorithm allows updating probabilities, in linear time, when a single piece of evidence is observed. The main result is given in the following theorem, which characterizes the CPDs obtained from a Gaussian JPD (see, for example, Anderson (1984)).

Theorem 8.1 Conditionals of a Gaussian distribution. *Let Y and Z be two sets of random variables having a multivariate Gaussian distribution with mean vector and covariance matrix given by*

$$\mu = \begin{pmatrix} \mu_Y \\ \mu_Z \end{pmatrix} \quad \text{and} \quad \Sigma = \begin{pmatrix} \Sigma_{YY} & \Sigma_{YZ} \\ \Sigma_{ZY} & \Sigma_{ZZ} \end{pmatrix},$$

where μ_Y and Σ_{YY} are the mean vector and covariance matrix of Y, μ_Z and Σ_{ZZ} are the mean vector and covariance matrix of Z, and Σ_{YZ} is the covariance of Y and Z. Then the CPD of Y given $Z = z$ is multivariate Gaussian with mean vector $\mu_{Y|Z=z}$ and covariance matrix $\Sigma_{Y|Z=z}$ that are given by

$$\mu_{Y|Z=z} = \mu_Y + \Sigma_{YZ} \Sigma_{ZZ}^{-1} (z - \mu_Z), \tag{8.52}$$

$$\Sigma_{Y|Z=z} = \Sigma_{YY} - \Sigma_{YZ} \Sigma_{ZZ}^{-1} \Sigma_{ZY}. \tag{8.53}$$

Note that the conditional mean $\mu_{Y|Z=z}$ depends on z but the conditional variance $\Sigma_{Y|Z=z}$ does not.

Theorem 8.1 suggests an obvious procedure to obtain the means and variances of any subset of variables $Y \subset X$, given a set of evidential nodes $E \subset X$ whose values are known to be $E = e$. Replacing Z in (8.52) and (8.53) by E, we obtain the mean vector and covariance matrix of the conditional distribution of the nodes in Y. Note that considering $Y = X \setminus E$ we get the joint distribution of the remaining nodes, and then we can answer questions involving the joint distribution of nodes instead of the usual information that refers only to individual nodes. The above introduced methods for evidence propagation in Gaussian Bayesian network models use the same idea, but perform local computations by taking advantage of the factorization of the JPD as a product of CPDs.

384   8. Exact Propagation in Probabilistic Network Models

$$
\begin{aligned}
&Y \leftarrow X \\
&\mu \leftarrow E[X] \\
&\Sigma \leftarrow Var[X] \\
&\text{For } i \leftarrow 1 \text{ to the number of elements in } E, \text{ do:} \\
&\qquad z \leftarrow \text{the } i\text{th element of } e \\
&\qquad Y \leftarrow Y \setminus Z \\
&\qquad \mu \leftarrow \mu_y + \Sigma_{yz} \Sigma_z^{-1}(z - \mu_z) \\
&\qquad \Sigma \leftarrow \Sigma_y - \Sigma_{yz} \Sigma_z^{-1} \Sigma_{yz}^T \\
&f(y|E=e) \sim N(\mu, \Sigma)
\end{aligned}
$$

FIGURE 8.40. Pseudocode of an incremental algorithm for updating the joint probability of the nonevidential nodes Y given the evidence $E = e$.

In order to simplify the computations, it is more convenient to use an incremental method, updating one evidential node at a time (taking elements one by one from E). In this case we do not need to calculate the inverse of a matrix because it degenerates to a scalar. Moreover, μ_y and Σ_{yz} are column vectors, and Σ_z is also a scalar. Then the number of calculations needed to update the probability distribution of the nonevidential variables given a single piece of evidence is linear in the number of variables in X. Thus, this algorithm provides a simple and efficient method for evidence propagation in Gaussian Bayesian network models.

Due to the simplicity of this incremental algorithm, the implementation of this propagation method in the inference engine of an expert system is an easy task. Figure 8.40 shows the corresponding pseudocode. The algorithm gives the CPD of the nonevidential nodes Y given the evidence $E = e$. The performance of this algorithm is illustrated in the following example.

Example 8.11 Propagation in Gaussian Bayesian network models.
Consider the Gaussian Bayesian network given in Figure 8.41. This network was used in Example 6.18. The JPD of the variables can be factorized as follows:
$$f(a, b, c, d) = f(a)f(b)f(c|a)f(d|a, b),$$
where
$$
\begin{aligned}
f(a) &\sim N(\mu_A, v_A), \\
f(b) &\sim N(\mu_B, v_B), \\
f(c) &\sim N(\mu_C + \beta_{CA}(a - \mu_A), v_C), \\
f(d) &\sim N(\mu_D + \beta_{DA}(a - \mu_A) + \beta_{DB}(b - \mu_B), v_D).
\end{aligned}
$$

8.9 Exact Propagation in Gaussian Networks 385

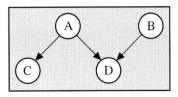

FIGURE 8.41. Example of a DAG used to define a Gaussian Bayesian network model.

For example, when all the means are equal to zero, all the variances are equal to one, and the regression parameters are $\beta_{CA} = 1$, $\beta_{DA} = 0.2$, $\beta_{DB} = 0.8$, we get

$$f(a) \sim N(0,1),$$
$$f(b) \sim N(0,1),$$
$$f(c) \sim N(a,1),$$
$$f(d) \sim N(0.2a + 0.8b, 0).$$

This set of CPDs constitutes one of two equivalent representations of the Gaussian Bayesian network model. An alternative representation can be obtained using the covariance matrix. Equation (6.40) provides a formula to obtain this matrix from the above set of CPDs. Using this formula we get the covariance matrix (see Example 6.18)

$$\Sigma = \begin{pmatrix} 1.0 & 0.0 & 1.0 & 0.20 \\ 0.0 & 1.0 & 0.0 & 0.80 \\ 1.0 & 0.0 & 2.0 & 0.20 \\ 0.2 & 0.8 & 0.2 & 1.68 \end{pmatrix}. \tag{8.54}$$

Then, we apply the algorithm given in Figure 8.40 to (8.54) and the mean vector $\mu = (0,0,0,0)$ to propagate evidence in the resulting Gaussian network model. Suppose we have the evidence $\{A = 1, B = 3, C = 2\}$. Figure 8.42 shows a *Mathematica* program implementing this incremental algorithm to calculate the conditional mean vector and covariance matrix for the different iteration steps.

In the first iteration step, we consider the first evidential node $A = 1$. We obtain the following mean vector and covariance matrix for the rest of the nodes $Y = \{B, C, D\}$:

$$\mu_{Y|A=1} = \begin{pmatrix} 0.0 \\ 1.0 \\ 0.2 \end{pmatrix}; \Sigma_{Y|A=1} = \begin{pmatrix} 1.0 & 0.0 & 0.80 \\ 0.0 & 1.0 & 0.00 \\ 0.8 & 0.0 & 1.64 \end{pmatrix}.$$

The second step of the algorithm adds evidence $B = 3$; we obtain the following mean vector and covariance matrix for the rest of the nodes $Y =$

```
(* Definition of the JPD *)
M={0,0,0,0};
V={{1.0, 0.0, 1.0, 0.2},
   {0.0, 1.0, 0.0, 0.8},
   {1.0, 0.0, 2.0, 0.2},
   {0.2, 0.8, 0.2, 1.68}};
(* Nodes and evidence *)
X={A,B,C,D};
Ev={A,B,C};
ev={1,3,2};
(* Incremental updating of M and V *)
NewM=Transpose[List[M]];
NewV=V;
For[k=1, k<=Length[Ev], k++,
(* Position of the ith element of E[[k]] in X *)
    i=Position[X,Ev[[k]]][[1,1]];
    My=Delete[NewM,i];
    Mz=NewM[[i,1]];
    Vy=Transpose[Delete[Transpose[Delete[NewV,i]],i]];
    Vz=NewV[[i,i]];
    Vyz=Transpose[List[Delete[NewV[[i]],i]]];
    NewM=My+(1/Vz)*(ev[[k]]-Mz)*Vyz;
    NewV=Vy-(1/Vz)*Vyz.Transpose[Vyz];
(* Delete ith element *)
    X=Delete[X,i];
(* Printing results *)
    Print["Iteration step = ",k];
    Print["Remaining nodes = ",X];
    Print["M = ",Together[NewM]];
    Print["V = ",Together[NewV]];
    Print["--------------------"];
]
```

FIGURE 8.42. *Mathematica* code for exact propagation of evidence in a Gaussian Bayesian network model.

$\{C, D\}$:

$$\mu_{Y|A=1,B=3} = \begin{pmatrix} 1.0 \\ 2.6 \end{pmatrix}; \ \Sigma_{Y|A=1,B=3} = \begin{pmatrix} 1.0 & 0.0 \\ 0.0 & 1.0 \end{pmatrix}.$$

From the structure of this covariance matrix, we observe that the two remaining variables are independent given the evidential nodes A and B (this fact can actually be verified in the DAG given in Figure 8.41 using the D-separation criterion). Therefore, the last iteration step will not modify the current values. Finally, after considering evidence $C = 2$ we get the conditional mean and variance of D, which are given by $\mu_{D|A=1,B=3,C=2} = 2.6$, $\sigma_{D|A=1,B=3,C=2} = 1.0$. ∎

Exercises

8.1 Consider the Bayesian network given by the DAG in Figure 8.43 and the CPDs shown in Table 8.11.

(a) Apply the polytrees Algorithm 8.1 to obtain the MPDs of the nodes.

(b) Draw the necessary messages and the order in which they are computed.

(c) Suppose now that the evidence $\{D = 0, C = 1\}$ has been observed. Repeat the calculations to obtain the conditional probabilities of the nodes given this evidence.

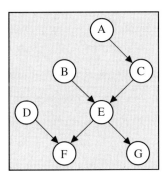

FIGURE 8.43. A polytree Bayesian network.

8.2 Repeat the previous exercise using the polytree in Figure 8.44 and the CPDs in Table 8.12.

8.3 Repeat the calculations in Example 8.4 using the graph in Figure 8.19(b).

8.4 Given the multiply-connected graph in Figure 8.45 and the corresponding CPDs in Table 8.11 with $p(b)$ and $p(d)$ replaced by $p(b|a)$ and $p(d|b)$ as shown in Table 8.13,

(a) Obtain a loop-cutset that renders this graph the polytree in Figure 8.43 and use the conditioning algorithm to obtain the MPDs of the nodes and the conditional probabilities when evidence $\{D = 0, C = 1\}$ has been observed.

388   8. Exact Propagation in Probabilistic Network Models

a	p(a)
0	0.3
1	0.7

b	p(b)
0	0.6
1	0.4

e	p(d)
0	0.7
1	0.3

a	c	p(c|a)
0	0	0.15
0	1	0.85
1	0	0.25
1	1	0.75

e	g	p(g|e)
0	0	0.10
0	1	0.90
1	0	0.30
1	1	0.70

b	c	e	p(e|b,c)
0	0	0	0.40
0	0	1	0.60
0	1	0	0.45
0	1	1	0.55
1	0	0	0.60
1	0	1	0.40
1	1	0	0.30
1	1	1	0.70

d	e	f	p(f|d,e)
0	0	0	0.25
0	0	1	0.75
0	1	0	0.60
0	1	1	0.40
1	0	0	0.10
1	0	1	0.90
1	1	0	0.20
1	1	1	0.80

TABLE 8.11. The CPDs for the Bayesian network in Figure 8.43.

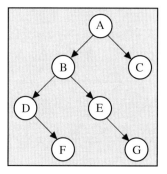

FIGURE 8.44. A polytree Bayesian network.

(b) Is this a minimal loop-cutset? Compute all possible minimum loop-cutsets for this graph.

(c) Apply the conditioning algorithm with some of the new minimum loop-cutsets.

8.5 Consider the triangulated graph in Figure 8.46.

(a) Obtain a chain of cliques using Algorithm 4.3.

8.9 Exact Propagation in Gaussian Networks

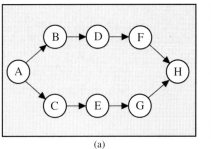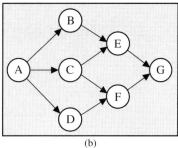

FIGURE 8.47. Two multiply-connected graphs.

8.10 Consider the graph in Figure 8.36.

(a) Construct the moralized ancestral graph that shows that Θ_E is D-separated from C by D.

(b) Construct the moralized ancestral graph that shows that Θ_F is D-separated from C by D.

8.11 Given the DAG in Figure 8.43 and supposing that E is the goal node,

(a) Apply Algorithm 8.6 to reduce the graph to the relevant set of nodes.

(b) Obtain the new numerical values for the CPDs from the values given in Table 8.11.

(c) Use a suitable method to compute the initial (no evidence) MPD of node E.

8.12 Repeat the previous exercise using the graph in Figure 8.45 and the CPDs in Table 8.12.

8.13 Given that $(X_1, X_2, X_3, X_4, X_5)$ are multivariate normally distributed random variables with mean vector and covariance matrix given by

$$\mu = \begin{pmatrix} \mu_1 \\ \mu_2 \\ \mu_3 \\ \mu_4 \end{pmatrix} \quad \text{and} \quad \Sigma = \begin{pmatrix} 1 & 0.3 & 0 & 0.4 & 0 \\ 0.3 & 1 & 0 & 0.2 & 0 \\ 0 & 0 & 1 & 0 & 0.1 \\ 0.4 & 0.2 & 0 & 1 & 0 \\ 0 & 0 & 0.1 & 0 & 1 \end{pmatrix},$$

apply the algorithm for exact propagation in Gaussian networks (see Figure 8.40) to calculate the initial MPDs of the nodes, and the CPDs given the evidence $X_2 = 1$, $X_4 = 2$.

Chapter 9
Approximate Propagation Methods

9.1 Introduction

In Chapter 8, we presented several algorithms for exact propagation of evidence in Markov and Bayesian network models. However, there are some problems associated with these methods. On one hand, some of these algorithms are not generally applicable to all types of network structures. For example, the polytrees algorithm (Section 8.3) applies only to networks with simple polytree structure. On the other hand, general exact propagation methods that apply to any Bayesian or Markov network become increasingly inefficient with certain types of network structures. For example, conditioning algorithms (Section 8.5) suffer from a combinatorial explosion when dealing with large sets of cutset nodes, and clustering methods (Section 8.6) require building a join tree, which can be an expensive computational task; and they also suffer from a combinatorial explosion when dealing with networks with large cliques. This is not surprising because as we have seen in Chapter 8, the task of exact propagation has been proven to be NP-hard (see Cooper (1990)). Thus, from the practical point of view, exact propagation methods may be restrictive or even inefficient in situations where the type of network structure requires a large number of computations and huge amount of memory and computer power.

In this chapter we introduce some approximate methods for the propagation of evidence that are applicable to all types of network structures. We remind the reader that by *approximate propagation algorithms* we mean algorithms that compute the conditional probability distribution of nodes approximately. The basic idea behind these methods is to generate a sample of size N from the joint probability distribution (JPD) of the variables, and then use the generated sample to compute approximate values for the probabilities of certain events given the evidence. The probabilities are approximated by the ratios of frequency of events in the sample to the sample size.

Approximate propagation methods can be classified into two types: stochastic simulation methods and deterministic search methods. The methods in the first class generate the sample from the JPD using some random mechanism, whereas the deterministic search methods generate the sample in a systematic way.

Section 9.2 gives the intuitive and theoretical bases of the simulation methods. Section 9.3 gives a general framework for these methods. Sections 9.4 to 9.8 present five stochastic simulation methods. These are the acceptance-rejection sampling, uniform sampling, likelihood weighing, backward-forward sampling, and Markov sampling methods. Sections 9.9 and 9.10 introduce the systematic sampling and the maximum probability search methods, which fall in the class of deterministic search methods. Finally, in Section 9.11 we analyze the complexity of approximate propagation.

9.2 Intuitive Basis of Simulation Methods

In this section we illustrate a general simulation scheme by a simple example. Consider an urn that contains six balls numbered $\{1, \ldots, 6\}$. Suppose we want to conduct the following experiment. We select a ball at random from the urn, record its number, put it back in the urn, and mix the balls before selecting the next ball. This sampling scheme is called *sampling with replacement*. Each selection of a ball is called a *trial* or an *experiment*. In this case each trial has six possible outcomes, $\{1, \ldots, 6\}$.

Let X_i denote the outcome (the number of the selected ball) of the ith trial. Because sampling is done with replacement, the trials are independent (the outcome of one trial does not affect the outcome of the next trial). Clearly, X_i is a *uniform* random variable with probability mass function $p(X_i = x_i) = 1/6$, for $x_i = 1, \ldots, 6$ and $i = 1, \ldots, N$, where N is the number of trials (the sample size).

As in the previous chapters, we write $p(X_i = x_i)$ as $p(x_i)$, for simplicity. In this case, the JPD of $X = \{X_1, \ldots, X_N\}$, $p(x)$, is the product of the individual probability distributions, that is,

$$p(x) = \prod_{i=1}^{N} p(x_i). \tag{9.1}$$

Using this JPD, we can calculate the exact probabilities of certain events such as $p(X_1 = 1, \ldots, X_N = 1)$, $p(\textit{odd outcomes} = \textit{even outcomes})$, etc.

9.2 Intuitive Basis of Simulation Methods

These computations are easy in this case because the distribution is uniform (there is exactly one ball for each of the numbers $\{1, \ldots, 6\}$), the trials are identical (we use the same urn), and the outcome of each trial is independent of the outcomes of the others (we sample with replacement). The calculations of the exact probabilities become complicated and time-consuming when the distribution is not uniform (e.g., unequal numbers of balls of different types), the trials are not identical (e.g, sampling from urns with different numbers of balls), and/or the trials are not independent (e.g., sampling without replacement).

In these complicated situations, we can compute the probabilities of certain events approximately using simulation approaches. We can, for example, perform the experiment N times. We obtain what we call a *sample* of size N. Then the probability of an event can be approximated by the ratio of the number of times the event occurs to the total number of simulations N. Clearly, the larger the sample size, the more accurate the approximation. The methods we present in this chapter differ mainly in the way the sample is generated from the JPD of the variables.

An equivalent way of obtaining a sample of size N balls with replacement from Urn 1 is rolling a six-sided fair die N times. Let X denote the number of a ball selected at random from Urn 1 and Y_i denote the number observed when we roll a die. Then Y has the same probability distribution as that of X. Let $p(x)$ denote the probability distribution of X and $h(y)$ denote the probability distribution of Y; then $p(x) = h(x)$, as can be seen in Figure 9.1. Thus, drawing N balls with replacement from Urn 1 can be simulated by rolling a six-sided die N times.

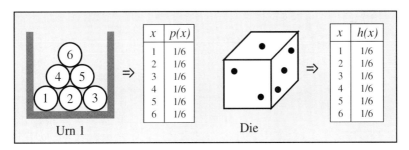

FIGURE 9.1. Simulating drawing balls with replacement from Urn 1 using a six-sided fair die.

It is helpful at this point to distinguish between the probability distribution $p(x)$ (generated by Urn 1) and the probability distribution $h(y)$ (generated by the die). We shall refer to the distribution of X, which is the distribution that we wish to generate the sample from, as the *population distribution*. The distribution of Y is called the *simulation distribution* because it is the one that we use to generate (simulate) the sample. In this case, the simulation distribution happens to be the same as the population distribution. But as we shall see below, the simulation distribution can be different from the population distribution.

The reason we use a die (a simulation distribution) to simulate sampling balls from Urn 1 (a population distribution) is that it is easier to roll a die than to withdraw a ball from an urn and mix the balls before the next withdrawal. In other words, if the population distribution is not easy to sample from, we may choose an easier distribution to sample from. But can we always do that? For example, can we always use a die to simulate withdrawing balls from urns with different numbers of balls? The answer is, fortunately, yes. For example, suppose that the urn contains only five balls numbered $\{1,\ldots,5\}$ as shown in Figure 9.2 (Urn 2). Let X denote the number of the ith ball drawn at random with replacement from Urn 2. Then X is a random variable whose probability distribution, $p(x)$, is shown in Figure 9.2 (Urn 2). In this case, the simulation distribution (the die) is not the same as the population distribution (Urn 2), that is, $p(x) \neq h(x)$ (the columns labeled $s(x)$ will be explained shortly). Despite the fact that Urn 2 and the die do not lead to the same distribution, we can still use the six-sided die to simulate withdrawing balls from Urn 2, but we have to account for the fact that the simulation and population distributions are not the same.

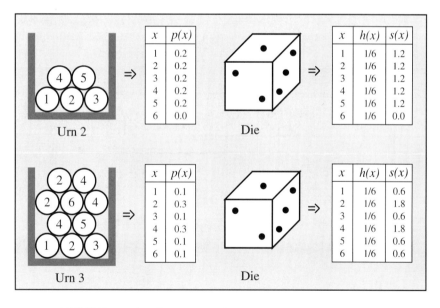

FIGURE 9.2. An illustration of a general simulation scheme.

One way to account for this difference is the following: when the die shows up a 6, we ignore the roll and try again until the die shows up a number less than 6, in which case we let y be the obtained number and accept y as a value generated from the population distribution, $p(x)$. This example is actually a special case of a method known as the *acceptance-rejection* method. The theoretical basis is given by the following theorem, which is due to Von Neumann (1951) (see also Rubinstein (1981), Devroye (1986), and Ripley (1987)).

Theorem 9.1 Acceptance-rejection method. *Let X be a random variable with probability distribution $p(x)$. Suppose that $p(x)$ can be expressed as*

$$p(x) = c\,g(x)\,h(x), \tag{9.2}$$

where $c \geq 1, 0 \leq g(x) \leq 1$, and $h(x)$ is a probability distribution. Let U be a standard uniform $U(0,1)$ random variable and let Y be a random variable with probability density function $h(y)$ independent of U. Then, the conditional probability distribution of Y given that $u \leq g(y)$ is the same as the probability distribution of X. Furthermore, the probability of acceptance (efficiency) is $1/c$.

For instance, in the Urn 2 example shown in Figure 9.2, we can write $p(x) = cg(x)h(x)$, where $p(x)$ and $h(x)$ are shown in Figure 9.2, $c = 6/5$ and

$$g(x) = \begin{cases} 0, & \text{if } x = 6, \\ 1, & \text{otherwise.} \end{cases} \tag{9.3}$$

Thus, using the above theorem, we can obtain a sample from $p(x)$ (Urn 2) by using $h(x)$ (the die) and checking the condition $u \leq g(x)$ for every value x simulated from $h(x)$, where u is a number generated from the uniform $U(0,1)$ distribution. Therefore, in this case, the outcome $x = 6$ is always rejected, because $g(6) = 0$, and all other outcomes are always accepted.

Before illustrating the acceptance-rejection method by an example, we show how to simulate a sample from a probability distribution $h(x)$.

Example 9.1 Simulating from a probability distribution. To generate a sample from a probability distribution $h(x)$, we first compute the cumulative distribution function (CDF),

$$H(x) = p(X \leq x) = \int_{-\infty}^{x} h(x)dx.$$

We then generate a sequence of random numbers $\{u_1, \ldots, u_N\}$ from $U(0,1)$ and obtain the corresponding values $\{x_1, \ldots, x_N\}$ by solving $H(x_i) = u_i$, $i = 1 \ldots, N$, which gives $x_i = H^{-1}(u_i)$, where $H^{-1}(u_i)$ is the inverse of the CDF evaluated at u_i. For example, Figure 9.3 shows the CDF $H(x)$ and the two values x_1 and x_2 corresponding to the uniform $U(0,1)$ numbers u_1 and u_2. ■

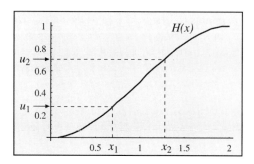

FIGURE 9.3. Sampling from a probability distribution $h(x)$ using the corresponding cumulative distribution function $H(x)$.

Example 9.2 The Acceptance-rejection method. Suppose that $p(x) = 3x(2-x)/4$, $0 \leq x \leq 2$, and we wish to obtain a sample of size N from $p(x)$. In this case, $p(x)$ can be factorized in the form (9.2) by taking $h(x) = 1/2$, $0 \leq x \leq 2$, $g(x) = x(2-x)$, and $c = 3/2$. These functions are shown in Figure 9.4(a). The CDF corresponding to $h(x)$ is

$$H(x) = \int_0^x h(x)dx = \int_0^x (1/2)dx = x/2.$$

Note that this function is simpler to simulate from than the CDF of the population distribution $P(x)$. The functions $P(x)$, $H(x)$, and $g(x)$ are drawn in Figure 9.4(b).

Suppose we generate a random number y from $h(y)$ and a random number u from $U(0,1)$. We compute $g(y)$, and if $u \leq g(y)$, we accept y as a number generated from $p(x)$. Otherwise, we reject both u and y, and try again. For example, if $y = 1.5$, then $g(1.5) = 0.75$, which means that the probability of accepting $y = 1.5$ as a random number generated from $p(x)$ is 0.75. On the other hand, if $y = 1$, $g(1) = 1$, which means that the probability of accepting $y = 1$ as a random number generated from $p(x)$ is 1. This should not be surprising, because as can be seen from Figure 9.4(b), when $x = 1$, $P(x) = H(x)$. It can also be seen from Figure 9.4(b) that $g(x)$ attains its maximum at $x = 1$; hence, as y deviates from 1, the probability of acceptance decreases. The probability of accepting a random number generated from $h(y)$ is equal to $1/c = 2/3$. ■

According to Theorem 9.1, the probability of acceptance is $1/c$. Thus, the probability of acceptance is high when c is close to 1, and this usually happens when $h(x)$ is close to $p(x)$. Thus, on the one hand, we want $h(x)$ to be as close to $p(x)$ as possible so that the probability of acceptance becomes large. On the other hand, we want $h(x)$ to be as easy to simulate from as possible.

Theorem 9.1 suggests the following algorithm for generating a random sample of size N from $p(x)$ but using $h(x)$.

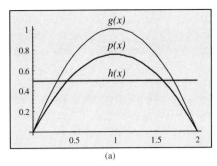
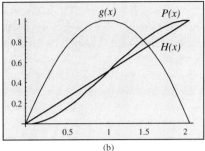

FIGURE 9.4. An illustration of the acceptance-rejection method.

Algorithm 9.1 Acceptance-Rejection Method.

- **Input:** The population distribution $p(x)$, the simulation distribution $h(x)$, and the sample size N.

- **Output:** A sample $\{x_1, \ldots, x_N\}$ from $p(x)$.

1. Let $i = 1$.

2. Generate u from $U(0, 1)$.

3. Generate y from $h(y)$.

4. If $u \leq g(y)$, let $x_i = y$. Otherwise go to Step 2.

5. If $i < N$, increase increase i by 1 and go to Step 2. Otherwise return $\{x_1, \ldots, x_N\}$. ∎

When c is large, the efficiency of the above algorithm is low and the rejection rate is high. Consequently, we may have to generate a huge number of random variables from $h(y)$ to obtain a small sample from $p(x)$. The acceptance-rejection algorithm, however, can be made more efficient by the following modification. Write $p(x)$ as

$$p(x) = \frac{p(x)}{h(x)} h(x) = s(x) h(x), \qquad (9.4)$$

where

$$s(x) = \frac{p(x)}{h(x)}, \qquad (9.5)$$

is a *score function*. Thus, the score of the event x is the ratio of the population distribution, $p(x)$, to the simulation distribution, $h(x)$.

From (9.5) and (9.4), we see that $s(x) = c g(x)$, that is, the score is proportional to $g(x)$. Therefore, instead of rejecting a number x generated from $h(x)$, we assign it a probability proportional to $s(x)$ or $g(x)$. Then at the end we normalize the scores (by dividing each score by the sum of all scores) and use the normalized scores to estimate the probability of any event of interest. This leads to a much higher efficiency of the simulation process.

For example, the scores in our urn-die example are shown in Figure 9.2. Note that in the case of Urn 2 the score associated with the outcome $x = 6$ is zero, and the the rest of the outcomes have associated the same score. Thus, in this case, using the scores produces the same result as applying the acceptance-rejection method. However, the situation for Urn 3 is more complicated, and the scores are more efficient than the acceptance-rejection method.

9.3 General Frame for Simulation Methods

The above discussion suggests a general framework for simulation methods. Let $X = \{X_1, \ldots, X_n\}$ be a set of variables with JPD $p(x)$. Suppose that a subset E of the variables in X have known values e. The variables in E are called the *evidence* or *evidential* variables and the remaining variables are called the *nonevidential* variables. Our objective is to compute the probability distribution for each nonevidential variable given the evidence. In general, given a set of variables $Y \subset X$, we wish to calculate the conditional probability of y given e, the values of the evidential variables. This conditional probability can be written as

$$p(y|e) = \frac{p_e(y)}{p(e)} \propto p_e(y), \qquad (9.6)$$

where $p(e)$ is the probability of e and

$$p_e(y) = \begin{cases} p(y \cup e), & \text{if } y \text{ is consistent with } e, \\ 0, & \text{otherwise.} \end{cases} \qquad (9.7)$$

Note that if $Y \cap E = \phi$, then $p_e(y) = p(y, e)$. It can be seen from (9.6) that to compute $p(y|e)$, we need only to compute and normalize $p_e(y)$. Note also that when $E = \phi$ (no evidence is available), $p(y|e) = p(y)$ is simply the marginal distribution of Y.

As we mentioned earlier, the exact computations of $p(x_i|e)$ can be time-consuming and often infeasible. Thus, we turn to simulation methods to approximate $p(x_i|e)$. We first generate a sample of size N from $p(x)$ but using a simulation distribution $h(x)$. Next, we compute and normalize the scores. Then, the probability $p(x_i|e)$ can be approximated by adding up the normalized scores of all instantiations that are consistent with the event x_i and e.

After a sample of N instantiations, $x^j = \{x_1^j, \ldots, x_n^j\}$, $j = 1, \ldots, N$, is obtained, the conditional probability distribution of any subset $Y \subset X$

given the evidence $E = e$ is then estimated by the sum of the normalized scores of all instantiations in which y occurs,

$$p(y) \approx \frac{\sum_{y \in x^j} s(x^j)}{\sum_{j=1}^{N} s(x^j)}. \qquad (9.8)$$

The above procedure is described in the following general algorithm:

Algorithm 9.2 General Simulation Framework.

- **Input:** *The population distribution $p(x)$, the simulation probability distribution $h(x)$, the sample size N, and a subset $Y \subset X$.*
- **Output:** *An approximation of $p(y)$ for every possible value y of Y.*

1. for $j = 1$ to N
   - Generate $x^j = (x_1^j, \ldots, x_n^j)$ using $h(x)$.
   - Calculate $s(x^j) = \dfrac{p(x^j)}{h(x^j)}$.

2. For every possible value y of Y, approximate $p(y)$ using (9.8). ■

The accuracy of the approximation obtained using Algorithm 9.2 depends on the following factors:

- The simulation distribution $h(x)$ chosen to obtain the sample.
- The method used to generate instantiations from $h(x)$.
- The sample size N.

Although the selection of the simulation distribution influences the quality of the approximation considerably, sampling schemes leading to similar scores for all instantiations are associated with high quality and those leading to substantially unequal scores have low quality.

The next example illustrates this simulation framework.

Example 9.3 Bayesian network model. Suppose that the directed graph shown in Figure 9.5 is given as a graphical model for the independence structure of the JPD of six binary variables $X = \{X_1, \ldots, X_6\}$ that take values in $\{0, 1\}$. This graph defines a Bayesian network whose JPD can be factorized as

$$p(x) = p(x_1)p(x_2|x_1)p(x_3|x_1)p(x_4|x_2)p(x_5|x_2,x_3)p(x_6|x_3). \qquad (9.9)$$

The numerical values of the set of conditional probability distributions (CPDs) needed to specify the JPD, $p(x)$, are given in Table 9.1.

402   9. Approximate Propagation Methods

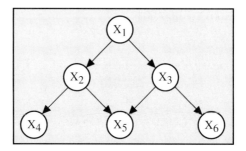

FIGURE 9.5. Example of a Bayesian network.

x_1	$p(x_1)$
0	0.3
1	0.7

x_1	x_2	$p(x_2|x_1)$
0	0	0.4
0	1	0.6
1	0	0.1
1	1	0.9

x_1	x_3	$p(x_3|x_1)$
0	0	0.2
0	1	0.8
1	0	0.5
1	1	0.5

x_2	x_4	$p(x_4|x_2)$
0	0	0.3
0	1	0.7
1	0	0.2
1	1	0.8

x_3	x_6	$p(x_6|x_3)$
0	0	0.1
0	1	0.9
1	0	0.4
1	1	0.6

x_2	x_3	x_5	$p(x_5|x_2,x_3)$
0	0	0	0.4
0	0	1	0.6
0	1	0	0.5
0	1	1	0.5
1	0	0	0.7
1	0	1	0.3
1	1	0	0.2
1	1	1	0.8

TABLE 9.1. Probability distributions required for specifying the JPD corresponding to the Bayesian network given in Figure 9.5.

9.3 General Frame for Simulation Methods

Instantiation x^j	$p(x^j)$	$h(x^j)$	$s(x^j)$
$x^1 = (0, 1, 1, 1, 0, 0)$	0.0092	1/64	0.5898
$x^2 = (1, 1, 0, 1, 1, 0)$	0.0076	1/64	0.4838
$x^3 = (0, 0, 1, 0, 0, 1)$	0.0086	1/64	0.5529
$x^4 = (1, 0, 0, 1, 1, 0)$	0.0015	1/64	0.0941
$x^5 = (1, 0, 0, 0, 1, 1)$	0.0057	1/64	0.3629

TABLE 9.2. Five instantiations randomly obtained from all 64 possible instantiations of the six binary variables.

We generate a sample of N observations from this distribution. Each observation $x^j = \{x_1^j, \ldots, x_6^j\}$ consists of an observed value for each of the six variables in X. An observation x^j is sometimes referred to as an *instantiation*. Since the six variables are binary, there are $2^6 = 64$ possible instantiations. For illustrative purpose, suppose we select five of the 64 instantiations at random with replacement.[1] This implies that $h(x^j) = 1/64$; $j = 1, \ldots, 5$. The selected instantiations and their corresponding $p(x)$, $h(x)$, and $s(x)$ are given in Table 9.2.

Then based on the instantiations in Table 9.2, we approximate the probability distribution of the variables by normalizing the scores by dividing each score by the sum of all scores. The desired probability is obtained by adding up the normalized scores of all instantiations where the event of interest occurs. For example,

$$p(X_1 = 0) \approx \frac{s(x^1) + s(x^3)}{\sum_{i=1}^{5} s(x^i)} = \frac{0.5898 + 0.5529}{2.0835} = 0.5485$$

and

$$p(X_2 = 0) \approx \frac{s(x^3) + s(x^4) + s(x^5)}{\sum_{i=1}^{5} s(x^i)} = \frac{0.5529 + 0.0941 + 0.3629}{2.0835} = 0.4847.$$

This is because $X_1 = 0$ appears in instantiations x^1 and x^3, while $X_2 = 0$ appears in instantiations x^3, x^4, and x^5. The conditional probabilities for other variables can be computed in a similar way.

The probability distributions that we dealt with so far are the marginal probability distributions of the nodes, that is, in the case where no evidence is available. If some evidence is available, we also proceed in a similar

---

[1] We wish to emphasize here that approximation based on only five observations is likely to be inaccurate. Furthermore, as we shall see later in this chapter, choosing the instantiations uniformly at random is not necessarily an optimal choice of a simulation distribution.

404   9. Approximate Propagation Methods

Instantiation x^j	$p(x^j)$	$h(x^j)$	$s(x^j)$
$x^1 = (0, 1, 1, 1, 0, 1)$	0.0138	1/16	0.2212
$x^2 = (1, 0, 1, 1, 0, 0)$	0.0049	1/16	0.0784
$x^3 = (1, 0, 1, 1, 1, 1)$	0.0073	1/16	0.1176
$x^4 = (0, 1, 1, 1, 1, 0)$	0.0369	1/16	0.5898
$x^5 = (1, 1, 1, 1, 0, 0)$	0.0202	1/16	0.3226

TABLE 9.3. Five instantiations randomly obtained from all 16 possible instantiations of the six binary variables when the evidential nodes have the values $X_3 = 1$ and $X_4 = 1$.

fashion. To illustrate, suppose $E = \{X_3, X_4\}$ are evidential nodes and their values are known to be $e = \{X_3 = 1, X_4 = 1\}$. Because nodes X_3 and X_4 are fixed, we only have $2^4 = 16$ possible instantiations. Again, suppose we select five of the 16 instantiations at random with replacement. This implies that $h(x^j) = 1/16$; $j = 1, \ldots, 5$. The selected instantiations and their corresponding $p(x)$, $h(x)$, and $s(x)$ are given in Table 9.3.

Based on the instantiations in Table 9.3, we approximate the conditional probability distribution of the nonevidential variables given the evidence using the normalized scores. For example,

$$p(X_1 = 0|X_3 = 1, X_4 = 1) \approx \frac{s(x^1) + s(x^4)}{\sum_{i=1}^{5} s(x^i)} = \frac{0.2212 + 0.5898}{1.3296} = 0.6099$$

and

$$p(X_2 = 0|X_3 = 1, X_4 = 1) \approx \frac{s(x^2) + s(x^3)}{\sum_{i=1}^{5} s(x^i)} = \frac{0.0784 + 0.1176}{1.3296} = 0.1474.$$

This is because $X_1 = 0$ appears in instantiations x^1 and x^4, while $X_2 = 0$ appears in instantiations x^2 and x^3. The conditional probabilities for other variables can be computed in a similar way.

It is worth noting here that the obtained instantiations can be used to approximate not only univariate but also multivariate probability distributions. This can be done if we store the frequencies of all feasible values of the discrete variables and all simulated values of the continuous variables. For example, using the instantiations in Table 9.2, we can compute

$$p(X_5 = 0, X_6 = 0) \approx \frac{s(x^1)}{\sum_{i=1}^{5} s(x^i)} = \frac{0.5898}{2.0835} = 0.2831,$$

$$p(X_5 = 0, X_6 = 1) \approx \frac{s(x^3)}{\sum_{i=1}^{5} s(x^i)} = \frac{0.5529}{2.0835} = 0.2654,$$

$$p(X_5 = 1, X_6 = 0) \approx \frac{s(x^2) + s(x^4)}{\sum_{i=1}^{5} s(x^i)} = \frac{0.4838 + 0.0941}{2.0835} = 0.2773,$$

$$p(X_5 = 1, X_6 = 1) \approx \frac{s(x^5)}{\sum_{i=1}^{5} s(x^i)} = \frac{0.3629}{2.0835} = 0.1742.$$

Note, however, that the sample size required to compute the probabilities with certain accuracy increases as dimensionality increases. Thus, the sample size required to compute multivariate probabilities is necessarily larger than the sample size required to compute the univariate probabilities with the same accuracy. ∎

Of special interest is the case in which the population and simulation distributions are given in terms of factorizations of the forms

$$p(x) = \prod_{i=1}^{n} p(x_i|s_i) \qquad (9.10)$$

and

$$h(x) = \prod_{i=1}^{n} h(x_i|s_i), \qquad (9.11)$$

where $S_i \subset X$ is a subset of variables and $h(x_i)$ is the simulation distribution of node X_i. In this situation, the simulation process can be simplified by sequentially simulating the nodes as, for example, in the example of rolling a die. Then, using (9.10) and (9.11), the score of a given instantiation $x = (x_1, \ldots, x_n)$ can be obtained as the product of the scores of the variables

$$s(x) = \frac{p(x)}{h(x)} = \prod_{i=1}^{n} \frac{p(x_i|s_i)}{h(x_i|s_i)} = \prod_{i=1}^{n} s(x_i|s_i). \qquad (9.12)$$

Note that all Bayesian network models, all decomposable Markov network models, and various other probabilistic models can be expressed as in (9.10) (see Chapter 6). For the sake of simplicity, we shall use Bayesian network models to illustrate the different methodologies.

When a set of evidential nodes E is known to take the values $E = e$, we can also use Algorithm 9.2 to calculate the CPD, $p(y|e)$, but taking into account the known evidence, that is, the population distribution in this case is

$$p_e(x) \propto \prod_{i=1}^{n} p_e(x_i|\pi_i), \qquad (9.13)$$

where

$$p_e(x_i|\pi_i) = \begin{cases} p(x_i|\pi_i), & \text{if } x_i \text{ and } \pi_i \text{ are consistent with } e, \\ 0, & \text{otherwise,} \end{cases} \qquad (9.14)$$

that is, $p_e(x_i|\pi_i) = 0$ if X_i or any of its parents are inconsistent with the evidence, otherwise $p_e(x_i|\pi_i) = p(x_i|\pi_i)$. Thus, the CPD $P(y|e)$ can be approximated using Algorithm 9.2 with the new unnormalized JPD $p_e(x)$.

From the above discussion we see that a simulation method consists of three components:

1. A simulation distribution, $h(x)$, used to generate the sample.
2. A method to obtain instantiations from $h(x)$.
3. A scoring function.

Many of the existing methods are variants of the above methods; they differ only in one or both of the last two components. In this chapter we discuss the following methods:

- Acceptance-rejection sampling method.
- Uniform sampling method.
- Likelihood weighing method.
- Backward-forward sampling method.
- Markov sampling method.
- Systematic sampling method.
- Maximum probability search method.

Given a population distribution $p(x)$, each of these methods generates a sample of size N from $p(x)$. They differ only in how the sample is generated and in the choice of the scoring function. The first five of these methods belong to the class of stochastic simulation methods, and the last two are deterministic search methods.

For the sake of simplicity, we illustrate the different algorithms in the case of discrete random variables, though the same scheme is applicable to continuous or mixed discrete-continuous cases (Castillo, Gutiérrez, and Hadi (1995b)).

9.4 Acceptance-Rejection Sampling Method

Henrion (1988) suggests a simulation method that generates the instantiations one variable at a time in a forward manner, that is, a variable is sampled only after all of its parents have been sampled. According to this method, all variables, including the evidence variables, if any, are sampled. The simulation distribution for X_i, is its associated CPD in (9.10), that is

$$h(x_i|\pi_i) = p(x_i|\pi_i), \; i \in \{1,\ldots,n\}. \qquad (9.15)$$

```
Initialize
    Arrange the nodes in ancestral ordering
Main loop
    for j ← 1 to N do
        for i ← 1 to n do
            x_i ← generate a value from p(x_i|π_i)
            if X_i ∈ E and x_i ≠ e_i then, repeat loop i
```

FIGURE 9.6. Pseudocode for the acceptance-rejection sampling method.

Thus, the variables must be ordered in such a way that all parents of X_i must be assigned values before X_i is sampled. An ordering of the nodes satisfying such a property is called an *ancestral ordering*. This simulation strategy is called forward sampling because it goes from parents to children. Once the parents of X_i are assigned their values, we simulate a value for X_i using the simulation distribution $h(x_i|\pi_i)$, which in this case is the same as the population distribution $p(x_i|\pi_i)$. Hence, the score becomes

$$s(x) = \frac{p_e(x)}{h(x)} = \frac{\prod_{X_i \notin E} p_e(x_i|\pi_i) \prod_{X_i \in E} p_e(x_i|\pi_i)}{\prod_{X_i \notin E} p(x_i|\pi_i) \prod_{X_i \in E} p(x_i|\pi_i)}. \quad (9.16)$$

From (9.14) and (9.16), it follows that

$$s(x) = \begin{cases} 1, & if\ x_i = e_i,\ for\ all\ X_i \in E, \\ 0, & otherwise. \end{cases} \quad (9.17)$$

Note that if $x_i \neq e_i$ for some $X_i \in E$, the score is zero; thus, as soon as the simulated value for an evidential node does not coincide with the observed value, the sample is discarded. Thus, the method is a variant of the acceptance-rejection method (see Theorem 9.1). Henrion (1988), however, refers to this method as the *logic sampling* method. The pseudocode for this algorithm is given in Figure 9.6.

The simulation proceeds sequentially, one variable at a time. Consequently, it is not possible to account for evidence known to have occurred, until the evidential variables are sampled. When the evidential variables are sampled, if the simulated values match the evidence, the simulated sample is counted; otherwise, it is discarded. Then, conditional probabilities are approximated by averaging the frequency of events over those cases in which the evidence variables agree with the data observed.

In some cases, this method leads to a very high rejection rate of simulated samples and may require too many simulation runs, especially in cases where probabilities corresponding to the evidences are small, that is, in networks with extreme probabilities. The following example illustrates the method.

408   9. Approximate Propagation Methods

Example 9.4 Acceptance-rejection sampling. Consider the network with six nodes in Figure 9.5 and the corresponding CPDs in Table 9.1. The JPD of the six variables can be factorized as in (9.9). Given the observed evidence $X_3 = 1, X_4 = 1$ (see Figure 9.7), we wish to compute the posterior probability distribution for each of the other four variables using the acceptance-rejection sampling method.

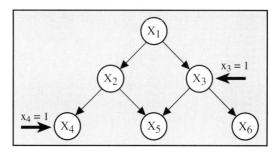

FIGURE 9.7. Adding evidence to the network of Figure 9.5.

We must first choose an ancestral numbering of the nodes. Figure 9.8 shows the ancestral structure associated with this example. Accordingly, the variables have to be numbered from top to bottom, choosing the numbers for the variables in the same row arbitrarily. For example, we take the ordering $(X_1, X_2, X_3, X_4, X_5, X_6)$.

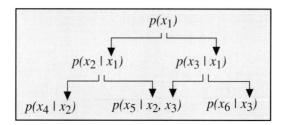

FIGURE 9.8. Ancestral structure of the CPDs associated with the network given in Figure 9.7.

Then, variables are sampled using the CPDs in Table 9.1 as follows:

1. Sampling variable X_1: Based on $p(x_1)$, a random number generator that gives a zero with probability 0.3 and a one with probability 0.7 is consulted. For illustrative purposes, suppose that the sampled value for X_1 happens to be $x_1^1 = 1$. We use the generated value x_1^1 to calculate the probabilities for the other nodes in this run.

2. Sampling variable X_2: Given $X_1 = 1$, from Table 9.1 the probabilities of X_2 taking the values 0 or 1 are $p(X_2 = 0|x_1^1) = p(X_2 = 0|X_1 = 1) = 0.1$ and $p(X_2 = 1|X_1 = 1) = 0.9$, respectively. Thus, we use a

random number generator with these values to obtain the simulated X_2 value. Suppose that the generated value of X_2 is $x_2^1 = 0$.

3. Sampling variable X_3: Given $X_1 = 1$, from Table 9.1 X_3 takes the value 0 with probability 0.5 and 1 with probability 0.5. We generate a value for X_3 from this probability distribution. If the sampled value is 0, then this simulation run is rejected because it does not match the evidence $X_3 = 1$ and we start with Step 1 again. Otherwise, X_3 takes the value 1 and the simulation continues. We therefore, keep simulating values for X_1, X_2, and X_3 until X_3 turns out to be $x_3^3 = 1$.

4. Sampling variables X_4, X_5, X_6: the situation is similar to the previous steps. If the simulated value for node X_4 does not match the evidence $X_4 = 1$, we reject the whole sample and start over again; otherwise, we simulate nodes X_5 and X_6. Suppose we obtain $x_5^1 = 1$ and $x_6^1 = 0$.

The run concludes with the first instantiation

$$x^1 = (x_1^1, x_2^1, x_3^1, x_4^1, x_5^1, x_6^1) = (1, 0, 1, 1, 1, 0).$$

We repeat this process until we obtain N instantiations. The probability distribution of any variable can then be estimated by the percentage of samples in which the associated events occur, as shown in Example 9.3. ∎

The acceptance-rejection sampling method can be written in the notation of Theorem 9.1 as

$$p_e(x) = c\, g(x)\, h(x) = c\, g(x) \prod_{i=1}^{n} p(x_i|\pi_i),$$

where $p_e(x)$ is given in (9.13), and

$$g(x_i) = \begin{cases} 1, & \text{if } x_i = e_i,\ \text{for all } X_i \in E, \\ 0, & \text{otherwise}. \end{cases}$$

This means that the condition of acceptance given by Theorem 9.1 will always be met if the instantiation is consistent with the evidence and will always fail if not.

A clear disadvantage of the acceptance-rejection sampling method is that evidences that are known to have occurred cannot be accounted for until the corresponding variables are sampled. In response to this problem, other simulation methods that account for the evidence have been developed. These are presented in the following sections.

9.5 Uniform Sampling Method

The population distribution for the variable X_i is given by (9.13). Thus, $p_e(x_i|\pi_i)$ is $p(x_i|\pi_i)$ with the evidence variables instantiated according to

```
Initialize
    for X_i ∈ E do
        x_i = e_i
Main loop
    for j ← 1 to N do
        for X_i ∉ E do
            x_i ← generate a value from h(x_i) in (9.18)
        s_j = ∏_{i=1}^{n} p(x_i|π_i)
Normalize scores
```

FIGURE 9.9. Pseudocode for the Uniform sampling method.

their observed values or zero if the values x_i or π_i are inconsistent with the evidence. The simulation distribution of the variable X_i, $h(x_i)$, is uniform, that is,

$$h(x_i) = \begin{cases} \frac{1}{card(X_i)}, & if\ X_i \notin E, \\ 1, & if\ X_i \in E\ and\ x_i = e_i, \\ 0, & otherwise, \end{cases} \quad (9.18)$$

where $card(X_i)$ denotes the cardinality of X_i. This guarantees that the evidential variables will always take their observed values and all instantiations are compatible with the evidence (no instantiations are rejected). Therefore, we need to simulate only the nonevidential variables.

Once an instantiation $x = \{x_1, \ldots, x_n\}$ is generated, the associated score becomes

$$s(x) = \frac{p_e(x)}{h(x)}$$
$$= \frac{p_e(x)}{\prod_{X_i \notin E} \frac{1}{card(X_i)} \prod_{X_i \in E} 1} \propto p_e(x) = p(x), \quad (9.19)$$

where the factor $\prod_{X_i \notin E} card(X_i)$ need not be considered because it is a constant for all instantiations. The last equality in (9.19) follows because evidential variables are assigned their corresponding evidential values in each instantiation. Figure 9.9 gives the pseudocode for the uniform sampling method.

This method can be applied to the set of nodes in any order because $h(x_i)$ does not depend on the value of any other variable. This method is simple, but unfortunately, it gives unsatisfactory results in cases where the instantiations are far from being uniformly distributed. If this is the case, many nonrepresentative samples can be generated, which leads to inaccurate approximations.

Example 9.5 Uniform sampling method. Consider again the situation given in Example 9.4, where we have six variables, two of which are evidential nodes: $e = \{X_3 = 1, X_4 = 1\}$. To obtain an instantiation $x = (x_1, \ldots, x_6)$ of X, we first fix evidential variables to their evidence values: $x_3^1 = 1, x_4^1 = 1$. We then select an arbitrary order for nonevidential nodes, for example, (X_1, X_2, X_5, X_6) and generate values for each of these variables randomly with equal probabilities because the cardinality of the variables in this case is 2. Suppose in the first trial we obtain the values $x_1^1 = 0, x_2^1 = 1, x_5^1 = 0$, and $x_6^1 = 1$. Then we calculate the score associated with the resulting instantiation,

$$x^1 = (x_1^1, x_2^1, x_3^1, x_4^1, x_5^1, x_6^1) = (0, 1, 1, 1, 0, 1),$$

using (9.9), (9.19), and the CPDs in Table 9.1:

$$\begin{aligned} s(x^1) &= p(x_1^1)p(x_2^1|x_1^1)p(x_3^1|x_1^1)p(x_4^1|x_2^1)p(x_5^1|x_2^1,x_3^1)p(x_6^1|x_3^1) \\ &= 0.3 \times 0.6 \times 0.8 \times 0.8 \times 0.2 \times 0.6 = 0.0138. \end{aligned}$$

The process is repeated until the desired number of instantiations is reached. ∎

9.6 The Likelihood Weighing Sampling Method

The likelihood weighing was developed independently by Fung and Chang (1990) and Shachter and Peot (1990). It deals with the problems of high rejection rate (as in the acceptance-rejection sampling method) and of nonuniform probabilities of instantiations (as in the uniform sampling method). The population distribution is (9.13) with (9.14). The simulation distribution is given by

$$h(x_i) = \begin{cases} p_e(x_i|\pi_i), & \text{if } X_i \notin E, \\ 1, & \text{if } X_i \in E \text{ and } x_i = e_i, \\ 0, & \text{otherwise}. \end{cases} \quad (9.20)$$

Because $h(x_i)$ depends on π_i, the ordering of the nodes must be such that the parents are sampled before the children (forward sampling). Thus, we order the nodes according to an *ancestral ordering*.

The score of an instantiation $x = (x_1, \ldots, x_n)$ becomes

$$\begin{aligned} s(x) &= \frac{p_e(x)}{h(x)} \\ &= \prod_{X_i \notin E} \frac{p_e(x_i|\pi_i)}{p_e(x_i|\pi_i)} \prod_{X_i \in E} \frac{p_e(x_i|\pi_i)}{1} \\ &= \prod_{X_i \in E} p_e(x_i|\pi_i) = \prod_{X_i \in E} p(e_i|\pi_i). \end{aligned} \quad (9.21)$$

> Initialize
>   Arrange the nodes in ancestral ordering
>   for $X_i \in E$ do
> $x_i = e_i$
> Main loop
>   for $j \leftarrow 1$ to N do
>   for $X_i \notin E$ do
> $x_i \leftarrow$ generate a value from $h(x_i)$ in (9.20)
> $s_j = \prod_{X_i \in E} p(e_i|\pi_i)$
> Normalize scores

FIGURE 9.10. Pseudocode for the likelihood weighing method.

The last equality follows because when $X_i \in E$, $x_i = e_i$ (all instantiations are consistent with the evidence). The pseudocode for the likelihood weighing method is given in Figure 9.10. This method gives representative instantiations as long as the probability of the observed evidence is not very close to zero.

Example 9.6 Likelihood weighing method. Consider again the Bayesian network in Example 9.5 with two evidential nodes: $e = \{X_3 = 1, X_4 = 1\}$. We arrange the nonevidential nodes in an ancestral ordering, for example, $X = (X_1, X_2, X_5, X_6)$. Then, we start by clamping the evidential nodes to the evidential values $x_3^1 = 1$ and $x_4^1 = 1$. Following the order chosen for the nonevidential nodes, we simulate X_1 using $p(x_1)$ in Table 9.1, that is, we use a random number generator that returns 0 with probability 0.3 and 1 with probability 0.7. Suppose that the sampled value for X_1 turned out to be $x_1^1 = 0$. We then simulate X_2 using $p(x_2|X_1 = x_1^1) = p(x_2|X_1 = 0)$, that is, we select a zero with probability 0.4 and a one with probability 0.6. Suppose that we get $x_2^1 = 1$. We then simulate X_5 using $p(x_5|x_2^1, x_3^1) = p(x_5|X_2 = 1, X_3 = 1)$, which assigns a probability 0.2 to zero and 0.8 to one. Then we simulate X_6 using $p(x_6|x_3^1) = p(x_6|x_3 = 1)$, which assigns 0.4 to zero and 0.6 to one. Suppose that we get $x_5^1 = 0$ and $x_6^1 = 1$. Then, using (9.21), the score of the obtained sample $x^1 = (x_1^1, x_2^1, x_3^1, x_4^1, x_5^1, x_6^1) = (0, 1, 1, 1, 0, 1)$ becomes

$$s(x^1) = p(x_3^1|x_1^1)p(x_4^1|x_2^1) = 0.8 \times 0.8 = 0.64.$$

Thus, we obtain the first instantiation, $x^1 = (0, 1, 1, 1, 0, 1)$, with the associated score $s(x^1) = 0.64$. We repeat this process until we reach the desired number of instantiations. ∎

Note that when the probabilities of the instantiations are not uniform, the likelihood weighing method will exhibit a better performance in the simulation process than both the acceptance-rejection sampling and the uniform sampling methods.

The likelihood weighing method is simple yet efficient and powerful method for uncertainty propagation, especially in cases when no exact propagation methods exist, for example, networks with discrete and continuous variables (see Castillo, Gutiérrez, and Hadi (1995b)). For these reasons, we use this method in a real-life application in Chapter 12.

9.7 Backward-Forward Sampling Method

Both the acceptance-rejection sampling and likelihood weighing methods can be thought of as forward sampling methods in the sense that a variable is sampled only after all of its parents have been sampled. In this way, $p(x_i|\pi_i)$ is readily available from the CPD tables (see, e.g., Table 9.1). A method that does not adhere to this structure has been recently introduced by Fung and Del Favero (1994). This method does not require the parents to be sampled before their children. The method uses $p_e(x)$ in (9.13) with (9.14) as the simulation distribution, but it considers a different method for obtaining the instantiations from this distribution. Fung and Del Favero (1994) refer to this method as *backward sampling*, but it actually combines backward sampling and forward sampling. Which one of the two methods is used to sample a node depends on the network topology. For nodes with no evidence from its descendant nodes, forward sampling is used. For nodes with an evidence node as a descendant, sampling from a node's conditional probability occurs after the node has been instantiated, and it determines the values for the node's predecessors (backward sampling).

The first step in the backward-forward simulation method is node ordering, referred to as a *valid ordering*. Valid node ordering must satisfy

1. A node must be instantiated before it is used for backward sampling,

2. A node's predecessors must be instantiated before the node is forward sampled, and

3. Each node in the network must be either a node in the ordering or a direct predecessor of a node in the ordering that is used for backward sampling.

Conditions 1 and 2 are imposed to guarantee that backward and forward sampling are possible. Condition 3 guarantees that all nodes get assigned values.

In backward sampling, values of the uninstantiated parents of a node X_i get assigned values under the condition that node X_i has already been instantiated. Node X_i may have been instantiated either because it is an evidence node or because it has been backward sampled before. The values of π_i are generated according to the assessment function of node X_i, namely with probability

$$h(\pi_i^*) = \frac{p(x_i|\pi_i)}{\alpha_i}, \tag{9.22}$$

where π_i^* is the set of uninstantiated parents of X_i, and

$$\alpha_i = \sum_{x_j \in \pi_i^*} p(x_j|\pi_j) \tag{9.23}$$

is a normalizing constant that makes the simulation distribution sum to unity. Since not all nodes can be backward sampled, for example, because there is no evidence yet, backward sampling must be mixed with another sampling method such as likelihood weighing sampling. Then, the remaining sampled nodes are forward sampled by this method. The score then becomes

$$\begin{aligned} s(x) &= \frac{p_e(x)}{h(x)} \\ &= \frac{p_e(x_i|\pi_i)}{\prod_{X_i \in B} \frac{p_e(x_i|\pi_i)}{\alpha_i} \prod_{X_i \in F} p_e(x_i|\pi_i)} \\ &= \prod_{X_i \notin B \cup F} p(x_i|\pi_i) \prod_{X_i \in B} \alpha_i, \end{aligned} \tag{9.24}$$

where B and F are the nodes backward and forward sampled, respectively.

The pseudocode for the backward-forward sampling method is shown in Figure 9.11. The following is an illustrative example.

Example 9.7 Backward-forward sampling method. Consider the Bayesian network in Figure 9.5, the corresponding CPDs in Table 9.1, and the observed evidence $X_3 = 1, X_4 = 1$. First we need to select a valid ordering satisfying the above three conditions. There are several possibilities, for example, $\{X_4, X_2, X_5, X_6\}$, $\{X_4, X_3, X_5, X_6\}$, and $\{X_6, X_4, X_5, X_3\}$ are valid orderings. Let us suppose that we choose the valid order $\{X_4, X_2, X_5, X_6\}$, where X_4 and X_2 are backward sampled and X_5 and X_6 are forward sampled. We proceed as follows:

1. First, we assign the evidence values to the evidence variables. Thus, we have, $x_3^1 = x_4^1 = 1$.

2. Backward sampling variable X_2 using node X_4: We use $p(x_4^1|x_2)$, that is, $p(X_4 = x_4^1|X_2 = 0) = 0.7$ and $p(X_4 = x_4^1|X_2 = 1) = 0.8$, which leads to $\alpha_4 = 1.5$. Then we use a random generator with $h(X_2 = 0) = 0.7/1.5$ and $h(X_2 = 1) = 0.8/1.5$. Suppose that we get $x_2^1 = 1$.

3. Backward sampling variable X_1 using node X_2: We use $p(x_2^1|x_1)$, that is, $p(X_2 = x_2^1|X_1 = 0) = 0.6$ and $p(X_2 = x_2^1|X_1 = 1) = 0.9$, which

```
Initialize
    Valid Node Ordering
    Create list of backward (B) and forward (F) sampled nodes
    for $X_i \in E$ do
        $x_i = e_i$
Main loop
    for $j \leftarrow 1$ to $N$ do
        for $i \leftarrow 1$ to $n$ do
            if $X_i \in F$ then
                $x_i \leftarrow$ generate a value from $p(x_i|\pi_i)$
            if $X_i \in B$ then
                $\pi_i^* \leftarrow$ generate a value from $p(x_i|\pi_i)/\alpha_i$
        $s_j = \prod_{X_i \notin B \cup F} p(x_i|\pi_i) \prod_{X_i \in B} \alpha_i$
Normalize scores
```

FIGURE 9.11. Pseudocode for the backward-forward sampling method.

leads to $\alpha_1 = 0.6 + 0.9 = 1.5$. Then we use a random generator with $h(X_1 = 0) = 0.6/1.5$ and $h(X_1 = 1) = 0.9/1.5$. Suppose that we get $x_1^1 = 0$.

4. Forward sampling variable X_5: We use $p(x_5|x_2^1, x_3^1)$, that is, $p(X_5 = 0|X_1 = x_2^1, X_3 = x_3^1) = 0.2$ and $p(X_5 = 1|X_1 = x_2^1, X_3 = x_3^1) = 0.8$. Then we use a random generator with these probabilities and get $x_5^1 = 0$, say.

5. Forward sampling variable X_6: We use $p(x_6|x_3^1)$, that is, $p(X_6 = 0|X_3 = x_3^1) = 0.4$ and $p(X_6 = 1|X_3 = x_3^1) = 0.6$ and we get $x_6^1 = 0$, say.

Thus, we end with the instantiation $x^1 = (0, 1, 1, 1, 0, 0)$. We then calculate the score using (9.24):

$$s(x^1) = p(x_1^1)p(x_3^1|x_1^1)\alpha_4\alpha_2 = 0.3 \times 0.8 \times 1.5 \times 1.5 = 0.54.$$

The same process can be repeated until we reach the desired number of instantiations. ∎

9.8 Markov Sampling Method

A stochastic sampling method developed by Pearl (1987b) accounts for the evidence by clamping the evidence variables to the values observed, and then conducting a stochastic simulation on the clamped network. This method is known as the *Markov*, or *stochastic, sampling method*.

416   9. Approximate Propagation Methods

Initialize
    for $X_i \in E$ **do**
        $x_i = e_i$
    for $X_i \notin E$ **do**
        $x_i \leftarrow$ uniformly generated value
Main loop
    for $j \leftarrow 1$ **to** N **do**
        for $X_i \notin E$ **do**
            for each value x_i of X_i **do**
$$q(x_i) \leftarrow p(x_i|\pi_i) \prod_{X_j \in C_i} p(x_j|\pi_j)$$
            Normalize $q(x_i)$
            $x_i \leftarrow$ generate from the normalized $q(x_i)$

FIGURE 9.12. Pseudocode for the Markov sampling method.

An initial instantiation is generated either randomly from among all possible instantiations or using one of the previous sampling methods. Then, to generate the next instantiation, all nonevidential variables are sampled one at a time (in any given order or in a random order). A random value for the selected variable is generated from the conditional probability distribution of the variable given the current state of all other variables. This conditional distribution is given by the following theorem (Pearl (1987b)).

Theorem 9.2 CPD of one variable given the rest. *The probability distribution of each variable X_i, conditioned on the state of all other variables, is given by*

$$h(x_i) = p(x_i|x \setminus x_i) \propto p(x_i|\pi_i) \prod_{X_j \in C_i} p(x_j|\pi_j), \qquad (9.25)$$

where C_i is the set of children of X_i and $X \setminus X_i$ denotes all variables in X but X_i.

An instantiation is obtained after all nonevidential variables have been sampled. This instantiation is then used as above to generate the next instantiation. The pseudocode for this method is shown in Figure 9.12. Note that for node X_i the only nodes playing a role are X_i and its *Markov blanket*, which is the set of parents of X_i, children of X_i, and parents of the children of X_i except X_i itself. Note also that for variables that have not been assigned a value in the current run, their values in the previous run are used. Thus, this method does not require ordering of the nodes.

Since we simulate with the real probabilities, the score associated with this method is constant and equal to 1.

After generating the desired number of instantiations, the conditional probability distribution of every node given the evidence is then estimated as shown in Section 9.3:

- By the proportion of the sample runs in which a given event has occurred, or
- By the average of the conditional probabilities of a given event in all simulation runs.

This method gives representative samples, as long as the CPDs do not contain values close to zero or one.

Example 9.8 Markov sampling method. Consider again the Bayesian network used in the previous examples. As before, the observed evidence is $X_3 = 1, X_4 = 1$. According to (9.25), the simulation distributions for the nonevidential variables are

$$
\begin{aligned}
h(x_1) &= p(x_1|x \setminus x_1) \propto p(x_1)p(x_2|x_1)p(x_3|x_1), \\
h(x_2) &= p(x_2|x \setminus x_2) \propto p(x_2|x_1)p(x_4|x_2)p(x_5|x_2, x_3), \\
h(x_5) &= p(x_5|x \setminus x_5) \propto p(x_5|x_2, x_3), \\
h(x_6) &= p(x_6|x \setminus x_6) \propto p(x_6|x_3).
\end{aligned}
\quad (9.26)
$$

The Markov sampling method starts by assigning all evidential variables to their observed evidence. Thus, $x_3^j = 1$ and $x_4^j = 1$, for all $j = 1, \ldots, N$. The method then proceeds as follows:

1. Instantiate all nonevidential variables to some arbitrary initial state. Suppose we obtained $X_1 = 0, X_2 = 1, X_5 = 0, X_6 = 1$. Thus, the initial instantiation for all six variables is $x^0 = (0, 1, 1, 1, 0, 1)$.

2. Choose an arbitrary order to select nonevidential nodes, for example $\{X_1, X_2, X_5, X_6\}$. For each variable in this list, generate a random value from the corresponding CPD in (9.26) given the current state of all other variables as follows:

   Variable X_1: The current state of all other variables is given by x^0. Thus, using Table 9.1, we calculate

$$
\begin{aligned}
p(X_1 = 0|x \setminus x_1) &\propto p(X_1 = 0)p(X_2 = 1|X_1 = 0)p(X_3 = 1|X_1 = 0) \\
&= 0.3 \times 0.6 \times 0.8 = 0.144,
\end{aligned}
$$

$$
\begin{aligned}
p(X_1 = 1|x \setminus x_1) &\propto p(X_1 = 1)p(X_2 = 1|X_1 = 1)p(X_3 = 1|X_1 = 1) \\
&= 0.7 \times 0.9 \times 0.5 = 0.315.
\end{aligned}
$$

Normalizing the above probabilities by dividing by their sum, $0.144 + 0.315 = 0.459$, we obtain $p(X_1 = 0|x \setminus x_1) = 0.144/0.459 = 0.314$ and $p(X_1 = 1|x \setminus x_1) = 0.315/0.459 = 0.686$. Therefore, we generate a value for X_1 using a random number generator that gives 0 with probability 0.314 and 1 with probability 0.686. Suppose that the value sampled is 0. Then the current state of all variables is still the same as in x^0.

Variable X_2: Using the current state of the variables and from (9.26) and Table 9.1 we get

$$p(X_2 = 0|x \setminus x_2) \propto 0.4 \times 0.3 \times 0.5 = 0.06,$$
$$p(X_2 = 1|x \setminus x_2) \propto 0.6 \times 0.2 \times 0.2 = 0.024.$$

Normalizing the above probabilities, we obtain $p(X_2 = 0|x \setminus x_2) = 0.06/0.084 = 0.714$ and $p(X_2 = 1|x \setminus x_2) = 0.024/0.084 = 0.286$. Thus, we generate a value for X_2 from this distribution. Suppose that the value sampled is 1.

Variables X_5 and X_6 are simulated in a similar way. Suppose we obtain $X_5 = 0$ and $X_6 = 1$. Thus, we obtain the first instantiation of the sample $x^1 = (0, 1, 1, 1, 0, 1)$ after the first run.

3. Repeat Step 2 for N simulation runs.

Now, the required conditional probability distribution is approximated by the percentage of samples in which the event is true, or by taking the average of its associated conditional probability over the simulation runs. For example, if we want to know the probability of event $X_2 = 1$, we can choose one of two alternatives:

1. Divide the number of samples in which $X_2 = 1$ by the total number of runs.

2. Compute the average of $p(X_2 = 1|x \setminus x_2)$ in all simulation runs. ∎

The convergence of the Markov sampling method is guaranteed, under certain conditions, by a theorem due to Feller (1968) regarding the existence of a limiting distribution for Markov chains. In each simulation run, the system's instantiation changes from state i to state j, and the change is governed by the transition probability of the simulated variable. In this situation, when transition probabilities are strictly positive, the probability that the system will be found at a given state approaches a stationary limit. The case where some of the transition probabilities are zero corresponds to reducible Markov chains and limits the applicability of stochastic simulation schemes.

The Markov sampling method clearly avoids the problem of sample rejection, but unfortunately, it has its own problems. The method will have convergence problems when the network contains variables with extreme probabilities. These problems are due to the fact that successive cycles in Markov simulation schemes are not independent and the simulation can get trapped in particular states or sets of states.

9.9 Systematic Sampling Method

Recently, Bouckaert (1994) and Bouckaert, Castillo, and Gutiérrez (1996) introduced a new method for generating the instantiations forming a sample in a systematic way. Unlike the algorithms introduced in the previous sections, which are stochastic in nature, this method proceeds in a deterministic way.

9.9.1 The Basic Idea

To illustrate the basic idea of this method, we need the following definition.

Definition 9.1 Instantiation ordering. *Let $X = \{X_1, \ldots, X_n\}$ be a set of discrete variables and let $\{0, \ldots, r_i\}$ be the possible values of X_i. Let $x^j = (x_1^j, \ldots, x_n^j)$, $j = 1, \ldots, m$, be the set of all possible instantiations of X. Suppose that the instantiations are given in an order satisfying*

$$x^j < x^{j+1} \Leftrightarrow \exists\, k \text{ such that } \forall\, i < k,\ x_i^j = x_i^{j+1} \text{ and } x_k^j < x_k^{j+1}. \qquad (9.27)$$

Then, we say that $(x_1^j, \ldots, x_n^j)$ precedes $(x_1^{j+1}, \ldots, x_n^{j+1})$ and refer to the ordering that satisfies (9.27) as an instantiation ordering.

Example 9.9 Instantiation ordering. Consider the Bayesian network in Figure 9.13. Suppose that X_1 and X_2 are binary and X_3 is ternary. The first column of Table 9.4 contains all possible instantiations in the order satisfying (9.27) from top to bottom. For example, for $j = 1$, we have $x^1 < x^2$ because $x_1^1 = x_1^2$, $x_2^1 = x_2^2$, and $x_3^1 < x_3^2$. The reader can verify that $x^j < x^{j+1}$, for $j = 2, \ldots, 12$, in the sense of the instantiation ordering in (9.27). ■

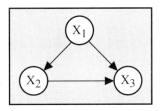

FIGURE 9.13. A Bayesian network with three variables.

Using the instantiations ordering, the unit interval $[0, 1]$ can be divided into subintervals associated with the different instantiations. Thus, we can associate each instantiation, x^j, with an interval $I_j = [l_j, u_j) \subset [0, 1)$, whose lower bound is given by the cumulative probabilities of all previous instantiations:

$$l_j = \sum_{x^i < x^j} p_e(x^i) \geq 0,$$

Instantiation (x_1, x_2, x_3)	Probability $p(x_1, x_2, x_3)$	Cumulative Probability	Interval $[l_i, u_i)$
(0,0,0)	$0.4 \times 0.6 \times 0.3 = 0.072$	0.072	[0.000, 0.072)
(0,0,1)	$0.4 \times 0.6 \times 0.3 = 0.072$	0.144	[0.072, 0.144)
(0,0,2)	$0.4 \times 0.6 \times 0.4 = 0.096$	0.240	[0.144, 0.240)
(0,1,0)	$0.4 \times 0.4 \times 0.3 = 0.048$	0.288	[0.240, 0.288)
(0,1,1)	$0.4 \times 0.4 \times 0.3 = 0.048$	0.336	[0.288, 0.336)
(0,1,2)	$0.4 \times 0.4 \times 0.4 = 0.064$	0.400	[0.336, 0.400)
(1,0,0)	$0.6 \times 0.4 \times 0.3 = 0.072$	0.472	[0.400, 0.472)
(1,0,1)	$0.6 \times 0.4 \times 0.4 = 0.096$	0.568	[0.472, 0.568)
(1,0,2)	$0.6 \times 0.4 \times 0.3 = 0.072$	0.640	[0.568, 0.640)
(1,1,0)	$0.6 \times 0.6 \times 0.3 = 0.108$	0.748	[0.640, 0.748)
(1,1,1)	$0.6 \times 0.6 \times 0.4 = 0.144$	0.892	[0.748, 0.892)
(1,1,2)	$0.6 \times 0.6 \times 0.3 = 0.108$	1.000	[0.892, 1.000)

TABLE 9.4. Ordered instantiations and associated absolute and cumulative probabilities and their upper and lower bounds.

and the upper bound is

$$u_j = l_j + p_e(x^j) \leq 1,$$

where

$$p_e(x^j) = \begin{cases} p(x^j), & \text{if } x^j \text{ is consistent with } e, \\ 0, & \text{otherwise}. \end{cases}$$

Note that in the case of no evidence, the function $p_e(x)$ is simply $p(x)$.

The systematic sampling method generates the instantiations systematically by choosing a sequence of equally spaced values in the unit interval $(0, 1)$ and finding the associated instantiations, that is, the instantiations whose corresponding intervals contain the given values. To generate a sample of size N we take the values

$$f_j = (j - 0.5)/N, \ j = 1, 2, \ldots, N, \tag{9.28}$$

from which it follows that $0 < f_j < 1, j = 1, \ldots, N$. The instantiation whose interval contains the value f_j is chosen as the jth instantiation in the sample. Note that due to the deterministic character of the procedure, random numbers are not used.

Example 9.10 Instantiations probability intervals. Consider again the Bayesian network in Figure 9.13. The JPD of the three variables can be factorized as (see Section 6.4.4)

$$p(x_1, x_2, x_3) = p(x_1)p(x_2|x_1)p(x_3|x_1, x_2). \tag{9.29}$$

x_1	x_2	x_3	$p(x_3\|x_1,x_2)$
0	0	0	0.3
0	0	1	0.3
0	0	2	0.4
0	1	0	0.3
0	1	1	0.3
0	1	2	0.4
1	0	0	0.3
1	0	1	0.4
1	0	2	0.3
1	1	0	0.3
1	1	1	0.4
1	1	2	0.3

x_1	x_2	$p(x_2\|x_1)$
0	0	0.6
0	1	0.4
1	0	0.4
1	1	0.6

x_1	$p(x_1)$
0	0.4
1	0.6

TABLE 9.5. CPDs required for specifying the JPD for the nodes in the Bayesian network of Figure 9.13.

The CPDs needed to specify the JPD in (9.29) are given in Table 9.5. The second column of Table 9.4 shows the probability of each instantiation, the third column gives the cumulative probabilities, and the last column shows the corresponding intervals. Figure 9.14 also illustrates the instantiations, the cumulative probabilities, and their associated intervals.

Let us now generate a sample of size $N = 4$ using the systematic sampling method. The sequence $f_j = (j - 0.5)/4$ is $(0.125, 0.375, 0.625, 0.875)$. From Table 9.4 or Figure 9.14, we see that the generated sample consists of the following instantiations: $\{(001), (012), (102), (111)\}$. The first instantiation (001) is generated because $f_1 = 0.125$ falls in the interval corresponding to this instantiation (the second interval in Table 9.4), the second instantiation (012) is generated because $f_2 = 0.375$ falls in the interval corresponding to this instantiation (the sixth interval in Table 9.4), and so on. ∎

Note that to generate a sample using the systematic sampling method, we enumerated all possible instantiations of the variables as in Figure 9.14. This is done only for the purpose of illustration because the method does not actually require the generation of all possible instantiations (see Section 9.9.2). This is fortunate because the number of different instantiations grows exponentially with the number of variables in a such a way that even with only ten binary variables there are 1024 different instantiations. Thus, several of the f_j's values may fall in the interval corresponding to a given instantiation. If the probability of an instantiation x is $p_e(x)$, the number of different values of f_j associated with this instantiation in a sample of size N is equal to $Np_e(x)$ if $p_e(x) \geq 1/N$. This means that the associated instantiation will appear $Np_e(x)$ times in the sample. Thus, we have $h(x) = p_e(x)$ as $N \to \infty$. Otherwise, i.e., if $p_e(x) < 1/N$, we have in the

422   9. Approximate Propagation Methods

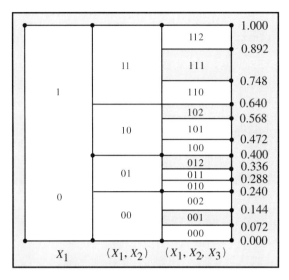

FIGURE 9.14. Ordered instantiations and associated cumulative probabilities.

sample either one instantiation, with probability $Np_e(x)$, or zero instantiations, with probability $1 - Np_e(x)$, which again implies that $h(x) = p_e(x)$, since

$$h(x) = \frac{1 \times Np_e(x) + 0 \times (1 - Np_e(x))}{N} = p_e(x).$$

It is clear that as N increases, the frequency of a given instantiation converges to the exact frequency, and hence, the approximated values obtained with this method converge to the exact values.

Since the method generates the instantiations systematically (in a similar way to systematic random sampling), we refer to this method as *systematic sampling*. Bouckaert (1994) refers to this method as *stratified sampling* because all possible sets of instantiations are divided into mutually exclusive and collectively exhaustive subsets, or strata,[2] and considers random sampling from each stratum. Stratified sampling is a well-known statistical technique that leads to more representative samples than those obtained by simple random sampling.

9.9.2 Implementation Aspects

The systematic simulation method is conceptually very simple, but its implementation is complicated. In general, we cannot generate or calculate

---

[2]These strata are obtained by ordering the instantiations and using the f_j values that divide the interval $(0, 1)$ into equal and disjoint subintervals, each of which contains a certain number of instantiations.

the probabilities of all instantiations because of the associated computational effort (for n binary variables, there are 2^n possible instantiations). Due to the fact that the method is supposed to be utilized when exact methods cannot be used, we assume that the number of instantiations is much larger than the sample size N. This implies that many of the instantiations (most of those with small probabilities) will not appear in the simulated sample. A computationally efficient method must be able to skip these instantiations, avoiding unnecessary calculations. The method proceeds in a systematic way, moving from the first instantiation $(0,\ldots,0)$ to the last instantiation $(r_1,\ldots,r_n)$ and taking advantage of the order and the deterministic character of the selected sequence to determine the instantiations corresponding to the sequence of the values f_j. The main advantage of this procedure is that for obtaining a new instantiation we only need to update the values of the last k variables. This allows a fast procedure that skips many instantiations at a time. However, we need to determine which variables to update.

To this end, suppose the variables $X_1,\ldots,X_n$ are given in an ancestral ordering and that we have a current instantiation, $(x_1^{j-1},\ldots,x_n^{j-1})$, corresponding to the value f_{j-1} of the sequence. Then, we define a lower bound $l(i)$ and an upper bound $u(i) \geq l(i)$ for each variable X_i, which indicate the probability values where each variable will undergo the next two value changes. For example, refer to Figure 9.15 and observe that in Step 4, which corresponds to a value of f_j in the interval $(0.240, 0.288)$, the associated instantiation is $(0, 1, 0)$. The next change in variable X_3 occurs at 0.288 (instantiation $(0, 1, 1)$) and the following occurs at 0.336 (instantiation $(0, 1, 2)$). Thus, to move to the next instantiation, we change $l(3)$ from 0.240 to 0.288 and $u(3)$ from 0.288 to 0.336. Note that the limit functions for variables X_1 and X_2 do not change. Figure 9.15 shows the values of the limit functions $l(i)$ and $u(i)$ for the three variables when the value f_j is in each of the shaded intervals.

Once the instantiation associated with f_{j-1} has been calculated and the limit functions for the nodes have been updated, the key problem to obtaining the instantiation associated with the next value, f_j, is the determination of the variable number k above which the rest of variables need to be updated. The lower and upper limits defined above are used to identify this variable in an efficient way.

From the above discussion, the systematic sampling algorithm can be summarized in the following steps:

1. Initialize the limit functions.

2. Generate the sequence of f_j values.

3. Compute the associated instantiations.

4. Compute and normalize scores.

424   9. Approximate Propagation Methods

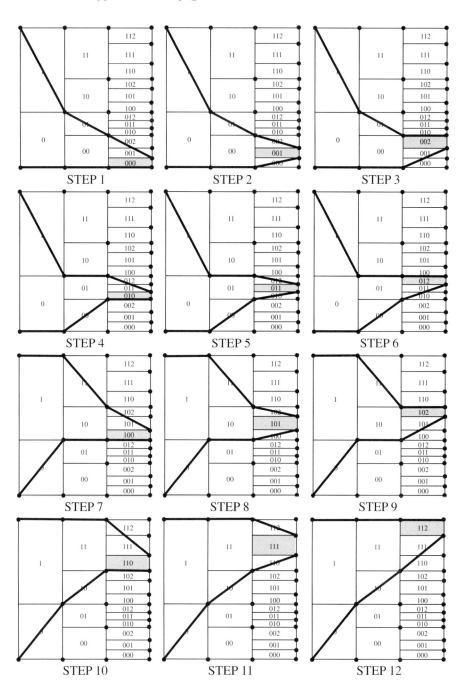

FIGURE 9.15. An Illustration of the $l(j)$ and $u(j)$ limits.

9.9 Systematic Sampling Method

> Initialize
>   Arrange the nodes in ancestral ordering
>   Initialize limit functions (Use the code in Figure 9.17)
> Main loop
>   Generate the sequence of f_j values using (9.28)
>   for $j \leftarrow 1$ to N do
> $x^j \leftarrow$ calculate instantiation associated with f_j
>   (Use the code in Figure 9.18)
> $s(x^j) \leftarrow \prod_{X_i \in E} p(e_i | \pi_i)$
> Normalize scores

FIGURE 9.16. Pseudocode for the systematic sampling algorithm to obtain a sample of size N.

Figure 9.16 gives the corresponding pseudocode. Note that no score is needed in the case that no evidence is available. The codes corresponding to the two main steps, initialization of the limit functions and computation of the instantiations associated with the values f_j, are given in Figures 9.17 and 9.18. In these figures $p(X_i = k | \pi_i^j)$ is the probability of node X_i taking value k given its parents as instantiated in x^j.

In the initialization process, we choose the first instantiation, $(0, \ldots, 0)$, to initialize the values of the limit functions $l()$ and $u()$. Step 1 of Figure 9.15 shows the limit values corresponding to this instantiation. These limits are calculated as follows:

- For node X_1 we have $x_1^0 = 0$, then the lower and upper bounds for the first variable are $l(1) = 0.0$ and $u(1) = p(X_1 = 0) = 0.4$.

- For X_2, we have $x_2^0 = 0$, and hence $l(2) = 0.0$ and $u(2) = u(1) \times p(X_2 = 0 | x_1^0) = 0.4 \times 0.6 = 0.24$.

- Finally, we have $l(3) = 0.0$ and $u(3) = u(2) \times p(X_3 = 0 | x_1^0, x_2^0) = 0.24 \times 0.3 = 0.072$.

Once the limit functions have been initialized, we use the obtained values to calculate the first instantiation, associated with the value f_1. We use a binary search function $Binsearch(f_1, l, u)$, which locates in $\log_2 n$ operations the largest index i such that $l(i) \leq f_1 \leq u(i)$. Therefore, the values $(x_1^0, \ldots, x_i^0)$ are the same for instantiation x^1 and we only need to update the values for the rest of the nodes, $(x_{i+1}, \ldots, x_n)$, to obtain the first instantiation. Then, the new values for the limit functions are used to obtain the instantiation associated with f_2, and so on until we obtain the N instantiations forming the sample.

Figure 9.18 gives a pseudocode of this algorithm optimized to reduce the number of multiplications, which are costly operations. This is especially useful for variables with high cardinalities.

426   9. Approximate Propagation Methods

```
l(0) ← 0;  u(0) ← 1
for i ← 1 to n do
    l(i) ← 0
    if X_i ∈ E then
        x_i^0 ← e_i
        u(i) ← u(i − 1)
    else
        x_i^0 ← 0
        u(i) ← u(i − 1) × p(X_i = 0|π_i^0)
```

FIGURE 9.17. Pseudocode for the initialization step of the limit functions in the systematic sampling algorithm.

```
i ← Binsearch (f_j, l, u)
while i <= n do
    if X_i ∈ E then
        l(i) ← l(i − 1)
        u(i) ← u(i − 1)
        x_i^j ← e_i
    else
        k ← 0
        l(i) ← l(i − 1)
        u(i) ← l(i) + (u(i − 1) − l(i − 1)) × p(X_i = k|π_i^j)
        while f_j > u(i) do
            k ← k + 1
            l(i) ← u(i)
            u(i) ← l(i) + (u(i − 1) − l(i − 1)) × p(X_i = k|π_i^j)
        x_i^j ← k
    i ← i + 1
return (x_1^{j−1}, … x_{i−1}^{j−1}, x_i^j, … x_n^j)
```

FIGURE 9.18. Pseudocode for the calculation of the instantiation associated with f_j in the systematic sampling algorithm.

A modification of this systematic sampling algorithm has been recently introduced by Bouckaert, Castillo, and Gutiérrez (1996). Since we work with a deterministic sequence, once we generate one instantiation associated with a given value of f_j, we can determine how many of the values in the sequence will lead to the same instantiation using the formula (see

Figure 9.19)

$$\delta = \left\lfloor \frac{u(n) - f_j}{N} \right\rfloor + 1, \qquad (9.30)$$

where $\lfloor . \rfloor$ is the integer part function and n is the number of variables. Then we increment the j counter of the f_j sequence δ units instead of one unit. In this way, we save the work of searching for the same instantiation again and again when the f_j values correspond to the same instantiation. With this technique the simulation time is greatly reduced.

Figure 9.20 gives a pseudocode for this modification of the systematic sampling method. The only difference between the codes in Figures 9.16 and 9.20 is that in the latter code the counter j is incremented by δ units instead of only one unit.

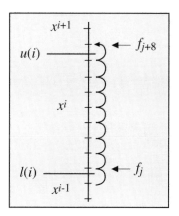

FIGURE 9.19. An Illustration of how the same values of the f_j sequence are skipped when they correspond to the same instantiation.

The efficiency of the modified systematic method increases in situations where a large number of values f_j fall in the interval associated with one instantiation. In probabilistic networks with extreme probabilities, most of the instantiations have very small probabilities, and only a few instantiations have large probabilities. In this situation, the above modification gives a substantial improvement of the efficiency of the method. For instance, consider the Bayesian network in Figure 9.13 and suppose we take the new values for the CPDs given in Table 9.6. The values contained in this table are more extreme that the values shown in Table 9.5.

The new structure of the instantiations is shown in Figure 9.21. From this figure, it can be seen that the probability of the instantiation $(0, 1, 0)$ alone is more than half of the total probability. Thus, the modified algorithm will save more than $N/2$ iteration steps to obtain a sample of size N.

428   9. Approximate Propagation Methods

> Initialize
>   Arrange the nodes in ancestral ordering
>   Initialize limit functions
> Main loop
>   Generate the sequence of f_j values using (9.28)
>   for $j \leftarrow 1$ to N do
> $x^j \leftarrow$ generate instantiation associated with f_j
> $\delta \leftarrow \lfloor \frac{u(n) - f_j}{N} \rfloor$
> $s(x^j) \leftarrow \delta$
> $i \leftarrow i + \delta$
> Normalize scores

FIGURE 9.20. The general sampling framework for the modified systematic sampling algorithm.

x_1	$p(x_1)$
0	0.9
1	0.1

x_1	x_2	$p(x_2\|x_1)$
0	0	0.2
0	1	0.8
1	0	0.1
1	1	0.9

X_1	X_2	X_3	$p(X_3\|X_1, X_2)$
0	0	0	0.2
0	0	1	0.2
0	0	2	0.6
0	1	0	0.8
0	1	1	0.1
0	1	2	0.1
1	0	0	0.3
1	0	1	0.4
1	0	2	0.3
1	1	0	0.7
1	1	1	0.1
1	1	2	0.2

TABLE 9.6. New values for the CPDs required for specifying the joint probability distribution for the nodes in the Bayesian network of Figure 9.13.

9.9.3 Comparison with Stochastic Algorithms

Bouckaert, Castillo, and Gutiérrez (1996) compare the following four of the approximate propagation methods presented in the previous sections: likelihood weighing, Markov sampling, the systematic sampling, and the modified systematic sampling. They try the methods on ten different random Bayesian network models and use two measures of performance: the average time to execute the algorithm and the average error of the approximation. They conclude the following:

9.10 Maximum Probability Search Method

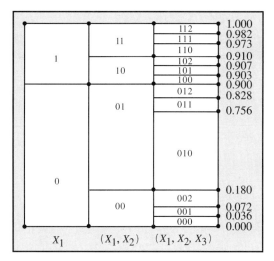

FIGURE 9.21. Instantiations and accumulated probabilities.

1. Overall, the modified systematic sampling method is the best of all four methods.

2. In terms of errors, the two systematic sampling methods have equal performance. This is because the calculated sample is the same for both algorithms. In terms of time, for a small N, the systematic and modified systematic sampling methods give similar performances when the probabilities in the Bayesian network are chosen from the unit interval. When the probabilities are extreme (chosen, for example, from $[0, 0.1] \cup [0.9, 1]$) the modified systematic sampling method performs better than the systematic sampling method. The reason for this behavior is that networks with extreme probabilities result in large strata, where skipping saves a lot of calculations. These large strata are less frequent in networks with nonextreme probabilities, so that skipping does not help.

3. Likelihood weighing and Markov methods becomes increasingly inefficient with networks including extreme probabilities (close to zero). However, the efficiency of the systematic sampling method increases in this type of networks. Thus, this algorithm eliminates the problem of extreme probabilities associated with stochastic methods.

9.10 Maximum Probability Search Method

In Section 9.9 we discussed a deterministic method that generates the instantiations systematically. In this section we describe a deterministic *search-based propagation algorithm*, which we refer to as the *maximum probability search method*. Search-based propagation methods compute the N instantiations of the sample by creating a tree whose branches are associated with partial instantiations of the variables. In every iteration step, the search process chooses one of the branches of the tree associated with an instantiation $(x_1^i, \ldots, x_m^i)$. If the associated instantiation is complete, that is, if $m = n$, then the branch is pruned from the tree and the instantiation is included in the sample. Otherwise, the tree is augmented with as many new branches as values of the next variable, x_{m+1}. Thus, the original branch, $(x_1^i, \ldots, x_m^i)$, is replaced by the branches $(x_1^i, \ldots, x_m^i, x_{m+1})$ for all possible values of X_m.

Several search-based methods have been proposed in the literature (see, for example, Pearl (1987a), Henrion (1991), Poole (1993a, 1993b) and Srinivas and Nayak (1996)). The main difference among them is the criterion to choose the branches in each iteration step. For example, the algorithm of maximum probability search (Poole (1993a)) uses the criterion of maximum probability to choose the branches in every iteration step. The algorithm proceeds as follows.

Given an ancestral ordering of the variables $X = \{X_1, \ldots, X_n\}$, we start with a tree that consists of as many branches as the number of possible values of X_1. The probability of each of these branches is calculated, and the branch with the maximum probability is chosen. We then augment this branch with a subtree that contains as many branches as the number of possible values of X_2. The probability of each of these branches is calculated by multiplying the probabilities of the nodes in the branch. The branch (in the augmented tree) with the maximum probability is chosen. The tree is expanded with the branches of X_3, and so on until we reach the last variable X_n. We shall refer to every branch with n nodes as a *complete* branch. The first instantiation consists of the values of the nodes associated with the complete branch with maximum probability. This branch will be trimmed from the tree.

The next instantiation is obtained by finding the branch with maximum probability from among the remaining branches of the tree. If the chosen branch is complete, the second instantiation will consist of the values of the variables associated with the branch. If the chosen branch is incomplete, the tree is augmented with the subtree corresponding to the next variable in the ancestral ordering. This process is continued until we generate the desired number of instantiations. This method is best explained by example.

Example 9.11 Maximum probability search method. Consider the Bayesian network in Figure 9.13 and the associated CPDs in Table 9.6. The variables $X = (X_1, X_2, X_3)$ have cardinalities $(2, 2, 3)$, respectively. The JPD of the three variables can be expressed as

$$p(x_1, x_2, x_3) = p(x_1)p(x_2|x_1)p(x_3|x_1, x_2).$$

9.10 Maximum Probability Search Method

The variables are already given in ancestral ordering. Since X_1 has two possible values, we start with a tree with two branches corresponding to $X_1 = 0$ and $X_1 = 1$. From Table 9.6, the probabilities of the two branches are $p(X_1 = 0) = 0.9$ and $p(X_1 = 1) = 0.1$. This is illustrated in Figure 9.22(a). Therefore, we have the initial branches

$$B_1^1 = (X_1 = 0), \quad p(B_1^1) = 0.9,$$
$$B_2^1 = (X_1 = 1), \quad p(B_2^1) = 0.1.$$

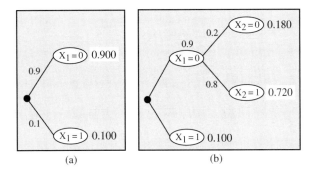

FIGURE 9.22. First two steps of the maximum probability search algorithm.

The first branch, B_1^1 has associated the maximum probability. Now we consider the next variable, X_2, which has two possible values. In Step 2, we augment the tree by a subtree with two branches, $X_2 = 0$, and $X_2 = 1$, as shown in Figure 9.22(b). The corresponding probabilities, obtained from Table 9.6, are $p(X_2 = 0|X_1 = 0) = 0.2$ and $p(X_2 = 1|X_1 = 0) = 0.8$. Now the tree contains three branches:

$$B_1^2 = (X_1 = 0, X_2 = 0), \quad p(B_1^2) = 0.9 \times 0.2 = 0.180,$$
$$B_2^2 = (X_1 = 0, X_2 = 1), \quad p(B_2^2) = 0.9 \times 0.8 = 0.720,$$
$$B_3^2 = (X_1 = 1), \quad p(B_3^2) = 0.100.$$

In Step 3, we select the branch with the maximum probability. In this case, it is the branch B_2^2. We enlarge the tree with the three branches corresponding to the three possible values of node X_3, as shown in Figure 9.23(a). The corresponding probabilities, obtained from Table 9.6, are $p(X3 = 0|X_1 = 0, X_2 = 1) = 0.8$, $p(X3 = 1|X_1 = 0, X_2 = 1) = 0.1$ and $p(X3 = 2|X_1 = 0, X_2 = 1) = 0.1$. The new tree contains five branches:

$$B_1^3 = (X_1 = 0, X_2 = 0), \quad p(B_1^3) = 0.180,$$
$$B_2^3 = (X_1 = 0, X_2 = 1, X_3 = 0), \quad p(B_2^3) = 0.720 \times 0.8 = 0.576,$$
$$B_3^3 = (X_1 = 0, X_2 = 1, X_3 = 1), \quad p(B_3^3) = 0.720 \times 0.1 = 0.072,$$
$$B_4^3 = (X_1 = 0, X_2 = 1, X_3 = 2), \quad p(B_4^3) = 0.720 \times 0.1 = 0.072,$$
$$B_5^3 = (X_1 = 1), \quad p(B_5^3) = 0.100.$$

432   9. Approximate Propagation Methods

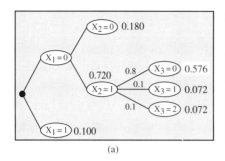
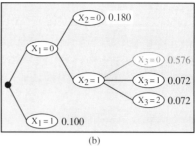

FIGURE 9.23. Maximum probability search method: (a) Step 3 and (b) Step 4.

The branch with the maximum probability is $B_2^3 = (X_1 = 0, X_2 = 1, X_3 = 0)$. Since this branch is complete, the first instantiation is obtained: $x^1 = (010)$, and the branch is pruned from the tree (see Figure 9.23(b)).

Now, of the remaining four branches, B_1^3 has the maximum probability. The branch is incomplete, so we augment the tree by three branches for X_3. This is shown in Figure 9.24(a). The corresponding probabilities are $p(X_3 = 0|X_1 = 0, X_2 = 0) = 0.2$, $p(X_3 = 1|X_1 = 0, X_2 = 0) = 0.2$, and $p(X_3 = 3|X_1 = 0, X_2 = 0) = 0.6$. The tree now contains six branches:

$$
\begin{array}{ll}
B_1^4 = (X_1 = 0, X_2 = 0, X_3 = 0), & p(B_1^4) = 0.180 \times 0.2 = 0.036, \\
B_2^4 = (X_1 = 0, X_2 = 0, X_3 = 1), & p(B_2^4) = 0.180 \times 0.2 = 0.036, \\
B_3^4 = (X_1 = 0, X_2 = 0, X_3 = 2), & p(B_3^4) = 0.180 \times 0.6 = 0.108, \\
B_4^4 = (X_1 = 0, X_2 = 1, X_3 = 1), & p(B_4^4) = 0.072, \\
B_5^4 = (X_1 = 0, X_2 = 1, X_3 = 2), & p(B_5^4) = 0.072, \\
B_6^4 = (X_1 = 1), & p(B_6^4) = 0.100.
\end{array}
$$

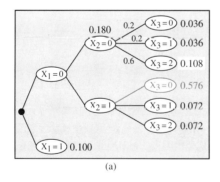
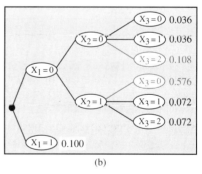

FIGURE 9.24. Maximum probability search method: (a) Step 5 and (b) Step 6.

Branch B_3^4 has the maximum probability and is complete. Therefore, $x^2 = (002)$. The process continues until we generate the desired number of instantiations. If we wish to compute exact, rather than approximate, probabilities, we continue the process until all possible instantiations are

9.10 Maximum Probability Search Method

generated. In this case, the complete tree contains $2 \times 2 \times 3 = 12$ possible branches (instantiations). This tree is shown in Figure 9.25. However, in most, if not all, practical cases, exact values are not needed, and we can stop the process once we have obtained N instantiations. To illustrate this point, we compute the approximate probabilities for each node based on the first j instantiations, for $j = 1, \ldots, 12$. The results are shown in Table 9.7. The last row in the table contains the exact probabilities because they are based on all 12 possible instantiations. It can be seen that as the number of instantiations increases, the approximation gets better.

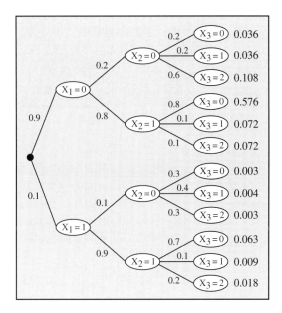

FIGURE 9.25. Search tree showing the probabilities associated with all instantiations.

The process can be viewed as one in which the total probability 1 is sequentially distributed among the instantiations or groups of instantiations. For example, in Step 1 we assign a probability 0.9 to the group of instantiations of the form $(0, \ldots)$, and a probability 0.1 to the group of instantiations of the form $(1, \ldots)$, and in Step 2 we distribute a probability 0.9 among the groups of instantiations of the forms $(0, 0, \ldots)$ and $(0, 1, \ldots)$, assigning them 0.18 and 0.72, respectively. ■

An important advantage of this method is that in the case of no evidence, it allows computing lower and upper bounds for the marginal probabilities of all variables at each step. If we have some evidence, these bounds can be calculated after the evidential nodes are included in all branches of the tree, that is, after the tree has been completed up to the last evidential node, since only then we can know the normalization constant. These bounds can

434   9. Approximate Propagation Methods

x^j	$p(x^j)$	$p(x_1)$		$p(x_2)$		$p(x_3)$		
		0	1	0	1	0	1	2
(0,1,0)	0.576	1.000	0.000	0.000	1.000	1.000	0.000	0.000
(0,0,2)	0.108	1.000	0.000	0.158	0.842	0.842	0.000	0.158
(0,1,2)	0.072	1.000	0.000	0.143	0.857	0.762	0.000	0.238
(0,1,1)	0.072	1.000	0.000	0.130	0.870	0.696	0.087	0.217
(1,1,0)	0.063	0.929	0.071	0.121	0.879	0.717	0.081	0.202
(0,0,1)	0.036	0.932	0.068	0.155	0.845	0.689	0.117	0.194
(0,0,0)	0.036	0.935	0.065	0.187	0.813	0.701	0.112	0.187
(1,1,2)	0.018	0.917	0.083	0.183	0.817	0.688	0.110	0.202
(1,1,1)	0.009	0.909	0.091	0.182	0.818	0.682	0.118	0.200
(1,0,1)	0.004	0.905	0.095	0.185	0.815	0.679	0.122	0.199
(1,0,2)	0.003	0.903	0.097	0.188	0.812	0.677	0.121	0.202
(1,0,0)	0.003	0.900	0.100	0.190	0.810	0.678	0.121	0.201

TABLE 9.7. The instantiations generated by the maximum probability search method and the approximate probability of each node based on the first j instantiations.

then be used to stop the sampling process when the bounds are within a previously specified tolerance. This results in a very large saving in computations. In fact, the method is optimal, in the sense that no better ordering of the instantiations can be obtained, that is, the instantiations with large probabilities of occurrence are generated first.

During the simulation process the probability bounds are calculated as follows. We start by the trivial bounds $(0, 1)$ and in each iteration we change only the bounds of the variable to which we have added new branches. We proceed in the following manner:

- **Lower bounds:** We add to the previous lower bound of the actual variable all the probabilities assigned to that variable in the new branches.

- **Upper bounds:** We calculate the width of the interval by summing all unassigned probabilities, that is, the sum of the probabilities in terminal nodes that do not correspond to the variable. Then, we add this width to the lower bound of the variable to obtain the upper bound. Note that this width is identical for all possible values of the variable, because the unassigned probability can be later assigned to any of the possible values. This is illustrated in the following example.

Example 9.12 Computing probability bounds. Consider the Bayesian network in Example 9.11. Initially, the lower and upper bounds are 0 and 1. In Step 1, we assign 0.9 to instantiations of the form $(0, x_2, x_3)$ and 0.1 to those of the form $(1, x_2, x_3)$. Thus, the minimum value of $p(X_1 = 0)$

		Step 1		Step 2		Steps 3 and 4	
Marginals	Exact	Lower	Upper	Lower	Upper	Lower	Upper
$X_1 = 0$	0.900	0.9	0.9	–	–	–	–
$X_1 = 1$	0.100	0.1	0.1	–	–	–	–
$X_2 = 0$	0.190	0.0	1.0	0.180	0.280	0.180	0.280
$X_2 = 1$	0.810	0.0	1.0	0.720	0.820	0.720	0.820
$X_3 = 0$	0.678	0.0	1.0	0.0	1.0	0.576	0.856
$X_3 = 1$	0.121	0.0	1.0	0.0	1.0	0.072	0.352
$X_3 = 2$	0.201	0.0	1.0	0.0	1.0	0.072	0.352

TABLE 9.8. Exact, lower, and upper bounds for marginal probabilities (Steps 1 to 4).

is 0.9 (lower bound) and the maximum is the same, because the remaining probability, 0.1 has been already assigned to instantiations with $X_1 = 1$, and no more probability is available. Similarly, the lower and upper bounds for $p(X_1 = 1)$ are equal to 0.1. Because of the zero width of the this interval, the bounds coincide with the exact values and they shall be no longer modified in subsequent steps.

In Step 2, we modify the bounds for variable X_2 because this variable has been used in the branching process. Probabilities of 0.18 and 0.72 (a total of 0.90) are assigned to groups of instantiations of the form $(0, 0, x_3)$ and $(0, 1, x_3)$, respectively, which are the lower bounds for $p(X_2 = 0)$ and $p(X_2 = 1)$, respectively. The remaining probability, 0.1, is still unassigned. Thus, it could be assigned to any of them. Thus, upper bounds can be obtained by adding 0.1 to the lower bounds. The same method is applied to obtain the bounds in the remaining steps.

Tables 9.8 and 9.9 show the exact marginal probabilities of the three nodes X_1, X_2, and X_3 and the lower and upper bounds corresponding to Steps 1–8. ∎

It is clear from the above discussion that using the maximum probability search method we can

- Control the error of the marginal probabilities of the nodes.
- Deal with different errors for different variables.
- Use the errors as a stopping rule.

9.10.1 Dealing With Evidence

If we are given a set of evidential nodes, E, we need to modify the process in the following way. We change the conditional probability $p(x_i|\pi_i)$ associated

		Steps 5 and 6		Step 7		Step 8	
Marginals	Exact	Lower	Upper	Lower	Upper	Lower	Upper
$X_1 = 0$	0.900	–	–	–	–	–	–
$X_1 = 1$	0.100	–	–	–	–	–	–
$X_2 = 0$	0.190	0.180	0.280	0.190	0.190	–	–
$X_2 = 1$	0.810	0.720	0.820	0.810	0.810	–	–
$X_3 = 0$	0.678	0.612	0.712	0.612	0.712	0.675	0.685
$X_3 = 1$	0.121	0.108	0.208	0.108	0.208	0.117	0.127
$X_3 = 2$	0.201	0.180	0.280	0.180	0.280	0.198	0.208

TABLE 9.9. Exact, lower, and upper bounds for marginal probabilities (Steps 5 to 8).

with the evidential nodes $X_i \in E$ by clamping the evidence variables to the evidence values. Then, we consider the CPDs $p_e(x_i|\pi_i)$, defined in (9.14), as

$$p_e(x_i|\pi_i) = \begin{cases} p(x_i|\pi_i), & \text{if } x_i \text{ and } \pi_i \text{ are consistent with } e, \\ 0, & \text{otherwise.} \end{cases}$$

Therefore, only the branch corresponding to the evidential value e_i is added to the tree. This is illustrated below by an example.

Example 9.13 Dealing with evidence. Consider the Bayesian network in Example 9.11 and suppose the evidence $X_2 = 0$ is known. Before we create the search tree, we need to change the CPD for X_2 to

$$p(X_2 = 0|X_1 = 0) = 0.2, \quad p(X_2 = 0|X_1 = 1) = 0.1,$$
$$p(X_2 = 1|X_1 = 0) = 0.0, \quad p(X_2 = 1|X_1 = 1) = 0.0.$$

All other CPDs remain unchanged. The corresponding search tree is shown in Figure 9.26. Note that the evidential node X_2 is clamped to its known value of 0.

Finally, we normalize the probabilities. ∎

9.10.2 Implementation Aspects

The pseudocode for the maximum probability search method is given in Figure 9.27. For the sake of simplicity, we take binary variables in the algorithm. A priority queue (a data structure on which the operations "insert" and "get maximum" are defined, both of which are at most log(queue size) operations) is used for storing the structure of the search tree and for retrieving the branch with maximum probability. For every *branch* of the tree, the queue is used to store the associated partial instantiation, *branch.ins*, and the corresponding probability, *branch.prob*.

9.10 Maximum Probability Search Method   437

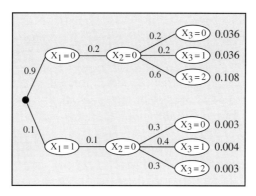

FIGURE 9.26. Incorporating evidence in the maximum probability search method.

```
Initialize
    queue ← {(X₁ = 0, p(X₁ = 0)), (X₁ = 1, p(X₁ = 1))}
    TotalProb ← 0

Main loop
    while TotalProb < δ do
        branch ← get maximum element from queue.
        if (|branch.ins| = n) then
            Add branch.ins to the sample
            Add branch.prob to TotalProb
            Remove branch from queue
        if (|branch.ins| < n) then
            expand branch in queue
```

FIGURE 9.27. Pseudocode for the maximum probability search method.

Note that for calculating the bounds, we use the sum of the logarithm of the probabilities instead of the product of the probabilities. This decreases the numerical instability that usually occurs when a multiplication of many small numbers is performed.

9.10.3 Experimental Results

We perform some experiments to evaluate the performance of the maximum probability search algorithm. A Bayesian network consisting of ten binary variables is randomly generated. In the first experiment, the random numbers associated with the CPDs are selected from the unit interval and in the second experiment the numbers are uniformly selected from $[0, 0.1] \cup [0.9, 1]$. The experiments are performed by running the al-

438   9. Approximate Propagation Methods

gorithms with values of the minimal accumulated probability δ from the values $\{0.8, 0.9, 0.95, 0.975, 0.98125, 0.99\}$. The performance is measured by (a) the number of complete generated instantiations, (b) the maximum size of the queue, and (c) the time to execute the approximation. The results are averaged over the runs of all ten joint probability distributions.

Figure 9.28 shows the results for the case where the probability tables have been selected from the unit interval. Figure 9.29 shows the results for the interval $[0, 0.1] \cup [0.9, 1]$. As could be expected, the execution time rises when the probability rises. When Figure 9.28 and 9.29 are compared, one sees that all measured criteria are about ten times larger for the distributions that have their probabilities from the unit interval. So, the maximum probability search method works better for cases where extreme probabilities are involved. This is caused by the large size of the largest intervals that appear when distributions contain extreme probabilities.

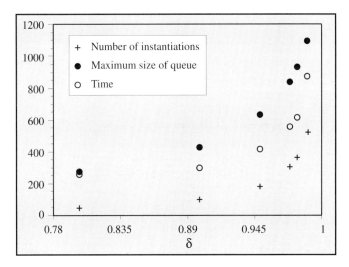

FIGURE 9.28. A scatter plot of three performance measures versus δ for the cases in which probability tables are chosen from $[0, 1]$.

In Figure 9.29 it is striking that the maximum queue size is constant for all values of δ. Apparently, after a given number of partial instantiations have been stored, the expansion of the queue is in balance with the number of instantiations taken from the queue. This is interesting, because it means that it is not necessary to store all possible instantiations in the queue at the same time.

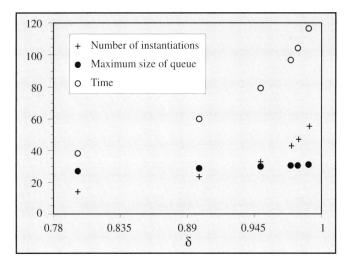

FIGURE 9.29. A scatter plot of three performance measures versus δ for the cases in which probability tables are chosen from $[0, 0.1] \cup [0.9, 1]$.

9.11 Complexity Analysis

Cooper (1990) shows that the complexity of exact inference in Bayesian networks is an NP-hard problem. Recently, Dagum, and Luby (1993) showed that the complexity of approximate propagation is NP-hard in the precision of the approximation. However, if the precision is kept fixed, then the runtime is linear in the number of variables. The fact that the worst case is NP-hard does not mean that this is true for most of the cases that are encountered in practice. The key advantage of simulation methods over exact methods is that, for a given level of precision, their complexities are linear in the number of nodes irrespective of the degree of connectedness of the graph.

Complexity problems are due to the fact that the absolute errors in p_i can be an exponential function of the number of nodes. This occurs for very small probabilities p_i such as those that arise in reliability problems. However, an absolute error can be fixed so that for the estimation above such precision is unnecessary. For example, in a nuclear power plant the probability of some critical events must be estimated with an absolute error of 10^{-7} or 10^{-8}, but this error is fixed and does not depend on the number of nodes. In addition, in many practical applications, simultaneous confidence intervals are needed only for a few cells.

440   9. Approximate Propagation Methods

Exercises

9.1 Consider the JPD $p(x) = 3x(2-x)/4$ given in Example 9.2. Obtain a sample of size 20 using

   (a) The associated cumulative distribution function $P(x) = (3x^2 - x^3)/4$ (see Figure 9.3).

   (b) The acceptance-rejection method with $h(x) = 1/2$ and $g(x) = x(2-x)$.

   Compare the obtained samples.

9.2 Consider the population distribution $p(x) = x^2$ and the simulation distribution $h(x) = x$. Calculate a function $g(x)$ and the associated constant c that satisfy (9.2) and repeat the same calculations shown in the previous exercise.

9.3 Calculate all the possible ancestral orderings for the Bayesian network given in Figure 9.5 (use Figure 9.8). What is the most efficient ordering to reject instantiations that do not match the evidence?

9.4 Consider the example of acceptance-rejection sampling given in Example 9.4:

   - Compute the posterior probability distribution for all four nonevidential nodes in Figure 9.5 using samples of sizes 10, 20, and 100, respectively.

   - Use any of the exact propagation methods introduced in Chapter 8 to compute the exact values for these probabilities.

   - Analyze the convergence of the approximated probabilities for the three different sample sizes.

   - What is the percentage of rejected instantiations in this example?

9.5 Repeat the above exercise considering the new CPDs for nodes X_3 and X_4 given in Table 9.10. What is the percentage of rejected instantiations?

9.6 Repeat the two previous exercises using the likelihood weighing algorithm. Compare the results obtained with the ones obtained using the acceptance-rejection method.

9.7 Repeat the steps of the backward-forward sampling algorithm shown in Example 9.7 considering the alternative orderings (X_4, X_3, X_5, X_6) and (X_6, X_4, X_5, X_3), where nodes X_2, X_3, X_4 are backward sampled and nodes X_5 and X_6 are forward sampled.

x_1	x_3	$p(x_3\|x_1)$
0	0	0.01
0	1	0.99
1	0	0.01
1	1	0.99

x_2	x_4	$p(x_4\|x_2)$
0	0	0.01
0	1	0.99
1	0	0.01
1	1	0.99

TABLE 9.10. New CPDs for nodes X_3 and X_4 for the Bayesian network given in Figure 9.5.

9.8 Generate two systematic samples, one of size five and the other of size ten, from the network in Figure 9.13 and the CPDs in Table 9.5.

9.9 Consider the Bayesian network given in Figure 9.13 with the values for the CPDs given in Table 9.6. Use the systematic sampling method to obtain a sample of size 10 (see Figure 9.21). Repeat the same process using the modified sampling method. Compare the performance of both methods.

9.10 Consider the Bayesian network with six nodes given in Example 9.3 and use the maximum probability search method, step by step, calculating the corresponding upper and lower bounds for the probabilities of all variables. Are the error bounds dependent on the order of the variables?

Chapter 10
Symbolic Propagation of Evidence

10.1 Introduction

In chapters 8 and 9 we introduced several methods for exact and approximate propagation of evidence in probabilistic network models. These methods require that the joint probability distribution (JPD) of the model be specified numerically, that is, all the parameters must be assigned fixed numeric values. However, numeric specification of these parameters may not be available, or it may happen that the subject-matter specialists can specify only ranges of values for the parameters rather than their exact values. In such cases, the numeric propagation methods must be replaced by symbolic propagation methods, which are able to deal with the parameters themselves without assigning them numeric values.

Symbolic propagation leads to solutions that are expressed as functions of the parameters in symbolic form. Thus, the answers to general queries can be given symbolically in terms of the parameters, and the answers to specific queries can then be obtained by plugging the values of the parameters into the solution given in symbolic form, without need to redo the propagation. Furthermore, symbolic propagation allows one to study the sensitivity of the results to changes in parameter values with little additional computational effort. Symbolic propagation is specially useful in the following cases:

1. When exact numeric specification of the probabilistic model is not available.

2. When the subject-matter specialists are able to specify only ranges of values for the parameters rather than the exact values. In this case, we show how to use symbolic propagation methods to obtain lower and upper bounds for the probabilities for all possible values of the parameters in the given intervals.

3. When a sensitivity analysis is required. A question that usually arises in this context is, How sensitive are the results to small changes in the parameters and evidence values?

Symbolic propagation algorithms have been recently introduced in the literature. For example, Castillo, Gutiérrez, and Hadi (1995c, 1995d) perform symbolic propagation by adapting some of the standard numeric propagation algorithms described in Chapter 8 for symbolic computations. These methods perform the necessary symbolic calculations by using computer packages with symbolic computational capabilities (such as *Mathematica* and *Maple*).

Another method with symbolic capabilities is the symbolic probabilistic inference algorithm (SPI) (Shachter, D'Ambrosio, and DelFabero (1990) and Li and D'Ambrosio (1994)). This method is goal oriented and it performs only those calculations that are required to respond to queries. Symbolic expressions can be obtained by postponing evaluation of expressions, maintaining them in symbolic form.

However, both of the above symbolic methods suffer from the same problem: They need to use either special programs or extra computational effort to carry out the symbolic computations. Furthermore, computing and simplifying symbolic expressions is a computationally expensive task, and it becomes increasingly inefficient when dealing with large networks or large sets of symbolic parameters. Recently, Castillo, Gutiérrez, and Hadi (1996c) introduced an efficient approach to symbolic propagation that takes advantage of the polynomial structure of the probabilities of the nodes (see Section 7.5.1) to avoid symbolic computations. The main idea of the method is obtaining the symbolic expressions through a numeric algorithm that computes the coefficients of the associated polynomials. Then, all the computations are carried out numerically, avoiding the use of the computationally expensive symbolic manipulations.

In Section 10.2 we introduce the notation and the basic framework of the symbolic methods. Section 10.3 discusses the automatic generation of symbolic codes. The algebraic structure of probabilities is analyzed in Section 10.4. Section 10.5 shows how numeric propagation methods can be used for symbolic propagation of evidence to obtain the polynomial structure of the probabilities. Section 10.6 presents an improvement of the above method for goal-oriented tasks. Section 10.7 deals with the problem of symbolic random evidence. Section 10.8 shows that a sensitivity analysis can be easily performed using symbolic methods. Finally, Section 10.9 analyzes the problem of symbolic propagation of evidence in Gaussian Bayesian networks.

10.2 Notation and Basic Framework

We have seen in Chapter 6 that the JPD associated with the two main types of probabilistic network models (Bayesian and decomposable Markov network models) can be given through a factorization of the JPD as a product of conditional probability distributions (CPDs)

$$p(x_1, \ldots, x_n) = \prod_{i=1}^{n} p(x_i | \pi_i). \tag{10.1}$$

In the case of Bayesian networks, the conditioning sets are the parents of the node, Π_i, $i = 1, \ldots, n$. In the case of decomposable Markov networks, these sets are obtained by applying the chain rule to the factorization obtained from a chain of cliques (see Chapter 6). Therefore, although some of the methods introduced in this chapter can be easily extended to deal with a potential representation of the JPD, for simplicity but without loss of generality, we use the set of CPDs in (10.1) as a basic parametric representation of the JPD.

Let $X = \{X_1, \ldots, X_n\}$ be a set of n discrete variables, each of which can take values in the set $\{0, 1, \ldots, r_i\}$, and $B = (D, P)$ a Bayesian network over X, where the directed acyclic graph (DAG) D determines the structure of the set of CPDs, and $P = \{p(x_1|\pi_1), \ldots, p(x_n|\pi_n)\}$ is the set of CPDs needed to specify the JPD. Note that $p(x_i|\pi_i)$ gives the probabilities of X_i given the values of the variables in its parent set Π_i.

Some of the CPDs in (10.1) can be specified numerically and others symbolically, that is, $p(x_i|\pi_i)$ can be either a parametric family or fully specified numerically.

Definition 10.1 Symbolic Node. *When $p(x_i|\pi_i)$ is a symbolic parametric family (i.e., it depends on at least one parameter in symbolic form), we refer to the node X_i as a symbolic node and we use Θ_i to denote its associated symbolic parameters.*

As we have already seen in Section 7.5.1, when $p(x_i|\pi_i)$ is a parametric family, that is, when X_i is a symbolic node, a convenient choice of the parameters is given by

$$\theta_{ij\pi} = p(X_i = j | \Pi_i = \pi), \ j \in \{0, \ldots, r_i\}, \tag{10.2}$$

where π is any possible instantiation of the parents of X_i, Π_i. Thus, the first subscript in $\theta_{ij\pi}$ refers to the node number, the second subscript refers to the state of the node, and the remaining subscripts refer to the parents' instantiations. Since $\sum_{j=0}^{r_i} \theta_{ij\pi} = 1$, for all i and π, not all parameters are free parameters. Because the parameters sum up to one, one parameter can be written as one minus the sum of the other parameters. For example, the first parameter can be written as

$$\theta_{i0\pi} = 1 - \sum_{j=1}^{r_i} \theta_{ij\pi}. \qquad (10.3)$$

To simplify the notation in cases where a variable X_i does not have parents, we use θ_{ij} to denote $p_i(X_i = j)$, $j \in \{0, \ldots, r_i\}$. We illustrate this notation using the following example.

Example 10.1 Symbolic nodes. Consider a discrete Bayesian network consisting of the variables $X = \{X_1, \ldots, X_8\}$ whose corresponding directed acyclic graph (DAG) is given in Figure 10.1. The structure of the graph implies that the joint probability of the set of nodes can be written, in the form of (10.1), as

$$p(x) = p(x_1)p(x_2|x_1)p(x_3|x_1)p(x_4|x_2,x_3)p(x_5|x_3)p(x_6|x_4)p(x_7|x_4)p(x_8|x_5). \qquad (10.4)$$

For simplicity, but without loss of generality, assume that all nodes represent binary variables with values in the set $\{0,1\}$. This and the structure of the probability distribution in (10.4) imply that the JPD of the eight variables depends on 34 parameters $\Theta = \{\theta_{ij\pi}\}$. Note, however, that only 17 of the parameters are free (because the probabilities in each conditional distribution must add up to one). These 17 parameters are given in Table 10.1.

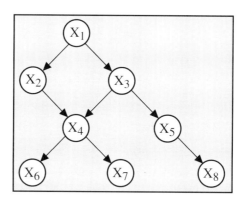

FIGURE 10.1. A directed acyclic graph.

In this example, only nodes X_3 and X_6 are symbolic nodes because their corresponding CPDs contain at least one symbolic parameter. We have the sets of parameters $\Theta_3 = \{\theta_{300}, \theta_{310}\}$ and $\Theta_6 = \{\theta_{600}, \theta_{610}\}$. Note that these sets include all the symbolic parameters, not only the free parameters. Thus, the set of symbolic parameters associated with the Bayesian network is $\Theta = \{\Theta_3, \Theta_6\}$. ∎

X_i	Π_i	Free Parameters
X_1	ϕ	$\theta_{10} = p(X_1 = 0) = 0.2$
X_2	X_1	$\theta_{200} = p(X_2 = 0\|X_1 = 0) = 0.3$ $\theta_{201} = p(X_2 = 0\|X_1 = 1) = 0.5$
X_3	X_1	$\theta_{300} = p(X_3 = 0\|X_1 = 0)$ $\theta_{301} = p(X_3 = 0\|X_1 = 1) = 0.5$
X_4	X_2, X_3	$\theta_{4000} = p(X_4 = 0\|X_2 = 0, X_3 = 0) = 0.1$ $\theta_{4001} = p(X_4 = 0\|X_2 = 0, X_3 = 1) = 0.8$ $\theta_{4010} = p(X_4 = 0\|X_2 = 1, X_3 = 0) = 0.3$ $\theta_{4011} = p(X_4 = 0\|X_2 = 1, X_3 = 1) = 0.4$
X_5	X_3	$\theta_{500} = p(X_5 = 0\|X_3 = 0) = 0.3$ $\theta_{501} = p(X_5 = 0\|X_3 = 1) = 0.1$
X_6	X_4	$\theta_{600} = p(X_6 = 0\|X_4 = 0)$ $\theta_{601} = p(X_6 = 0\|X_4 = 1) = 0.9$
X_7	X_4	$\theta_{700} = p(X_7 = 0\|X_4 = 0) = 0.3$ $\theta_{701} = p(X_7 = 0\|X_4 = 1) = 0.6$
X_8	X_5	$\theta_{800} = p(X_8 = 0\|X_5 = 0) = 0.2$ $\theta_{801} = p(X_8 = 0\|X_5 = 1) = 0.4$

TABLE 10.1. The set of free parameters associated with the CPDs in (10.4).

10.3 Automatic Generation of Symbolic Code

Dealing with symbolic parameters is the same as dealing with numeric values with the only difference being that all the required operations must be performed by a program with symbolic manipulation capabilities. Symbolic computations, however, are slower than numeric computations and they require more memory. We have seen in Examples 7.13 and 7.14 that symbolic propagation can be directly performed using the marginal and conditional probability formulas given in (3.4) and (3.5). However, this brute force method is computationally intensive and becomes inefficient even with a small number of variables.

An alternative to this method is to adapt some of the numeric propagation algorithms introduced in Chapter 8 to symbolic computation. Castillo, Gutiérrez, and Hadi (1995d) show that the symbolic adaptation of some of these methods requires only minor modifications. For example, the clustering propagation algorithm (Algorithm 8.4) can be easily adapted to symbolic propagation using a symbolic computer language, such as *Mathematica*. In this section we examine the symbolic capabilities of this propagation algorithm. Then, we illustrate its application using an example.

448   10. Symbolic Propagation of Evidence

From the clustering algorithms introduced in Chapter 8, in particular, the algorithm for propagation in probabilistic network models using a join tree (Algorithms 8.4 and 8.5), one can conclude the following:

1. The number of arguments of the functions involved (conditional probabilities, potential functions, and probability functions of the cliques) depend on the topology of the network. In addition, there are several possibilities for building the potential functions from the CPDs in (10.1).

2. Given two neighboring cliques C_i and C_j with separator set S_{ij}, the message $M_{ij}(s_{ij})$ that clique C_i sends to C_j is

$$M_{ij}(s_{ij}) = \sum_{c_i \setminus s_{ij}} \psi_i(c_i) \prod_{k \neq j} M_{ki}(s_{ki}), \qquad (10.5)$$

a function that depends on the symbolic parameters contained in C_i and all its neighboring cliques except C_j. Thus, the maximum number of symbolic parameters involved in this message is bounded by the maximum number of neighbors of a clique.

3. The messages among cliques have to be ordered in such a way that no message is used before it has been defined. Once all messages have been defined, the JPD of the cliques can be calculated using

$$p(c_i) = \psi_i(c_i) \prod_k M_{ki}(s_{ik}). \qquad (10.6)$$

This expression depends on the symbolic parameters contained in C_i and all its neighboring cliques C_k.

4. The marginal probability distribution of a node can be calculated in several ways by marginalizing the JPD of any clique containing the node. However, the smaller the size of the selected clique, the less the required computations.

Thus, writing the symbolic code for a given clique tree requires analyzing its topology in order to build an efficient join tree for symbolic propagation. To this purpose, a code generator can be easily written based on Algorithm 8.4. In this algorithm, the symbolic code is written in the order indicated by the algorithm in such a way that the sequential definitions of the functions, messages, and probabilities coincide with the order needed for the algorithm. As shown in the following example, the codes in Figures 10.3 and 10.4 are generated by a computer program written by the authors in the C++ language.[1]

---

[1] The program *X-pert Symbolic* for the automatic generation of symbolic code for *Mathematica* and *Maple* programs can be obtained from the World Wide Web site http://ccaix3.unican.es/~AIGroup.

10.3 Automatic Generation of Symbolic Code

Example 10.2 Symbolic code. In this example we illustrate a symbolic implementation of Algorithm 8.4 using *Mathematica* to solve the propagation of evidence problem in a given example. Consider again the directed graph in Figure 10.1 with the associated JPD in (10.4). Figure 10.2 shows a join tree associated with the cliques of the moralized and triangulated graph (see Section 4.5).

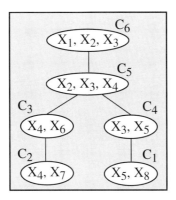

FIGURE 10.2. A Join tree for the moralized and triangulated graph associated with the DAG given in Figure 10.1.

Figures 10.3 and 10.4 give *Mathematica* code for the symbolic propagation. A similar code can be easily obtained using *Maple*, *Axiom*, or any other symbolic package. These symbolic codes have been automatically generated by the code generator mentioned above. This code first reads the graphical and the probabilistic structures of the Bayesian network, in terms of numeric and symbolic parameters. Then it generates the required symbolic propagation code.

Initially, we define the probability distributions required to specify $p(x)$ in (10.4) using a and b to denote the symbolic parameters θ_{300} and θ_{600}, contained in the tables for the CPDs given in Table 10.1.

For clarity of exposition, we select some numeric and some symbolic parameters. However, dealing only with symbolic parameters is identical and does not affect the rest of the code. Under the heading "Probability Tables," the numeric and symbolic values for the parameters are assigned, and a list of commands is given to build the conditional probability distributions for each node. The names of the functions that generate these tables are $P1, P2, \ldots, P8$. Note that the automatic code generator takes into account the different numbers of arguments for the different probability tables. Note also that only the free parameters are needed in the definition of the JPD. The next step in the algorithm is the definition of the potential functions ψ_i, one for each clique. The cliques associated with the graph in Figure 10.1 are shown in Figure 10.2. Then the functions $\{F1, F2, F3, F4, F5, F6\}$ are defined by assigning each of the above condi-

450   10. Symbolic Propagation of Evidence

```
(* Probability Tables *)

T={0.2,0.8};
n=1; Do[P1[i1]=T[[n]];n++,{i1,0,1}];
T={0.3,0.5,0.7,0.5};
n=1; Do[P2[i1,i2]=T[[n]];n++,{i1,0,1},{i2,0,1}];
T={a,0.5,1-a,0.5};
n=1; Do[P3[i1,i2]=T[[n]];n++,{i1,0,1},{i2,0,1}];
T={0.1,0.8,0.3,0.4,0.9,0.2,0.7,0.6};
n=1; Do[P4[i1,i2,i3]=T[[n]];n++,{i1,0,1},{i2,0,1},{i3,0,1}];
T={0.3,0.1,0.7,0.9};
n=1; Do[P5[i1,i2]=T[[n]];n++,{i1,0,1},{i2,0,1}];
T={b,0.9,1-b,0.1};
n=1; Do[P6[i1,i2]=T[[n]];n++,{i1,0,1},{i2,0,1}];
T={0.3,0.6,0.7,0.4};
n=1; Do[P7[i1,i2]=T[[n]];n++,{i1,0,1},{i2,0,1}];
T={0.2,0.4,0.8,0.6};
n=1; Do[P8[i1,i2]=T[[n]];n++,{i1,0,1},{i2,0,1}];

(* Potential Functions *)

F1[X8_,X5_]:=P8[X8,X5];
F2[X7_,X4_]:=P7[X7,X4];
F3[X6_,X4_]:=P6[X6,X4];
F4[X5_,X3_]:=P5[X5,X3];
F5[X4_,X2_,X3_]:=P4[X4,X2,X3];
F6[X3_,X1_,X2_]:=P1[X1]*P2[X2,X1]*P3[X3,X1];

(* Initialize Ranges *)

Do[inf[i]=0; sup[i]=1, {i,1,8}];
```

FIGURE 10.3. A *Mathematica* program for symbolic propagation of evidence (Part 1).

tional tables to a clique that includes its associated family. Next, we define the ranges, $\{(inf[i], sup[i]), i = 1, 2, \ldots, 8\}$, of all variables to be $\{0, 1\}$, because they are binary variables.

In the code of Figure 10.4, the messages between neighboring cliques are defined using (10.5). These messages are denoted by

$$\{M14, M23, M34, M45, M56, M65, M53, M54, M32, M41\},$$

where the indices refer to the cliques sending and receiving the messages, respectively. The order of these statements is partially imposed by the topology of the network, because some messages depend on previous mes-

10.3 Automatic Generation of Symbolic Code   451

(∗ *Definition of Messages* ∗)

M14[X5_]:=Sum[F1[X8,X5],{X8,inf[8],sup[8]}];
M23[X4_]:=Sum[F2[X7,X4],{X7,inf[7],sup[7]}];
M35[X4_]:=Sum[F3[X6,X4]*M23[X4],{X6,inf[6],sup[6]}];
M45[X3_]:=Sum[F4[X5,X3]*M14[X5],{X5,inf[5],sup[5]}];
M56[X2_,X3_]:=Sum[F5[X4,X2,X3]*M35[X4]*M45[X3],{X4,inf[4],sup[4]}];
M65[X2_,X3_]:=Sum[F6[X3,X1,X2],{X1,inf[1],sup[1]}];
M53[X4_]:=Sum[F5[X4,X2,X3]*M45[X3]*M65[X2,X3],{X2,inf[2],sup[2]},
  {X3,inf[3],sup[3]}];
M54[X3_]:=Sum[F5[X4,X2,X3]*M35[X4]*M65[X2,X3],{X4,inf[4],sup[4]},
  {X2,inf[2],sup[2]}];
M32[X4_]:=Sum[F3[X6,X4]*M53[X4],{X6,inf[6],sup[6]}];
M41[X5_]:=Sum[F4[X5,X3]*M54[X3],{X3,inf[3],sup[3]}];

(∗ *JPDs of the Cliques* ∗)

Q1[X8_,X5_]:=F1[X8,X5]*M41[X5];
Q2[X7_,X4_]:=F2[X7,X4]*M32[X4];
Q3[X6_,X4_]:=F3[X6,X4]*M23[X4]*M53[X4];
Q4[X5_,X3_]:=F4[X5,X3]*M14[X5]*M54[X3];
Q5[X4_,X2_,X3_]:=F5[X4,X2,X3]*M35[X4]*M45[X3]*M65[X2,X3];
Q6[X3_,X1_,X2_]:=F6[X3,X1,X2]*M56[X2,X3];

(∗ *Node Marginals* ∗)

P[1,X1_]:=Sum[Q6[X3,X1,X2],{X3,inf[3],sup[3]},{X2,inf[2],sup[2]}];
P[2,X2_]:=Sum[Q6[X3,X1,X2],{X3,inf[3],sup[3]},{X1,inf[1],sup[1]}];
P[3,X3_]:=Sum[Q4[X5,X3],{X5,inf[5],sup[5]}];
P[4,X4_]:=Sum[Q3[X6,X4],{X6,inf[6],sup[6]}];
P[5,X5_]:=Sum[Q4[X5,X3],{X3,inf[3],sup[3]}];
P[6,X6_]:=Sum[Q3[X6,X4],{X4,inf[4],sup[4]}];
P[7,X7_]:=Sum[Q2[X7,X4],{X4,inf[4],sup[4]}];
P[8,X8_]:=Sum[Q1[X8,X5],{X5,inf[5],sup[5]}];

(∗ *Normalization and Printing* ∗)

Do[c=Chop[Simplify[Sum[P[i,t],{t,inf[i],sup[i]}]]];
   Do[R[i,t]:=Simplify[Chop[P[i,t]]/c];
     Print["p(Node",i,"=",t,")=",R[i,t]],
   {t,inf[i],sup[i]}],
{i,1,8}]

FIGURE 10.4. A *Mathematica* program for symbolic propagation of evidence (Part 2).

X_i	$p(x_i = 0)$
X_1	0.2
X_2	0.46
X_3	$0.4 + 0.2\,\theta_{300}$
X_4	$0.424 - 0.056\,\theta_{300}$
X_5	$0.18 + 0.04\,\theta_{300}$
X_6	$0.5184 + 0.0504\,\theta_{300} + 0.424\,\theta_{600} - 0.056\,\theta_{300}\,\theta_{600}$
X_7	$0.4728 + 0.0168\,\theta_{300}$
X_8	$0.364 - 0.008\,\theta_{300}$

TABLE 10.2. Initial probabilities of nodes.

sages. However, the symbolic code generator takes this into consideration when deciding this order.

Now, using the potential and the message functions we calculate the JPD for all the cliques $\{Q1, Q2, Q3, Q4, Q5, Q6\}$, using (10.6). The required *Mathematica* statements are shown in Figure 10.4 under the heading "JPDs of the cliques."

Next, we calculate the unnormalized univariate marginals of each node by marginalizing in one of the cliques containing the corresponding node. Note that for a given node there are as many possible alternatives as the number of cliques it belongs to. Choosing the clique with the minimum size leads to a minimum computational effort. The code generator selects the optimal clique for each node.

Finally, we normalize the probabilities by dividing by the associated sum and printing the results. The initial (no evidence) marginal probabilities of the nodes obtained with this program are given in Table 10.2. Note that we obtain polynomials in the parameters and that the exponents of these parameters are always one.

We now consider the evidence $\{X_2 = 1, X_5 = 1\}$. In this case, propagation of evidence can be performed using the same code, but making the ranges of variables X_2 and X_5 equal to $(1, 1)$, that is, $inf[2] = inf[5] = 1$ and repeating the calculations. Table 10.3 gives the new probabilities of the nodes given this evidence. Note that we get rational functions, i.e., quotients of polynomial functions in the parameters with unit exponents.

Tables 10.2 and 10.3 can then be used to answer all queries regarding initial or evidential probabilities associated with the network in Figure 10.1 simply by plugging in specific values for the parameters.

In Figure 10.5 we show the conditional probabilities for the nodes, given $X_2 = 1, X_5 = 1$ for the values of the parameters $\theta_{300} = 0.4, \theta_{600} = 0.8$, which can be obtained from the information in Table 10.3 by simple substitution. ■

X_i	$p(x_i = 0 \mid X_2 = 1, X_5 = 1)$
X_1	$\dfrac{0.126 - 0.028\,\theta_{300}}{0.446 - 0.028\,\theta_{300}}$
X_2	0
X_3	$\dfrac{0.14 + 0.098\,\theta_{300}}{0.446 - 0.028\,\theta_{300}}$
X_4	$\dfrac{0.1644 - 0.021\,\theta_{300}}{0.446 - 0.028\,\theta_{300}}$
X_5	0
X_6	$\dfrac{0.25344 - 0.0063\,\theta_{300} + 0.1644\,\theta_{600} - 0.021\,\theta_{300}\theta_{600}}{0.446 - 0.028\,\theta_{300}}$
X_7	$\dfrac{0.21828 - 0.0105\,\theta_{300}}{0.446 - 0.028\,\theta_{300}}$
X_8	$\dfrac{0.178 - 0.011\,\theta_{300}}{0.446 - 0.028\,\theta_{300}} = 0.4$

TABLE 10.3. Conditional probabilities of the nodes given $\{X_2 = 1, X_5 = 1\}$.

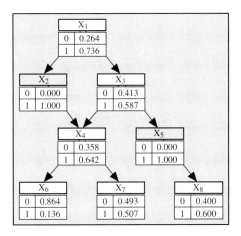

FIGURE 10.5. Conditional probabilities of the nodes given the evidence $X_2 = 1, X_5 = 1$.

10.4 Algebraic Structure of Probabilities

The marginal and conditional probabilities possess an interesting algebraic structure. The symbolic methods that we shall present in the next sections take advantage of this structure. In Section 7.5.1 we have analyzed the algebraic structure of marginal and conditional probabilities of the nodes in probabilistic network models. The following results are taken from Castillo, Gutiérrez, and Hadi (1995c).

Theorem 10.1 *The initial marginal probability of any given instantiation, $(x_1, \ldots, x_n)$, of the nodes in a Bayesian network is a polynomial in the symbolic parameters of degree less than or equal to the number of symbolic nodes. However, it is a first-degree polynomial in each parameter.*

Proof: According to (10.1) the probability of an instantiation $(x_1, \ldots, x_n)$ is

$$p(x_1, \ldots, x_n) = \prod_{i=1}^{n} p(x_i | \pi_i) = \prod_{i=1}^{n} \theta_{ix_i\pi_i}.$$

Note that all the parameters appearing in the above product are associated with different variables, and some of them may be specified numerically. Thus, $p(x_1, \ldots, x_n)$ is a monomial of degree less than or equal to the number of symbolic nodes. ■

Corollary 10.1 *The marginal probability of any set of nodes $Y \subset X$ is a polynomial in the parameters of degree less than or equal to the number of symbolic nodes. However, it is a first-degree polynomial in each parameter.*

For example, as can be seen in Table 10.2, the marginal probability of all nodes in the Bayesian network are polynomials of first-degree in each of the free symbolic parameters θ_{300} and θ_{600}.

Corollary 10.2 *The conditional probability of any set of nodes Y given some evidence $E = e$ is a rational function (a ratio of two polynomial functions) of the parameters. Furthermore, the denominator polynomial is the same for all nodes.*

For example, the probabilities in Table 10.3 show that the conditional probabilities of the nodes given the evidence $\{X_2 = 1, X_5 = 1\}$ is a ratio of two polynomials, and that the denominator polynomial is the same for all nodes.

These polynomial structures of the probabilities in probabilistic network models give rise to an efficient method for symbolic propagation. This method is presented in the next section.

10.5 Symbolic Propagation Through Numeric Computations

10.5.1 Polynomial Structure

Suppose that we are dealing with a set of symbolic nodes $\{X_{i_1}, \ldots, X_{i_s}\} \subset X$. Let $\Theta = \{\Theta_1, \ldots, \Theta_s\}$ be the set of corresponding symbolic parameters, where Θ_k stands for the symbolic parameters associated with the symbolic node X_{i_k}, with $k = 1, \ldots, s$.

Corollaries 10.1 and 10.2 guarantee that the conditional probabilities of a typical node X_i, given some evidence $E = e$, $p(X_i = j|E = e), j = 0, \ldots, r_i$, is either a polynomial or a ratio of two polynomials of the symbolic parameters. In addition, Theorem 10.1 guarantees that each monomial forming these polynomials contains at most one parameter of Θ_k, for each $k = 1, \ldots, s$. Therefore, we build the set of feasible monomials, M, by taking the Cartesian product of the sets of free symbolic parameters corresponding to the different symbolic nodes, including a numeric value 1 to account for the possible numeric values assigned to some of the parameters of the symbolic node. Then, we have

$$M = \{1, \Theta_1\} \times \ldots \times \{1, \Theta_s\}, \tag{10.7}$$

where Θ_i represents, in this case, the set of free symbolic parameters. Thus, the general form of these polynomials is

$$\sum_{m_r \in M} c_r m_r, \tag{10.8}$$

where c_r is the numeric coefficient associated with the monomial $m_r \in M$.

For example, the initial polynomial structure of the probabilities of the Bayesian network given in Example 10.1 is given by

$$M = \{1, \theta_{300}\} \times \{1, \theta_{600}\}.$$

Then, the general polynomials associated with the marginal and conditional probabilities of the the nodes have the form

$$c_1 + c_2 \theta_{300} + c_3 \theta_{600} + c_4 \theta_{300}\theta_{600}. \tag{10.9}$$

An alternative representation of this polynomial can be given by considering not only the free symbolic parameters, but all the symbolic parameters associated with the symbolic nodes. In the above example, the Cartesian product of all symbolic parameters associated with the different symbolic nodes is

$$M = \{\theta_{300}, \theta_{310}\} \times \{\theta_{600}, \theta_{610}\},$$

which defines the following structure for the polynomials

$$c_1 \theta_{300}\theta_{600} + c_2 \theta_{310}\theta_{600} + c_3 \theta_{300}\theta_{610} + c_4 \theta_{310}\theta_{610}. \tag{10.10}$$

The representations in (10.9) and (10.10) are equivalent. This is due to the fact that the parameters for each node add up to one. Thus, starting with (10.10), we have

$$c_1\theta_{300}\theta_{600} + c_2\theta_{310}\theta_{600} + c_3\theta_{300}\theta_{610} + c_4\theta_{310}\theta_{610} = c_1\theta_{300}\theta_{600}$$
$$+ c_2(1-\theta_{300})\theta_{600} + c_3\theta_{300}(1-\theta_{600}) + c_4(1-\theta_{300})(1-\theta_{600})$$
$$= c_4 + (c_3-c_4)\theta_{300} + (c_2-c_4)\theta_{600} + (c_4+c_1-c_2-c_3)\theta_{300}\theta_{600},$$

which is of the same form as (10.9). We shall see later that the representation given in (10.10) is more convenient than the one in (10.9) from a computational point of view. Thus, we shall use the representation

$$M = \Theta_1 \times \ldots \times \Theta_s, \tag{10.11}$$

where Θ_i represents the set of all the symbolic parameters associated with the symbolic node X_{i_s}, to represent the polynomials in (10.8).

In this section, we develop a method, which we refer to as the *numeric canonical components* method, for computing the coefficients c_r associated with a representation of the polynomials. The method first instantiates the symbolic parameters Θ and computes the resulting numeric probabilities. Then, once the coefficients have been obtained, the polynomials and hence, the probabilities $p(x_i|e)$ can be easily obtained.

We shall also show that there are some analogies between the canonical components symbolic method and the conditioning algorithms for evidence propagation presented in Section 8.5. Both methods perform several numeric propagations associated with different instantiations of certain parameters in the network to achieve the desired solution.

10.5.2 The Numeric Canonical Components Method

Let M be the set of monomials needed to compute $p(X_i = j|E = e)$ for $j = 0, \ldots, r_i$. Let m be the number of monomials in M. From (10.8), the polynomial needed to compute $p(X_i = j|E = e)$ is of the form

$$p(X_i = j|E = e) \propto \sum_{m_k \in M} c_k^{ij} m_k = p_{ij}(\Theta), \; j = 0, \ldots, r_i. \tag{10.12}$$

The term $p_{ij}(\Theta)$ represents the unnormalized probability $p(X_i = j|E = e)$. Thus, $p_{ij}(\Theta)$ can be written as a linear combination of the monomials in M. Our objective now is to compute the coefficients c_k^{ij}.

If the parameters Θ are assigned numeric values, say θ, then $p_{ij}(\theta)$ can be obtained by replacing Θ by θ and using any numeric propagation method to compute $p(X_i = j|E = e, \Theta = \theta)$. Similarly, the monomial m_k takes a numeric value, which is the product of the parameters involved in m_k. Thus, we have

$$p(X_i = j|E = e, \Theta = \theta) \propto \sum_{m_k \in M} c_k^{ij} m_k = p_{ij}(\theta). \tag{10.13}$$

10.5 Symbolic Propagation Through Numeric Computations

Note that in (10.13) all the monomials m_k and the unnormalized probability $p_{ij}(\theta)$ are known numbers, and the only unknowns are the coefficients c_k^{ij}, $k = 1,\ldots,m$. To compute these coefficients, we need to construct any set of m linearly independent equations each of the form (10.13). These equations can be obtained using m sets of distinct instantiations Θ. Let these values be denoted by $C = \{\theta_1,\ldots,\theta_m\}$. Let $\mathbf{T}_{ij}$ be the $m \times m$ nonsingular matrix whose rkth element is the value of the monomial m_k obtained by replacing Θ by θ_r, the rth instantiation of Θ. We refer to the matrix $\mathbf{T}_{ij}$ as the *canonical matrix* associated with the set of canonical components C. Let

$$\mathbf{c}_{ij} = \begin{pmatrix} c_1^{ij} \\ \vdots \\ c_m^{ij} \end{pmatrix} \quad \text{and} \quad \mathbf{p}_{ij} = \begin{pmatrix} p_{ij}(\theta_1) \\ \vdots \\ p_{ij}(\theta_m) \end{pmatrix}. \tag{10.14}$$

From (10.13) the m linearly independent equations can be written as

$$\mathbf{T}_{ij}\,\mathbf{c}_{ij} = \mathbf{p}_{ij}. \tag{10.15}$$

Since $\mathbf{T}_{ij}$ is nonsingular, $\mathbf{T}_{ij}^{-1}$ exists and the solution of (10.14) is given by

$$\mathbf{c}_{ij} = \mathbf{T}_{ij}^{-1}\,\mathbf{p}_{ij}. \tag{10.16}$$

The values of the coefficients in $\mathbf{c}_{ij}$ can then be substituted in (10.12) and the unnormalized probability $p_{ij}(\Theta)$ can be expressed as a function of Θ.

Therefore, Equations (10.12)–(10.16) provide an efficient algorithm for symbolic propagation that does not require any symbolic computation. We refer to this algorithm as the *numeric canonical components method* (NCCM). This algorithm is summarized as follows:

Algorithm 10.1 Numeric Canonical Components.

- **Input:** A Bayesian network model (D, P), where P is defined using numeric and symbolic parameters, and evidence $E = e$.

- **Output:** The symbolic probabilities $p(X_i = j|E = e)$, $i = 1,\ldots,n$.

1. Construct m sets of instantiations of Θ: $\theta_1,\ldots,\theta_m$, providing m linearly independent equations in $\mathbf{c}_{ij}$, after the substitution of the θ_i values in (10.13).

2. Calculate the $m \times m$ nonsingular matrix $\mathbf{T}_{ij}$ whose rkth element is the value of the monomial m_k obtained by replacing Θ by θ_r, the rth instantiation of Θ.

3. Compute the vector of probabilities $\mathbf{p}_{ij}$ in (10.14) using any standard numeric propagation method.

458   10. Symbolic Propagation of Evidence

4. Solve the linear system of equations (10.15) to obtain the desired coefficients c_{ij}.

5. Substitute the obtained values of c_{ij} in (10.12) and normalize to obtain the symbolic expression for the probabilities $p(X_i = j | E = e)$.   ∎

Note that Step 3 of Algorithm 10.1 requires the use of a numeric propagation method to propagate uncertainty as many times as the number of possible combinations of the symbolic parameters. This means that the number of numeric propagations increases combinatorially with the number of symbolic parameters. This problem is also present in other propagation algorithms. For example, the conditioning algorithm has this problem with respect to the number of nodes in the loop-cutset (see Chapter 8). Therefore, the role of symbolic nodes in the canonical components symbolic method is similar to the role of conditioning nodes in the conditioning algorithms. However, there we instantiated evidences and here we instantiate parameters.

Algorithm 10.1 requires calculating and solving a linear system of equations (see, for example, Press et al. (1992)). In the following, we show that by imposing certain conditions in the symbolic parameters, it is always possible to find a set of canonical components whose corresponding $\mathbf{T}_{ij}$ matrix is the identity matrix. Thus, the symbolic expressions associated with the probabilities can be obtained directly without the need for solving the linear system of equations (10.15) or inverting the matrix $\mathbf{T}_{ij}$ in (10.16).

Consider a typical symbolic node X_i with the associated parameters $\theta_{ij\pi}$. Some of these parameters may be specified numerically, and some may be given in symbolic form. Assume that a subset of the parameters $\theta_{ij\pi}$ is given in symbolic form for a given instantiation π of Π_i, and the parameters for all other instantiations are numeric. For example, the Bayesian network of Example 10.1 satisfies this assumption because the symbolic parameters corresponding to each of the symbolic nodes are associated with the same instantiation of the set of parents (see Table 10.1). For example, X_1 is the only parent of the symbolic node X_3, and both the symbolic parameters θ_{300} and θ_{310} are associated with the same instantiation of X_1, $X_1 = 0$.

In this situation, the canonical components resulting from considering extreme values for the symbolic parameters produce an identity canonical matrix $\mathbf{T}_{ij}$. The next theorem states this fact.

Theorem 10.2 *Given a set of symbolic nodes* $\{X_{i_1}, \ldots, X_{i_s}\}$ *with associated symbolic parameters* $\Theta = \{\Theta_1, \ldots, \Theta_s\}$, *where* $\Theta_k = \{\theta_{i_k j \pi_k}, j = 0, \ldots, r_{i_k}\}$, *and* π_k *is a given instantiation of* Π_{i_k}, *then the canonical matrix associated with the set of canonical components C defined by the Cartesian*

10.5 Symbolic Propagation Through Numeric Computations

product $C = C_1 \times \ldots \times C_n$, where

$$C_k = \begin{pmatrix} \{\theta_{i_k 0 \pi_k} = 1, \theta_{i_k 1 \pi_k} = 0, \ldots, \theta_{i_k r_{i_k} \pi_k} = 0\} \\ \vdots \\ \{\theta_{i_k 0 \pi_k} = 0, \theta_{i_k 1 \pi_k} = 0, \ldots, \theta_{i_k r_{i_k} \pi_k} = 1\}\end{pmatrix},$$

is the $m \times m$ identity matrix.

Proof: From (10.11), the set of monomials m_k is given by

$$M = \Theta_1 \times \ldots \times \Theta_s.$$

In this case, using the assumption $\Theta_k = \{\theta_{i_k j \pi_k}, j = 0, \ldots, r_{i_k}\}$, we have

$$M = \{\{\theta_{i_1 0 \pi_1}, \ldots, \theta_{i_1 r_{i_1} \pi_1}\} \times \ldots \times \{\theta_{i_s 0 \pi_s}, \ldots, \theta_{i_1 r_{i_s} \pi_s}\}\}.$$

Therefore, any instantiation of the symbolic parameters in C annihilates all the monomials in M but one. Let $\theta \in C$ be a typical canonical component. Then, all the parameters in Θ_k but one are zero, $\theta_{i_k j_k \pi_k}$, for $k = 1, \ldots, m$. Thus, substituting this numeric values in the monomial in M, the monomial $\theta_{i_1 j_1 \pi_1} \ldots \theta_{i_s j_s \pi_s}$ takes the value 1. The rest of monomials vanish. Therefore, every row in the matrix $\mathbf{T}_{ij}$ contains one element with the value 1 and all the others are 0. Finally, the process of the construction of M and C guarantees that the matrix $\mathbf{T}_{ij}$ is the identity matrix of order $m \times m$. ∎

It follows from Theorem 10.2 that the solution $\mathbf{c}_{ij}$ of the system of linear equations (10.15) becomes

$$\mathbf{c}_{ij} = \mathbf{p}_{ij}. \tag{10.17}$$

Therefore, in this situation, Algorithm 10.1 can be simplified as follows:

Algorithm 10.2 Modified Numeric Canonical Components.

- **Input:** A Bayesian network model (D, P), where P is defined using numeric and symbolic parameters satisfying the condition in Theorem 10.2, and evidence $E = e$.

- **Output:** The symbolic probabilities $p(X_i = j | E = e)$, $i = 1, \ldots, n$.

1. Construct m sets of instantiations of Θ in the canonical form indicated in Theorem 10.2, $C = \{\theta_1, \ldots, \theta_m\}$.

2. Compute the vector of probabilities $\mathbf{p}_{ij}$ in (10.14) using any standard numeric propagation method.

3. Let $\mathbf{c}_{ij} = \mathbf{p}_{ij}$, substitute $\mathbf{c}_{ij}$ in (10.12), and normalize to obtain the symbolic expression for the probabilities $p(X_i = j | E = e)$. ∎

460   10. Symbolic Propagation of Evidence

Note that with the assumption considered in Theorem 10.2, the canonical components allow us to perform symbolic propagation in a straightforward and efficient way. If the assumption given in Theorem 10.2 is not satisfied, then the resulting canonical matrix associated with the canonical set of parameters C may be different from the identity matrix, and it will be necessary to solve the system of equations (10.15) to obtain the polynomial coefficients.

We now illustrate this algorithm by an example.

Example 10.3 Numeric canonical components. Consider the network in Figure 10.1 and the evidence $e = \{X_2 = 1, X_5 = 1\}$. We wish to assess the influence of the symbolic parameters on the conditional probabilities of the remaining nodes. In this example the set of symbolic nodes is $\{X_3, X_6\}$ and the set of parameters is $\Theta = \{\Theta_3, \Theta_6\} = \{\{\theta_{300}, \theta_{310}\}, \{\theta_{600}, \theta_{610}\}\}$ (see Table 10.1). Then, the set of feasible monomials is given by

$$\begin{aligned}
M &= \Theta_3 \times \Theta_6 \\
&= \{\theta_{300}\theta_{600}, \theta_{300}\theta_{610}, \theta_{310}\theta_{600}, \theta_{310}\theta_{610}\} \\
&= \{m_1, m_2, m_3, m_4\}.
\end{aligned}$$

From (10.12), the unnormalized conditional probability $p(X_i = j|e)$ is a polynomial function of the form

$$p(X_i = j|e) \propto \sum_{k=1}^{4} c_k^{ij} \, m_k = p_{ij}(\Theta). \qquad (10.18)$$

Thus, our objective is to obtain the coefficients $\{c_k^{ij} \, ; \, k = 1, \ldots, 4\}$ for each node X_i and each possible value j. To this end, we consider the canonical components associated with the symbolic set of parameters Θ. In this case, given that we are dealing with binary variables and that we are in the conditions of Theorem 10.2, there are only two possible canonical combinations of the parameters in Θ_i, $\{1, 0\}$ and $\{0, 1\}$. Thus, we have the following set of canonical components:

$$\begin{aligned}
C &= \{\{1, 0\}, \{0, 1\}\} \times \{\{1, 0\}, \{0, 1\}\} \\
&= \{\{1, 0; 1, 0\}, \{1, 0; 0, 1\}, \{0, 1; 1, 0\}, \{0, 1; 0, 1\}\} \\
&= \{c_1, c_2, c_3, c_4\}.
\end{aligned}$$

Then, by instantiating the symbolic parameters to the corresponding values given in its canonical components, all the monomials appearing in (10.18) become either 0 or 1. Thus, we obtain an expression that depends only on the coefficients c^{ij}:

$$\begin{aligned}
p_{ij}(\Theta = c_1) &= c_1^{ij}, & p_{ij}(\Theta = c_2) &= c_2^{ij}, \\
p_{ij}(\Theta = c_3) &= c_3^{ij}, & p_{ij}(\Theta = c_4) &= c_4^{ij}.
\end{aligned}$$

10.5 Symbolic Propagation Through Numeric Computations

Thus, in this case, the matrix $\mathbf{T}_{ij}$ is the identity matrix because all the symbolic parameters of the symbolic nodes are associated with the same instantiation of the set of parents. Then, we have

$$\begin{pmatrix} c_1^{ij} \\ c_2^{ij} \\ c_3^{ij} \\ c_4^{ij} \end{pmatrix} = \begin{pmatrix} p_{ij}(c_1) \\ p_{ij}(c_2) \\ p_{ij}(c_3) \\ p_{ij}(c_4) \end{pmatrix}. \tag{10.19}$$

It is interesting to note here that the feasible set (the set generated by all feasible parameter values) for the probabilities of any set of nodes is the convex hull generated by the canonical probabilities.

In Figure 10.6 we show the unnormalized conditional probabilities $p_{ij}(c_k)$ of all nodes, given the evidence $e = \{X_2 = 1, X_5 = 1\}$, associated with the four possible canonical components. Using these values we can obtain all the rational functions in Table 10.3. For example, from Figure 10.6 we get the following values for the node X_6:

$$\begin{pmatrix} c_1^{60} \\ c_2^{60} \\ c_3^{60} \\ c_4^{60} \end{pmatrix} = \begin{pmatrix} 0.390 \\ 0.247 \\ 0.418 \\ 0.254 \end{pmatrix} \tag{10.20}$$

and

$$\begin{pmatrix} c_1^{61} \\ c_2^{61} \\ c_3^{61} \\ c_4^{61} \end{pmatrix} = \begin{pmatrix} 0.028 \\ 0.171 \\ 0.029 \\ 0.193 \end{pmatrix}, \tag{10.21}$$

that is, the coefficients of the numerator polynomials for $X_6 = 0$ and $X_6 = 1$, respectively. Then, substituting these values in (10.18) we obtain

$$p(X_6 = 0|e) \propto 0.390\,\theta_{300}\theta_{600} + 0.247\,\theta_{300}\theta_{610}$$
$$+ 0.418\,\theta_{310}\theta_{600} + 0.254\,\theta_{310}\theta_{610},$$

$$p(X_6 = 1|e) \propto 0.028\,\theta_{300}\theta_{600} + 0.171\,\theta_{300}\theta_{610}$$
$$+ 0.029\,\theta_{310}\theta_{600} + 0.193\,\theta_{310}\theta_{610}.$$

Adding both polynomials, we obtain the denominator normalizing polynomial, that is,

$$\begin{pmatrix} c_1^{60} \\ c_2^{60} \\ c_3^{60} \\ c_4^{60} \end{pmatrix} + \begin{pmatrix} c_1^{61} \\ c_2^{61} \\ c_3^{61} \\ c_4^{61} \end{pmatrix} = \begin{pmatrix} 0.418 \\ 0.418 \\ 0.447 \\ 0.447 \end{pmatrix}.$$

Thus, we have

$$p(X_6 = 0|e) = (0.390\theta_{300}\theta_{600} + 0.247\theta_{300}\theta_{610}$$
$$+ 0.418\theta_{310}\theta_{600} + 0.254\theta_{310}\theta_{610})/d,$$

$$p(X_6 = 1|e) = (0.028\theta_{300}\theta_{600} + 0.171\theta_{300}\theta_{610}$$
$$+ 0.029\theta_{310}\theta_{600} + 0.193\theta_{310}\theta_{610})/d,$$

where $d = 0.418\,\theta_{300}\theta_{600} + 0.418\,\theta_{300}\theta_{610} + 0.447\,\theta_{310}\theta_{600} + 0.447\,\theta_{310}\theta_{610}$.

Finally, eliminating the dependent parameters, θ_{310} and θ_{610}, we get the expression shown in Table 10.3. Note that the only symbolic operation in this process consists of simplifying the final expression to eliminate the dependent parameters. However, this is an optional operation, and in some cases, it is more convenient to keep the expression with all the parameters and simplify the numeric results after plugging in some specific values for the parameters. ∎

We should mention here that although we are using only exact propagation methods to compute $\mathbf{p}_{ij}$, the methodology remains valid using the approximate or simulation propagation methods described in Chapter 9. In this case, we shall obtain approximate symbolic solutions for the probabilities.

10.5.3 Efficient Computation of Canonical Components

The proposed symbolic propagation method requires several applications of a numeric, approximate or exact, propagation algorithm to calculate the numeric probabilities $\mathbf{p}_{ij}(c_k)$ associated with each one of the canonical components C_k. Therefore, the number of propagations increases combinatorially with the number of symbolic parameters. However, when propagating uncertainty in the canonical cases using some of the message-passing algorithms introduced in Chapter 8, we can save many calculations because some messages are common to several canonical components. In this section, we illustrate this fact in two of the main message-passing algorithms: the algorithm for polytrees (see Section 8.3), and the clustering algorithm (Section 8.6).

Figure 10.7 illustrates the message-passing process corresponding to a typical node X_i in a polytree when applying Algorithm 8.1. In this figure, Θ^k stands for the set of parameters contained in the connected component associated with node X_k when dropping the link $X_k - X_i$, while Θ_k is the set of parameters associated with node X_k. Note that the message from node X_k to node X_i depends only on those parameters, while the message passed from X_i to X_k depends on the remaining parameters in the Bayesian network. Note also that if Θ^k does not contain any symbolic parameter, then all the messages coming from this region of the graph to X_i need to be calculated only once, because they have the same value for every canonical component.

10.5 Symbolic Propagation Through Numeric Computations

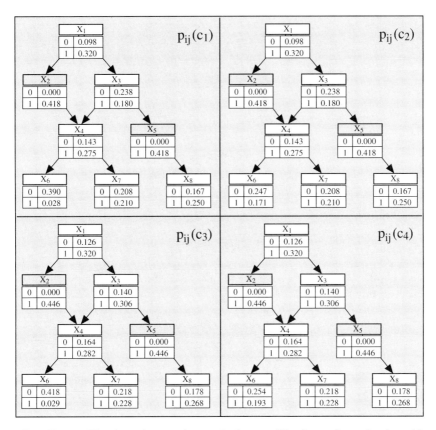

FIGURE 10.6. The four elemental canonical cases. The first column in the table for each node X_i is the state of X_i and the second column is the unnormalized marginal probabilities of X_i given the evidence $e = \{X_2 = 1, X_5 = 1\}$. Light rectangles indicate the evidence nodes.

The same computational savings are obtained when applying clustering message-passing algorithms. In this case, the situation is the same except that we deal with a tree of clusters (sets of nodes) instead of a tree of nodes. For example, consider the multiply-connected graph in Figure 10.1 with the numeric and symbolic probabilities in Table 10.1. Figure 10.8 shows all the messages needed to propagate evidence in a family tree associated with the Bayesian network of Figure 10.1, using the clustering method. In Figure 10.8, the cluster messages are indicated by arrows. We can distinguish between two types of messages:

1. Messages with no index. These messages are common messages for all canonical components. Hence, they need to be calculated only once.

2. Messages with one or more indices such as Θ_3, Θ_6, or $\Theta_3\Theta_6$. These messages depend on the parameters, and then we must calculate as

464   10. Symbolic Propagation of Evidence

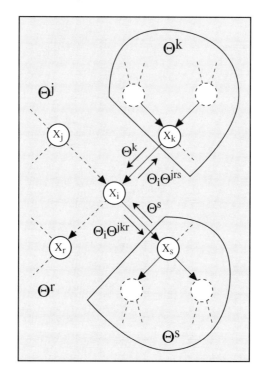

FIGURE 10.7. Parametric dependence of the messages in a general node in a polytree.

many different messages as the number of monomials associated with them.

Thus, in this example we can build the rational function associated with the node marginals by performing only half of the computations.

A further improvement of the canonical components method that reduces the number of feasible monomials by dealing only with the monomials formed by parameters consistent with the given evidence is suggested by Castillo, Gutiérrez, and Hadi (1996a). This method is illustrated in the next section using the case of goal-oriented propagation.

10.6 Goal-Oriented Symbolic Propagation

In Section 8.8 we introduced an algorithm to calculate the relevant set of nodes needed to perform a certain task. In the case of symbolic propagation, those reductions are of significant importance, since the complexity of symbolic propagation is given by the total number of symbolic parameters.

Suppose that we are interested in a given goal node X_i, and that we want to obtain the CDP $p(X_i = j | E = e)$, where E is a set of evidential nodes

10.6 Goal-Oriented Symbolic Propagation

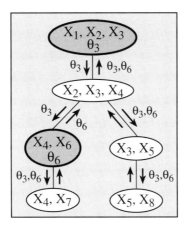

FIGURE 10.8. Join tree and messages affected by the relevant parameters. The clusters involving potential functions depending on parameters Θ_3 or Θ_6 are outlined.

with known values $E = e$. In this section we show the steps to be followed for goal-oriented symbolic propagation. The algorithm is organized in four main parts (see Castillo, Gutiérrez, and Hadi (1996a)):

- **Part I: Identify all relevant nodes.**
  As we have seen in Section 8.8, the CDP $p(X_i = j|E = e)$ does not necessarily involve parameters associated with all nodes. Thus, we can identify the set of nodes that are relevant to the calculation of $p(X_i = j|E = e)$ using Algorithm 8.6. Once the relevant set of nodes has been identified, the corresponding set of relevant parameters Θ is identified and the remaining nodes are removed from the graph. The calculations are performed on the reduced graph using the set of relevant parameters.

- **Part II: Identify sufficient parameters.**
  By considering the values of the evidence variables, the set of parameters Θ can be further reduced by identifying and eliminating the set of parameters that are in contradiction with the evidence. These parameters are eliminated using the following two rules:

    - **Rule 1:** Eliminate the parameters $\theta_{jk\pi}$ if $e_j \neq k$ for every $X_j \in E$.
    - **Rule 2:** Eliminate the parameters $\theta_{jk\pi}$ if parents' instantiations π are incompatible with the evidence.

- **Part III: Identify feasible monomials.**
  Once the minimal sufficient subsets of parameters have been identified, the parameters are combined in monomials by taking the Cartesian product of the minimal sufficient subsets of parameters and

466   10. Symbolic Propagation of Evidence

X_i	Π_i	Free Parameters
X_1	ϕ	$\theta_{10} = p(X_1 = 0)$
X_2	$\{X_1\}$	$\theta_{200} = p(X_2 = 0\|X_1 = 0) = 0.2$ $\theta_{201} = p(X_2 = 0\|X_1 = 1) = 0.5$
X_3	$\{X_1\}$	$\theta_{300} = p(X_3 = 0\|X_1 = 0)$ $\theta_{301} = p(X_3 = 0\|X_1 = 1)$
X_4	$\{X_2, X_3\}$	$\theta_{4000} = p(X_4 = 0\|X_2 = 0, X_3 = 0) = 0.1$ $\theta_{4001} = p(X_4 = 0\|X_2 = 0, X_3 = 1) = 0.2$ $\theta_{4010} = p(X_4 = 0\|X_2 = 1, X_3 = 0) = 0.3$ $\theta_{4011} = p(X_4 = 0\|X_2 = 1, X_3 = 1) = 0.4$
X_5	$\{X_3\}$	$\theta_{500} = p(X_5 = 0\|X_3 = 0)$ $\theta_{501} = p(X_5 = 0\|X_3 = 1)$
X_6	$\{X_4\}$	$\theta_{600} = p(X_6 = 0\|X_4 = 0)$ $\theta_{601} = p(X_6 = 0\|X_4 = 1)$
X_7	$\{X_4\}$	$\theta_{700} = p(X_7 = 0\|X_4 = 0)$ $\theta_{701} = p(X_7 = 0\|X_4 = 1)$
X_8	$\{X_5\}$	$\theta_{800} = p(X_8 = 0\|X_5 = 0)$ $\theta_{801} = p(X_8 = 0\|X_5 = 1)$

TABLE 10.4. A set of free parameters associated with the CPDs in (10.4).

eliminating the set of all infeasible combinations of the parameters using

– **Rule 3:** Parameters associated with contradictory conditioning instantiations cannot appear in the same monomial.

• **Part IV: Calculate the coefficients of all polynomials.**
This part calculates the coefficients by applying the numeric canonical components method described in Section 10.5.2.

To illustrate the algorithm we use the Bayesian network introduced in Example 10.1.

Example 10.4 Consider again the network in Figure 10.1. For illustrative purposes, Table 10.4 gives the numeric and symbolic values for the parameters associated with the CPDs in (10.4). Thus, the symbolic nodes are X_1, X_3, X_5, X_6, X_7, and X_8. Suppose that the target node is X_7 and that the evidence is $e = \{X_1 = 1\}$. Then, we wish to compute the conditional probabilities $p(x_7|X_1 = 1)$. We proceed as follows:

10.6 Goal-Oriented Symbolic Propagation

Part I:

- **Step 1:** Following the steps of Algorithm 8.6, we need to add to the initial graph in Figure 10.1 the dummy nodes $V_1, \ldots, V_8$. This gives the augmented graph in Figure 10.9.

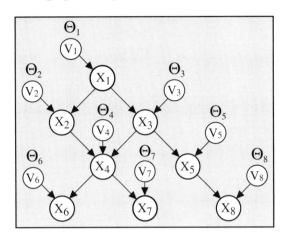

FIGURE 10.9. The augmented graph obtained by adding a dummy node V_i for every node X_i.

- **Step 2:** The set V of dummy nodes D-separated from X_7 (the goal node) by X_1 (the evidential node) is found to be $V = \{V_5, V_6, V_8\}$. The new graph obtained by removing the corresponding nodes X_5, X_6, and X_8 is shown in Figure 10.10. Thus, the set of all parameters associated with the symbolic dummy nodes that are not included in V is
$$\Theta = \{\{\theta_{300}, \theta_{301}, \theta_{310}, \theta_{311}\}; \{\theta_{700}, \theta_{701}, \theta_{710}, \theta_{711}\}\}.$$
This is the relevant set of parameters. Note that the parameters of node X_1 (θ_{10} and θ_{11}) are not included in Θ because X_1 is an evidential node. Note also that at this step we have reduced the number of symbolic parameters from 22 to 8 (or the number of free parameters from 11 to 4).

Part II:

- **Step 3:** The set Θ does not contain parameters associated with the evidential node X_1. Thus, no reduction is possible applying Rule 1.

- **Step 4:** Since θ_{300} and θ_{310} are inconsistent with the evidence (because they indicate that $X_1 = 0$), we can remove from Θ these parameters, obtaining the minimum set of sufficient parameters:
$$\Theta = \{\{\theta_{301}, \theta_{311}\}; \{\theta_{700}, \theta_{701}, \theta_{710}, \theta_{711}\}\}.$$

468   10. Symbolic Propagation of Evidence

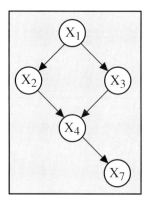

FIGURE 10.10. The graph containing the set of nodes with associated dummy nodes not D-separated from X_7 by X_1.

M_0	M_1
$\theta_{301}\theta_{700}$	$\theta_{301}\theta_{710}$
$\theta_{301}\theta_{701}$	$\theta_{301}\theta_{711}$
$\theta_{311}\theta_{700}$	$\theta_{311}\theta_{710}$
$\theta_{311}\theta_{701}$	$\theta_{311}\theta_{711}$

TABLE 10.5. Required monomials to determine the indicated probabilities.

Part III:

- **Step 5:** The initial set of candidate monomials is given by taking the Cartesian product of the minimal sufficient subsets, that is, $M = \{\theta_{301}, \theta_{311}\} \times \{\theta_{700}, \theta_{701}, \theta_{710}, \theta_{711}\}$. Thus, the candidate monomials are shown in Table 10.5.

- **Step 6:** The parents of nodes X_3 and X_7 do not have common elements; hence all monomials shown in Table 10.5 are feasible monomials.

Part IV:

- **Step 7:** The sets of monomials M_0 and M_1 needed to calculate $p(X_7 = 0|X_1 = 1)$ and $p(X_7 = 1|X_1 = 1)$, respectively, are shown in Table 10.5.

- **Step 8:** For $j = 0$ we have the following polynomial equation:

$$\begin{aligned} p_0(\Theta) &= c_{01}m_{01} + c_{02}m_{02} + c_{03}m_{03} + c_{04}m_{04} \\ &= c_{01}\theta_{301}\theta_{700} + c_{02}\theta_{301}\theta_{701} \\ &\quad + c_{03}\theta_{311}\theta_{700} + c_{04}\theta_{311}\theta_{701}. \end{aligned} \qquad (10.22)$$

10.6 Goal-Oriented Symbolic Propagation

Thus, taking the canonical components

$$\theta_1 = (1,0,1,0,1,0), \quad \theta_2 = (1,0,0,1,1,0),$$
$$\theta_3 = (0,1,1,0,1,0), \quad \theta_4 = (0,1,0,1,1,0)$$

for the set $\Theta = \{\theta_{301}, \theta_{311}, \theta_{700}, \theta_{701}, \theta_{710}, \theta_{711}\}$ of sufficient parameters, we get the following system of equations:

$$c_0 = \begin{pmatrix} 1 & 0 & 0 & 0 \\ 0 & 1 & 0 & 0 \\ 0 & 0 & 1 & 0 \\ 0 & 0 & 0 & 1 \end{pmatrix} \begin{pmatrix} p_0(\theta_1) \\ p_0(\theta_2) \\ p_0(\theta_3) \\ p_0(\theta_4) \end{pmatrix} = \begin{pmatrix} 0.15 \\ 0.85 \\ 0.35 \\ 0.65 \end{pmatrix}. \quad (10.23)$$

Similarly, for $j = 1$ we get

$$c_1 = \begin{pmatrix} p_1(\theta_1) \\ p_1(\theta_2) \\ p_1(\theta_3) \\ p_1(\theta_4) \end{pmatrix} = \begin{pmatrix} 0.15 \\ 0.85 \\ 0.35 \\ 0.65 \end{pmatrix}. \quad (10.24)$$

Table 10.6 shows the results of calculating the numeric probabilities needed in the above expressions. Combining (10.22) and (10.23) we get the final polynomial expressions

$$p(X_7 = 0 | X_1 = 1) \propto 0.15\theta_{301}\theta_{700} + 0.85\theta_{301}\theta_{701}$$
$$+ 0.35\theta_{311}\theta_{700} + 0.65\theta_{311}\theta_{701}. \quad (10.25)$$

Similarly, for $X_7 = 1$ we get

$$p(X_7 = 1 | X_1 = 1) \propto 0.15\theta_{301}\theta_{710} + 0.85\theta_{301}\theta_{711}$$
$$+ 0.35\theta_{311}\theta_{710} + 0.65\theta_{311}\theta_{711}. \quad (10.26)$$

- **Step 9:** Adding the unnormalized probabilities in (10.25) and (10.26) we get the normalizing constant. Because $\theta_{i0\pi} + \theta_{i1\pi} = 1$, for all i, the normalizing constant is found to be 1.

- **Step 10:** The expressions in (10.25) and (10.26) can be simplified by replacing $\theta_{311}, \theta_{710}$, and θ_{711} by $(1 - \theta_{301})$, $(1 - \theta_{700})$, and $(1 - \theta_{701})$, respectively. We obtain

$$p(X_7 = 0 | X_1 = 1) \propto 0.15\theta_{301}\theta_{700} + 0.85\theta_{301}\theta_{701}$$
$$+ (1 - \theta_{301})(0.35\theta_{700} + 0.65\theta_{701})$$
$$= 0.35\theta_{700} - 0.2\theta_{301}\theta_{700}$$
$$+ 0.65\theta_{701} + 0.2\theta_{301}\theta_{701}$$

$X_7 = 0$			
$(\theta_{301}, \theta_{311}, \theta_{700}, \theta_{701}, \theta_{710}, \theta_{711})$	$p_0(\theta)$	Monomials	Coefficients
(1,0,1,0,1,0)	0.15	$\theta_{301}\theta_{700}$	$c_{01} = 0.15$
(1,0,0,1,1,0)	0.85	$\theta_{301}\theta_{701}$	$c_{02} = 0.85$
(0,1,1,0,1,0)	0.35	$\theta_{311}\theta_{700}$	$c_{03} = 0.35$
(0,1,0,1,1,0)	0.65	$\theta_{311}\theta_{701}$	$c_{04} = 0.65$
$X_7 = 1$			
$(\theta_{301}, \theta_{311}, \theta_{700}, \theta_{701}, \theta_{710}, \theta_{711})$	$p_1(\theta)$	Monomials	Coefficients
(1,0,1,0,1,0)	0.15	$\theta_{301}\theta_{710}$	$c_{11} = 0.15$
(1,0,0,1,1,0)	0.85	$\theta_{301}\theta_{711}$	$c_{12} = 0.85$
(0,1,1,0,1,0)	0.35	$\theta_{301}\theta_{710}$	$c_{13} = 0.35$
(0,1,0,1,1,0)	0.65	$\theta_{311}\theta_{711}$	$c_{14} = 0.65$

TABLE 10.6. Monomial coefficients and their corresponding values of $p_j(\theta)$.

and

$$\begin{aligned}
p(X_7 = 1|X_1 = 1) &\propto 0.15\theta_{301}(1 - \theta_{700}) + 0.85\theta_{301}(1 - \theta_{701}) \\
&\quad + 0.35(1 - \theta_{301})(1 - \theta_{700}) \\
&\quad + 0.65(1 - \theta_{301})(1 - \theta_{701}) \\
&= 1 + 0.2\theta_{301}\theta_{700} - 0.35\theta_{700} \\
&\quad - 0.65\theta_{701} - 0.2\theta_{301}\theta_{701}.
\end{aligned}$$

∎

10.7 Symbolic Treatment of Random Evidence

In the previous sections, we assumed that the available evidence is deterministic, that is, the evidential set E is known to have values e. In some situations the available evidence may be stochastic, that is, it may involve some degree of uncertainty. For example, we may say that $E = e$ with probability $q(e)$, where $\sum_e q(e) = 1$. Thus, when $q(e) = 1$, we have a deterministic evidence. In this section we deal with stochastic evidence.

Suppose that we have a stochastic evidence $E = e$ with probability $q(e)$. We wish to compute the conditional probability $p(x_i|(e \text{ with } q(e)))$, for all nonevidential nodes X_i. This conditional probability is

$$p(x_i|(E = e \text{ with } q(e))) = \sum_e q(e)p(x_i|e). \tag{10.27}$$

10.7 Symbolic Treatment of Random Evidence

Thus, the conditional probability in (10.27) can be computed by applying the symbolic canonical components method once for each combination of values of the evidential variables.

It is important to note that $p(x_i|(e \text{ with } q(e)))$ is also a rational function of two polynomials because it is a linear convex combination of $p(x_i|e)$, which is a rational function (see Corollary 10.2). However, in this case the parameters can appear with exponents larger than one, which implies polynomials of degree larger than one. This result is stated in the following theorem (Castillo, Gutiérrez, and Hadi (1995c)).

Theorem 10.3 *The conditional probability of any nonevidential node given a stochastic evidence is a ratio of two polynomial functions, where the degree of the polynomials involved is at most equal to the sum of the cardinalities of the evidential nodes.*

Proof: The polynomials in the denominator of the rational function are in general different for different combinations of the evidence set e. Thus, the common denominator is the product of different rational functions. The number of these rational functions, and hence the degree of the polynomial, cannot exceed the sum of the cardinalities of the evidential nodes. ■

Example 10.5 Let us consider the Bayesian network given in Example 10.1 and suppose that we know a deterministic evidence $X_2 = 1$ and a stochastic evidence $X_5 = x_5$ with probability $q(x_5)$, where $q(0) = p$ and $q(1) = 1 - p$. With this information, we wish to compute $p(X_6 = 0|(X_2 = 1, X_5 = x_5 \text{ with } q(x_5)))$. In this case, we use the symbolic canonical components method twice (once with the evidence $(X_2 = 1, X_5 = 0)$ and once with the evidence $(X_2 = 1, X_5 = 1)$) and obtain

$$\begin{aligned} p(X_6 = 0|X_2 = 1, X_5 = 0) &= s_0, \\ p(X_6 = 0|X_2 = 1, X_5 = 1) &= s_1, \end{aligned} \quad (10.28)$$

where

$$s_0 = \frac{0.056 + 0.019\,\theta_{300} + 0.032\,\theta_{600} + 0.007\,\theta_{300}\,\theta_{600}}{0.094 + 0.028\,\theta_{300}},$$

$$s_1 = \frac{0.253 - 0.006\,\theta_{300} + 0.164\,\theta_{600} - 0.021\,\theta_{300}\,\theta_{600}}{0.446 - 0.028\,\theta_{300}}.$$

Substituting s_0, s_1, and $q(x_5)$ in (10.27), we obtain

$$p(x_i|(X_2 = 1, X_5 = 0 \text{ with } q(x_5))) = p s_0 + (1-p) s_1 = a/b,$$

where

$$\begin{aligned} a =\ & -30.334 - 1.522\,p - 8.316\,\theta_{300} - 0.492\,p\,\theta_{300} + 0.214\,\theta_{300}^2 \\ & + 0.464\,p\,\theta_{300}^2 - 19.663\,\theta_{600} + 1.459\,p\,\theta_{600} - 3.339\,\theta_{300}\,\theta_{600} \\ & + 0.5\,p\,\theta_{300}\,\theta_{600} + 0.75\,\theta_{300}^2\,\theta_{600} - 0.5\,p\,\theta_{300}^2\,\theta_{600}, \\ b =\ & -53.474 - 12.571\,\theta_{300} + \theta_{300}^2. \end{aligned}$$

Thus, the probability of the event $X_6 = 0$ in this case is a linear convex combination of s_0 and s_1 which is a rational function of two polynomials a and b. Note that the polynomials involved are second-degree in θ_{300}, as would be expected by Theorem 10.3. ∎

Note also that in cases where the linear convex combinations are complicated expressions, they may be kept in the expanded form, and the simplification process may be realized after considering specific numeric values.

10.8 Sensitivity Analysis

Sensitivity analysis means studying the effects of changing some of the parameter values on the conditional probabilities of the nonevidential nodes given the evidence. One way of performing sensitivity analysis is to change the parameter values and then monitor the effects of these changes on the conditional probabilities by redoing the computations using an appropriate method for evidence propagation. Clearly, this brute force method is computationally intensive.

Other ways of performing sensitivity analysis are found in the literature. For example, Breese and Fertig (1991) and Tessem (1992) propagate intervals of probabilities instead of single values, and Laskey (1995) measures the impact of small changes in one parameter on a target probability of interest using the partial derivative of $p(x_i|e)$ with respect to the parameters.

In this section we show that sensitivity analysis can also be performed by using symbolic propagation methods with little additional computational effort (see Castillo, Gutiérrez, and Hadi (1996d)). We show that the numeric canonical components method allows us to obtain lower and upper bounds for probabilities obtained from symbolic propagation. These bounds can provide valuable information for performing sensitivity analysis of a Bayesian network.

The symbolic expressions of conditional probabilities obtained by Algorithm 10.1 can be used to obtain lower and upper bounds for the marginal probabilities. To compute these bounds, we first need the following result, due to Martos (1964):

Theorem 10.4 Bounds for the probabilities. *If the linear fractional functional of a vector u,*

$$f(\mathbf{u}) = \frac{\mathbf{c}\,\mathbf{u} - c_0}{\mathbf{d}\,\mathbf{u} - d_0}, \qquad (10.29)$$

where $\mathbf{c}$ and $\mathbf{d}$ are vector coefficients and c_0 and d_0 are real constants, is defined in the convex polyhedral set $A\mathbf{u} \leq a_0, \mathbf{u} \geq 0$, where A is a constant matrix and a_0 is a constant vector and the denominator in (10.29) does

10.8 Sensitivity Analysis

Parameter	Lower	Upper
θ_{10}	0.7	0.9
θ_{300}	0.3	0.5
θ_{301}	0.1	0.4
θ_{500}	0.0	0.2
θ_{501}	0.8	1.0
θ_{600}	0.3	0.7
θ_{601}	0.4	0.5
θ_{700}	0.2	0.3
θ_{701}	0.5	0.7
θ_{800}	0.0	0.3
θ_{801}	0.4	0.6

TABLE 10.7. Lower and upper bounds for the parameters provided by the human expert.

not vanish in the polyhedral set, then the maximum of $f(\mathbf{u})$ occurs at one of the vertices of the polyhedron.

In our case, $\mathbf{u}$ is the set of symbolic parameters and $f(\mathbf{u})$ is the set of symbolic expressions associated with the probabilities $p(x_i|e)$. In this case, the convex polyhedral set is defined by $\mathbf{u} \leq 1, \mathbf{u} \geq 0$, that is, A is the identity matrix. Then, using Theorem 10.4, the lower and upper bounds of the symbolic expressions associated with the probabilities are attained at the vertices of this polyhedron, that is, at some of the canonical components associated with the symbolic set of parameters.

Thus, the lower and upper bounds for the ratio of polynomial probabilities $p(x_i|e)$ are given by the minimum and maximum, respectively, of the numeric values attained by this probability over all the possible extreme canonical components associated with the parameters contained in Θ, i.e., for all possible combinations of extreme values of the parameters (the vertices of the set of parameters).

Example 10.6 Probability bounds. Let us compute the lower and upper bounds associated with all the variables in the Bayesian network in Example 10.4, first for the case of no evidence and then for the case of evidence $X_1 = 0$. We assume that the human expert gives, as initial information, lower and upper bounds for all the parameters, as shown in Table 10.7.

For comparison purpose, we compute the probabilities for two models:

- The model specified by the eleven free symbolic parameters in Table 10.4.

| | Case 1: Eleven Parameters ||| Case 2: Seven Parameters |||
Node	Lower	Upper	Range	Lower	Upper	Range
X_1	0.7000	0.9000	0.2000	0.7000	0.9000	0.2000
X_2	0.2300	0.2900	0.0600	0.2300	0.2900	0.0600
X_3	0.2400	0.4900	0.2500	0.3400	0.3800	0.0400
X_4	0.2770	0.3230	0.0460	0.3010	0.3030	0.0020
X_5	0.4080	0.8080	0.4000	0.4960	0.7280	0.2320
X_6	0.3677	0.5646	0.1969	0.4301	0.4303	0.0002
X_7	0.4031	0.5892	0.1861	0.4091	0.5796	0.1705
X_8	0.0768	0.4776	0.4008	0.1088	0.4512	0.3424

TABLE 10.8. Lower and upper bounds for the initial marginal probabilities $p(X_i = 0)$ (the case of no evidence).

- The reduced model obtained from the above model by replacing the parameters of variables X_3 and X_6 by fixed numeric values, that is,

$$\theta_{300} = 0.4, \quad \theta_{301} = 0.2, \quad \theta_{600} = 0.5, \quad \theta_{601} = 0.4.$$

This model has seven free symbolic parameters.

For each of the above cases, we compute the bounds for the no-evidence case and for the evidence $X_1 = 0$. Tables 10.8 and 10.9 show the lower and upper bounds for the four different cases. We can make the following remarks:

1. By comparing the seven- and eleven-parameter models, we see that the range (the difference between upper and lower bounds) is non-decreasing in the number of symbolic parameters. For example, the ranges for the seven-parameter case are no larger than those for the eleven-parameter case (actually with the exception of X_1 and X_2, the ranges in the seven-parameter case are smaller than the corresponding ranges in the eleven-parameter case). This is expected, because fewer symbolic parameters means less uncertainty.

2. By comparing the no-evidence case (Table 10.8) with the evidence case (Table 10.9), we see that with the bounds in the former case are generally larger than those in the latter case. Again, this is expected because more evidence means less uncertainty. ∎

10.9 Symbolic Propagation in Gaussian Bayesian Networks

In Section 8.9 we presented several methods for exact propagation in Gaussian Bayesian networks. Some of these methods have been extended for

10.9 Symbolic Propagation in Gaussian Bayesian Networks

Node	Case 1: Eleven Parameters			Case 2: Seven Parameters		
	Lower	Upper	Range	Lower	Upper	Range
X_1	1.0000	1.0000	0.0000	1.0000	1.0000	0.0000
X_2	0.2000	0.2000	0.0000	0.2000	0.2000	0.0000
X_3	0.3000	0.5000	0.2000	0.4000	0.4000	0.0000
X_4	0.2800	0.3200	0.0400	0.3000	0.3000	0.0000
X_5	0.4000	0.7600	0.3600	0.4800	0.6800	0.2000
X_6	0.3680	0.5640	0.1960	0.4300	0.4300	0.0000
X_7	0.4040	0.5880	0.1840	0.4100	0.5800	0.1700
X_8	0.0960	0.4800	0.3840	0.1280	0.4560	0.3280

TABLE 10.9. Lower and upper bounds for the conditional probabilities $p(X_i = 0|X_1 = 0)$.

symbolic computation (see, for example, Chang and Fung (1991) and Lauritzen (1992)). In this section we illustrate symbolic propagation in Gaussian Bayesian networks using the conceptually simple method given in Section 8.9. When dealing with symbolic computations, all the required operations must be performed by a program with symbolic manipulation capabilities. Figure 10.11 shows the *Mathematica* code for the symbolic implementation of the method given in Section 8.9. The code calculates the mean and variance of all nodes given the evidence in the evidence list.

Example 10.7 Consider the set of variables $X = \{X_1, \ldots, X_5\}$ with mean vector and covariance matrix

$$\mu = \begin{pmatrix} m \\ 0 \\ 0 \\ 1 \\ 0 \end{pmatrix} \text{ and } \Sigma = \begin{pmatrix} a & 1 & 1 & 2 & 1 \\ 1 & 2 & 1 & -1 & 1 \\ 1 & 1 & b & -1 & c \\ 2 & -1 & -1 & 4 & -2 \\ 1 & 1 & c & -2 & 6 \end{pmatrix}. \quad (10.30)$$

Note that the mean of X_1, the variances of X_1 and X_3, and the covariance of X_3 and X_5 are specified in symbolic form.

We use the *Mathematica* code in Figure 10.11 to calculate the conditional mean and variance of all the nodes. The first part of the code defines the mean vector and covariance matrix of the Bayesian network. Table 10.10 shows the initial marginal probabilities of the nodes (no evidence) and the conditional probabilities of the nodes given each of the evidences $\{X_3 = x_3\}$ and $\{X_3 = x_3, X_5 = x_5\}$. An examination of the results in Table 10.10 shows that the conditional means and variances are rational expressions, that is, ratios of polynomials in the parameters. Note, for example, that for the case of evidence $\{X_3 = x_3, X_5 = x_5\}$, the polynomials are first-degree in m, a, b, x_3, and x_5, that is, in the mean and variance parameters and in the evidence variables, and second-degree in c, the covariance parameters.

476   10. Symbolic Propagation of Evidence

```
(* Mean vector and covariance matrix *)
mean={m,0,0,1,0};
var={{a,1,1,2,1},{1,2,1,-1,1},{1,1,b,-1,c},
     {2,-1,-1,4,-2},{1,1,c,-2,6}};
(* Evidence order *)
evidence={3,5}

(* Marginal Probabilities *)
For[k=0,k<=Length[evidence],k++,
  For[i=1,i<=Length[mean],i++,
    If[MemberQ[Take[evidence,k],i],
      cmean=x[i];
      cvar=0,
      meany=mean[[i]];
      meanz=Table[{mean[[evidence[[j]]]]},{j,1,k}];
      vary=var[[i,i]];
      If[k==0,
        cmean=Together[meany];
        cvar=Together[vary],
        varz=Table[Table[
               var[[evidence[[t]],evidence[[j]]]],
             {t,1,k}],{j,1,k}];
        covaryz=Table[{var[[evidence[[t]]]][[i]]},{t,1,k}];
        zaux=Table[{x[evidence[[t]]]},{t,1,k}];
        aux=Inverse[varz];
        cmean=meany+Transpose[covaryz].aux.(zaux-meanz);
        cvar=vary-Transpose[covaryz].aux.covaryz;
      ]
    ];
    Print["Evidential nodes ",k," Node ",i];
    Print["Mean ",Together[cmean],"Var ",Together[cvar]];
  ]
]
```

FIGURE 10.11. A *Mathematica* program for symbolic propagation of evidence in a Gaussian Bayesian network.

Note also that the common denominator for the rational functions gives the conditional mean and the conditional variance. ∎

The fact that the mean and variances of the conditional probability distributions of the nodes are rational functions of polynomials is given by the following theorem (see Castillo, Gutiérrez, Hadi, and Solares (1996)).

Theorem 10.5 *Consider a Gaussian Bayesian network over a set of variables $X = \{X_1, \ldots, X_n\}$ with mean vector μ and covariance matrix Σ.*

10.9 Symbolic Propagation in Gaussian Bayesian Networks   477

	No Evidence	
X_i	Mean	Variance
X_1	m	a
X_2	0	2
X_3	0	b
X_4	1	4
X_5	0	6
	Evidence $X_3 = x_3$	
X_i	Mean	Variance
X_1	$(bm + x_3)/b$	$(ab - 1)/b$
X_2	$(x_3)/b$	$(2b - 1)/b$
X_3	x_3	0
X_4	$(b - x_3)/b$	$(4b - 1)/b$
X_5	$(cx_3)/b$	$(6b - c^2)/b$
	Evidence $X_3 = x_3$ and $X_5 = x_5$	
X_i	Mean	Variance
X_1	$\dfrac{6bm - c^2 m + (6-c)x_3 + (b-c)x_5}{6b - c^2}$	$\dfrac{6ab + 2c - ac^2 - b - 6}{6b - c^2}$
X_2	$\dfrac{(6-c)x_3 + (b-c)x_5}{6b - c^2}$	$\dfrac{11b + 2c - 2c^2 - 6}{6b - c^2}$
X_3	x_3	0
X_4	$\dfrac{6b - c^2 + (2c - 6)x_3 + (c - 2b)x_5}{6b - c^2}$	$\dfrac{20b + 4c - 4c^2 - 6}{6b - c^2}$
X_5	x_5	0

TABLE 10.10. Means and variances of the marginal probability distributions of nodes, initially and after evidence.

Partition X, μ, and Σ as $X = \{Y, Z\}$,

$$\mu = \begin{pmatrix} \mu_Y \\ \mu_Z \end{pmatrix}, \quad \text{and} \quad \Sigma = \begin{pmatrix} \Sigma_{YY} & \Sigma_{YZ} \\ \Sigma_{ZY} & \Sigma_{ZZ} \end{pmatrix},$$

where μ_Y and Σ_{YY} are the mean vector and covariance matrix of Y, μ_Z and Σ_{ZZ} are the mean vector and covariance matrix of Z, and Σ_{YZ} is the covariance of Y and Z. Suppose that Z is the set of evidential nodes. Then the conditional probability distribution of any variable $X_i \in Y$ given Z is normal, with mean and variance that are ratios of polynomial functions in the evidential variables and the related parameters in μ and Σ. The polynomials involved are at most of degree one in the conditioning variables and in the mean and variance parameters and are of degree two in the covariance parameters. Finally, the polynomial in the denominator is the same for all nodes.

Proof: From Theorem 8.1,

$$\mu_{Y|Z=z} = \mu_Y + \Sigma_{YZ}\Sigma_{ZZ}^{-1}(z - \mu_Z). \tag{10.31}$$

Note that $\Sigma_{YZ}\Sigma_{ZZ}^{-1}(z - \mu_Z)$ is a rational function because it can be written as the quotient of the polynomials $\Sigma_{YZ}adj(\Sigma_{ZZ})(z - \mu_Z)$ and $det(\Sigma_{ZZ})$, where $adj(\Sigma_{ZZ})$ is the adjoint matrix of Σ_{ZZ} and $det(\Sigma_{ZZ})$ is the determinant of Σ_{ZZ}. Therefore, the conditional expectation $\mu_{Y|Z=z}$ in (10.31) is μ_Y plus a rational function, which implies that $\mu_{Y|Z=z}$ is a rational function with polynomial denominator $det(\Sigma_{ZZ})$. Note also that each parameter appears in only one of the three factors above, which implies linearity in each parameter.

Similarly, from Theorem 8.1 the conditional variance is

$$\Sigma_{Y|Z=z} = \Sigma_{YY} - \Sigma_{YZ}\Sigma_{ZZ}^{-1}\Sigma_{ZY}, \tag{10.32}$$

which is Σ_{YY} minus the rational function $\Sigma_{YZ}\Sigma_{ZZ}^{-1}\Sigma_{ZY}$. This implies that $\Sigma_{Y|Z=z}$ is a rational function with polynomial denominator $|\Sigma_{ZZ}|$. Note also that all parameters except those in Σ_{YZ} appear in only one of the factors, which implies linearity in these parameters. On the contrary, the parameters in Σ_{YZ} appear in two factors, and hence they can generate second-degree terms in the polynomials.

Finally, the denominator polynomial can be of second degree in the covariance parameters because of the symmetry of the covariance matrix. ■

Note that because the denominator polynomial is identical for all nodes, for implementation purposes it is more convenient to calculate and store all the numerator polynomials for each node and calculate and store the common denominator polynomial separately.

Exercises

10.1 Consider the directed graph in Figure 10.12 and suppose that all three variables are given in symbolic form. Write the free parameters using the notation in (10.2) in each of the following cases:

(a) All three variables are binary.

(b) X_1 and X_2 are binary but X_3 is ternary.

(c) X_1 and X_3 are ternary but X_2 is binary.

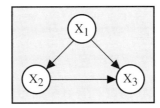

FIGURE 10.12. A directed graph with three variables.

10.2 Consider the directed graph in Figure 9.5 and the corresponding CPDs in Table 9.1. Write the free parameters using the notation in (10.2) in each of the following cases:

(a) The CPDs for X_3 and X_4 are replaced by symbolic parameters.

(b) The CPDs for X_1 and X_5 are replaced by symbolic parameters.

10.3 Consider the Bayesian network given in Figure 10.13 and the corresponding numerical and symbolical CPDs given in table 10.11. Write a *Mathematica* program (using Figures 10.3 and 10.4 as a reference) for the symbolic propagation of uncertainty in each of the following cases:

(a) No evidence is given.

(b) Evidence $\{X_2 = 1, X_5 = 1\}$ is given.

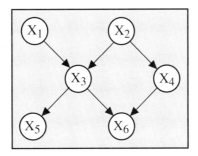

FIGURE 10.13. A multiply-connected DAG.

10.4 Consider the Bayesian network given in Figure 10.13 and Table 10.11. Write the general form of the polynomial structure of the initial marginal probabilities of the nodes (no evidence), as in (10.8).

$p(X_1 = 0) = 0.3$	$p(X_2 = 0) = \theta_{20}$
$P(X_3 = 0\|X_1 = 0, X_2 = 0) = 0.1$	$P(X_6 = 0\|X_3 = 0, X_4 = 0) = 0.1$
$P(X_3 = 0\|X_1 = 0, X_2 = 1) = 0.3$	$P(X_6 = 0\|X_3 = 0, X_4 = 1) = 0.4$
$P(X_3 = 0\|X_1 = 1, X_2 = 0) = 0.4$	$P(X_6 = 0\|X_3 = 1, X_4 = 0) = 0.3$
$P(X_3 = 0\|X_1 = 1, X_2 = 1) = 0.5$	$P(X_6 = 0\|X_3 = 1, X_4 = 1) = 0.2$
$p(X_4 = 0\|X_2 = 0) = 0.3$	$p(X_5 = 0\|X_3 = 0) = \theta_{500}$
$p(X_4 = 0\|X_2 = 1) = \theta_{401}$	$p(X_5 = 0\|X_3 = 1) = 0.8$

TABLE 10.11. Set of parameters for the Bayesian network given in Figure 10.13

10.5 Consider the Bayesian network given in Figure 10.13 and Table 10.11. Use the numeric canonical components method to compute $p(X_6 = 0|E = e)$ in each of the following cases:

(a) No evidence is available ($E = \phi$).

(b) The evidence is $X_2 = 0$.

10.6 Consider the Bayesian network given in Example 10.1. Use the numeric canonical components method to compute the initial marginal probability of X_6 in each of the following cases:

(a) The CPDs for X_3 and X_4 are replaced by symbolic parameters.

(b) The CPDs for X_1 and X_2 are replaced by symbolic parameters.

10.7 Consider the situation given in Example 10.6. Compute the upper and lower bounds for the probabilities for the following reduced models:

(a) The model obtained from the seven-parameter model by replacing the parameters of X_8 by $\theta_{300} = 0.2$ and $\theta_{801} = 0.4$.

(b) The model obtained from the above model by replacing the parameters of variable X_7 by $\theta_{700} = 0.1$ and $\theta_{701} = 0.6$.

10.8 Repeat the calculations in Example 10.7 for each of the following cases:

(a) The evidential nodes are $X_1 = 0$ and $X_2 = 0$.

(b) The evidential nodes are $X_1 = 0$ and $X_5 = 0$.

10.9 Given the Bayesian network in Figure 9.5, suppose that nodes X_4 and X_8 are symbolic nodes. Identify the messages that are affected by both parameters, those affected by only one of them, and those affected by none of them.

Chapter 11
Learning Bayesian Networks

11.1 Introduction

In the previous chapters we assumed that both the dependency structure of the model and the associated conditional probability distributions (CPDs) are provided by the human experts. In many practical situations, this information may not be available. In addition, different experts can give different and sometimes conflicting assessments due to the subjective nature of the process. In these situations, the dependency structure and the associated CPDs can be estimated from the data. This is referred to as *learning*.

An interesting book edited by Fisher and Lenz (1996) has come to our attention just before our book is sent to press. The articles contained in this book deal primarily with the learning problem, but they cover other areas related to the interface of artificial intelligence and statistics.

Learning can be performed for both Markov and Bayesian probabilistic network models. Because of space limitation, we concentrate in this chapter on learning Bayesian network models. For learning Markov network models, the reader is referred to Dawid and Lauritzen (1993), Madigan and Raftery (1994), and the references therein.

As we have seen in Chapter 6, a Bayesian network model

$$B(\theta) = (D, P(\theta))$$

over a set of n variables $X = \{X_1, \ldots, X_n\}$ consists of a directed acyclic graph (DAG) D and a set $P(\theta)$ of n CPDs each of the form

$$p(x_i|\pi_i; \theta_i); \quad i = 1, \ldots, n,$$

where π_i is a realization of the set of parents of node X_i, θ_i is a set of unknown parameters of the CPD associated with node X_i, and $p(x_i|\pi_i; \theta_i)$ is a known parametric family that depends on θ_i. Thus, $\theta = \{\theta_1, \ldots, \theta_n\}$ is the set of parameters for the entire joint probability distribution (JPD) over X, and $\theta \in \Theta = \{\Theta_1, \ldots, \Theta_n\}$. When explicit reference to θ is not necessary, we shall write $B = (D, P)$ instead of $B(\theta) = (D, P(\theta))$. The JPD over X is defined as a product of the n CPDs in $P(\theta)$. This factorization of the JPD is implied by D (see Section 6.4.4).

When D and/or $P(\theta)$ are unknown, they can be estimated or learned using prior information and/or a data set S. The prior information reflects the human expert's opinion as to which graphical and/or probabilistic structure is more likely than others to represent the relationships among the variables. This information is sometimes given in the form of prior probability distributions for the graphical structure and/or the parameters (the probabilistic structure).

The data set S consists of N observations each of which consists of n values, one value for each of the variables $\{X_1, \ldots, X_n\}$. The data set S may contain missing values. When S contains no missing values, it is referred to as *complete data*; otherwise the data set is referred to as *incomplete data*.

When dealing with the problem of learning Bayesian networks, it is important to note that different directed graphs can represent the same independence structures and/or the same JPD for the set of associated variables. In other words, when solving the learning problem over the set of all Bayesian networks, one may obtain many, apparently different, solutions, but some of these solutions, in fact, represent the same independence structures and/or the same JPD. From the learning point of view, directed graphs with the same independence structures and/or the same JPD are equivalent. Consequently, if the links have no causal interpretations, considerable computational savings can be obtained by considering only one graphical structure from each set of equivalent graphical structures. It may be helpful to the reader at this point to review the material in Section 6.5 on equivalent Bayesian networks.

Thus, using the notion of independence, or distribution equivalence, one can divide the set of all possible directed graphs over a set of n variables or nodes into a number of equivalence classes as discussed in Section 6.5.

The practical importance of the concept of independence equivalence is that we need select only a single directed graph from a set of equivalent ones, because they lead to the same solution. Thus, in this chapter we take D to be the class of all directed graphs leading to the same dependence structure, that is, the class of all independence equivalent DAGs. This class can, of course, be represented by any one of them. We have selected this definition of D in order to avoid problems with quality measures. In the following, however, D can be any of these graphs, the results being independent of the one being selected.

We shall differentiate in this chapter between two different conceptual types of learning:

1. **Structural learning:** This refers to learning the graphical (dependency) structure of D, that is, determining which links to be included in D.

2. **Parametric learning:** This refers to learning the parametric (probability) structure of P. In the language of statisticians, parametric learning is called *parameter estimation*.

A learning method consists of two elements:

1. A *quality measure*, which is used for deciding which one of a set of candidate Bayesian networks is the best. This is a global measure of quality, since it measures the quality of the Bayesian network as a whole, that is, both the quality of the graphical structure and the quality of the estimated parameters. Several quality measures are presented and discussed in Sections 11.2–11.8.

2. A *search algorithm*, which is used to select a small subset of high quality Bayesian networks, from which the best is selected. Note that the number of all possible networks, even for a small number of variables, is extremely large, and trying all possible networks is virtually impossible. Search methods are presented in Section 11.9.

Thus, learning both the graphical and the probabilistic structures of Bayesian networks involves the following steps:

1. Choose a quality measure and a search algorithm.

2. Use the search algorithm to select a subset of Bayesian networks with high qualities. This requires estimating the parameters of the selected networks using an estimation method, and evaluating the qualities of all Bayesian networks in the chosen subset.

3. Select the network structure with the highest quality from the above subset.

These steps are discussed in this chapter. Sections 11.2–11.8 introduce and discuss some quality measures. This material requires some results from Bayesian statistics. These results are included in an appendix to this chapter. Two search algorithms are given in Section 11.9. Finally, the problem of missing observations is treated in Section 11.10.

11.2 Measuring the Quality of a Bayesian Network Model

A quality measure, $Q(B|S,\xi)$, is a criterion by which one can order a set of possible Bayesian networks, where B is a Bayesian network, ξ the prior information, and S a set of data. Thus, given the prior information ξ and/or a set of data S, our objective is to find a Bayesian network model of high quality. A good quality measure should satisfy some desirable properties. For example, networks leading to the same independency structure should be assigned the same quality value. This important property is defined below.

Definition 11.1 Score equivalence. *Given a data set S, a quality measure $Q(B|S,\xi)$ is said to be score equivalent if it assigns the same value to every pair of independence equivalent Bayesian networks B_1 and B_2, that is, if $Q(B_1|S,\xi) = Q(B_2|S,\xi)$.*

Other desirable properties of quality measures include

- Networks recommended by experts should be assigned quality values higher than those rejected by experts.

- Perfect maps should receive quality values higher than networks which are nonperfect maps.

- Minimal I-maps should receive quality values higher than networks that are not minimal I-maps.

- A network with a small number of parameters should be assigned a quality value higher than a network with a large number of parameters.

- Networks in agreement with the data should be assigned quality values higher than those contradicting the data.

For some theorems related to these desirable properties see Bouckaert (1995).

Quality measures depend on the uncertainty of the available information. Two possible situations are

1. A situation in which the graphical and probabilistic structures both involve uncertainty. In this case, we are given prior information ξ and a data set S, and our objective is to find the best Bayesian network $B(\theta) = (D, P(\theta))$ using some quality criterion. Note that ξ contains prior information regarding both the graphical and parametric structures. Given ξ and S, the quality of a Bayesian network model $B(\theta)$ depends on the quality of its subcomponents, D and $P(\theta)$. We use

$$Q(B(\theta)|S,\xi) \text{ or } Q(D, P(\theta)|S,\xi) \qquad (11.1)$$

11.2 Measuring the Quality of a Bayesian Network Model

to denote the quality measure of the entire Bayesian network model and to emphasize that the measure depends on both S and ξ. In some cases, however, we may be interested only in structural learning. In these cases, a measure for the quality of the graphical structure can be obtained by maximizing the quality of its associated Bayesian network $Q(B(\theta)|S,\xi)$ with respect to θ, that is,

$$Q(D|S,\xi) = Q(D, P(\hat{\theta})|S,\xi), \qquad (11.2)$$

where $\hat{\theta}$ is the value of θ that maximizes $Q(D, P(\theta)|S,\xi)$. Alternatively, one can use any other appropriate estimate of θ such as the *maximum likelihood estimate*, a *Bayesian estimate*, etc.

2. A situation in which the graphical structure D is known and only the probabilistic structure involves uncertainty. In this case, we are interested only in parametric learning, and our objective is to find the best probabilistic structure $P(\theta)$ using some quality criterion. Given S, D, and ξ, the quality of $P(\theta)$ depends on the quality of the estimated parameters. We use

$$Q(P(\theta)|D, S, \xi) \qquad (11.3)$$

to denote the quality measure of the probabilistic structure of a Bayesian network model and to emphasize that it is conditional on D, S, and ξ. Note that ξ here contains prior information about the parametric structure only because the graphical structure is known with certainty.

Some quality measures are defined as the sum of three terms, or components:

$$Q = f(\text{prior information}) + g(\text{available data}) + h(\text{network complexity}),$$

where $f(.), g(.)$, and $h(.)$ are functions. These terms are explained below:

1. The *prior information:* The function $f(\text{prior information})$ assigns a high quality to networks that have been indicated as likely by the prior information, and a low quality to those which have been indicated as unlikely. The larger the contribution of this term to the quality measure, the larger is the weight of the prior knowledge, relative to data knowledge. This term has an important contribution to the quality when the data are lacking, but it is negligible when the data are abundant. A common choice for this term is $\log p(B)$, where $p(B) = p(D, \theta)$ is the prior probability assigned to the network model B, where θ is used instead of P to show explicitly the dependence of P on the parameter θ. If no prior knowledge is available, this term can be replaced by zero, which is equivalent to assuming that $p(B)$ is a uniform distribution.

2. The *available data*: The function g(available data) is a goodness-of-fit term that measures how well a given network fits the data S. It gives a high quality to networks that are in agreement with the data and a low quality to those that contradict the data. This term increases when links are added to the network structure. In that case we have more parameters or degrees of freedom, and we can get a better fit to the data. Usual choices for this term are

    (a) The log likelihood of the data: $\log p(S|D, \theta)$.
    (b) The log of the posterior probability of θ given the structure D and the data S: $\log p(\theta|S, D)$.

3. The *complexity of the model*: The function h(network complexity) penalizes networks with complex structure (e.g., networks with large numbers of links and/or large numbers of parameters). Thus, the function $h(.)$ leads to a high quality for simple networks with a small number of links and a small number of parameters, and a low quality for networks with a large number of links and/or a large number of parameters.

To measure the complexity of a Bayesian network it is important to know its dimension.

Definition 11.2 Dimension of a Bayesian network. *Let X be a set of variables and $B = (D, P)$ be a Bayesian network defined over X. The dimension of this Bayesian network, $Dim(B)$, is the number of free parameters required to completely specify the JPD of X.*

Chickering (1995a) shows that independence equivalent Bayesian networks have the same dimension.

Several quality measures have been proposed in the literature. These measures can be categorized into the following classes:

- Bayesian quality measures.
- Minimum description length measures.
- Information measures.

These classes of quality measures are discussed in the following sections.

11.3 Bayesian Quality Measures

Bayesian quality measures rely on Bayesian statistics philosophy (Bayes' theorem and conjugacy in particular). For readers who are not familiar with Bayesian statistical theory, the appendix to this chapter gives an introduction to the necessary concepts (e.g., prior distributions, posterior distributions, conjugate distributions, etc.) of Bayesian statistics. In this appendix we also include several interesting conjugate families of probability distributions. For a more detailed presentation of Bayesian statistics the reader is referred to any of the standard books, e.g., DeGroot (1970), Press (1992), or Bernardo and Smith (1994).

In Bayesian statistical theory, we initially assume that a prior probability distribution $p(B) = p(D,\theta)$ is given by experts. This probability distribution reflects the experts' opinion about the relative frequency of occurrence of the different Bayesian networks $B = (D, P(\theta))$. To improve our knowledge, a data set S is obtained, and using Bayes' theorem, the posterior probability distribution $p(B,\theta|S)$ is calculated as follows:

$$\begin{aligned} p(D,\theta|S) &= \frac{p(D,\theta,S)}{p(S)} = \frac{p(D,\theta,S)}{\sum_{D,\theta} p(D,\theta,S)} \\ &= \frac{p(D,\theta)p(S|D,\theta)}{\sum_{D,\theta} p(D,\theta)p(S|D,\theta)} \\ &= \frac{p(S,D,\theta)}{\sum_{D,\theta} p(D,\theta)p(S|D,\theta)}. \end{aligned} \quad (11.4)$$

The basic idea of Bayesian quality measures is to assign to every network a quality value that is a function of the posterior probability distribution (see Heckerman (1995)). For example, a quality measure can be proportional to the posterior probability distribution $p(B,\theta|S)$.

Note that the denominator in (11.4) is a normalizing constant, so we can write

$$p(D,\theta|S) \propto p(S,D,\theta) = p(D)p(\theta|D)p(S|D,\theta), \quad (11.5)$$

where the symbol $\propto$ means "proportional to". The equality in (11.5) is obtained using the chain rule (see Section 5.5). Using the chain rule, we can also write

$$p(S,D,\theta) = p(D)p(\theta|S,D)p(S|D) \propto p(D)p(\theta|S,D). \quad (11.6)$$

This indicates that we can use either $p(S,D,\theta)$ or $p(D)p(\theta|S,D)$ instead of $p(D,\theta|S)$ to measure the quality of the network. The meanings of the factors in (11.5) and (11.6) are as follows:

- The factor $p(D)$ is the prior probability distribution of the graphical structure that is given by the human experts.

- The factor $p(\theta|D)$ is the prior probability distribution of the parameters of the network given its graphical structure D. This prior probability distribution is also given by the human experts.

- The factor $p(S|D,\theta)$ is the likelihood of the data given the Bayesian network model $B = (D, P(\theta))$. This factor is computed using $P(\theta)$.

- The factor $p(\theta|S, D)$ is the posterior probability distribution of the parameters of the network given its graphical structure D and the data S. This factor is computed based on the data and the prior probability distribution.

Thus, to compute Bayesian quality measures we need to perform the following steps:

1. Assess a prior probability distribution $p(D)$ for the graphical structure.

2. Assess a prior probability distribution $p(\theta|D)$ for the parameters.

3. Calculate the posterior probability distribution $p(\theta|S, D)$ using the standard Bayesian approach (see the appendix to this chapter).

4. Obtain an estimate $\hat{\theta}$ of θ.

5. Calculate the desired measure of quality, $Q(B(\theta), S)$.

The prior probability distribution, $p(D)$, in Step 1 is provided by the human experts based on their knowledge in the areas of their expertise. They assign a probability to each network structure D. Specification of the prior probability distribution, $p(\theta|D)$, in Step 2, for all possible network structures is very complicated because of the huge amount of information required. One way to avoid this problem is to start with a prior probability distribution, $p(\theta|D_c)$, for the complete Bayesian network $B_c = (D_c, P_c)$. From this complete prior probability distribution, we can calculate $p(\theta|D)$ for any network structure D. Geiger and Heckerman (1995) make two important assumptions that allow computing $p(\theta|D)$ for any graphical structure D by specifying only the prior probability distribution of the parameters for the complete network B_c. The two assumptions are

1. **Parameter Independence:** For every Bayesian network $B = (D, P)$, assume that

$$p(\theta|D) = \prod_{i=1}^{n} p(\theta_i|D). \qquad (11.7)$$

Sometimes it is also assumed that θ_i can be partitioned as $(\theta_{i1}, \ldots, \theta_{is_i})$ with $\theta_{ik} \in \Theta_{ik}$ and

$$p(\theta_i|D) = \prod_{k=1}^{s_i} p(\theta_{ik}|D). \qquad (11.8)$$

Assumption (11.7) says that the parameters associated with each node or variable in D are independent. This is referred to as the *global parameter independence* assumption. Assumption (11.8) says that the parameters associated with each node can be partitioned into

s_i independent parameters. This is referred to as the *local parameter independence* assumption. If both global and local parameters independence assumptions hold, we simply say that we have *parameter independence*. This assumption means that the parameters defining different CPDs are mutually independent. Note that these CPDs can be given independently, with no more restrictions other than those imposed by the probability axioms. As we shall see, in some cases, such as the multinomial case with a Dirichlet prior probability distribution and the normal case with the normal-Wishart prior probability distribution (see the appendix to this chapter), the parameter independence assumption is automatically satisfied; hence it does not need to be assumed explicitly.

2. **Parameter Modularity:** If two Bayesian networks D_1 and D_2 have the same parents for a given node X_i, then

$$p(\theta_i|D_1) = p(\theta_i|D_2). \qquad (11.9)$$

This assumption means that the uncertainty in the parameters of different moduli, or families,[1] is independent of how these moduli are assembled in the network.

Parameter independence and modularity assumptions allow constructing the prior probability distribution $p(\theta|D)$ for every network structure D, given a single prior probability distribution $p(\theta|D_c)$ for a complete network structure D_c. Assume then that we are given a prior probability distribution $p(\theta|D_c)$ for the complete structure D_c and that we wish to obtain the prior probability distribution $p(\theta|D)$ for the structure D. This implies computing $p(\theta_i|D)$ for all nodes $X_i; i = 1, \ldots, n$. These computations are performed using the following algorithm:

Algorithm 11.1 Computes the prior probability distribution of the parameters for a given network from the prior probability distribution of a complete Bayesian network.

- **Input:** *A complete Bayesian network $B_c = (D_c, P_c)$, a prior probability distribution $p(\theta|D_c)$, and another Bayesian network $B = (D, P)$.*

- **Output:** *The prior probability distribution $p(\theta|D)$ associated with D.*

1. Let $i = 1$, i.e., select the first node.

2. Find a complete structure D_c^i having node X_i and the same parents Π_i, using an ordering of the nodes in which the nodes in Π_i come before X_i.

---

[1] The family of a node consists of the node and its parents.

3. Determine the prior probability distribution $p(\theta|D_c^i)$ associated with this structure from the prior probability distribution $p(\theta|D_c)$ by a change of variables. Note that the parameter independence assumption guarantees that we obtain $p(\theta_i|D_c^i)$.

4. Use parameter modularity to obtain $p(\theta_i|D) = p(\theta_i|D_c^i)$ and $p(\theta_{ik}|D) = p(\theta_{ik}|D_c^i), k = 1, \ldots, s_i$.

5. If $i < n$, let $i = i+1$ and go to Step 2. Otherwise continue.

6. Use parameter independence to obtain the prior JPD

$$p(\theta|D) = \prod_{i=1}^{n} p(\theta_i|D) = \prod_{i=1}^{n} \prod_{k=1}^{s_i} p(\theta_{ik}|D). \qquad (11.10)$$

∎

Although parameter independence and modularity assumptions are useful in deriving the prior probability distribution of any network from the prior probability distribution of the complete network, a clear disadvantage of this approach is that the specification of the prior probability distribution of the complete network can itself be a difficult task, especially when the number of variables is large.

In Sections 11.4 and 11.5 we develop Bayesian quality measures for the multinomial Bayesian network (as an example of discrete networks) and the multinormal Bayesian network (as an example of continuous networks), respectively.

11.4 Bayesian Measures for Multinomial Networks

In this section we present Bayesian quality measures for multinomial Bayesian networks, which are one of the most commonly assumed discrete networks. Let $X = (X_1, \ldots, X_n)$ be a multinomial random variable, i.e., a set of n discrete variables, where each variable X_i can take one of r_i distinct values $0, 1, \ldots, r_i - 1$. Thus, we assume that $p(S|D, \theta)$ in (11.5) is a multinomial likelihood with unknown parameters θ. Let S be a set of data (e.g., a sample) of N cases, that is, each case consists of a value for every variable, which represents a random sample $(x_{1j}, \ldots, x_{nj}), j = 1, \ldots, N$, from a multinomial distribution. Let D be a network structure defined over X and $N_{x_1,\ldots,x_n}$ be the number of cases in S such that $X_1 = x_1, \ldots, X_n = x_n$. For every variable X_i, let

$$s_i = \prod_{X_j \in \Pi_i} r_j \qquad (11.11)$$

be the number of all possible instantiations of the parents of X_i.

11.4.1 Assumptions and Derivations

To derive Bayesian quality measures for a multinomial Bayesian network, the following assumptions are made:

- **Assumption 1:** The data set S is *complete*, that is, there are no missing data. The case where we have *missing values* is treated in Section 11.10.

- **Assumption 2:** Parameter modularity is as described in Section 11.3.

- **Assumption 3:** For a complete network, D_c, the prior probability distribution of the parameters $\theta_{x_1,\ldots,x_n}$, where $\theta_{x_1,\ldots,x_n} = p(X_i = x_i, \ldots, X_n = x_n)$, is a Dirichlet distribution

$$p(\theta_{x_1,\ldots,x_n}|D_c) \propto \prod_{x_1,\ldots,x_n} \theta_{x_1,\ldots,x_n}^{\eta_{x_1,\ldots,x_n}-1}, \qquad (11.12)$$

where $\eta_{x_1,\ldots,x_n}$ are its parameters (hyperparameters).

One reason for using the Dirichlet distribution is that it is the natural conjugate for the multinomial distribution, i.e., the posterior JPD, $p(\theta_{x_1,\ldots,x_n}|S, D_c)$, given the data and the prior information is also a Dirichlet distribution

$$p(\theta_{x_1,\ldots,x_n}|S, D_c) \propto \prod_{x_1,\ldots,x_n} \theta_{x_1,\ldots,x_n}^{\eta_{x_1,\ldots,x_n}+N_{x_1,\ldots,x_n}-1}, \qquad (11.13)$$

where $N_{x_1,\ldots,x_n}$ is the number of cases in S with $X_i = x_i; i = 1,\ldots,n$.

As indicated above, one way of specifying a prior probability distribution of the parameters θ is by giving the prior probability distribution for the complete Bayesian network (D_c, P_c). From this complete prior probability distribution, we can calculate the prior values of the hyperparameters $\eta_{x_1,\ldots,x_n}$ by

$$\eta_{x_1,\ldots,x_n} = \eta \prod_{i=1}^{n} p(x_i|x_1,\ldots,x_{i-1}, D_c). \qquad (11.14)$$

The hyperparameter η measures the importance (weight) of the prior probability distribution and can be given the interpretation of an equivalent sample size.

Due to the fact that we factorize the JPD into a product of CPDs, it is more convenient to choose the new parameters

$$\theta_{x_i|x_1,\ldots,x_{i-1}}, \; \forall \, x_1,\ldots,x_{i-1}, \; x_i = 0, 1, \ldots, r_i - 1, \; i = 1,\ldots,n, \qquad (11.15)$$

where

$$\theta_{x_1,\ldots,x_n} = \prod_{i=1}^{n} \theta_{x_i|x_1,\ldots,x_{i-1}}, \; \forall \, (x_1,\ldots,x_n) \neq (r_1,\ldots,r_n), \qquad (11.16)$$

and
$$\theta_{r_i|x_1,\ldots,x_{i-1}} = 1 - \sum_{x_i=0}^{r_i-1} \theta_{x_i|x_1,\ldots,x_{i-1}}.$$

For the sake of simplicity, we shall also use θ_{ijk} to denote $\theta_{x_i|x_1,\ldots,x_{i-1}}$. Thus, θ_{ijk} refers to the probability that $X_i = j$ given the kth instantiation of the parent set Π_i. We refer to $\theta_{x_1,\ldots,x_n}$ as the initial parameters and to θ_{ijk} as the conditional parameters. Consequently, the prior probability distribution of the conditional parameters is also Dirichlet. This is stated in the following theorem (Geiger and Heckerman (1995)).

Theorem 11.1 Distribution of conditional parameters. *If the probability density of the initial parameters in a complete network is Dirichlet of the form (11.12), the probability density of the conditional parameters for each node in (11.15) is independent and jointly distributed as a Dirichlet distribution with probability density function*

$$p(\theta_{ijk}) \propto \prod_{j=0}^{r_i} \theta_{ijk}^{\eta_{ijk}-1}; \ k=1,\ldots,s_i; \ i=1,\ldots,n, \qquad (11.17)$$

where

$$\eta_{ijk} = \eta_{x_i|x_1\ldots x_{i-1}} = \sum_{x_{i+1},\ldots,x_n} \eta_{x_1,\ldots,x_n}, \qquad (11.18)$$

and r_i is the number of distinct values of X_i.

Theorem 11.1 shows that parameter independence, which does not hold for the initial parameters, holds for the new parameters. This independence comes as a consequence of the assumed Dirichlet distribution for the initial parameters and is not an additional assumption.

Theorem 11.1, combined with the conjugate character of the Dirichlet distribution for the multinomial likelihood, shows that the prior and posterior probability distributions for both the initial and conditional parameters are also Dirichlet. Then the posterior probability distribution of all the parameters $\theta = (\theta_1, \ldots, \theta_n)$ is a Dirichlet distribution with

$$p(\theta|S, D_c) \propto \prod_{i=1}^{n} \prod_{j=0}^{r_i} \prod_{k=1}^{s_i} \theta_{ijk}^{\eta_{ijk}+N_{ijk}-1}, \qquad (11.19)$$

where $N_{ik} = \sum_{j=0}^{r_i} N_{ijk}$, and N_{ijk} is the number of cases in S where $X_i = j$ and the set of parents of X_i takes values associated with the kth instantiation of Π_i.

According to (11.5), to calculate $p(S, D, \theta)$ we need

1. The prior probability distribution $p(D)$, which is given by the human experts.

2. The likelihood of the data set S, which is given by

$$p(S|D,\theta) \propto \prod_{i=1}^{n}\prod_{j=0}^{r_i}\prod_{k=1}^{s_i} \theta_{ijk}^{N_{ijk}}. \qquad (11.20)$$

3. The prior probability distribution, $p(\theta|D)$, for the parameters θ which is calculated using the independence assumption and Algorithm 11.1. Thus, according to (11.10), we get

$$p(\theta|D) = \prod_{i=1}^{n}\prod_{j=0}^{r_i}\prod_{k=1}^{s_i} \theta_{ijk}^{\eta_{ijk}-1}. \qquad (11.21)$$

4. The posterior probability distribution then becomes

$$p(\theta|S,D) \propto \prod_{i=1}^{n}\prod_{j=0}^{r_i}\prod_{k=1}^{s_i} \theta_{ijk}^{N_{ijk}+\eta_{ijk}-1}. \qquad (11.22)$$

The above results are used to derive three different measures:

- The Geiger and Heckerman measure.
- The Cooper-Herskovits measure.
- The standard Bayesian measure.

11.4.2 The Geiger and Heckerman Measure

Geiger and Heckerman (1995) propose the Bayesian quality measure

$$\begin{aligned} Q_{GH}(D,S) &= \log p(D) + \log \int p(S|D,\theta)p(\theta|D)d\theta \\ &= \log p(D) + \sum_{i=1}^{n}\left[\sum_{k=1}^{s_i}\left[\log \frac{\Gamma(\eta_{ik})}{\Gamma(\eta_{ik}+N_{ik})} \right.\right. \\ &\left.\left. + \sum_{j=0}^{r_i} \log \frac{\Gamma(\eta_{ijk}+N_{ijk})}{\Gamma(\eta_{ijk})}\right]\right], \quad (11.23) \end{aligned}$$

where $\Gamma(.)$ is the *gamma* function. Note that (11.23) does not include an explicit penalizing term for the complexity of the network.

11.4.3 The Cooper-Herskovits Measure

Cooper and Herskovits (1992) propose the following quality measure:

$$Q_{CH}(D, S) = \log p(D) + \sum_{i=1}^{n} \left[\sum_{k=1}^{s_i} \left[\log \frac{\Gamma(r_i)}{\Gamma(N_{ik} + r_i)} \right. \right.$$
$$\left. \left. + \sum_{j=0}^{r_i} \log \Gamma(N_{ijk} + 1) \right] \right]. \quad (11.24)$$

This measure does not also include an explicit penalizing term for the complexity of the network.

11.4.4 The Standard Bayesian Measure

As an alternative measure to (11.23) and (11.24), we introduce a quality measure that is obtained by evaluating the likelihood at the posterior mode and by considering an explicit penalizing term. The posterior mode, which is the maximum of (11.22) with respect to θ, is

$$\hat{\theta}_{ijk} = \frac{\eta_{ijk} + N_{ijk} - 1}{\eta_{ik} + N_{ik} - r_i}. \quad (11.25)$$

Asymptotically, (11.25) goes to

$$\hat{\theta}_{ijk} = \frac{N_{ijk}}{N_{ik}}. \quad (11.26)$$

Thus, substituting (11.25) in the likelihood, we obtain the standard Bayesian measure

$$Q_{SB}(D, S) = \log p(D)$$
$$+ \sum_{i=1}^{n} \sum_{j=0}^{r_i} \sum_{k=1}^{s_i} (N_{ijk} + \eta_{ijk} - 1) \log \left(\frac{\eta_{ijk} + N_{ijk} - 1}{\eta_{ik} + N_{ik} - r_i} \right)$$
$$- \frac{1}{2} Dim(B) \log N, \quad (11.27)$$

where $Dim(B)$ is the dimension of the Bayesian network model B (the number of free parameters required to completely specify the JPD of X) and N is the sample size. The last term in (11.27) is a penalty term that penalizes networks with large number of parameters.

Alternatively, we can use the asymptotic estimate (11.26) instead of (11.25) and get

$$Q_{SB}(D, S) = \log p(D) + \sum_{i=1}^{n} \sum_{j=0}^{r_i} \sum_{k=1}^{s_i} N_{ijk} \log \frac{N_{ijk}}{N_{ik}}$$
$$- \frac{1}{2} Dim(B) \log N. \quad (11.28)$$

11.4.5 Examples and Computer Code

To illustrate the above quality measures by an example and to provide a limited evaluation, we first need the following definition.

Definition 11.3 Perfect sample. *Assume that we have a set of discrete variables $(X_1, X_2, \ldots, X_n)$ with a JPD $p(x_1, x_2, \ldots, x_n)$. A sample S from $p(x_1, x_2, \ldots, x_n)$ is said to be perfect iff every possible combination of the variables, $(x_1, x_2, \ldots, x_n)$, is included in S with frequency proportional to its probability $p(x_1, x_2, \ldots, x_n)$.*

In the following we shall use perfect samples to evaluate the performance of different quality measures.

Example 11.1 Perfect sample. Consider the Bayesian network in Figure 11.1, where each of the variables (X_1, X_2, X_3) is binary, taking values 0 or 1. The network implies that the JPD can be factorized as

$$p(x_1, x_2, x_3) = p(x_1)p(x_2)p(x_3|x_1). \tag{11.29}$$

Suppose also that the CPDs required to specify $p(x_1, x_2, x_3)$ are given in Table 11.1. The eight possible combinations of the three variables together with the probability of each are shown in the first two columns of Table 11.2. A perfect sample of size 1,000 from the above JPD consists of all the combinations in Table 11.2 with the frequencies given in the last column of the same table. Thus, the frequency of each combination of the variables is proportional to its probability, $p(x_1, x_2, x_3)$. ∎

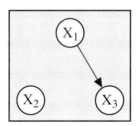

FIGURE 11.1. An example of a Bayesian network.

Example 11.2 Quality measures for a multinomial network. For the three binary variables in Example 11.1, Figure 11.2 shows the 11 possible network structures leading to distinct independence equivalence classes. Suppose we have the perfect sample S shown in Table 11.2. Recall that this sample is drawn from the JPD given by (11.29) and Table 11.1. This means that the true model that generated S is the Bayesian network number 4 in Figure 11.2.

TABLE 11.1. CPDs required for specifying the JPD for the nodes of the Bayesian network in Figure 11.1.

x_1	$p(x_1)$
0	0.4
1	0.6

x_2	$p(x_2)$
0	0.2
1	0.8

x_1	x_3	$p(x_3\|x_1)$
0	0	0.1
0	1	0.9
1	0	0.6
1	1	0.4

$\{x_1, x_2, x_3\}$	$p(x_1, x_2, x_3)$	Frequency
$\{0,0,0\}$	0.008	8
$\{0,0,1\}$	0.072	72
$\{0,1,0\}$	0.032	32
$\{0,1,1\}$	0.288	288
$\{1,0,0\}$	0.072	72
$\{1,0,1\}$	0.048	48
$\{1,1,0\}$	0.288	288
$\{1,1,1\}$	0.192	192
Total	1.000	1,000

TABLE 11.2. Perfect sample associated with the network in Example 11.1.

Figure 11.3 gives a *Mathematica* program for computing the quality measure in (11.27). Small modifications of this program allow computing the other measures in (11.23), (11.24), and (11.28).

We now use perfect samples of sizes 1,000, 5,000, and 10,000 to compute quality measures for each of the 11 structures in Figure 11.2. We also use a prior probability distribution that assigns the same frequency to each cell (we selected a value of 20, which implies that $\eta = 160$).

Table 11.3 shows the posterior probability distributions of the 11 network structures for different sample sizes for the Geiger and Heckerman quality measure in (11.23). For a perfect sample of size 1,000, this measure fails to give network 4, which is the true network that generated the sample, the highest posterior probability. But for sample sizes 5,000 and 10,000, Network 4 has the highest posterior probability. One reason for this is due to the use of the uniform prior probability distribution. Another reason is that the weight for the prior probability distribution dominates the measure. For example, when the weight is set to 1 instead of 20 (that is $\eta = 8$), network 4 is identified as the network with the highest posterior probability, as shown in Table 11.4.

Tables 11.5, 11.6 and 11.7 show the posterior probability distributions of the 11 network structures for different sample sizes for the three quality measures in (11.24), (11.27), and (11.28), respectively. As we can see, all

11.4 Bayesian Measures for Multinomial Networks 497

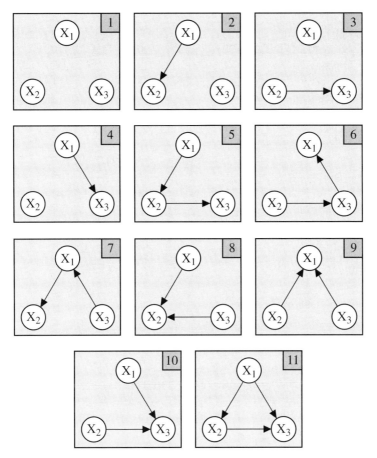

FIGURE 11.2. All distinct (leading to different dependence structures) Bayesian networks for three variables.

three measures correctly identify network 4 as the most likely network for all three sample sizes. The asymptotic standard Bayesian measure in (11.28), however, gives the highest posterior probability for this network. This is due to the fact that the η has no effect in this asymptotic case. Note also that as the sample size increases, the posterior probability for the asymptotic measure approaches the standard Bayesian measure, as would be expected.

The results in Tables 11.3–11.7 can be summarized as follows:

1. Network 4 (which is the structure that was used to generate S) has the highest quality measure, as would be expected. The other networks have much lower measures of quality.

2. The quality measures for network 4 increase with the sample size, as would be expected. This means that the chance for the quality measure to select the correct network increases with the sample size.

```
TI={{{0,0,0},{20}},{{0,0,1},{20}},{{0,1,0},{20}},{{0,1,1},{20}},
{{1,0,0},{20}},{{1,0,1},{20}},{{1,1,0},{20}},{{1,1,1},{20}}};
TD={{{0,0,0},{8}},{{0,0,1},{72}},{{0,1,0},{32}},{{0,1,1},{288}},
{{1,0,0},{72}},{{1,0,1},{48}},{{1,1,0},{288}},{{1,1,1},{192}}};

SampleSize=Sum[TotalData[[i,2,1]],{i,1,Length[TotalData]}];
Netw[1]={{1},{2},{3}}; Netw[2]={{1},{2,1},{3}};
Netw[3]={{1},{2},{3,2}}; Netw[4]={{1},{2},{3,1}};
Netw[5]={{1},{2,1},{3,2}}; Netw[6]={{1,3},{2},{3,2}};
Netw[7]={{1,3},{2,1},{3}}; Netw[8]={{1},{2,1,3},{3}};
Netw[9]={{1,2,3},{2},{3}}; Netw[10]={{1},{2},{3,1,2}};
Netw[11]={{1},{2,1},{3,2,1}};

Card={2,2,2};
For[Net=1,Net<=11,Net++,Nnodes=Length[Netw[Net]];
For[i=0,i<Nnodes,i++,For[k=0,k<Card[[i+1]],k++,
Nsons[i]=Length[Netw[Net][[i+1]]]-1;
For[jj=0;fact=1;q1=0,jj<Nsons[i],jj++,
fact1=Card[[Netw[Net][[i+1]][[jj+2]]]];
q1+=fact*(fact1-1);fact*=fact1;q[i]=q1;
For[j=0,j<=q1,j++,N2[i][j][k]=0;N21[i][j][k]=0]]];
KK=Sum[(Card[[i+1]]-1)*(q[i]+1),{i,0,Nnodes-1}];
For[Ndat=1,Ndat<=Length[TD],Ndat++,
Data=TD[[Ndat]];DatInit=TI[[Ndat]];
For[i=0,i<Nnodes,i++,k=Data[[1]][[i+1]];
kInit=DatInit[[1]][[i+1]];
For[jj=0;j=0;jInit=0;fact=1,jj<Nsons[i],jj++,
fact1=Card[[Netw[Net][[i+1]][[jj+1]]]];
j+=fact*Data[[1]][[Netw[Net][[i+1]][[jj+2]]]];
jInit+=fact*DatInit[[1]][[Netw[Net][[i+1]][[jj+2]]]];fact*=fact1];
N2[i][j][k]+=Data[[2]][[1]];N21[i][jInit][kInit]+=DatInit[[2]][[1]]]];
For[i=0,i<Nnodes,i++,
For[j=0,j<=q[i],j++,For[k=0;N3[i][j]=0;N31[i][j]=0,k<Card[[i+1]],
k++,N3[i][j]+=N2[i][j][k];N31[i][j]+=N21[i][j][k] ]]];
PNetw[Net]=N[Exp[-KK*Log[SampleSize]/2
+Sum[(N2[i][j][k]+N21[i][j][k]-1)*Log[(N2[i][j][k]+N21[i][j][k]-1)/
(N3[i][j])+N31[i][j][k]-Card[i+1]],
{i,0,Nnodes-1},{j,0,q[i]},{k,0,Card[[i+1]]-1}]]];
Print["Net=",Net," p=",PNetwork[Net]]]
For[S=0;jj=1,jj<=11,jj++,S+=PNetw[jj]]
For[L={};jj=1,jj<=11,jj++,L=Append[L,{PNetw[jj]/S,jj}]];Sort[L]
```

FIGURE 11.3. A *Mathematica* program for computing the standard Bayesian quality measure in (11.27) using the perfect sample in Table 11.2.

Sample size					
1,000		5,000		10,000	
Network	Prob.	Network	Prob.	Network	Prob.
10	0.310	4	0.505	4	0.671
9	0.310	10	0.117	6	0.109
11	0.160	9	0.117	7	0.109
4	0.108	7	0.117	10	0.051
6	0.056	6	0.117	9	0.051
7	0.056	11	0.027	11	0.008
8	0.000	1	0.000	1	0.000
1	0.000	2	0.000	3	0.000
3	0.000	3	0.000	2	0.000
2	0.000	8	0.000	5	0.000
5	0.000	5	0.000	8	0.000

TABLE 11.3. The Geiger and Heckerman measure in (11.23) with $\eta = 160$: Posterior probabilities of the 11 network structures in Figure 11.2 for different sample sizes.

3. The asymptotic standard Bayesian measure in (11.28) gives better results than the other three measures. The performance of a measure, however, may depend on other factors not considered in the above analysis, such as the complexity of the network and/or the JPD that was used to generate the data. Therefore, more thorough comparisons and investigations of the properties of the various quality measures are needed to determine which one of these measures is to be recommended. ∎

11.5 Bayesian Measures for Multinormal Networks

In this section we consider quality measures for multinormal Bayesian networks as a special case of continuous Bayesian networks.

11.5.1 Assumptions and Derivations

To obtain quality measures for multinormal Bayesian networks, the following assumptions are made:

1. **Assumption 1:** The data set S contains no missing data. The case where we have missing values is treated in Section 11.10.

2. **Assumption 2:** Parameter modularity, as described in Section 11.3.

	Sample size		
Network	1,000	5,000	10,000
4	0.808	0.910	0.936
6	0.083	0.042	0.031
7	0.083	0.042	0.031
10	0.012	0.003	0.001
9	0.012	0.003	0.001
11	0.001	0.000	0.000
1	0.000	0.000	0.000
3	0.000	0.000	0.000
2	0.000	0.000	0.000
5	0.000	0.000	0.000
8	0.000	0.000	0.000

TABLE 11.4. The Geiger and Heckerman measure in (11.23) with $\eta = 8$: Posterior probabilities of the 11 network structures in Figure 11.2 for different sample sizes.

	Sample size		
Network	1,000	5,000	10,000
4	0.844	0.928	0.949
6	0.082	0.040	0.029
7	0.054	0.027	0.019
10	0.010	0.002	0.001
9	0.010	0.002	0.001
11	0.001	0.000	0.000
1	0.000	0.000	0.000
3	0.000	0.000	0.000
2	0.000	0.000	0.000
5	0.000	0.000	0.000
8	0.000	0.000	0.000

TABLE 11.5. The Cooper and Herskovits measure in (11.24): Posterior probabilities of 11 network structures in Figure 11.2 for three perfect samples.

3. **Assumption 3:** The data set S is a random sample from an n-variate normal distribution $N(\mu, \Sigma)$ with unknown mean vector μ and covariance matrix Σ. Thus, $p(S|D, \theta)$ in (11.5) is assumed to be a multivariate normal likelihood. Instead of the covariance matrix Σ, it is sometimes convenient to work with the closely related precision matrix $W = \Sigma^{-1}$, which is the inverse of Σ. Thus, here $\theta = \{\mu, W\}$.

4. **Assumption 4:** The prior probability distribution on the parameters $\theta = \{\mu, W\}$,
$$p(\theta|D_c) = p(\mu, W|D_c),$$

11.5 Bayesian Measures for Multinormal Networks

Sample size					
1,000		5,000		10,000	
Network	Prob.	Network	Prob.	Network	Prob.
4	0.488	4	0.831	4	0.891
10	0.153	7	0.075	7	0.056
9	0.153	6	0.051	6	0.038
7	0.099	10	0.020	10	0.007
6	0.075	9	0.020	9	0.007
11	0.031	11	0.002	11	0.000
1	0.000	1	0.000	1	0.000
2	0.000	2	0.000	3	0.000
3	0.000	3	0.000	2	0.000
8	0.000	8	0.000	5	0.000
5	0.000	5	0.000	8	0.000

TABLE 11.6. The standard Bayesian measure in (11.28): Posterior probabilities of the 11 network structures in Figure 11.2 for different sample sizes.

	Sample size		
Network	1,000	5,000	10,000
4	0.938	0.972	0.980
6	0.030	0.014	0.010
7	0.030	0.014	0.010
10	0.001	0.000	0.000
9	0.001	0.000	0.000
11	0.000	0.000	0.000
1	0.000	0.000	0.000
3	0.000	0.000	0.000
2	0.000	0.000	0.000
5	0.000	0.000	0.000
8	0.000	0.000	0.000

TABLE 11.7. The asymptotic standard Bayesian measure in (11.27): Posterior probabilities of the 11 network structures in Figure 11.2 for different sample sizes.

where D_c is a complete network, is a normal-Wishart distribution (see the appendix to this chapter), that is, the conditional distribution of μ given W is $N(\mu_0, \nu W)$, where ν is a constant and μ_0 is a vector of constants, and the marginal distribution of W is a Wishart distribution with $\alpha > n - 1$ degrees of freedom and precision matrix W_0.

One reason for using the normal-Wishart distribution is that it is the natural conjugate for the multivariate normal distribution, i.e.,

Prior Hyperparameters	Posterior Hyperparameters
μ_0	$\frac{\nu\mu_0+N\bar{x}}{\nu+N}$
ν	$\nu+N$
α	$\alpha+N$
W_0	$W_N = W_0 + S + \frac{\nu N}{\nu+N}(\mu_0-\bar{x})(\mu_0-\bar{x})^T$

TABLE 11.8. Prior and posterior hyperparameters for the multivariate normal case.

- The posterior JPD, $p(\mu, W|D_c, S)$, of μ and W given the data is also a normal-Wishart distribution,

- The conditional distribution of μ given W is $N(\mu_N, ((\nu+N)W)^{-1})$ and the marginal distribution of W is a Wishart distribution with $\alpha + N$ degrees of freedom and precision matrix

$$W_N = W_0 + S + \frac{\nu N}{\nu+N}(\mu_0-\bar{x})(\mu_0-\bar{x})^T, \qquad (11.30)$$

where

$$\bar{x} = N^{-1}\sum_{i=1}^{N} x_i \qquad (11.31)$$

and

$$S = \sum_{i=1}^{N}(x_i-\bar{x})(x_i-\bar{x})^T \qquad (11.32)$$

are the sample mean and covariance matrix, respectively.

The derivation of these results is given in the appendix to this chapter, Example 11.9. Table 11.8 gives the prior and posterior hyperparameters and shows how they can be updated from the sample values.

The Bayes' parameter estimates are

$$\hat{\mu}_N = \frac{\nu\mu_0 + N\bar{x}}{\nu+N} \quad \text{and} \quad \hat{W}_N = (\alpha+N)W_N^{-1},$$

which are asymptotically equivalent to the estimates $\bar{x}$ and NS^{-1}, respectively.

The prior probability distribution is specified, in a similar way as in the multinomial case, by giving a complete normal Bayesian network. From this we can calculate μ_0, W_0, and the two weights ν and α that measure the

11.5 Bayesian Measures for Multinormal Networks

relative importance of the information in the prior probability distribution distribution with respect to information in the data.

Since we use conditional probabilities, and for convenience, instead of using the parameters μ (mean) and W (precision matrix), we use the new parameters m_i, $\beta_i = \{\beta_{ij}, j = 1, \ldots, i-1\}$, and v_i, where the β_{ij} are the regression coefficients when the variable X_i is regressed on the variables $X_1, \ldots, X_{i-1}$, m_i is the intercept in this regression, and v_i is the inverse of the residual variance, that is,

$$p(x_i|x_1, \ldots, x_{i-1}) = N(\mu_i, 1/v_i),$$

where

$$\mu_i = m_i + \sum_{j=1}^{i-1} \beta_{ij}\mu_j. \tag{11.33}$$

This implies that we may interpret a multivariate normal distribution as a Bayesian network, where there is no arc from X_j to X_i if $\beta_{ij} = 0, i < j$.

Thus, we are interested in knowing the prior and posterior probability distributions of the new parameters, as functions of the corresponding distributions for the old parameters. Heckerman and Geiger (1995) give the following theorem:

Theorem 11.2 Global independence of normal parameters. *If the old parameters (μ, W) have a normal-Wishart distribution, then the new parameters satisfy*

$$p(m, v, D) = \prod_{i=1}^{n} p(m_i, v_i, \beta_i). \tag{11.34}$$

Thus, the parameter independence automatically holds and we can use the prior probability distribution of a complete Bayesian network to obtain the prior probability distribution of any other network (see Section 11.3). The modularity assumption becomes

$$p(m_i, v_i\beta_i|D_1) = p(m_i, v_i\beta_i|D_2). \tag{11.35}$$

Once we know the prior probability distribution for a given Bayesian network, we can calculate the posterior probability distribution and define quality measures based on it. Here two measures are available: the Geiger and Heckerman measure and the standard Bayesian measure.

11.5.2 The Geiger and Heckerman Measure

Geiger and Heckerman (1994) propose a Bayesian quality measure similar to that in (11.23), that is,

$$Q_{GH}(D, S) = \log p(D) + \log \int p(S|D, \theta)p(\theta|D)d\theta, \tag{11.36}$$

where $\theta = (\mu, W)$. It is well known (see DeGroot (1970)) that the Bayesian or mixed distribution,

$$p(x|D_c) = \int p(x|\mu, W, D_c) p(\mu, W|D_c) d\mu dW$$

is an n-dimensional t-distribution with $\gamma = \alpha - n + 1$ degrees of freedom, location vector μ_0, and precision matrix $W_0' = (\nu\gamma/(\nu+1))W_0^{-1}$. This gives the quality measure

$$Q_{GH}(D, S) \propto \log p(D)$$
$$+ \log \left[(\pi)^{-nN/2} \left(\frac{\nu}{\nu + N} \right)^{n/2} \right]$$
$$+ \log \left[\frac{c(n, \alpha) d(N, \alpha)}{c(n, \alpha + N)} \right], \qquad (11.37)$$

where

$$c(n, \alpha) = \left[\prod_{i=1}^{n} \Gamma\left(\frac{\alpha - i + 1}{2} \right) \right]^{-1}$$

and

$$d(N, \alpha) = [det(W_0)]^{\alpha/2} [det(W_N)]^{-(\alpha+N)/2}.$$

Here $det(A)$ denotes the determinant of the matrix A. Like the quality measure in (11.23), this quality measure does not include a penalizing term for the complexity of the network.

11.5.3 The Standard Bayesian Measure

A quality measure for multinormal Bayesian networks, similar to that in (11.27) for multinomial Bayesian networks, can be obtained by evaluating the likelihood at the posterior mode of θ and by considering an extra penalizing term. This gives

$$Q_B(B(\theta), S) = \log p(D) + \log p(S|D, \hat{\theta})$$
$$- \frac{1}{2} Dim(B) \log N, \qquad (11.38)$$

where $\hat{\theta}$ is the posterior mode of θ, that is,

$$\hat{\theta} = \arg\max_{\theta} \log p(\theta|D, S). \qquad (11.39)$$

11.5.4 Examples and Computer Codes

We now illustrate the quality measures for normal Bayesian networks by an example. We also give the corresponding computer codes in *Mathematica*.

11.5 Bayesian Measures for Multinormal Networks

Example 11.3 Quality measures for normal Bayesian networks.
Consider a three-dimensional random variable (X_1, X_2, X_3). Suppose that the prior probability distribution of the parameters is a normal-Wishart distribution with

$$\nu = \alpha = 6; \ \hat{\mu}_0 = (0.1, -0.3, 0.2), v = (1, 1, 1), \beta_2 = (0); \ \beta_3 = (1, 1),$$

where β_2 and β_3 refer to the regression coefficients β_{ij} in (11.33) and

$$W_0 = \begin{pmatrix} 12/7 & 0 & 12/7 \\ 0 & 12/7 & 12/7 \\ 12/7 & 12/7 & 36/7 \end{pmatrix}.$$

Using the sample given in Table 11.9, and according to (11.30), (11.32), and (11.31), we get

$$\bar{x} = \begin{pmatrix} 0.510 \\ -0.362 \\ -0.785 \end{pmatrix}, \quad S_N = \begin{pmatrix} 11.337 & 11.439 & 6.806 \\ 11.439 & 34.014 & 25.717 \\ 6.806 & 25.717 & 31.667 \end{pmatrix},$$

$$\mu_N = \begin{pmatrix} 0.415 \\ -0.348 \\ -0.558 \end{pmatrix}, \quad W_N = \begin{pmatrix} 13.825 & 11.322 & 6.659 \\ 11.322 & 35.746 & 27.713 \\ 6.659 & 27.713 & 41.288 \end{pmatrix}.$$

Figure 11.4 lists a *Mathematica* program for computing the Geiger and Heckerman measure for normal networks. Table 11.10 shows the 11 distinct possible Bayesian network structures and their associated normalized qualities. Note that network structure number 5 has the highest quality measure of 0.763. The result is not surprising because the sample was generated from this network structure with parameters $\mu_0 = (0.5, 0.2, -0.5)$, $v = (1, 1, 1)$, $\beta_2 = (1)$, and $\beta_3 = (0, 1)$. Then, the estimates are

$$\hat{\mu} = \mu_N = \begin{pmatrix} 0.415 \\ -0.348 \\ -0.558 \end{pmatrix},$$

$$\hat{W} = (\alpha + N)W_N^{-1} = \begin{pmatrix} 2.597 & -1.038 & 0.278 \\ -1.038 & 1.931 & -1.129 \\ 0.278 & -1.129 & 1.343 \end{pmatrix},$$

and the asymptotic estimates are

$$\hat{\mu} = \bar{x} = \begin{pmatrix} 0.510 \\ -0.362 \\ -0.785 \end{pmatrix},$$

$$\hat{W} = NS^{-1} = \begin{pmatrix} 2.773 & -1.249 & 0.418 \\ -1.249 & 2.086 & -1.426 \\ 0.418 & -1.425 & 1.699 \end{pmatrix},$$

Case	X_1	X_2	X_3
1	−0.78	−1.55	0.11
2	0.18	−3.04	−2.35
3	1.87	1.04	0.48
4	−0.42	0.27	−0.68
5	1.23	1.52	0.31
6	0.51	−0.22	−0.60
7	0.44	−0.18	0.13
8	0.57	−1.82	−2.76
9	0.64	0.47	0.74
10	1.05	0.15	0.20
11	0.43	2.13	0.63
12	0.16	−0.94	−1.96
13	1.64	1.25	1.03
14	−0.52	−2.18	−2.31
15	−0.37	−1.30	−0.70
16	1.35	0.87	0.23
17	1.44	−0.83	−1.61
18	−0.55	−1.33	−1.67
19	0.79	−0.62	−2.00
20	0.53	−0.93	−2.92

TABLE 11.9. A data set generated at random from the network structure number 5 in Figure 11.3, assuming normality.

which are close to those above but not very close, indicating that the prior information has some weight, due to the fact that the sample size $N = 20$ is small. ∎

11.6 Minimum Description Length Measures

The Bayesian quality measures described in the previous sections require specification of prior information on both the graphical and probabilistic structures. This information may not be readily available. The minimum description length measure is an alternative way of measuring the quality of a network structure. This concept comes from *coding theory*, where a string is encoded with as few bits as possible. The basic idea consists of compressing one string by dividing it into substrings; the most frequent strings are encoded by short messages and the least frequent are encoded by longer messages, leading to as short as possible average message length. The encoding consists of two parts: the description of the transformation used in the compression and the compressed string. The description length of a

```
Needs["Statistics`ContinuousDistributions`"];
dist=NormalDistribution[0,1];
x=Table[Random[dist],{i,1,3},{j,1,2000}];
cc={{1,0,0},{0,1,0},{1,0,1}};
X=Transpose[cc.x];
mu0={0.0,0.0,0.0};v={1,1,2};b2={0};b3={0,0};
nu=6;alp=6;
c[n_,v_]:=N[Product[Gamma[(v+1-i)/2],{i,1,n}]),1000];
L=Length[X];n=Length[X[[1]]];
XL={};;For[i=1,i<=n,i++,M=0;For[j=1,j<=L,j++,M+=X[[j]][[i]]];
AppendTo[XL,{M/L}]];
muL=(nu*mu0+L*XL)/(nu+L);
SL=Sum[(X[[i]]-XL).Transpose[(X[[i]]-XL)],{i,1,L}];
W={{1/v[[1]]+b3[[1]]^ 2/v[[3]],b3[[1]]*b3[[2]]/v[[3]],-b3[[1]]/v[[3]]},
{b3[[1]]*b3[[2]]/v[[3]],1/v[[2]]+b3[[2]]^ 2/v[[3]],-b3[[2]]/v[[3]]},
{-b3[[1]]/v[[3]],-b3[[2]]/v[[3]],1/v[[3]]}};
W1=Inverse[W]; T0=(nu*(alp-n-1))/(nu+1)*W1;
TL=T0+SL+(nu*L)/(nu+L)*(mu0-XL).Transpose[(mu0-XL)];
P[n_,L_,T0_,TL_]:=N[(Pi)^ (-n*L/2)*(nu/(nu+L))^ (n/2)*
Det[T0]^ (alp/2)*Det[TL]^ (-(alp+L)/2)*c[n,alp]/c[n,alp+L],1000];
T012={{T0[[1]][[1]],T0[[1]][[2]]},{T0[[2]][[1]],T0[[2]][[2]]}};
TL12={{TL[[1]][[1]],TL[[1]][[2]]},{TL[[2]][[1]],TL[[2]][[2]]}};
T023={{T0[[2]][[2]],T0[[2]][[3]]},{T0[[3]][[2]],T0[[3]][[3]]}};
TL23={{TL[[2]][[2]],TL[[2]][[3]]},{TL[[3]][[2]],TL[[3]][[3]]}};
T013={{T0[[1]][[1]],T0[[1]][[3]]},{T0[[3]][[1]],T0[[3]][[3]]}};
TL13={{TL[[1]][[1]],TL[[1]][[3]]},{TL[[3]][[1]],TL[[3]][[3]]}};
T01={{T0[[1]][[1]]}};TL1={{TL[[1]][[1]]}};
T02={{T0[[2]][[2]]}};TL2={{TL[[2]][[2]]}};
T03={{T0[[3]][[3]]}};TL3={{TL[[3]][[3]]}};
P12=P[2,L,T012,TL12];P13=P[2,L,T013,TL13];
P23=P[2,L,T023,TL23];P1=P[1,L,T01,TL1];
P2=P[1,L,T02,TL2];P3=P[1,L,T03,TL3];P123=P[3,L,T0,TL];
PNetw[1]=P1*P2*P3;PNetw[2]=P12*P3;PNetw[3]=P23*P1;
PNetw[4]=P13*P2;PNetw[5]=P12*P23/P2;PNetw[6]=P13*P23/P3;
PNetw[7]=P12*P13/P1;PNetw[8]=P123*P1*P3/P13;
PNetw[9]=P123*P2*P3/P23;
PNetw[10]=P123*P1*P2/P12;PNetw[11]=P123;
SS=Sum[PNetw[i],{i,1,11}];
Sort[Table[PNetw[i]/SS,{i,1,11}]]
```

FIGURE 11.4. A *Mathematica* program simulating a normal Bayesian network and computing the Geiger and Heckerman measure in (11.37).

Network	Probability
5	0.763
8	0.136
11	0.049
3	0.036
6	0.013
10	0.002
2	0.000
7	0.000
9	0.000
1	0.000
4	0.000

TABLE 11.10. The Geiger and Heckerman measure in (11.37): Posterior probabilities associated with the 11 networks in Figure 11.2.

string of symbols is the sum of the lengths of these two parts. The minimum description length principle selects the code transformation leading to the minimal description length.

In the case of Bayesian networks, the description length includes

1. The length required to store the structure of the Bayesian network. Since the maximum number of links in a Bayesian network with n nodes is $n(n-1)/2$, and we can store 1 if the link exists and 0 otherwise,[2] then the maximum number of links (length) required is $n(n-1)/2$. Note that this number does not depend on the particular structure. Thus, it need not be included in the quality measure.

2. The length required to store the parameters θ. It is well known that the mean length to store a number from 0 to N is $\frac{1}{2}\log N$ (see Rissanen (1983, 1986)). Thus, to store the parameters of the JPD, we need a length of $\frac{1}{2}Dim(B)\log N$, where N is the sample size and

$$Dim(B) = \sum_{i=0}^{n}(r_i - 1)\prod_{X_j \in \Pi_i} r_j = \sum_{i=1}^{n}(r_i - 1)s_i \qquad (11.40)$$

is the number of free parameters (degrees of freedom) associated with the JPD. Note that s_i here is the number of free parameters (degrees of freedom) associated with the CPD for X_i, $p(x_i|\pi_i)$.

3. The length of the description of the data set S compressed using the distribution associated with (D, P), which in the case of multinomial

---

[2]This may not be the optimal way of storing the structure. It has been chosen for the sake of simplicity.

Bayesian networks becomes

$$\sum_{i=1}^{n} \sum_{j=0}^{r_i} \sum_{k=1}^{s_i} N_{ijk} \log \frac{N_{ijk}}{N_{ik}},$$

which is $-N$ times the entropy

$$H(S,B) = \sum_{i=1}^{n} \sum_{j=0}^{r_i} \sum_{k=1}^{s_i} -\frac{N_{ijk}}{N} \log \frac{N_{ijk}}{N_{ik}}. \quad (11.41)$$

Bouckaert (1995) adds an extra term to incorporate the prior information and proposes the minimum description length (MDL) measure

$$Q_{MDL}(B,S) = \log p(D) + \sum_{i=1}^{n} \sum_{j=0}^{r_i} \sum_{k=1}^{s_i} N_{ijk} \log \frac{N_{ijk}}{N_{ik}}$$
$$- \frac{1}{2} Dim(B) \log N. \quad (11.42)$$

Note that the quality measures (11.28) and (11.42) are the same, and that the quality measures (11.27) and (11.42) are asymptotically equivalent.

11.7 Information Measures

Another way of measuring the quality of a network is by the information measure. The basic idea is to select the network structure that best fits the data, penalized by the number of values that need to be specified to define the probability associated with the network. This leads to the information measure

$$Q_I(B,S) = \log p(D) + \sum_{i=1}^{n} \sum_{j=0}^{r_i} \sum_{k=1}^{s_i} N_{ijk} \log \frac{N_{ijk}}{N_{ik}} - Dim(B) f(N), \quad (11.43)$$

where $f(N)$ is a nonnegative penalizing function.

Many different penalizing functions have been used in the literature, such as the *maximum likelihood information criterion* ($f(N) = 0$), the *Akaike information criterion* (Akaike (1974)), ($f(N) = 1$) and the *Schwarz information criterion* (Schwarz (1978)), ($f(N) = \log(N)/2$). Note that the minimum description length measure (11.42) is a particular case of this measure.

11.8 Further Analyses of Quality Measures

The quality measures presented in the previous sections are capable of distinguishing I-maps and perfect maps. Bouckaert (1995) studies these quality measures for discrete Bayesian networks and gives the following theorems:

Theorem 11.3 Quality measures and minimal I-maps. *Let X be a set of variables and let $\alpha(X)$ be a total ordering on X. Assume that the prior probability distribution over all network structures over X is positive. Let P be a JPD over X such that D is a minimal I-map of P obeying $\alpha(X)$ and no other network structure obeying $\alpha(X)$ is a minimal I-map of P. Let S be a sample of size N obtained from P. Let Q be the Cooper-Herskovits Bayesian measure, the minimum description length measure, or the information measure with nonzero penalty function f, where $\lim_{N\to\infty} f(N) = \infty$ and $\lim_{N\to\infty} f(N)/N = 0$. Then, for any network structure D' over X that obeys the $\alpha(X)$ ordering, we have*

$$\lim_{N\to\infty} [N(Q(D',S) - Q(D,S))] = -\infty$$

if and only if D' is not a minimal I-map of P.

This theorem shows that asymptotically, I-maps are given higher quality values than non-I-maps.

Theorem 11.4 Quality measures and perfect maps. *Let X be a set of variables. Let the prior probability distribution over all network structures over X be positive. Let P be a positive distribution over X such that a perfect map exists for P. Let B be such a perfect map for P. Now, let S be a sample of size N from P. Let Q be either the Bayesian measure, the MDL measure, or an information measure with nonzero penalty function f, where $\lim_{N\to\infty} f(N) = \infty$ and $\lim_{N\to\infty} f(N)/N = 0$. Then, for any network structure D' over X we have*

$$\lim_{N\to\infty} [N(Q(D,S) - Q(D',S))] = -\infty$$

if and only if D' is not a perfect map of P.

This theorem shows that asymptotically perfect maps are given higher quality values than nonperfect maps.

Finally, the following theorem shows that there exist optimal directed graphs with a limited number of links:

Theorem 11.5 *Let X be a set of variables. Let the prior probability distribution over all network structures over X be uniform. Let S be a sample of size N from X. Let B be a network structure over X with a parent set containing more than $\log(N/f(N) + 1)$ variables. Then, a network structure D' exists such that $Q_f(D',S) > Q_f(D,S)$ with fewer links, where Q_f is an information quality measure with nonzero penalty function f.*

11.9 Bayesian Network Search Algorithms

In the previous sections we developed several measures of quality of Bayesian networks. These measures are used by search algorithms to find Bayesian networks with high quality. The number of all possible different network structures is so large that it is virtually impossible to evaluate each of them. This section presents two search algorithms that attempt to find the highest quality Bayesian network given a prior information and a set of data.

We note that there are search methods that work in the space of all Bayesian network structures and others that work on the equivalence class of Bayesian network structures. For details, see Spirtes and Meek (1995), or Chickering (1995a).

11.9.1 The K2-Algorithm

Cooper and Herskovits (1992) propose an algorithm for finding a high quality Bayesian network. They refer to it as the $K2$-algorithm. This algorithm starts with the simplest network, that is, a network without links, and assumes that the nodes are ordered. For each variable X_i, the algorithm adds to its parent set Π_i the node that is lower numbered than X_i and leads to a maximum increment in the chosen quality measure. The process is repeated until either adding new nodes does not increase the quality or a complete network is attained.

The pseudocode for this algorithm is given in Figure 11.5, where $q_i(\Pi_i)$ is the contribution of the variable X_i with parent set Π_i to the quality of the network. For example, using the Geiger and Heckerman quality measure in (11.23), we have

$$q_i(\Pi_i) = \sum_{k=1}^{s_i} \log \frac{\Gamma(\eta_{ik})}{\Gamma(\eta_{ik} + N_{ik})} + \sum_{j=0}^{r_i} \log \frac{\Gamma(\eta_{ijk} + N_{ijk})}{\Gamma(\eta_{ijk})}, \quad (11.44)$$

whereas using the Bouckaert (1995) minimum description length measure in (11.42), we get

$$q_i(\Pi_i) = \sum_{j=0}^{r_i} \sum_{k=1}^{s_i} N_{ijk} \log \frac{N_{ijk}}{N_{ik}} - \frac{1}{2} s_i (r_i - 1) \log N. \quad (11.45)$$

One problem with the $K2$-algorithm is that it requires a previous ordering of the nodes. The following algorithm does not require any ordering of the nodes.

11.9.2 The B-Algorithm

This algorithm is proposed by Buntine (1991). It starts with empty parent sets as in the $K2$-algorithm. At each step a new link is added that does

```
Initialization Step:
    Order the variables
    for i ← 1 to n do
        Π_i ← φ
Iteration Step:
    for i ← 1 to n do
    repeat
        select Y ∈ {X_1, ..., X_{i-1}} \ Π_i that
            maximizes g = q_i(Π_i ∪ {Y})
        δ ← g - q_i(Π_i)
        if δ > 0 then
            Π_i ← Π_i ∪ {Y}
    until δ ≤ 0 or Π_i = {X_1, ..., X_{i-1}}
```

FIGURE 11.5. Pseudocode for the $K2$-algorithm.

```
Initialization Step:
    for i ← 1 to n do
        Π_i ← φ
    for i ← 1 to n and j ← 1 to n do
        if i ≠ j then
            A[i, j] ← m_i(X_j) - m_i(φ)
        else
            A[i, j] ← -∞   {obstruct X_i → X_i}
Iteration Step:
    repeat
        select i, j, that maximize A[i, j]
        if A[i, j] > 0 then
            Π_i ← Π_i ∪ {X_j}
            for X_a ∈ Pred_i, X_b ∈ Desc_i do
                A[a, b] ← -∞   {obstruct loops}
            for k ← 1 to n do
                if A[i, k] > -∞ then
                    A[i, k] ← m_i(Π_i ∪ {X_k}) - m_i(Π_i)
    until A[i, j] ≤ 0 or A[i, j] = -∞, ∀ i, j
```

FIGURE 11.6. Pseudocode for the B-algorithm.

not lead to a cycle and that maximizes the quality increment. The process is repeated until no more increase of the quality is possible or a complete network is attained.

Figure 11.6 shows the pseudocode for this algorithm, where $Pred_i$ is the set of predecessors of node X_i and $Desc_i$ is the set of descendants of X_i.

11.10 The Case of Incomplete Data

In the previous sections we assumed that the available data set S is complete, that is, it contains no missing values. In this section we present the following two methods for dealing with the case where S contains missing values:

- The Gibbs' sampling.
- The EM-algorithm.

11.10.1 The Gibbs' Sampling

The Gibbs' sampling approach (see, for example, Gelfand and Smith (1990), Casella and George (1992), and Gilks and Wild (1992) for detailed descriptions) is based on sequentially simulating the univariate conditional distributions given the remaining variables, as shown in the following algorithm.

Algorithm 11.2 The Standard Gibbs' Sampling.

- **Input:** A function $f(x_1, \ldots, x_n)$, a set of conditional probabilities $p(x_i|x_1, \ldots, x_{i-1}, x_{i+1}, \ldots, x_n), i = 1, \ldots, n$, and two integers m_1 and m_2, which are the numbers of iterations of the initialization and estimation phases, respectively.

- **Output:** The expected value of $f(x_1, \ldots, x_n)$ with respect to the JPD corresponding to $p(x_i|x_1, \ldots, x_{i-1}, x_{i+1}, \ldots, x_n)$.

1. **Initialization phase:** Set $mean = 0$ and $(X_1, \ldots, X_n)$ to arbitrary values $(x_1^0, \ldots, x_n^0)$.

2. For $j = 1$ to m_1 do: For $i = 1$ to n do: simulate x_i^j using the conditional probability $p(x_i|x_1^{j-1}, \ldots, x_{i-1}^{j-1}, x_{i+1}^{j-1}, \ldots, x_n^{j-1})$.

3. **Estimation phase:** For $j = m_1$ to $m_1 + m_2$ do:

    (a) For $i = 1$ to n do: simulate x_i^j using the conditional probability $p(x_i|x_1^{j-1}, \ldots, x_{i-1}^{j-1}, x_{i+1}^{j-1}, \ldots, x_n^{j-1})$.

    (b) $mean = mean + f(x_1^j, \ldots, x_n^j)$.

4. Return $mean/m_2$. ∎

In the case of incomplete data we apply the Gibbs' sampling algorithm to obtain the missing values and replace them by their expectations. Thus, our $f(x_1,\ldots,x_n)$ functions are $f(x_1,\ldots,x_n) = x_m$, where m takes the values corresponding to the missing data values only. However, since the conditional probabilities depend on parameters to be estimated, we need to iterate until convergence. The following algorithm, where we use $X_{(k)}$ to denote all variables but X_k, illustrates this process.

Algorithm 11.3 The Modified Gibbs' Sampling.

- **Input:** An incomplete data set S, a prior probability distribution $p(D)$, a set of conditional probabilities $p(x_k|x_{(k)},\theta)$, $k = 1,\ldots,n$, and two integers m_1 and m_2.

- **Output:** The expected value of $p(\theta|S,D)$.

1. Initialization: Set $meantheta = 0$ and θ^1 to its prior mean estimate.

2. For $j = 1$ to $j = m_1$ do:

    (a) Use the standard Gibbs' sampling algorithm to simulate the missing data using the conditional probability $p(x_k|x_{(k)}^j,\theta^j)$.

    (b) Use the complete sampling method to obtain the new estimate $\theta^{j+1} = \hat{\theta}^j$ of θ.

3. For $j = m_1$ to $j = m_1 + m_2$ do:

    (a) Use the standard Gibbs' sampling algorithm to simulate the missing data using the conditional probability $p(x_k|x_{(k)}^j,\theta^j)$.

    (b) Use the complete sampling method to obtain the new estimate $\theta^{j+1} = \hat{\theta}^j$ of θ.

    (c) $meantheta = meantheta + \theta^{j+1}$.

4. Return $meantheta/m_2$. ∎

11.10.2 The EM-Algorithm for Incomplete Data

The *Expectation Maximization* or *EM*-algorithm consists of two iteration steps. The first is the expectation step, in which we calculate an expectation of the missing statistic. The second is the maximization step, in which we maximize a certain function. The two steps are performed until convergence, which is guaranteed under some conditions (see Dempster, Laird, and Rubin (1977)).

We illustrate the method using multinomial Bayesian networks. Thus, we initialize θ to some θ^1, and for the kth iteration we have

1. **Expectation Step:** Calculate the expectation of N_{ijk} using the expression

$$E[N_{ijk}|\theta^k] = \sum_{\ell=1}^{N} \sum_{y_\ell} p(X_i = k, (\pi_i^j|S_\ell)|\theta^k, D), \qquad (11.46)$$

where $\pi_i^j|S_\ell$ means that in π_i^j we have assigned to the nonmissing sample values their corresponding sample values in the jth instantiation of Π_i, and y_ℓ is the set of missing values of S_ℓ. The terms in (11.46) are calculated as follows:

- If X_i and its parents are not missing variables in S_ℓ, $p(X_i = k, (\pi_i^j|S_\ell)|\theta, D)$ is zero or one depending on whether or not the sample values coincide with those in the term.
- Otherwise, we use a Bayesian network inference algorithm to evaluate this term.

2. **Maximization Step:** We obtain the θ^k values that maximize

$$p(\theta|S, D), \qquad (11.47)$$

which leads to

$$\theta_{ijk}^k = \frac{E[N_{ijk}|\theta^{k-1}] + \eta_{ijk}}{E[N_{ik}|\theta^{k-1}] + \eta_{ik}}. \qquad (11.48)$$

Thus, we initially choose some value of θ^1, and we repeat this process until convergence to get the final estimates of θ. Note that instead of the standard posterior mean estimates for the parameters θ we use the mode posteriors.

Appendix to Chapter 11: Bayesian Statistics

The material presented in this chapter requires some knowledge of Bayesian statistical theory. In this appendix we introduce the most important concepts of Bayesian statistical theory. For more detailed treatments of Bayesian statistics, the reader is referred to books such as DeGroot (1970), Press (1992), or Bernardo and Smith (1994).

Suppose that a random variable X belongs to some parametric family $f(x|\theta)$ that depends on a set of parameters θ. In Bayesian statistical theory θ is assumed to be a random variable with a probability density function (pdf) $g(\theta;\eta)$, where η are known parameters. The pdf $g(\theta;\eta)$ is called the *prior probability distribution* of the parameter θ. In order to improve our knowledge about θ, a sample (data x) is obtained from $f(x|\theta)$ by means of a sampling process. Based on the prior distribution $g(\theta;\eta)$ and the sample data x, the *posterior probability distribution* of θ, $p(\theta|x,\eta)$ is obtained. Under the above settings, the random variable X is a mixed random variable and we can use the total probability theorem to obtain its distribution

$$u(x;\eta) = \int f(x|\theta)g(\theta;\eta)d\theta. \qquad (11.49)$$

A key step in Bayesian statistics is the selection of the prior distribution $g(\theta;\eta)$. This prior distribution must be selected without consideration of the data; however, the information contained in previous data can be incorporated into the prior. One problem in Bayesian statistics is that unless the prior family is carefully selected, the resulting posterior probabilities may not belong to known families and the mathematical treatment gets considerably complicated. To overcome this problem, one possibility is to choose the prior from among the class of *conjugate* distributions. Another possibility is to use *convenient* posterior distributions. The conjugate and convenient distributions are discussed below.

11.11.1 Conjugate Prior Distributions

Definition 11.4 Natural conjugate priors. *If the prior and posterior distributions are in the family of distributions for a given sampling process, we say that they are* natural conjugate *to the given sampling process.*

Thus, the most important property of conjugate families is that they are *closed* under sampling, that is, if a sample is used to update a prior probability distribution in a conjugate family, the resulting posterior probability distribution will also belong to the same family. The idea of parameterized families of distributions conjugate to a given sampling process was proposed by Raiffa and Schlaifer (1961).

The benefits of conjugacy include mathematical tractability; we obtain closed formulas for many important quantities used in Bayesian inference. In fact, we can derive simple formulas giving the parameters of the posterior probability distribution as a function of the parameters of the prior probability distribution and the available data. Consequently, obtaining the posterior probability distribution consists of evaluating these formulas instead of using Bayes' formula. When a conjugate prior distribution is used, the posterior probability distribution becomes

$$g(\theta, h(\eta;x)) \propto g(\theta,\eta)f(x,\theta),$$

where $f(x,\theta)$ is the likelihood function, $g(\theta;\eta)$ is the family of prior (and posterior) probability distribution, $h(\eta;x)$ is the posterior parameters as a function of the prior parameters and the sample x, and the parameters η and θ are vector quantities. The parameters η are called *hyperparameters* to be distinguished from the parameters θ.

The most commonly used probability distributions belong to a very general class of distributions known as the *exponential family*. The probability density function of any multidimensional random variable X which belongs to the exponential family can be written in the form

$$f(x;\theta) = \exp\left[\sum_{j=0}^{m} g_j(\theta) T_j(x)\right], \qquad (11.50)$$

where $T_j(x)$ is a function of x and $g_j(\theta)$ is a function of the set of m parameters θ; see e.g., Bickel and Doksum (1977) and Brown (1986). Examples of probability distributions that belong to the exponential family are the normal, the binomial, the Poisson, and exponential distributions. In this section we present examples of conjugate prior distributions which belong to the exponential family.

We start with the following theorem by Arnold, Castillo, and Sarabia (1993), which gives the most general exponential family of prior probability distributions that is conjugate to the exponential family.

Theorem 11.6 Conjugate priors for exponential likelihoods. *Suppose that the likelihood can be expressed in the exponential family form,*

$$f(x;\theta) = \exp\left[Na(\theta) + \sum_{j=0}^{m} g_j(\theta) T_j(x)\right], \quad m < t, \qquad (11.51)$$

where by convention $g_0(\theta) = 1$, $\{T_j(x); j = 0, 1, \ldots, m\}$ is a set of linearly independent functions, N is the sample size, and $a(\theta)$ is a function of θ. Then the most general t-parameter exponential family of prior distributions for $\theta = (\theta_1, \ldots, \theta_m)$ that is conjugate with respect to (11.51) is

$$q(\theta;\eta) = \exp\left[\nu(\eta) + u(\theta) + \sum_{j=1}^{m} \eta_i g_i(\theta) + \eta_{m+1} a(\theta) + \sum_{j=m+2}^{t} \eta_i s_i(\theta)\right], \qquad (11.52)$$

where $s_{m+2}(\theta), \ldots, s_t(\theta)$; $\nu(\eta)$, and $u(\theta)$ are arbitrary functions of θ; and $\eta_1, \ldots, \eta_m$ are the hyperparameters. The posterior hyperparameter vector is

$$(\eta_1 + T_1(x), \ldots, \eta_m + T_m(x), \eta_{m+1} + N, \eta_{m+2}, \ldots, \eta_t), \qquad (11.53)$$

where the hyperparameters $\eta_{m+1}, \ldots, \eta_t$ do not depend on the sample data.

Note that (11.53) gives the posterior hyperparameters as a function of the prior hyperparameters and the sample values. In the following we give several important examples of applications of this theorem.

Example 11.4 Conjugate of a Bernoulli sampling process. Suppose that X is a Bernoulli random variable that takes the value $X = 1$ with probability θ and $X = 0$ with probability $1-\theta$. We obtain an independently and identically distributed (iid) random sample $x = (x_1, \ldots, x_N)$ of size N. Let $r = \sum_{i=1}^{N} x_i$, which is the number of ones in the sample. Then r is a binomial random variable $B(N, \theta)$, and the likelihood function becomes

$$f(x;\theta) = \exp\left\{\log\binom{N}{r} + r\log\left(\frac{\theta}{1-\theta}\right) + N\log(1-\theta)\right\},$$

where

$$\binom{N}{r} = \frac{N!}{r!(N-r)!}.$$

Identifying terms with (11.51) we get

$$a(\theta) = \log(1-\theta),$$
$$T_0(x) = \log\binom{N}{r},$$
$$T_1(x) = r,$$
$$g_1(\theta) = \log\left(\frac{\theta}{1-\theta}\right),$$

and substitution into (11.52) leads to

$$q(\theta,\eta) = \exp\left\{\nu(\eta) + u(\theta) + \sum_{i=3}^{t} \eta_i s_i(\theta) + \eta_1 \log\left(\frac{\theta}{1-\theta}\right) + \eta_2 \log(1-\theta)\right\},$$

which includes, but is not limited to, the *Beta* distribution. This shows that the Beta distribution is a natural conjugate of the Bernoulli sampling process above. It follows from (11.53) that the posterior hyperparameters in this case are

$$(\eta_1 + r, \eta_2 + N),$$

which shows that a $Beta(\eta_1, \eta_2)$ prior probability distribution leads to a $Beta(\eta_1 + r, \eta_2 + N)$ posterior probability distribution. ∎

Example 11.5 Conjugate of a Poisson sampling process. Consider a *Poisson process* with intensity θ events per unit time. We count the number of events X occurring in a period of N time units. Then X is a Poisson random variable with mean $N\theta$. The likelihood function in this case is

$$f(x;\theta) = \exp\left\{-N\theta + \log\left(\frac{N^x}{x!}\right) + x\log\theta\right\}.$$

Identifying terms with (11.51) we get

$$a(\theta) = -\theta,$$
$$T_0(x) = \log\left(\frac{N^x}{x!}\right),$$
$$T_1(x) = x,$$
$$g_1(\theta) = \log\theta,$$

and substitution into (11.52) leads to

$$q(\theta, \eta) = \exp\left\{\nu(\eta) + u(\theta) + \sum_{i=3}^{t} \eta_i s_i(\theta) + \eta_1 \log\theta - \eta_2\theta\right\},$$

which includes, but is not limited to, the *gamma* distribution. This shows that the gamma distribution is a natural conjugate of the Poisson sampling process.

The formula for the hyperparameters (11.53) shows that a $Gamma(\eta_1, \eta_2)$ prior probability distribution leads to a $Gamma(\eta_1 + x, \eta_2 + N)$ posterior probability distribution. ■

Example 11.6 Conjugate of a normal sampling process. Suppose that X is a univariate normal random variable $N(\theta, \sigma^2)$ with unknown mean θ and known variance σ^2. We obtain an iid random sample $(x_1, \ldots, x_N)$ from $N(\theta, \sigma^2)$. Then, the likelihood function becomes

$$f(x;\theta) = \exp\left\{-\log\left(\sigma\sqrt{2\pi}\right) - \sum_{i=1}^{N}\frac{x_i^2}{2\sigma^2} - N\frac{\theta^2}{2\sigma^2} + \theta\sum_{i=1}^{N}\frac{x_i}{\sigma^2}\right\}.$$

Identifying terms with (11.51) we get

$$a(\theta) = -\frac{\theta^2}{2\sigma^2},$$
$$T_0(x) = -\log\left(\sigma\sqrt{2\pi}\right) - \sum_{i=1}^{N}\frac{x_i^2}{2\sigma^2},$$
$$T_1(x) = \sum_{i=1}^{N}\frac{x_i}{\sigma^2},$$
$$g_1(\theta) = \theta,$$

and substitution into (11.52) leads to

$$q(\theta; \eta) = \exp\left\{\nu(\eta) + u(\theta) + \sum_{i=3}^{t}\eta_i s_i(\theta) + \eta_1\theta - \eta_2\frac{\theta^2}{2\sigma^2}\right\},$$

which includes, but is not limited to, the normal $N(\eta_1\sigma^2/\eta_2, \sigma^2/\eta_2)$ distribution. This shows that the normal distribution is a natural conjugate of the normal sampling process above.

It follows from (11.53) that a $N(\eta_1\sigma^2/\eta_2, \sigma^2/\eta_2)$ prior probability distribution leads to a $N((\eta_1\sigma^2 + N\bar{x})/(\eta_2 + N), \sigma^2/(\eta_2 + N))$ posterior probability distribution. ■

Example 11.7 Conjugate of a multinomial sampling process (I). Let $(x_1, \ldots, x_k)$ be an iid random sample from a *multinomial* distribution $M(N; p_1, \ldots, p_{k-1})$, where $\sum_{i=1}^{k} x_i = N$. Then, the likelihood function is

$$f(x;\theta) = \exp\left\{\log\left(\frac{N!}{x_1!\ldots x_k!}\right) + \sum_{i=1}^{k-1} x_i\log\theta_i + \left(N - \sum_{i=1}^{k-1} x_i\right)\log\left(1 - \sum_{i=1}^{k-1}\theta_i\right)\right\}.$$

520   11. Learning Bayesian Networks

Identifying terms with (11.51) we get

$$a(\theta) = \log(1 - \theta_1 - \ldots - \theta_{k-1}),$$
$$T_0(x) = \log\left(\frac{N!}{x_1!\ldots x_k!}\right),$$
$$T_j(x) = x_j;\ j = 1,\ldots k-1,$$
$$g_j(\theta) = \log\left(\frac{\theta_j}{1-\theta_1-\ldots-\theta_{k-1}}\right);\ j = 1,\ldots,k-1,$$

and substitution into (11.52) leads to

$$q(\theta;\eta) = \exp\left\{\nu(\eta) + u(\theta) + \sum_{i=k+1}^{t} \eta_i s_i(\theta)\right\}$$
$$\times \exp\left\{\sum_{i=1}^{k-1} \eta_i \log\left(\frac{\theta_i}{1-\theta_1-\ldots-\theta_{k-1}}\right) + \eta_k \log(1-\theta_1-\ldots-\theta_{k-1})\right\},$$

which includes, but is not limited to, the *Dirichlet* distribution. This shows that the Dirichlet distribution is a natural conjugate of the multinomial sampling process.

The hyperparameter formula (11.53) shows that a $Dirichlet(k;\eta_1,\ldots,\eta_k)$ prior probability distribution leads to a $Dirichlet(k;\eta_1 + x_1,\ldots,\eta_{k-1} + x_{k-1},\eta_k + N)$ posterior probability distribution. ■

Example 11.8 Conjugate of a multinomial sampling process (II).
Consider again the multinomial sampling process of Example 11.7, but with a new reparametrization. The new parameters are

$$\pi_i = \log\left(\frac{\theta_i}{1 - \theta_1 - \ldots - \theta_{k-1}}\right);\ i = 1,\ldots,k-1.$$

With this reparametrization, the likelihood function becomes

$$f(x;\pi) = \exp\{-N\log(1 + \exp(\pi_1) + \ldots + \exp(\pi_{k-1})) + \log(N!)\}$$
$$\times \exp\left\{-\log\left(x_1!\ldots x_{k-1}!\left(N - \sum_{i=1}^{k-1} x_i\right)!\right) + \sum_{i=1}^{k-1} \pi_i x_i\right\}.$$

Identifying terms with (11.51) we get

$$a(\pi) = -\log(1 + \exp(\pi_1) + \ldots + \exp(\pi_{k-1})),$$
$$T_0(x) = +\log(N!) - \log\left[x_1!\ldots x_{k-1}!\left(N - \sum_{i=1}^{k-1} x_i\right)!\right],$$
$$T_j(x) = x_j;\ j = 1,\ldots,k-1,$$
$$g_j(\pi) = \pi_j;\ j = 1,\ldots,k-1,$$

and substitution into (11.52) leads to

$$q(\pi;\eta) = \exp\left\{\nu(\eta) + u(\pi) + \sum_{i=k+1}^{t} \eta_i s_i(\pi)\right\}$$

$$\times \exp\left\{\sum_{i=1}^{k-1} \eta_i \pi_i - \eta_k \log\left(1 + \exp(\pi_1) + \ldots + \exp(\pi_{k-1})\right)\right\}.$$

The formula for the hyperparameters (11.53) shows that a $q(\pi; \eta_1, \ldots, \eta_k)$ prior probability distribution leads to a $q(\pi; \eta_1+x_1, \ldots, \eta_{k-1}+x_{k-1}, \eta_k+N)$ posterior distribution. ∎

Example 11.9 Conjugate of a multivariate normal sampling process. Let $(x_1, \ldots, x_N)$ be an iid random sample from the multivariate normal distribution $N(\mu, W^{-1})$, where the parameters μ and W, are the mean vector and the precision matrix (the inverse of the covariance matrix), respectively. Note that x_i is a vector. Then, the likelihood function is

$$f(x; \mu, W) =$$
$$\exp\left\{\log\left[\sqrt{det(W)}\right] - k\log(2\pi)/2 - \frac{1}{2}\left[\sum_{r=1}^{N}\sum_{i,j=1}^{k} w_{ij} z_{ir} z_{jr}\right]\right\}$$
$$= \exp\left\{\log\left[\sqrt{det(W)}\right] - k\log(2\pi)/2 - \frac{N}{2}\sum_{i,j=1}^{k} w_{ij}\mu_i\mu_j\right\}$$
$$\times \exp\left\{\sum_{i=1}^{k}\left(\sum_{j=1}^{k} w_{ij}\mu_j\right)\sum_{r=1}^{N} x_{ir} - \frac{1}{2}\sum_{i,j=1}^{k} w_{ij}\sum_{r=1}^{N} x_{ir}x_{jr}\right\}.$$

where $z_{ir} = x_{ir} - \mu_i$. Identifying terms with (11.51) and using double indices we get

$$a(\theta) = -\frac{1}{2}\sum_{i,j=1}^{k} w_{ij}\mu_i\mu_j,$$
$$W_0(x) = -k\log(2\pi)/2 + \log\left(\sqrt{det(W)}\right),$$
$$W_i(x) = \sum_{r=1}^{N} x_{ir}; \; i = 1, \ldots k,$$
$$W_{ij}(x) = -\frac{1}{2}\sum_{r=1}^{N} x_{ir}x_{jr}; \; i, j = 1, \ldots k,$$
$$g_i(\mu, W) = \sum_{j=1}^{k} w_{ij}\mu_j; \; i = 1, \ldots, k,$$
$$g_{ij}(\mu, W) = w_{ij}; \; i, j = 1, \ldots, k,$$

and substitution into (11.52) leads to

$$q(\mu, W; \eta) = \exp\left\{\nu(\eta) + u(\mu, W) + \sum_{i=k+k^2+2}^{t} \eta_i s_i(\mu, W)\right\}$$
$$\times \exp\left\{\sum_{i=1}^{k}\eta_i\left(\sum_{j=1}^{k} w_{ij}\mu_j\right) + \sum_{i,j=1}^{k}\eta_{ij}w_{ij} + \eta_{k+k^2+1}\left[-\frac{1}{2}\sum_{i,j=1}^{k} w_{ij}\mu_i\mu_j\right]\right\},$$

which includes, but is not limited to, the *normal-Wishart* distribution.

Therefore, with a reparametrization, if the prior probability distribution is a normal-Wishart$(\nu, \mu_0, \alpha, W_0)$ distribution with density

$$g(\mu, W; \nu, \mu_0, \alpha, W_0) \propto$$
$$det(W)^{(\alpha-k)} \exp\left\{-\tfrac{1}{2}\left[tr(W_0 W) + \sum_{i,j=1}^{k} \nu w_{ij}(\mu_i - \mu_{0i})(\mu_j - \mu_{0j})\right]\right\},$$

where $tr(A)$ denotes the trace of matrix A (the sum of the diagonal elements of A), then the posterior probability distribution is normal-Wishart$(\nu + N, \mu_N, \alpha + N, W_N)$, where

$$\bar{x} = \tfrac{1}{N}\sum_{i=1}^{N} x_i,$$
$$\mu_N = \tfrac{\nu \mu_0 + N \bar{x}}{\nu + N},$$
$$S = \sum_{i=1}^{N}(x_i - \bar{x})(x_i - \bar{x})^T,$$
$$W_N = W_0 + S + \tfrac{\nu N}{\nu + N}(\mu_0 - \bar{x})(\mu_0 - \bar{x})^T.$$

Therefore, the normal-Wishart distribution is a natural conjugate of the multivariate normal sampling process. ∎

Example 11.10 Conjugate of a Dirichlet sampling process. Let us consider $(x_1, \ldots, x_N)$ to be an iid random sample from a Dirichlet distribution with parameters $\theta_1, \ldots, \theta_{k+1}$. Noe that x_i is a k-dimensional vector. The probability density function of the Dirichlet distribution is

$$f(x_1, \ldots, x_k; \theta_1, \ldots, \theta_{k+1}) = \frac{\Gamma(\sum_{j=1}^{k+1} \theta_j)}{\prod_{j=1}^{k+1} \Gamma(\theta_j)} \left(\prod_{j=1}^{k} x_j^{\theta_j - 1}\right)\left(1 - \sum_{j=1}^{k} x_j\right)^{\theta_{k+1}-1},$$

where $\Gamma(x)$ is the gamma function. Note that if x is an integer, we have $\Gamma(x) = (x-1)!$. The parameters $\theta_i, i = 1, \ldots, k$ are proportional to the mean values of $X_1, \ldots, X - k$ and such that the variances decrease with their sum. Then the likelihood of the sample is

$$L = \prod_{i=1}^{N} f(x_i; \theta) = \prod_{i=1}^{N} \left[\frac{\Gamma\left(\sum_{j=1}^{k+1} \theta_j\right)}{\prod_{j=1}^{k+1} \Gamma(\theta_j)} \left(\prod_{j=1}^{k} x_{ij}^{\theta_j - 1}\right)\left(1 - \sum_{j=1}^{k} x_{ij}\right)^{\theta_{k+1}-1}\right],$$

which can be written in the form of (11.51) as

$$L = \exp\left[N\log\left[\Gamma\left(\sum_{j=1}^{k+1}\theta_j\right) - \sum_{j=1}^{k+1}\log\Gamma(\theta_j)\right]\right.$$
$$\left. + \sum_{j=1}^{k}(\theta_j - 1)\sum_{i=1}^{N}\log x_{ij} + (\theta_{k+1} - 1)\sum_{i=1}^{N}\log\left(1 - \sum_{j=1}^{k}x_{ij}\right)\right].$$
(11.54)

Then, we have

$$m = k + 1,$$

$$a(\theta) = \log\Gamma\left(\sum_{j=1}^{k+1}\theta_j\right) - \sum_{j=1}^{k+1}\log\Gamma(\theta_j),$$

$$g_j(\theta) = \theta_j, \quad j = 1,\ldots,k,$$

$$g_{k+1}(\theta) = \theta_{k+1},$$

$$T_0(x) = -\sum_{j=1}^{k}\sum_{i=1}^{N}\log x_{ij} - \sum_{i=1}^{N}\log\left(1 - \sum_{j=1}^{k}x_{ij}\right),$$

$$T_j(x) = \sum_{j=1}^{N}\log x_{ij}, \quad j = 1,\ldots,k,$$

$$T_{k+1}(x) = \sum_{i=1}^{N}\log\left(1 - \sum_{j=1}^{k}x_{ij}\right).$$

It follows from (11.52) that the prior probability distribution of θ, $q(\theta;\eta)$, is proportional to

$$\exp\left(\nu(\eta) + u(\theta) + \sum_{i=1}^{k+1}\eta_i\theta_i + \eta_{k+2}\log\frac{\Gamma\left(\sum_{j=1}^{k+1}\theta_j\right)}{\prod_{j=1}^{k+1}\Gamma(\theta_j)} + \sum_{i=k+3}^{t}\eta_i s_i(\theta)\right),$$

where $\nu(\eta)$, $u(\theta)$, and $s_i(\theta)$; $i = k+3,\ldots,t$, are arbitrary functions. The posterior hyperparameters become

$$(\eta_1 + T_1(x),\ldots,\eta_{k+1} + T_{k+1}(x),\eta_{k+2} + N, \eta_{k+3},\ldots,\eta_t).$$

Due to the fact that the hyperparameters η_i, $i = k+3,\ldots,t$, are static parameters (they are not altered by the information), we can take

$$q(\theta;\eta) \propto \exp\left(\sum_{i=1}^{k+1}\eta_i\theta_i + \eta_{k+2}\log\Gamma\left(\sum_{j=1}^{k+1}\theta_j\right) - \eta_{k+2}\sum_{j=1}^{k+1}\log(\Gamma(\theta_j))\right).$$
(11.55)

Since the mean of the conjugate distribution in (11.55) is difficult to obtain analytically, we can use the mode of (11.55) to estimate the Dirichlet parameters. The mode can be obtained by maximizing (11.55) with respect to θ. ∎

11.11.2 Convenient Posterior Probability Distributions

When selecting a family of prior probability distributions to be combined with a given likelihood, the prime consideration is that the resulting posterior probability distributions should be members of tractable families of distributions. Thus we do not need both, priors and posterior probability distributions to belong to the same family. It may suffice to have posterior probability distributions which belong to a given exponential family. These distributions are referred to as *convenient* posterior distributions.

The following theorem by Arnold, Castillo, and Sarabia (1994) identifies the family of prior probability distributions that lead to convenient posterior probability distributions.

Theorem 11.7 Convenient posterior distributions *Consider an iid random sample of N possibly vector valued observations $x_1, \ldots, x_N$ from an m-parameter exponential family of the form*

$$f(x; \theta) = \exp\left[a(\theta) + \sum_{j=0}^{m} \theta_j T_j(x)\right], \quad (11.56)$$

where $\theta_0 = 1$. Then, the most general class of prior probability distributions on Θ that leads to a posterior probability distribution for θ that belongs to the t-parameter exponential family of the form

$$f(\theta, \eta) = \exp\left[\nu(\eta) + \sum_{s=0}^{t} \eta_s g_s(\theta)\right], \quad t \leq m,$$

where $\eta_0 = 1$, is of the form

$$f(\theta; c) = \exp\left[c_{00} - a(\theta) + g_0(\theta) + \sum_{i=1}^{t} c_{i0} g_i(\theta)\right], \quad (11.57)$$

where

$$\begin{bmatrix} \theta_1 \\ \theta_2 \\ \ldots \\ \theta_m \end{bmatrix} = \begin{bmatrix} c_{01} & c_{11} & \ldots & c_{t1} \\ c_{02} & c_{12} & \ldots & c_{t2} \\ \ldots & \ldots & \ldots & \ldots \\ c_{0m} & c_{1m} & \ldots & c_{tm} \end{bmatrix} \begin{bmatrix} g_0(\theta) \\ g_1(\theta) \\ g_2(\theta) \\ \ldots \\ g_t(\theta) \end{bmatrix}. \quad (11.58)$$

11.10 The Case of Incomplete Data

The hyperparameters become

$$\begin{bmatrix} \eta_1(x) \\ \eta_2(x) \\ \vdots \\ \eta_t(x) \end{bmatrix} = \begin{bmatrix} c_{10} & c_{11} & \cdots & c_{1m} \\ c_{20} & c_{21} & \cdots & c_{2m} \\ \vdots & \vdots & \cdots & \vdots \\ c_{t0} & c_{t1} & \cdots & c_{tm} \end{bmatrix} \begin{bmatrix} T_0(x) \\ T_1(x) \\ T_2(x) \\ \vdots \\ T_m(x) \end{bmatrix}. \tag{11.59}$$

The coefficients $\{c_{ij}; i = 0, \ldots t; j = 0, \ldots, m\}$ should make the function (11.57) integrable.

In addition they have found that this problem has a solution only in the exponential family. Note, however, that (11.58) imposes severe restrictions on the candidate posterior probability distribution.

Exercises

11.1 For each of the 11 directed graphs in Figure 11.2

(a) Write the JPD in factorized form as suggested by the graph.

(b) Write the corresponding parameters using the notation $\theta_{ij\pi} = p(X_i = j | \Pi_i = \pi)$.

(c) Find the dimension of each of the Bayesian networks implied by the graph and its corresponding JPD.

11.2 Consider the Bayesian network defined by the complete directed graph in Figure 11.2 (graph number 11) and the JPD

$$p(x_1, x_2, x_3) = p(x_1)p(x_2|x_1)p(x_3|x_1, x_2). \tag{11.60}$$

Let this graph be denoted by D_{11}. Suppose that all variables are binary.

(a) Show that the JPD in (11.60) depends on seven free parameters.

(b) Provide a numerical example of the CPDs in (11.60) by assigning each of the parameters to an appropriate value of your choice.

(c) Let θ_k denote the kth parameter. Suppose that a prior probability distribution $p(\theta_1, \ldots, \theta_7 | D_c)$ is specified to be uniform, that is, $p(\theta_1, \ldots, \theta_7 | D_c) = 1; 0 \le \theta_k \le 1, k = 1, \ldots, 7$. Assume parameter independence and modularity. Use Algorithm 11.1 to compute the prior probability distribution for the Bayesian network (D_5, P_5), where D_5 is the graph number 5 in Figure 11.2 and P_5 is the set of CPDs suggested by D_5.

x_1	$p(x_1)$
0	0.7
1	0.3

x_2	$p(x_2)$
0	0.6
1	0.4

x_1	x_2	x_3	$p(x_3\|x_1,x_2)$
0	0	0	0.1
0	0	1	0.6
0	0	2	0.3
0	1	0	0.4
0	1	1	0.2
0	1	2	0.4
1	0	0	0.7
1	0	1	0.1
1	0	2	0.2
1	1	0	0.1
1	1	1	0.1
1	1	2	0.8

TABLE 11.11. A set of CPDs.

(d) Under the above conditions, use Algorithm 11.1 to compute the prior probability distribution for the Bayesian network (D_8, P_8), where D_8 is the graph number 8 in Figure 11.2 and P_8 is the set of CPDs suggested by D_8.

11.3 Repeat the previous exercise when X_1 and X_2 are binary but X_3 is ternary.

11.4 Consider the Bayesian network defined by graph number 10 in Figure 11.2 and the CPDs in Table 11.11. Note that one of the variables is ternary and the other two are binary.

(a) What is the dimension of this Bayesian network model?
(b) Compute the frequency of each possible instantiation of the three variables in a perfect sample of size 1,000.

11.5 Assume a prior probability distribution that assigns the same probability to each of the free parameters in a Bayesian network model. Use the perfect sample of size 1,000 from the previous exercise to compute the following quality measures for each of the 11 networks in Figure 11.2 and compare the various measures using the obtained results:

(a) The Geiger and Heckerman measure in (11.23) with $\eta = 160$.

Case	X_1	X_2	X_3	Case	X_1	X_2	X_3
1	4.3	10.3	16.0	16	4.8	10.3	16.3
2	6.5	10.8	16.3	17	4.5	6.2	10.5
3	4.3	8.6	13.1	18	4.8	14.2	17.8
4	3.0	10.5	14.6	19	4.3	7.6	12.5
5	4.9	8.9	14.4	20	5.4	10.5	15.1
6	5.5	8.3	14.0	21	3.3	11.4	15.1
7	5.2	12.6	17.4	22	6.0	9.8	16.2
8	2.9	9.8	13.5	23	5.9	8.6	15.7
9	6.2	9.3	17.2	24	4.4	6.5	9.5
10	5.7	10.1	16.4	25	3.4	7.2	11.5
11	4.6	12.1	16.6	26	4.2	7.6	12.1
12	4.4	13.8	16.9	27	5.2	11.4	15.4
13	5.7	8.8	13.2	28	5.2	11.4	17.0
14	5.6	9.2	16.3	29	2.2	9.4	10.9
15	2.8	8.4	11.4	30	4.6	9.0	11.3

TABLE 11.12. A data set drawn from a trivariate Gaussian distribution.

(b) The Geiger and Heckerman measure in (11.23) with $\eta = 8$.

(c) The Cooper and Herskovits measure in (11.24).

(d) The standard Bayesian measure in (11.27).

(e) The asymptotic standard Bayesian measure in (11.28).

(f) The minimum description length measure in (11.42).

(g) The information measure in (11.43) for any penalizing function $f(N)$ of your choice.

11.6 The data set in Table 11.12 was generated from a trivariate Gaussian distribution. With the assumptions listed in Section 11.5.1 and the prior probability distribution of the parameters as given in Example 11.3, compute the Geiger and Heckerman measure in (11.37) for each of the 11 networks in Figure 11.2.

11.7 Assume a prior probability distribution that assigns the same probability to each of the free parameters in a Bayesian network model. Consider the perfect sample of size 1,000 generated from the JPD defined by the set of CPDs in Table 11.11. Find the highest quality Bayesian network using each of the following search algorithms:

(a) The $K2$-algorithm using the natural ordering X_1, X_2, X_3.

(b) The $K2$-algorithm using the reverse natural ordering X_3, X_2, X_1.

(c) The B-algorithm.

Chapter 12
Case Studies

12.1 Introduction

In this chapter we apply the methodology presented in the previous chapters to three case studies of real-life applications:

- The pressure tank problem (Section 12.2).

- The power distribution system problem (Section 12.3).

- The damage of reinforced concrete structure of buildings (Sections 12.4 and 12.5).

As one might expect, real-life applications are more complicated than the textbook examples that are usually used to illustrate a given method. Furthermore, many of the assumptions that we usually make to simplify matters do not hold in practice. For example,

- Variables can be discrete (binary, categorical, etc.), continuous, or mixed (some are discrete and others are continuous).

- The relationships among the variables can be very complicated. For example, the model may include feedback or cycles. As a result, specification of the probabilistic models can be difficult and may lead to problems.

- Propagation of evidence can be time-consuming due to the large number of parameters and to the complexity of the network structures.

530   12. Case Studies

We shall encounter many of the above problems in the three case studies we present in this chapter. We use these real-life examples to illustrate and further reinforce the steps to be followed when modeling probabilistic networks using the different models introduced in this book.

In some practical situations, the relationships among the variables in the model allow us to define an associated probabilistic model in an easy way. In some cases, Markov and Bayesian networks (see Chapter 6) provide appropriate methodologies to define a consistent probabilistic model based on a graphical representation. In Sections 12.2 and 12.3 we use this methodology to define a probabilistic network model for the *pressure tank* and the *power distribution* problems. However, in the general case, the relationships among the variables in a given problem can be very complex, and therefore, graphically specified models (see Chapter 6) may not be appropriate for the definition of a consistent probabilistic model associated with the problem. In these situations more general probabilistic frameworks are needed to analyze the relationships among the variables involved in the problem. For example, in Section 12.4 we apply the conditionally specified models (see Section 7.7) in the problem of the *damage of reinforced structures*. We shall see that this methodology provides a consistent probabilistic model in this case.

All calculations are done using computer programs written by the authors. [1]

12.2   Pressure Tank System

12.2.1   Definition of the Problem

Figure 12.1 shows a diagram of a pressure tank with its main elements. It is a tank used for the storing of a pressurized fluid, which is introduced with the help of a pump activated by an electric motor. The tank is known not to have problems if the pump is working for periods of less than one minute. Therefore, a security mechanism, based on a time relay, F, interrupts the electric current after 60 seconds. In addition, a pressure switch, A, also interrupts the current if the pressure in the tank reaches a certain threshold value. The system includes a switch, E, which initiates the operation of the system; one relay, D, which supplies the current after the initiation step and interrupts it after the activation of relay F; and relay C, which starts the operation of the electrical circuit of the motor. We are interested in knowing the probability of failure of the pressure tank.

---

[1] These programs can be obtained from the World Wide Web (WWW) site http://ccaix3.unican.es/~AIGroup.

12.2 Pressure Tank System

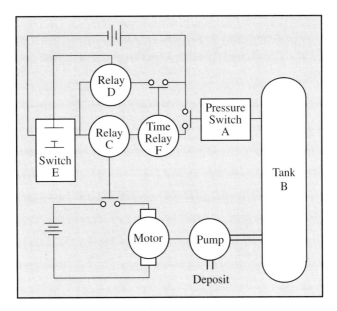

FIGURE 12.1. A diagram for a pressure tank system.

12.2.2 Variable Selection

Since we are interested in the analysis of all possible causes of the failure of tank B, we introduce a new variable K to denote this event. We shall use the notation $K = k$ to denote the failure of the tank, and $K = \bar{k}$ to denote its complementary event of no failure. Similarly, $a, \ldots, f$ represent failures of the respective components $A, \ldots, F$, and $\bar{a}, \ldots, \bar{f}$ represent the events of no failures.

Based on the previous description of the problem, we can write the following logical expression for the failure of the tank:

$$\begin{aligned} k &= b \vee c \vee (a \wedge e) \vee (a \wedge d) \vee (a \wedge f) \\ &= b \vee c \vee (a \wedge (e \vee d \vee f)), \end{aligned} \quad (12.1)$$

where the symbols $\vee$ and $\wedge$ are used for *or* and *and*, respectively. This expression is obtained by combining all of the possible partial failures of the components of the system that produce a failure of the tank. An alternative logical expression for the correct functioning of the system can be obtained by taking the complement of both sides of (12.1). We get

$$\begin{aligned} \bar{k} &= \overline{b \vee c \vee (a \wedge e) \vee (a \wedge d) \vee (a \wedge f)} \\ &= \bar{b} \wedge \bar{c} \wedge \overline{(a \wedge e)} \wedge \overline{(a \wedge d)} \wedge \overline{(a \wedge f)} \\ &= \bar{b} \wedge \bar{c} \wedge (\bar{a} \vee \bar{e}) \wedge (\bar{a} \vee \bar{d}) \wedge (\bar{a} \vee \bar{f}) \\ &= \left(\bar{b} \wedge \bar{c} \wedge \bar{a}\right) \vee \left(\bar{b} \wedge \bar{c} \wedge \bar{e} \wedge \bar{d} \wedge \bar{f}\right). \end{aligned} \quad (12.2)$$

Equations (12.1) and (12.2) form the basis for deriving the set of rules (for a rule-based expert system) or the dependency structure (of a probabilistic network model). These equations can be expressed in a more intuitive way using a *failure tree*. For example, Figure 12.2 shows a failure tree associated with (12.1). In this tree, the failures of relays D and F are combined into an intermediate cause of failure, G; then G is combined with E to define another intermediate cause of failure, H, and so on. This tree includes the initial variables $A, \ldots, F$ as well as the intermediate failures $\{G, \ldots, J\}$ that imply the failure of the tank. Therefore, the final set of variables used in this example is $X = \{A, \ldots, K\}$.

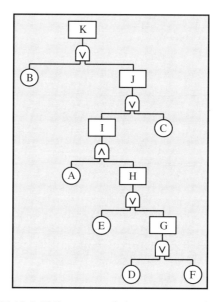

FIGURE 12.2. Failure tree of the pressure tank system.

12.2.3 Model Selection

The pressure tank example described above can be analyzed from a deterministic point of view using the rule-based expert systems introduced in Chapter 2. Figure 12.3 shows the rules derived from the failure tree in Figure 12.2 as well as the chaining among the premises and conclusions of different rules. The formal definition of the knowledge base (the rules) is left as an exercise for the reader.

Thus, once some evidence is introduced, the algorithms for rule chaining can be used to draw conclusions.

The rules state the conditions under which the truthfulness of the conclusion objects can be derived from those of the premise objects. However, from the point of view of the randomness paradigm, the objects can have

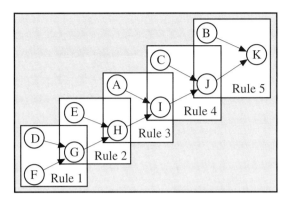

FIGURE 12.3. Chained rules for the pressure tank system.

an associated measure of uncertainty, such as the probability of being true. In addition, the rules can contain some extra information allowing the uncertainty of the conclusion objects to be obtained from that of the premise objects involved in the same rule. In fact, the key problem is how to derive the uncertainty associated with some objects (objects in the conclusion of a rule) when that of other objects (objects in the premise of a rule) are known. As we have seen in Chapter 3, from a statistical point of view, we can give a much wider interpretation to these rules by giving conditional probability tables, and even obtain aggregation formulas to combine the uncertainty of the premise objects with the uncertainty of the rule, in order to obtain the uncertainty of the conclusion objects. For example, the uncertainty of the rules "If A and B and C then D" and "If A or B or C then D" can be measured by the conditional probability $p(d|a, b, c)$. This includes the deterministic rules with

$$p(c|a,b,c) = \begin{cases} 1, & if\ A = B = C = true, \\ 0, & otherwise, \end{cases} \qquad (12.3)$$

for the first rule, and

$$p(d|a,b,c) = \begin{cases} 1, & if\ A\ or\ B\ or\ C\ is\ true, \\ 0, & otherwise, \end{cases} \qquad (12.4)$$

for the second.

The study of this example from a deterministic viewpoint is left as an exercise for the reader[2].

The structure given by the chained rules can be used to define the graphical structure of a probabilistic expert system. For example, since the failures

---

[2]We encourage the reader to use the *X-pert Rules* program for the analysis of this problem. The program can be obtained from the WWW site http://ccaix3.unican.es/~AIGroup.

of the different components of the system are the causes of the intermediate failures and finally, of the failure of the tank, we can obtain a directed graph reproducing the dependencies among the variables forming the model (see Figure 12.4). As we have seen in Chapter 6, this graph is the dependency structure of a Bayesian network. From this graph, the joint probability distribution (JPD) of all nodes can be written as

$$p(x) = p(a)p(b)p(c)p(d)p(e)p(f)p(g|d,f)p(h|e,g) \\ p(i|a,h)p(j|c,i)p(k|b,j). \qquad (12.5)$$

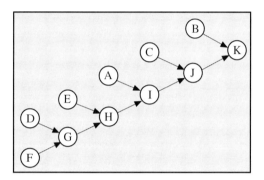

FIGURE 12.4. Directed graph for the pressure tank system.

On the one hand, the conditional probability distributions (CPDs) associated with the intermediate causes in the failure tree are defined using (12.3) and (12.4) as shown in Table 12.1, where we give only the conditional probabilities of failures, because $p(no\ failure) = 1 - p(failure)$. On the other hand, the marginal probabilities associated with the components of the system represent the initial information about the prior failure of these components. Suppose that these probabilities are

$$\begin{array}{lll} p(a) = 0.002, & p(b) = 0.001, & p(c) = 0.003, \\ p(d) = 0.010, & p(e) = 0.001, & p(f) = 0.010. \end{array} \qquad (12.6)$$

The graph in Figure 12.4 together with the probability tables shown in (12.6) and in Table 12.1 define a Bayesian network that corresponds to the above example of the pressure tank. The corresponding JPD is given in (12.5).

12.2.4 Propagation of Evidence

The graph in Figure 12.4 is a polytree, which means that we can use the efficient algorithm for polytrees (Algorithm 8.1) for evidence propagation. Suppose first that no evidence is available. In this case, Algorithm 8.1 gives

12.2 Pressure Tank System

D	F	$p(g|D,F)$
d	f	1
d	$\bar{f}$	1
$\bar{d}$	f	1
$\bar{d}$	$\bar{f}$	0

E	G	$p(h|E,G)$
e	g	1
e	$\bar{g}$	1
$\bar{e}$	g	1
$\bar{e}$	$\bar{g}$	0

A	H	$p(i|A,H)$
d	h	1
a	$\bar{h}$	0
$\bar{a}$	h	0
$\bar{a}$	$\bar{h}$	0

C	I	$p(j|C,I)$
e	i	1
c	$\bar{i}$	1
$\bar{c}$	i	1
$\bar{c}$	$\bar{i}$	0

B	J	$p(k|B,J)$
b	j	1
b	$\bar{j}$	1
$\bar{b}$	j	1
$\bar{b}$	$\bar{j}$	0

TABLE 12.1. Conditional probabilities of failures of the intermediate causes for the pressure tank system.

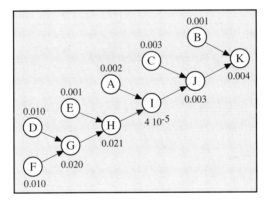

FIGURE 12.5. The initial marginal probabilities of the nodes (when no evidence is available) for the pressure tank system.

the initial marginal probabilities of the nodes, which are shown in Figure 12.5. Note that the initial probability of failure of the tank is $p(k) = 0.004$.

Suppose now that the components F and D fail, that is, we have the evidence $F = f, D = d$. We can use Algorithm 8.1 to propagate this evidence. The new conditional probabilities for the nodes $p(x_i|f, d)$ are shown in Figure 12.6. Note that the failure of relays F and D only induce the failure of the intermediate nodes G and H, but the probability of failure of the tank is still small ($p(k) = 0.006$).

To continue the illustration, suppose the pressure switch A also fails ($A = a$). Propagating the cumulative evidence ($F = f, D = d, A = a$) using Algorithm 8.1, we obtain the new conditional probabilities of the nodes as shown in Figure 12.7. Now, because $p(k) = 1$, the failures of these

536   12. Case Studies

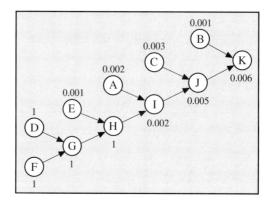

FIGURE 12.6. The conditional probabilities of the nodes given the evidence $F = f$ and $D = d$ for the pressure tank system.

components F, D, and A, imply the failure of all the intermediate causes and the failure of the tank.

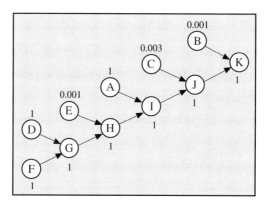

FIGURE 12.7. The conditional probabilities of the nodes given the evidence $F = f$, $D = d$, and $A = a$ for the pressure tank system.

12.2.5   Considering a Common Cause of Failure

Now let us suppose that there is a common cause of failure for the relays C, D, and F. For example, suppose that these relays have been built in the same circumstances. So, one possibility is to draw new links between them, to indicate the new dependency relationships in the model (see Castillo et al. (1994)), as shown in Figure 12.8, which is obtained from Figure 12.4 after linking the nodes C, D, and F. Note that the nodes have been rearranged to make the new graph visually uncluttered. Now, the graph in Figure 12.8 is a multiply-connected graph, and the polytree Algorithm 8.1 no longer applies. In this case, we have to use a more general propagation algorithm

12.2 Pressure Tank System

D	$p(c|D)$
d	0.750
$\bar{d}$	0.001

C	D	$p(f|C,D)$
c	d	0.99
c	$\bar{d}$	0.75
$\bar{c}$	d	0.75
$\bar{c}$	$\bar{d}$	0.01

D	F	$p(g|D,F)$
d	f	1
d	$\bar{f}$	1
$\bar{d}$	f	1
$\bar{d}$	$\bar{f}$	0

E	G	$p(h|E,G)$
e	g	1
e	$\bar{g}$	1
$\bar{e}$	g	1
$\bar{e}$	$\bar{g}$	0

A	H	$p(i|A,H)$
d	h	1
a	$\bar{h}$	0
$\bar{a}$	h	0
$\bar{a}$	$\bar{h}$	0

C	I	$p(j|C,I)$
e	i	1
c	$\bar{i}$	1
$\bar{c}$	i	1
$\bar{c}$	$\bar{i}$	0

B	J	$p(k|B,J)$
b	j	1
b	$\bar{j}$	1
$\bar{b}$	j	1
$\bar{b}$	$\bar{j}$	0

TABLE 12.2. Conditional probabilities of failures for the pressure tank system with a common cause of failure.

such as the clustering Algorithm 8.5 to propagate evidence in a join tree associated with the graph.

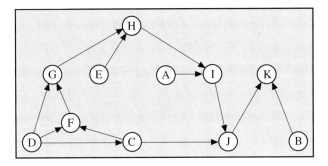

FIGURE 12.8. Directed graph for the pressure tank system with a common cause of failure.

According to the graph in Figure 12.8, the JPD of the nodes can now be factorized as

$$p(x) = p(d)p(c|d)p(f|c,d)p(g|d,f)p(e)$$
$$p(h|e,g)p(a)p(i|h,a)p(j|c,i)p(b)p(k|b,j). \quad (12.7)$$

The corresponding CPDs are given in Table 12.2.

538   12. Case Studies

To use the clustering Algorithm 8.5, we first need to moralize and triangulate the graph in Figure 12.8. The resultant moralized and triangulated undirected graph is shown in Figure 12.9.

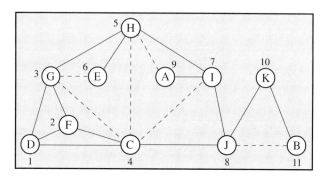

FIGURE 12.9. A moralized and triangulated graph associated with the directed graph in Figure 12.8. A perfect numbering of the nodes is shown.

The cliques associated with this graph are

$C_1 = \{A, H, I\}$, $C_2 = \{C, H, I\}$, $C_3 = \{C, G, H\}$, $C_4 = \{C, D, F, G\}$,
$C_5 = \{B, J, K\}$, $C_6 = \{C, I, J\}$, $C_7 = \{E, G, H\}$.

Thus, (12.7) can also be written using the potential representation

$$p(x) = \psi(a,h,i)\psi(c,h,i)\psi(c,g,h)$$
$$\psi(c,d,f,g)\psi(b,j,k)\psi(c,i,j)\psi(e,g,h), \qquad (12.8)$$

where

$\psi(a,h,i) = p(a)p(i|h,a)$,
$\psi(c,h,i) = 1$,
$\psi(c,g,h) = 1$,
$\psi(c,d,f,g) = p(d)p(c|d)p(f|c,d)p(g|d,f)$,
$\psi(b,j,k) = p(b)p(k|j,b)$,
$\psi(j,c,i) = p(j|c,i)$,
$\psi(e,g,h) = p(e)p(h|e,g)$.

The join tree obtained using Algorithm 4.4 is shown in Figure 12.10.

Suppose first that no evidence is available. Applying the clustering Algorithm 8.5 to this join tree, we obtain the initial marginal probabilities of the nodes as shown in Figure 12.11. The initial probability of failure for the tank is $p(k) = 0.009$. Note that the initial probability obtained when no common causes of failure are considered is $p(k) = 0.004$.

We now consider the evidence $F = f$ and $D = d$. Using Algorithm 8.5 to propagate this evidence, we obtain the conditional probabilities shown in Figure 12.12. The updated conditional probability of failure is now $p(k) = 0.799$. Note that with this evidence the conditional probability of failure

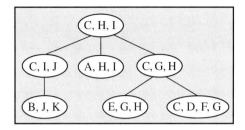

FIGURE 12.10. A join tree obtained from the undirected moralized and triangulated graph in 12.9.

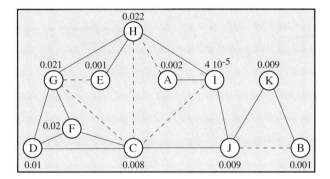

FIGURE 12.11. The initial marginal probabilities of the nodes (when no evidence is available) for the pressure tank system with a common cause of failure.

for the tank in the case of no common cause of failure was $p(k) = 0.006$. The reason for this difference is that relay C has been considered to have a common cause of failure with relays F and D, and hence, the failure of this relay implies a considerable probability of failure for C.

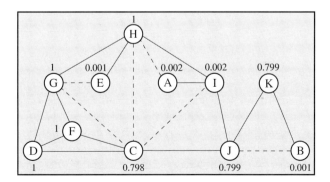

FIGURE 12.12. The conditional probabilities of the nodes given the evidence $F = f$ and $D = d$ for the pressure tank system with a common cause of failure.

540   12. Case Studies

Finally, when we consider the additional evidence $A = a$, we obtain $p(k) = 1$, indicating the failure of the tank in this situation (see Figure 12.13).

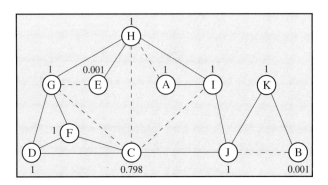

FIGURE 12.13. The conditional probabilities of the nodes given the evidence $F = f$, $D = d$, and $A = a$ for the pressure tank system with a common cause of failure.

12.2.6 Symbolic Propagation of Evidence

In this section we apply the symbolic propagation method introduced in Chapter 10 to perform sensitivity analysis, that is, we wish to study the effect of changing the probabilities of some nodes on the probabilities of the other nodes in the network. As an example, let us modify some of the conditional probabilities in (12.6) and Table 12.2 by including some symbolic parameters for nodes C and E. We replace the probabilities for nodes C and E by

$$p(c|d) = 0.75, \quad p(c|\bar{d}) = \theta_C,$$
$$p(\bar{c}|d) = 0.25, \quad p(\bar{c}|\bar{d}) = 1 - \theta_C;$$
$$p(e) = \theta_E, \quad p(\bar{e}) = 1 - \theta_E,$$

where $0 < \theta_C < 1$ and $0 < \theta_E < 1$.

For the case of no evidence, using the symbolic method discussed in Chapter 10 (Algorithm 10.1), we obtain the initial marginal probabilities of nodes shown in Table 12.3. From this table, we see that the marginal probabilities for nodes C, F, and G depend on θ_C but not on θ_E. We can also see that the marginal probabilities of nodes H to K depend on both θ_C and θ_E. However, the dependence of the marginal probabilities of nodes J and K depends more on θ_C than on θ_E (the coefficients of θ_C are very large as compared to those of θ_E). Also, the dependence of the marginal probabilities of node I on θ_C and θ_E is weak.

X_i	x_i	$p(x_i)$
A	a	0.002
B	b	0.001
C	c	$0.0075 + 0.99\theta_C$
D	d	0.01
E	e	θ_E
F	f	$0.0192 + 0.7326\theta_C$
G	g	$0.0199 + 0.7326\theta_C$
H	h	$0.0199 + 0.7326\theta_C + 0.9801\theta_E - 0.7326\theta_C\theta_E$
I	i	$0.0001 + 0.0015\theta_C + 0.0019\theta_E - 0.0014\theta_C\theta_E$
J	j	$0.0075 + 0.9900\theta_C + 0.0019\theta_E - 0.0019\theta_C\theta_E$
K	k	$0.0085 + 0.9890\theta_C + 0.0019\theta_E - 0.0019\theta_C\theta_E$

TABLE 12.3. The initial marginal probabilities of the nodes (no evidence) as a function of the parameters θ_C and θ_E.

X_i	x_i	$p(x_i)$
A	a	0.002
B	b	0.001
C	c	$(0.007 + 0.743\theta_C)/(0.019 + 0.733\theta_C)$
D	d	$(0.009)/(0.019 + 0.733\theta_C)$
E	e	θ_E
F	f	1
G	g	1
H	h	1
I	i	0.002
J	j	$(0.007 + 0.742\theta_C)/(0.019 + 0.733\theta_C)$
K	k	$(0.007 + 0.742\theta_C)/(0.019 + 0.733\theta_C)$

TABLE 12.4. The conditional probabilities of the nodes given the evidence $F = f$ as a function of the parameters θ_C and θ_E.

The symbolic methods can also be used to compute the conditional probabilities of the nodes given any evidence. For example, Table 12.4 gives the conditional probabilities of the nodes given the evidence $F = f$.

We have also seen in Chapter 10 how symbolic expressions, such as those in Table 12.3, can be used to obtain bounds for the marginal and conditional

X_i	x_i	p_{00}	p_{01}	p_{10}	p_{11}	Lower	Upper	Range
A	a	0.002	0.002	0.002	0.002	0.002	0.002	0.000
B	b	0.001	0.001	0.001	0.001	0.001	0.001	0.000
C	c	0.0075	0.007	0.997	0.997	0.007	0.997	0.990
D	d	0.010	0.010	0.010	0.010	0.010	0.010	0.000
E	e	0.000	1.000	0.000	1.000	0.000	1.000	1.000
F	f	0.019	0.019	0.752	0.752	0.019	0.752	0.733
G	g	0.020	0.020	0.752	0.752	0.020	0.753	0.733
H	h	0.020	1.000	0.752	1.000	0.020	1.000	0.980
I	i	10^{-4}	0.002	0.002	0.002	10^{-4}	0.002	0.002
J	j	0.008	0.009	0.997	0.997	0.008	0.998	0.990
K	k	0.008	0.010	0.997	0.997	0.008	0.997	0.989

TABLE 12.5. The initial marginal probabilities of the nodes and the corresponding and upper and lower bounds for the canonical cases in (12.9).

probabilities of the nodes. For the case of no evidence, Table 12.5 shows the initial marginal probabilities of the nodes and the corresponding lower and upper bounds that are obtained when the symbolic parameters are set to their extreme values (the so-called *canonical cases*):

$$p_{00} = (\theta_C = 0, \theta_E = 0), \quad p_{01} = (\theta_C = 0, \theta_E = 1), \\ p_{10} = (\theta_C = 1, \theta_E = 0), \quad p_{11} = (\theta_C = 1, \theta_E = 1). \quad (12.9)$$

Note that the range of the variable, i.e., the difference between the upper and lower bounds, can be used as a sensitivity factor to measure the dependence of the probabilities on the symbolic parameters (a small range means insensitive).

A similar table can be obtained when some evidence is available. For example, the conditional probabilities of nodes and the new upper and lower bounds are shown in Table 12.6 for two cases of evidence: $F = f$ and $\{F = f, E = e\}$. It can be concluded, for example, that given $\{F = f, E = e\}$, the conditional probabilities of the remaining nodes do not depend on the parameters θ_C and θ_E.

12.3 Power Distribution System

12.3.1 Definition of the Problem

Figure 12.14 displays a power distribution system with three motors, 1, 2, and 3, and three timers, A, B, and C, which are normally closed. A mo-

12.3 Power Distribution System

X_i	x_i	$F = f$			$(F = f, E = e)$		
		Lower	Upper	Range	Lower	Upper	Range
A	a	0.002	0.002	0.000	0.002	0.002	0.000
B	b	0.001	0.001	0.000	0.001	0.001	0.000
C	c	0.387	0.997	0.610	0.798	0.798	0.000
D	d	0.012	0.484	0.472	1.000	1.000	0.000
E	e	0.000	1.000	1.000	1.000	1.000	0.000
F	f	1.000	1.000	0.000	1.000	1.000	0.000
G	g	1.000	1.000	0.000	1.000	1.000	0.000
H	h	1.000	1.000	0.000	1.000	1.000	0.000
I	i	0.002	0.002	0.000	0.002	0.002	0.000
J	j	0.388	0.997	0.609	0.799	0.799	0.000
K	k	0.388	0.997	0.608	0.799	0.799	0.000

TABLE 12.6. Upper and lower bounds for the conditional probabilities for the canonical cases given two cases of evidence, $F = f$ and $\{F = f, E = e\}$.

mentary depression of pushbutton F applies power from a battery to the coils of *cutthroat* relays G and I. Thereupon G and I close and remain electrically latched. To check whether the three motors are operating properly, a 60-second test signal is impressed through K. Once K has been closed, power from battery 1 is applied to the coils of relays R and M. The closure of R starts motor 1. The closure of T applies power from battery 1 to coil S. The closure of S starts motor 3.

After an interval of 60 seconds, K is supposed to open, shutting down the operation of all three motors. Should K fail to be closed after the expiration of 60 seconds, all three timers A, B, and C open, de-energizing the coil of G, thus shutting down the system. Suppose K opens to de-energize G, and motor 1 stops. B and C act similarly to stop motor 2 or motor 3 should either M or S fail to be closed. In the following only the effect on motor 2 is analyzed. The analyses of motors 1 and 3 are left as an exercise for the reader.

12.3.2 Variable Selection

We are interested in knowing the operating status of motor 2. Let this random variable be denoted by Q. Hence q means failure and $\bar{q}$ means no failure. We wish to compute $p(q)$. Figure 12.15 shows the failure tree and the sets that lead to the failure of the system. Note that the failure of motor

544   12.   Case Studies

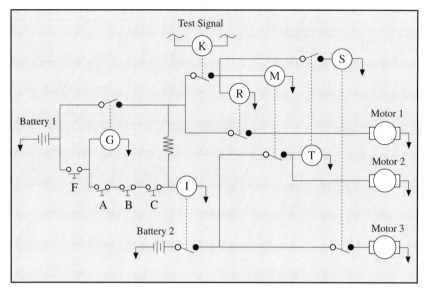

FIGURE 12.14. A diagram representing a power distribution system.

2 is equal to the logical expression

$$q = [(m \vee (k \wedge g) \vee k \wedge (a \wedge b \wedge c) \vee (k \wedge f)] \wedge (i \vee g \vee b \vee f), \quad (12.10)$$

where the symbols $\vee$ and $\wedge$ are used for *or* and *and*, respectively.

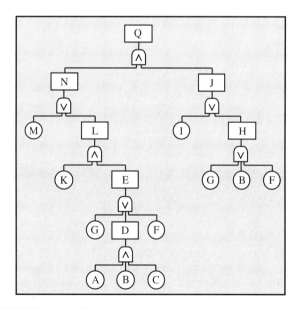

FIGURE 12.15. The failure tree for the overrun of motor 2.

Equation (12.10) can be used to derive the set of rules of a deterministic expert system. Figure 12.16 shows the set of chained rules obtained from (12.10). The set of variables used in this example is

$$X = \{A, B, C, D, E, F, G, H, I, J, K, L, M, N, Q\},$$

where $D, E, H, J, L,$ and N are intermediate failures.

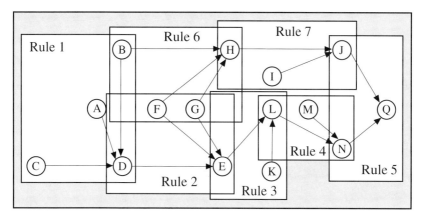

FIGURE 12.16. The chained rules for motor 2.

The study of this example from a deterministic viewpoint is left as an exercise for the reader.

As in the previous example, the chained structure given by the rules allows us to define a more powerful model for this example. Thus, we use the directed graph in Figure 12.17 as the graphical model for a Bayesian network whose JPD can be factorized as

$$\begin{aligned}p(x) =\ & p(a)p(b)p(c)p(d|a,b,c)p(f)p(g)p(e|d,f,g)p(h|b,f,g)\\ & p(i)p(j|h,i)p(k)p(l|e,k)p(m)p(n|l,m)p(q|j,n).\end{aligned}$$

The CPDs needed to define this JPD are given in Table 12.7 (we only give the probabilities of failures because $p(no\ failure) = 1 - p(failure)$). The marginals for the nonintermediate variables A, B, C, F, G, I, K, and M are

$$\begin{array}{llll}p(a) = 0.010, & p(b) = 0.010, & p(c) = 0.010, & p(f) = 0.011\\ p(g) = 0.011, & p(i) = 0.001, & p(k) = 0.002, & p(m) = 0.003.\end{array}$$

For illustrative purposes, we use one exact and one approximate method for the propagation of evidence in this multiply-connected Bayesian network.

12.3.3 Exact Propagation of Evidence

Figure 12.18 shows the undirected moralized and triangulated graph that corresponds to the directed graph in Figure 12.17. A perfect numbering of nodes is also shown in Figure 12.18.

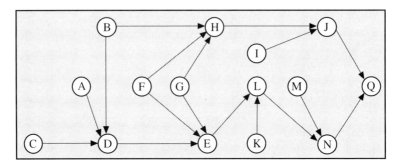

FIGURE 12.17. Multiply-connected directed graph for the power distribution system (motor 2).

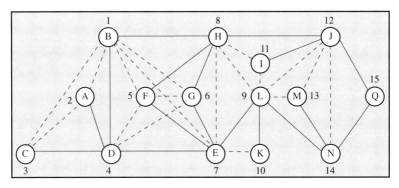

FIGURE 12.18. A moralized and triangulated graph associated with the directed graph in Figure 12.17. A perfect numbering of the nodes is shown.

The cliques obtained from the graph in Figure 12.18 are

$$
\begin{array}{lll}
C_1 = \{A, B, C, D\}, & C_2 = \{B, D, E, F, G\}, & C_3 = \{B, E, F, G, H\} \\
C_4 = \{E, H, L\}, & C_5 = \{H, I, J, L\}, & C_6 = \{E, K, L\}, \\
C_7 = \{J, L, M, N\} & C_8 = \{J, N, Q\},
\end{array}
$$

which implies that the JPD of the nodes can be written as a function of evidence potentials as follows:

$$
\begin{aligned}
p(x) \;=\; & \psi(a,b,c,d)\;\psi(b,d,e,f,g)\;\psi(b,e,f,g,h)\;\psi(e,h,l) \\
& \times \psi(h,i,j,l)\;\psi(e,k,l)\;\psi(j,l,m,n)\;\psi(j,n,q),
\end{aligned} \quad (12.11)
$$

12.3 Power Distribution System 547

I	H	$p(j\|I,H)$
i	h	1
i	$\bar{h}$	1
$\bar{i}$	h	1
$\bar{i}$	$\bar{h}$	0

E	K	$p(l\|E,K)$
e	k	1
e	$\bar{k}$	0
$\bar{e}$	k	0
$\bar{e}$	$\bar{k}$	0

L	M	$p(n\|L,M)$
l	m	1
l	$\bar{m}$	1
$\bar{l}$	m	1
$\bar{l}$	$\bar{m}$	0

J	N	$p(q\|J,N)$
j	n	1
j	$\bar{n}$	0
$\bar{j}$	n	0
$\bar{j}$	$\bar{n}$	0

A	B	C	$p(d\|A,B,C)$
a	b	c	1
a	b	$\bar{c}$	0
a	$\bar{b}$	c	0
a	$\bar{b}$	$\bar{c}$	0
$\bar{a}$	b	c	0
$\bar{a}$	b	$\bar{c}$	0
$\bar{a}$	$\bar{b}$	c	0
$\bar{a}$	$\bar{b}$	$\bar{c}$	0

D	F	G	$p(e\|D,F,G)$
d	f	g	1
d	f	$\bar{g}$	1
d	$\bar{f}$	g	1
d	$\bar{f}$	$\bar{g}$	1
$\bar{d}$	f	g	1
$\bar{d}$	f	$\bar{g}$	1
$\bar{d}$	$\bar{f}$	g	1
$\bar{d}$	$\bar{f}$	$\bar{g}$	0

B	F	G	$p(h\|B,F,G)$
b	f	g	1
b	f	$\bar{g}$	1
b	$\bar{f}$	g	1
b	$\bar{f}$	$\bar{g}$	1
$\bar{b}$	f	g	1
$\bar{b}$	f	$\bar{g}$	1
$\bar{b}$	$\bar{f}$	g	1
$\bar{b}$	$\bar{f}$	$\bar{g}$	0

TABLE 12.7. Conditional probabilities of failures of the variables in the power distribution system (motor 2).

where
$$\begin{aligned}
\psi(a,b,c,d) &= p(a)p(b)p(c)p(d|a,b,c), \\
\psi(b,d,e,f,g) &= p(f)p(g)p(e|d,f,g), \\
\psi(b,e,f,g,h) &= p(h|b,f,g), \\
\psi(e,h,l) &= 1, \\
\psi(h,i,j,l) &= p(i)p(j|i,h), \\
\psi(e,k,l) &= p(k)p(l|e,k), \\
\psi(j,l,m,n) &= p(m)p(n|l,m), \\
\psi(j,n,q) &= p(q|j,n).
\end{aligned} \qquad (12.12)$$

548   12. Case Studies

The corresponding join tree is drawn in Figure 12.19.

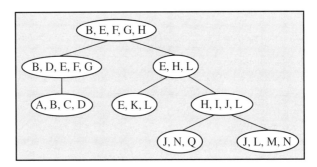

FIGURE 12.19. A join tree obtained from the undirected moralized and triangulated graph in Figure 12.18.

We use the clustering Algorithm 8.5 to obtain the initial marginal probabilities of the nodes when no evidence is available. These probabilities are shown in Figure 12.20. Suppose now we have the evidence $K = k$. The conditional probabilities of the nodes given this evidence are obtained using Algorithm 8.5 and are shown in Figure 12.21. In this case, the probability of failure increases from the initial value of $p(q) = 0.0001$ to $p(q|K = k) = 0.022$.

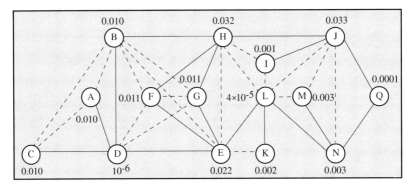

FIGURE 12.20. The marginal probabilities of the nodes when no evidence is available, for the power distribution system (motor 2).

When we introduce the additional evidence $E = e$, then L and N also fail. Consequently, the power system fails: $p(q|E = e, K = k) = 1$ (see Figure 12.22).

12.3.4   Approximate Propagation of Evidence

In Chapter 9 we introduced several algorithms for approximate propagation of evidence. We have seen that the *likelihood weighing* method was

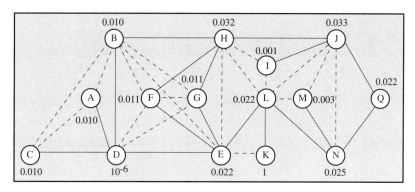

FIGURE 12.21. The conditional probabilities of the nodes given the evidence $K = k$, for the power distribution system (motor 2).

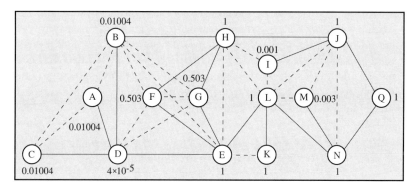

FIGURE 12.22. The conditional probabilities of the nodes given the evidence $E = e$ and $K = k$, for the power distribution system (motor 2).

the most efficient stochastic algorithm and that the *systematic sampling* and *maximum probability search* algorithms are the most efficient deterministic search algorithms when dealing with networks with extreme probabilities. In this case, we have a Bayesian network with probability tables that contain several extreme values (zeros and ones), a situation in which the likelihood weighing method is known to be inefficient. Nevertheless, we compare the likelihood weighing and the systematic sampling methods using the above Bayesian network model (initially and after the evidence $E = e, K = k$).

Table 12.8 gives the error of approximation,

$$\text{error} = |\text{exact} - \text{approximate}|,$$

for both methods for different numbers of simulation runs. Clearly, the systematic sampling algorithm (see Section 9.9) outperforms the likelihood algorithm, leading to much smaller errors for the same number of simulations. The inefficiency of the likelihood weighing algorithm is partially due

| Number of Simulations | Error ||||
| | No Evidence || $E = e, K = k$ ||
	Likelihood	Systematic	Likelihood	Systematic
100	0.00205	0.00023	0.19841	0.00650
1,000	0.00021	5.25×10^{-6}	0.04300	0.00292
2,000	6.26×10^{-5}	3.91×10^{-6}	0.01681	0.00109
10,000	1.49×10^{-5}	4.35×10^{-7}	0.00302	3.34×10^{-5}
20,000	9.36×10^{-6}	1.22×10^{-7}	0.00265	1.78×10^{-5}
50,000	5.79×10^{-6}	3.08×10^{-8}	0.00053	7.66×10^{-6}
100,000	1.26×10^{-6}	3.06×10^{-9}	0.00011	2.08×10^{-6}

TABLE 12.8. Performance of likelihood weighing and systematic sampling approximation algorithms with different number of simulations.

to the extreme values of the probabilities. Since most of the instantiations have associated zero probability, the most efficient approximate method is the *maximum probability search* algorithm. For example, even by considering a number of instantiations as small as 10, the error obtained (not shown) is smaller than 3×10^{-6}.

12.4 Damage of Concrete Structures

In Sections 12.2 and 12.3 we used probabilistic network models to define a consistent JPD for two real-life problems in a simple and straightforward way. Bayesian network models were used in these cases because of the simple structure of the dependency relationships among the variables. For example, there was no feedback or cycles. In this section we present a real-life problem that cannot be directly modeled as a Bayesian network due to the presence of cycles. We use the more general conditionally specified models presented in Section 7.7 to model the relationships among the variables in this problem. The problems of compatibility and elimination of redundant information, which do not arise in Bayesian network models, are now important and need to be addressed during the modeling process.

12.4.1 Definition of the Problem

In this case study, the objective is to assess the damage of reinforced concrete structures of buildings. In this section, we illustrate this problem using a mixed model with discrete and continuous variables. Alternatively, in Section 12.5, we use a Gaussian Bayesian network model in which all variables are continuous. This example, which is taken from Liu and Li (1994) (see also Castillo, Gutiérrez, and Hadi (1995b)), is slightly modified

12.4.2 Variable Selection

The model formulation process usually starts with the selection or specification of a set of variables of interest. This specification is dictated by the subject-matter specialists (civil engineers, in this case). In our example, the goal variable (the damage of a reinforced concrete beam) is denoted by X_1. A civil engineer initially identifies 16 variables $(X_9, \ldots, X_{24})$ as the main variables influencing the damage of reinforced concrete structures. In addition, the engineer identifies seven intermediate unobservable variables $(X_2, \ldots, X_8)$ that define some partial states of the structure. Table 12.9 shows the list of variables and their definitions. The table also shows whether each variable is continuous or discrete and the possible values that each variable can take. The variables are measured using a scale that is directly related to the goal variable, that is, the higher the value of the variable the more the possibility for damage. Let the set of all variables be denoted by $X = \{X_1, \ldots, X_{24}\}$.

12.4.3 Identification of Dependencies

The next step in model formulation is the identification of the dependency structure among the selected variables. This identification is also given by a civil engineer and is usually done by identifying the minimum set of variables, $Nbr(X_i)$, for each variable X_i such that

$$p(x_i | x \setminus x_i) = p(x_i | Nbr(X_i)), \qquad (12.13)$$

where the set $Nbr(X_i)$ is referred to as the *neighbors* of X_i. Equation (12.13) indicates that the variable X_i is conditionally independent of the set $R_i = X \setminus \{X_i, Nbr(X_i)\}$ given $Nbr(X_i)$. Thus, using the notation of conditional independence (see Section 5.1), we can write $I(X_i, R_i | Nbr(X_i))$. The variables and their corresponding neighbors are shown in the first two columns of Table 12.10. It follows that if $X_j \in Nbr(X_i)$, then $X_i \in Nbr(X_j)$.

Additionally, but optionally, the engineer can impose certain cause-effect relationships among the variables, that is, specifying which variables among the set $Nbr(X_i)$ are direct causes of X_i and which are direct effects of X_i. The set of direct causes of X_i is referred to as the *parents* of X_i and is denoted by Π_i.

In our example, the engineer specifies the following cause-effect relationships, as depicted in Figure 12.23. The goal variable X_1, depends primarily on three factors: X_9, the weakness of the beam available in the form of a damage factor; X_{10}, the deflection of the beam; and X_2, its cracking state. The cracking state, X_2, is influenced in turn by four variables: X_3,

552   12. Case Studies

X_i	Type	Values	Definition
X_1	Discrete	$\{0,1,2,3,4\}$	Damage assessment
X_2	Discrete	$\{0,1,2\}$	Cracking state
X_3	Discrete	$\{0,1,2\}$	Cracking state in shear domain
X_4	Discrete	$\{0,1,2\}$	Steel corrosion
X_5	Discrete	$\{0,1,2\}$	Cracking state in flexure domain
X_6	Discrete	$\{0,1,2\}$	Shrinkage cracking
X_7	Discrete	$\{0,1,2\}$	Worst cracking in flexure domain
X_8	Discrete	$\{0,1,2\}$	Corrosion state
X_9	Continuous	$(0-10)$	Weakness of the beam
X_{10}	Discrete	$\{0,1,2\}$	Deflection of the beam
X_{11}	Discrete	$\{0,1,2,3\}$	Position of the worst shear crack
X_{12}	Discrete	$\{0,1,2\}$	Breadth of the worst shear crack
X_{13}	Discrete	$\{0,1,2,3\}$	Position of the worst flexure crack
X_{14}	Discrete	$\{0,1,2\}$	Breadth of the worst flexure crack
X_{15}	Continuous	$(0-10)$	Length of the worst flexure cracks
X_{16}	Discrete	$\{0,1\}$	Cover
X_{17}	Continuous	$(0-100)$	Structure age
X_{18}	Continuous	$(0-100)$	Humidity
X_{19}	Discrete	$\{0,1,2\}$	PH value in the air
X_{20}	Discrete	$\{0,1,2\}$	Content of chlorine in the air
X_{21}	Discrete	$\{0,1,2,3\}$	Number of shear cracks
X_{22}	Discrete	$\{0,1,2,3\}$	Number of flexure cracks
X_{23}	Discrete	$\{0,1,2,3\}$	Shrinkage
X_{24}	Discrete	$\{0,1,2,3\}$	Corrosion

TABLE 12.9. Definitions of the variables related to damage assessment of reinforced concrete structures.

the cracking state in the shear domain; X_6, the evaluation of the shrinkage cracking; X_4, the evaluation of the steel corrosion; and X_5, the cracking state in the flexure domain. Shrinkage cracking, X_6, depends on shrinkage, X_{23}, and the corrosion state, X_8. Steel corrosion, X_4, is influenced by X_8, X_{24}, and X_5. The cracking state in the shear domain, X_3, depends on four factors: X_{11}, the position of the worst shear crack; X_{12}, the breadth of the worst shear crack; X_{21}, the number of shear cracks; and X_8. The cracking state in the flexure domain, X_5 is affected by three variables: X_{13}, the position of the worst flexure crack; X_{22}, the number of flexure cracks; and X_7, the worst cracking state in the flexure domain. The variable X_{13} is influenced by X_4. The variable X_7 is a function of five variables: X_{14}, the breadth of the worst flexure crack; X_{15}, the length of the worst flexure crack; X_{16}, the cover; X_{17}, the structure age; and X_8, the corrosion state.

12.4 Damage of Concrete Structures   553

X_i	$Nbr(X_i)$	Π_i
X_1	$\{X_9, X_{10}, X_2\}$	$\{X_9, X_{10}, X_2\}$
X_2	$\{X_3, X_6, X_5, X_4, X_1\}$	$\{X_3, X_6, X_5, X_4\}$
X_3	$\{X_{11}, X_{12}, X_{21}, X_8, X_2\}$	$\{X_{11}, X_{12}, X_{21}, X_8\}$
X_4	$\{X_{24}, X_8, X_5, X_2, X_{13}\}$	$\{X_{24}, X_8, X_5\}$
X_5	$\{X_{13}, X_{22}, X_7, X_2, X_4\}$	$\{X_{13}, X_{22}, X_7\}$
X_6	$\{X_{23}, X_8, X_2\}$	$\{X_{23}, X_8\}$
X_7	$\{X_{14}, X_{15}, X_{16}, X_{17}, X_8, X_5\}$	$\{X_{14}, X_{15}, X_{16}, X_{17}, X_8\}$
X_8	$\{X_{18}, X_{19}, X_{20}, X_7, X_4, X_6, X_3\}$	$\{X_{18}, X_{19}, X_{20}\}$
X_9	$\{X_1\}$	ϕ
X_{10}	$\{X_1\}$	ϕ
X_{11}	$\{X_3\}$	ϕ
X_{12}	$\{X_3\}$	ϕ
X_{13}	$\{X_5, X_4\}$	$\{X_4\}$
X_{14}	$\{X_7\}$	ϕ
X_{15}	$\{X_7\}$	ϕ
X_{16}	$\{X_7\}$	ϕ
X_{17}	$\{X_7\}$	ϕ
X_{18}	$\{X_8\}$	ϕ
X_{19}	$\{X_8\}$	ϕ
X_{20}	$\{X_8\}$	ϕ
X_{21}	$\{X_3\}$	ϕ
X_{22}	$\{X_5\}$	ϕ
X_{23}	$\{X_6\}$	ϕ
X_{24}	$\{X_4\}$	ϕ

TABLE 12.10. Variables and their corresponding neighbors, $Nbr(X_i)$, and parents, Π_i, for the damage assessment of reinforced concrete structure.

The variable X_8 is affected by three variables: X_{18}, the humidity; X_{19}, the PH value in the air; and X_{20}, the content of chlorine in the air.

The set of parents Π_i of each of the variables in Figure 12.23 is shown in the third column of Table 12.10. If no cause-effect relationships are given, the relationships are represented by undirected links (a line connecting two nodes).

12.4.4 Specification of Conditional Distributions

Once the graphical structure is specified, the engineer specifies a set of CPDs that is suggested by the graph. To simplify the probability assignment of the CPDs, the engineer assumes that the conditional probabilities belong to some parametric families (e.g., Binomial, Beta, etc.). The set of CPDs is given in Table 12.11, where the four continuous variables are as-

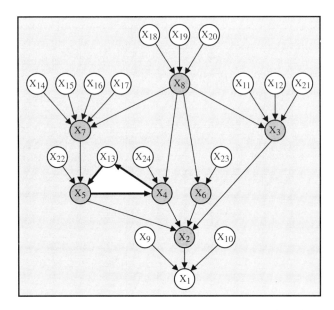

FIGURE 12.23. Directed graph for the damage assessment of reinforced concrete structure. Shadowed nodes represent unobservable (auxiliary) variables of the model.

sumed to have a $Beta(a,b)$ distribution with the specified parameters, and the discrete variables are assumed to be $Binomial\ B(n,p)$. The reason for this choice is that the beta distribution has finite bounds and also has a variety of shapes depending on the choice of the parameters.

The variable X_1 can assume only one of five values (states): $0, 1, 2, 3, 4$, with 0 meaning the building is free of damage and 4 meaning the building is seriously damaged. The values between 0 and 4 are intermediate states of damage. All other variables are defined similarly using a scale that is directly related to the goal variable, that is, the higher the value of the variable the more possibility for damage.

All discrete variables are assumed to have a binomial distribution with parameters N and p, with $N+1$ being the number of possible states of each variable. These distributions, however, can be replaced by any other suitable distributions. The parameter $0 \leq p \leq 1$ is specified as follows. Let π_i be the observed values of the parents of any given node X_i. The function $p_i(\pi_i)$, $i = 1, \ldots, 8$, in Table 12.11 is a function that takes π_i as an input and produces a probability value associated with the node X_i, that is, $p_i(\pi_i) = h(\pi_i)$. For the sake of simplicity let us consider $\Pi_i = \{X_1, \ldots, X_m\}$. Then, examples of $h(\pi_i)$ are given by

$$h(\pi_i) = \sum_{j=1}^{m} \frac{x_j/u_j}{m} \qquad (12.14)$$

12.4 Damage of Concrete Structures

X_i	$p(x_i\|u_i)$	Family
X_1	$p(x_1\|x_9, x_{10}, x_2)$	$B(4, p_1(x_9, x_{10}, x_2))$
X_2	$p(x_2\|x_3, x_6, x_4, x_5)$	$B(2, p_2(x_3, x_6, x_4, x_5))$
X_3	$p(x_3\|x_{11}, x_{12}, x_{21}, x_8)$	$B(2, p_3(x_{11}, x_{12}, x_{21}, x_8))$
X_4	$p(x_4\|x_{24}, x_8, x_5)$	$B(2, p_4(x_{24}, x_8, x_5))$
X_5	$p(x_5\|x_{13}, x_{22}, x_7)$	$B(2, p_5(x_{13}, x_{22}, x_7))$
X_6	$p(x_6\|x_{23}, x_8)$	$B(2, p_6(x_{23}, x_8))$
X_7	$p(x_7\|x_{14}, x_{15}, x_{16}, x_{17}, x_8)$	$B(2, p_7(x_{14}, x_{15}, x_{16}, x_{17}, x_8))$
X_8	$p(x_8\|x_{18}, x_{19}, x_{20})$	$B(2, p_8(x_{18}, x_{19}, x_{20}))$
X_9	$f(x_9)$	$10 * Beta(0.5, 8)$
X_{10}	$p(x_{10})$	$B(2, 0.1)$
X_{11}	$p(x_{11})$	$B(3, 0.2)$
X_{12}	$p(x_{12})$	$B(2, 0.1)$
X_{13}	$p(x_{13}\|x_4)$	$B(3, p_{13}(x_4))$
X_{14}	$p(x_{14})$	$B(2, 0.1)$
X_{15}	$f(x_{15})$	$10 * Beta(1, 4)$
X_{16}	$p(x_{16})$	$B(1, 0.1)$
X_{17}	$f(x_{17})$	$100 * Beta(2, 6)$
X_{18}	$f(x_{18})$	$100 * Beta(2, 6)$
X_{19}	$p(x_{19})$	$B(2, 0.2)$
X_{20}	$p(x_{20})$	$B(2, 0.2)$
X_{21}	$p(x_{21})$	$B(3, 0.2)$
X_{22}	$p(x_{22})$	$B(3, 0.2)$
X_{23}	$p(x_{23})$	$B(3, 0.1)$
X_{24}	$p(x_{24})$	$B(3, 0.1)$

TABLE 12.11. The marginal and conditional probability distributions associated with the network in Figure 12.23.

and

$$h(\pi_i) = 1 - \prod_{j=1}^{m}(1 - x_j/u_j), \quad (12.15)$$

where u_j is an upper bound (e.g., the maximum value) of the random variable X_j. The functions $h(\pi_i)$ in (12.14) and (12.15) are increasing with increasing values of Π_i. They also satisfy the probability axiom $0 \leq h(\pi_i) \leq 1$. We should emphasize here that these functions are only examples, given for illustrative purposes, and that they can be replaced by other suitable functions.

Table 12.12 gives the functions $h(\pi_i)$ used for calculating the conditional probabilities in Table 12.11. Alternatively, a marginal or conditional probability distribution table for each discrete variable can be given.

556   12. Case Studies

X_i	$p(\pi_i)$	$h(\pi_i)$
X_1	$p_1(x_9, x_{10}, x_2)$	(12.15)
X_2	$p_2(x_3, x_6, x_5, x_4)$	(12.14)
X_3	$p_3(x_{11}, x_{12}, x_{21}, x_8)$	(12.14)
X_4	$p_4(x_{24}, x_8, x_5, x_{13})$	(12.14)
X_5	$p_5(x_{13}, x_{22}, x_7)$	(12.14)
X_6	$p_6(x_{23}, x_8)$	(12.14)
X_7	$p_7(x_{14}, x_{15}, x_{16}, x_{17}, x_8)$	(12.14)
X_8	$p_8(x_{18}, x_{19}, x_{20})$	(12.14)

TABLE 12.12. Probability functions required for the computations of conditional probabilities in Table 12.11.

12.4.5 Diagnosing the Model

Once the set of CPDs in Table 12.11 is given, the task of the statistical expert begins. Before propagation of evidence can start, the given set of conditional probabilities has to be checked for uniqueness and compatibility because the directed graph in Figure 12.23 contains a cycle.

Given a set of nodes $X = \{X_1, \ldots, X_n\}$ and a set of conditional probabilities of the nodes of X, the statistical expert determines whether the set of conditional probabilities corresponds to a well-defined JPD of the variables in the network.

In some particular cases (when the model do not present cycles), we can set an ordering of the nodes, say $X_1, \ldots, X_n$, such that

$$\Pi_i = Nbr(X_i) \cap A_i, \qquad (12.16)$$

where $A_i = \{X_{i+1}, \ldots, X_n\}$, that is, the set of parents of a node is the set of neighbors with larger number than the given node.[3] In this case, the associated graph is a directed acyclic graph (DAG), and it defines a Bayesian network whose JPD can be factorized as

$$p(x_1, \ldots, x_n) = \prod_{i=1}^{24} p(x_i | a_i) = \prod_{i=1}^{24} p(x_i | \pi_i). \qquad (12.17)$$

However, when the nodes cannot be ordered as in (12.16), the JPD cannot be directly defined as in (12.17). For instance, in our example of the damage of concrete structures, the variables cannot be ordered as in (12.16) because the nodes X_5, X_4, and X_{13} form a cycle (see Figure 12.23). In this situation, the set of CPDs

$$P = \{p(x_i | \pi_i) : i = 1, \ldots, 24\}$$

---

[3]Note that an equivalent definition can be given by considering $\Pi_i = B_i \cap Nbr(X_i)$, where $B_i = \{X_1, \ldots, X_{i-1}\}$.

provided by the engineer in Table 12.11 has to be checked for consistency and uniqueness in order to define a consistent probabilistic model.

Note that if no cause-effect relationships were given, an ordering satisfying (12.16) could always be found because we have no restriction on the choice of parents, that is, any subsets of $Nbr(x_i)$ can serve as π_i. On the other hand, if some cause-effect relationships are given, the ordering of the nodes must satisfy (12.16), that is, a child must receive a lower number than all of its parents. It follows then that if the given cause-effect relationships contain cycles, there exists no ordering that satisfies (12.16). However, as we have seen in Section 7.7, cycles always lead to redundant information, and therefore, they can be removed from the model. In order to obtain a reduced model, the given set of CPDs has to be checked for uniqueness and consistency to detect the links associated with redundant information.

Given an ordering for the variables $\{X_1, \ldots, X_n\}$, in Section 7.7 we showed that a set of CPDs contains enough information to define at most one JPD if the set contains a sequence of CPDs of the form $p(x_i|a_i, u_i)$, $i = 1, \ldots, n$, where $A_i = \{X_{i+1}, \ldots, X_n\}$ and $U_i \subset B_i = \{X_1, \ldots, X_{i-1}\}$. We have arranged the variables in Table 12.11 in the order required to check uniqueness so that a permutation of the variables is not necessary. It can be easily seen that the CPDs given in Table 12.11 form the required sequence, since we are assuming (12.13).

The uniqueness theorem ensures only uniqueness but it does not guarantee the existence of a JPD for the set X. In Section 7.7 we provided a theorem by which one can determine whether a given set of conditionals defines a feasible JPD for X (compatibility). In this section we also showed that the redundant information is associated with a certain structure of the CPDs. More precisely, the CPDs of the form $p(x_i|a_i, u_i)$, with $U_i \neq \phi$ may be inconsistent with the previous CPDs, and hence, they contain redundant information. Therefore, any link $Y_j \to X_i$, with $Y_j \in U_i$, can be removed without restricting the JPDs associated with the graphical structure. In addition, by removing the cycles redundant information is eliminated from the model and the assessment process can be done with no restrictions on the numerical values assigned to the parameters (apart from the probability axioms that each individual distribution must satisfy).

From Table 12.11 it can be seen that all CPDs but $p(x_{13}|x_4)$ satisfy the compatibility condition. Thus, the link $X_4 \to X_{13}$ can be eliminated (removed or reversed) from the graph. This redundant information is graphically displayed in the diagram in Figure 12.23 by the cycle involving nodes X_5, X_4, and X_{13}. This cycle implies that in the engineer's mind, the cracking state in the flexure domain influences the steel corrosion, the steel corrosion influences position of the worst flexure crack, and the worst flexure crack influences the cracking state in the flexure domain.

Thus, we can eliminate this cycle without affecting the joint probability assignment. We have reversed the direction of the link $X_4 \to X_{13}$, thus obtaining another graph without cycles (Figure 12.24), which allows us to define the JPD of all nodes without restrictions in the selection of the conditional probabilities. Note that reversing this link requires changing the CPDs for X_4 and X_{13} in Table 12.11. The CPD of X_4 has to be changed from $p(x_4|x_{24}, x_8, x_5)$ to

$$p(x_4|x_{24}, x_8, x_5, x_{13}) = B(2, p_4(x_{24}, x_8, x_5, x_{13})). \qquad (12.18)$$

The CPD of X_{13} has to be changed from $p(x_{13}|x_4)$ to

$$p(x_{13}) = B(3, 0.2). \qquad (12.19)$$

Thus, we arrive at a set of conditionals in a standard canonical form and the probability assignment does not offer any compatibility problems. In other words, the conditional probabilities can be chosen freely.

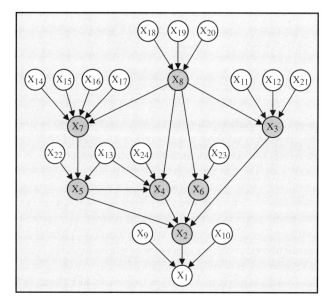

FIGURE 12.24. The network in Figure 12.23 after reversing the link from X_4 to X_{13}.

12.4.6 Propagation of Evidence

In this example we deal with both continuous and discrete random variables in the same network. Thus, we need an evidence propagation method to deal with this mixed type of network. The case of continuous variables complicates things because sums need to be replaced by integrals and the number of possible values becomes infinite. Exact propagation methods cannot be used here because they are applicable only when the variables

Available Evidence	$p(X_1 = x_1 \mid evidence)$				
	$x_1 = 0$	$x_1 = 1$	$x_1 = 2$	$x_1 = 3$	$x_1 = 4$
None	0.3874	0.1995	0.1611	0.1235	0.1285
$X_9 = 0.01$	0.5747	0.0820	0.1313	0.1002	0.1118
$X_{10} = 0$	0.6903	0.0651	0.0984	0.0606	0.0856
$X_{11} = 3$	0.6154	0.0779	0.1099	0.0783	0.1185
$X_{12} = 2$	0.5434	0.0914	0.1300	0.0852	0.1500
$X_{13} = 3$	0.3554	0.1033	0.1591	0.1016	0.2806
$X_{14} = 2$	0.3285	0.1052	0.1588	0.1043	0.3032
$X_{15} = 9.99$	0.3081	0.1035	0.1535	0.1096	0.3253
$X_{16} = 1$	0.2902	0.1054	0.1546	0.1058	0.3440
$X_{17} = 99.9$	0.2595	0.1029	0.1588	0.1064	0.3724
$X_{18} = 99.9$	0.2074	0.1027	0.1513	0.1010	0.4376
$X_{19} = 2$	0.1521	0.0937	0.1396	0.0908	0.5238
$X_{20} = 2$	0.1020	0.0813	0.1232	0.0786	0.6149
$X_{21} = 3$	0.0773	0.0663	0.1062	0.0698	0.6804
$X_{22} = 3$	0.0325	0.0481	0.0717	0.0437	0.8040
$X_{23} = 3$	0.0000	0.0000	0.0000	0.0001	0.9999
$X_{24} = 3$	0.0000	0.0000	0.0001	0.0000	0.9999

TABLE 12.13. The approximate probability distribution of the damage, X_1, given the cumulative evidence of $x_9, \ldots, x_{24}$ as indicated in the table. The results are based on 10,000 replications.

are discrete or belong to simple families (such as the normal distribution, see Chapter 8), and no general exact methods exist for networks with mixed variables (for a particular case see Lauritzen and Wermouth (1989)).

However, we can use one of the approximate propagation methods of Chapter 9. Because of its computational efficiency and generality, we choose the likelihood weighing method to propagate evidence in this example. The propagation of evidence is carried out using the set of marginal and conditional probabilities in Table 12.11 as modified in (12.18) and (12.19). To illustrate evidence propagation and to answer certain queries prompted by the engineer, we assume that the engineer examines a given concrete beam and obtains the values $x_9, \ldots, x_{24}$ corresponding to the observable variables $X_9, \ldots, X_{24}$. Note that these values can be measured sequentially. In this case, the inference can also be made sequentially. Table 12.13 shows the probabilities of the damage X_1 to a given beam for various types of evidence ranging from no knowledge at all to the knowledge of all the observed values $x_9, \ldots, x_{24}$. The values in Table 12.13 are explained and interpreted below.

As an illustrative example, suppose we wish to assess the damage (the goal variable X_1) in each of the following situations:

- **No evidence is available.** The row in Table 12.13 corresponding to the cumulative evidence "None" gives the initial marginal probability for each of the states of the goal variable X_1. For example, the probability that a randomly selected building has no damage ($X_1 = 0$) is 0.3874 and the probability that the building is seriously damaged ($X_1 = 4$) is 0.1285. These probabilities can be interpreted as 39% of the buildings in the area are safe and 13% are seriously damaged.

- **Evidence of high damage.** Suppose that we have the data for all the observable variables as given in Table 12.13, where the evidence is obtained sequentially in the order given in the table. The probabilities in the ith row in Table 12.13 are computed using $x_9, \ldots, x_i$, that is, they are based on cumulative evidences. Except for the key variables X_9 and X_{10}, the values of all other variables attain high values, resulting in high probabilities of damage. For example, as can be seen in the last row of the table, when all the evidences are considered, $p(X_1 = 4) \simeq 1$, an indication that the building is seriously damaged.

- **Evidence of low damage.** Now, suppose that we have the data for all the observable variables as given in Table 12.14, where the data are measured sequentially in the order given in the table. In this case all the variables attain low values, indicating that the building is in good condition. When considering all evidence, for example, the probability of no damage is as high as 1.

- **Observing a key evidence.** Suppose that we observe the value of only one key variable X_9, the weakness of the beam, and it turned out to be $X_9 = 8$, an indication that the beam is very weak. Propagating the evidence $X_9 = 8$ gives

$$p(X_1 = 0|X_9 = 8) = 0.0012, \quad p(X_1 = 1|X_9 = 8) = 0.0168,$$
$$p(X_1 = 2|X_9 = 8) = 0.0981, \quad p(X_1 = 3|X_9 = 8) = 0.3320,$$
$$p(X_1 = 4|X_9 = 8) = 0.5519.$$

Note that after observing the evidence $X_9 = 8$, $p(X_1 = 4)$ has increased from 0.1285 to 0.5519 and $p(X_1 = 0)$ has decreased from 0.3874 to 0.0012. The reason is that (12.15) is used to evaluate the conditional probability distribution

$$p(X_1 = x_1|X_9 = x_9, X_{10} = x_{10}, X_2 = x_2).$$

The function in (12.15) is similar to an OR-gate, which means that a very high value of one of the three parents of X_1 is sufficient for X_1 to be very high.

12.4 Damage of Concrete Structures

Available Evidence	$p(X_1 = x_1\|evidence)$				
	$x_1 = 0$	$x_1 = 1$	$x_1 = 2$	$x_1 = 3$	$x_1 = 4$
None	0.3874	0.1995	0.1611	0.1235	0.1285
$X_9 = 0$	0.5774	0.0794	0.1315	0.1002	0.1115
$X_{10} = 0$	0.6928	0.0630	0.0984	0.0603	0.0855
$X_{11} = 0$	0.7128	0.0550	0.0872	0.0615	0.0835
$X_{12} = 0$	0.7215	0.0571	0.0883	0.0551	0.0780
$X_{13} = 0$	0.7809	0.0438	0.0685	0.0469	0.0599
$X_{14} = 0$	0.7817	0.0444	0.0686	0.0466	0.0587
$X_{15} = 0$	0.7927	0.0435	0.0680	0.0441	0.0517
$X_{16} = 0$	0.7941	0.0436	0.0672	0.0421	0.0530
$X_{17} = 0$	0.8030	0.0396	0.0630	0.0428	0.0516
$X_{18} = 0$	0.8447	0.0330	0.0525	0.0316	0.0382
$X_{19} = 0$	0.8800	0.0243	0.0434	0.0269	0.0254
$X_{20} = 0$	0.9079	0.0217	0.0320	0.0217	0.0167
$X_{21} = 0$	0.9288	0.0166	0.0274	0.0172	0.0100
$X_{22} = 0$	0.9623	0.0086	0.0125	0.0092	0.0074
$X_{23} = 0$	0.9857	0.0030	0.0049	0.0037	0.0027
$X_{24} = 0$	1.0000	0.0000	0.0000	0.0000	0.0000

TABLE 12.14. The approximate probability distribution of the damage, X_1, given the cumulative evidence of $x_9, \ldots, x_{24}$ as indicated in the table. The results are based on 10,000 replications.

Available Evidence	$p(X_1 = x_1\|evidence)$				
	$x_1 = 0$	$x_1 = 1$	$x_1 = 2$	$x_1 = 3$	$x_1 = 4$
None	0.3874	0.1995	0.1611	0.1235	0.1285
$X_{11} = 2$	0.3595	0.1928	0.1711	0.1268	0.1498
$X_{12} = 2$	0.3144	0.1868	0.1700	0.1427	0.1861
$X_{21} = 2$	0.2906	0.1748	0.1784	0.1473	0.2089
$X_{18} = 80$	0.2571	0.1613	0.1764	0.1571	0.2481
$X_{19} = 2$	0.2059	0.1434	0.1797	0.1549	0.3161
$X_{20} = 1$	0.1835	0.1431	0.1716	0.1575	0.3443

TABLE 12.15. The approximate probability distribution of the damage, X_1, given the cumulative evidence of the variables as indicated in the table. The results are based on 10,000 replications.

- **Observing partial evidence.** Finally, suppose that the evidence we have available is only for a subset of the observable variables, as given in Table 12.15. The probabilities are reported in Table 12.15 and can be interpreted in a similar way.

It can be seen from the above examples that any query posed by the engineers can be answered simply by propagating the available evidence using the likelihood weighing method. Note also that it is possible for the inference to be made sequentially. An advantage of the sequential inference is that we may be able to make a decision concerning the state of damage of a given building immediately after observing only a subset of the variables. Thus, for example, once a very high value of X_9 or X_{10} is observed, the inspection can stop at this point and the building declared to be seriously damaged.

12.5 Damage of Concrete Structures: The Gaussian Model

12.5.1 Model Specification

In this section we present an alternative formulation of the damage of concrete structures example introduced in Section 12.4. Here we assume that all variables are continuous and have a multivariate normal distribution.

It is important to notice that in practice different subject-matter specialists can develop different dependency structures for the same problem. Moreover, it is a hard task to develop a consistent and not redundant probabilistic network, unless the problem can be described using Bayesian or Markov networks, which automatically lead to consistency. In Section 12.4 we studied this problem from a practical point of view, describing the steps to be followed in order to generate a unique and consistent cause-effect diagram. We have seen in Figure 12.23 that the initial network given by the engineer contains a cycle $X_4 - X_{13} - X_5 - X_4$. The cycle is broken by reversing the link from $X_4 \to X_{13}$. Now we assume that the JPD of $X = \{X_1, X_2, \ldots, X_{24}\}$ is a multivariate normal distribution $N(\mu, \Sigma)$, where μ is the 24-dimensional mean vector, Σ is the 24×24 covariance matrix, and the variables $X_1, \ldots, X_{24}$ are measured using a continuous scale that is consistent with the normality assumption. Then, as we have seen in Chapter 6, the JPD of X can be written as

$$f(x_1, \ldots, x_{24}) = \prod_{i=1}^{24} f_i(x_i | \pi_i), \qquad (12.20)$$

where

$$f_i(x_i | \pi_i) \sim N\left(m_i + \sum_{j=1}^{i-1} \beta_{ij}(x_j - \mu_j); v_i\right), \qquad (12.21)$$

m_i is the conditional mean of X_i, v_i is the conditional variance of X_i given values for Π_i, and β_{ij} is the regression coefficient associated with X_i and X_j (see Section 6.4.4). Note that if $X_j \notin \Pi_i$ then $\beta_{ij} = 0$.

12.5 Damage of Concrete Structures: The Gaussian Model   563

Alternatively, we can define the JPD by giving its mean vector and its covariance matrices. In Chapter 6 we introduced a method to transfer between these two representations.

Thus, we can consider the DAG in Figure 12.24 as the network structure of a Gaussian Bayesian network. Then, the next step is the definition of the JPD using (12.20). Suppose that the initial means of all variables are zeros, the coefficients β_{ij} in (12.21) are defined as shown in Figure 12.25, and the conditional variances are given by

$$v_i = \begin{cases} 10^{-4}, & \text{if } X_i \text{ is unobservable,} \\ 1, & \text{otherwise.} \end{cases}$$

Then the Gaussian Bayesian network is given by (12.20). Below we give examples illustrating both numeric and symbolic propagation of evidence.

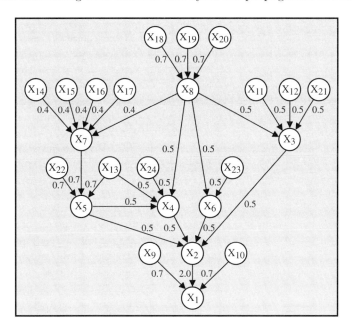

FIGURE 12.25. Directed graph for the damage assessment of a reinforced concrete structure. The numbers near the links are the coefficients β_{ij} in (12.21) used to build the Bayesian network.

12.5.2 Numeric Propagation of Evidence

To propagate evidence in the above Gaussian Bayesian network model, we use the incremental algorithm described in Chapter 8. For illustrative purpose, we assume that the engineer examines a given concrete beam and sequentially obtains the values $\{x_9, x_{10}, \ldots, x_{24}\}$ corresponding to the observable variables $X_9, X_{10}, \ldots, X_{24}$. For the sake of simplicity, suppose that the obtained evidence is $e = \{X_9 = 1, \ldots, X_{24} = 1\}$, which indicates serious damage to the beam.

Again, we wish to assess the damage (the goal variable, X_1). The conditional mean vector and covariance matrix of the remaining (unobservable and goal) variables $Y = (X_1, \ldots, X_8)$ given e, obtained using the incremental algorithm, are

$$E(y|e) = (1.8, 2.32, 1.4, 3.024, 3.412, 2.4, 5.118, 11.636),$$

$$Var(y|e) = \begin{pmatrix} 0.00010 & \ldots & 0.00006 & 0.00005 & 0.00009 & 0.00019 \\ 0.00004 & \ldots & 0.00006 & 0.00002 & 0.00009 & 0.00018 \\ 0.00005 & \ldots & 0.00003 & 0.00003 & 0.00010 & 0.00020 \\ 0.00003 & \ldots & 0.00009 & 0.00001 & 0.00014 & 0.00028 \\ 0.00006 & \ldots & 0.00018 & 0.00003 & 0.00017 & 0.00033 \\ 0.00005 & \ldots & 0.00003 & 0.00012 & 0.00010 & 0.00020 \\ 0.00010 & \ldots & 0.00017 & 0.00010 & 0.00035 & 0.00070 \\ 0.00013 & \ldots & 0.00033 & 0.00019 & 0.00072 & 1.00100 \end{pmatrix}.$$

Thus, the conditional distribution of the variables in Y is normal with the above mean vector and variance matrix.

Note that in this case, all elements in the covariance matrix except for the conditional variance of X_1 are close to zero, indicating that the mean values are very good estimates for $E(X_2, \ldots, X_8)$ and a reasonable estimate for $E(X_1)$.

We can also consider the evidence sequentially. Table 12.16 shows the conditional mean and variance of X_1 given that the evidence is obtained sequentially in the order given in the table. The evidence ranges from no information at all to complete knowledge of all the observed values $x_9, x_{10}, \ldots, x_{24}$. Thus, for example, the initial mean and variance of X_1 are $E(X_1) = 0$ and $Var(X_1) = 12.861$, respectively; and the conditional mean and variance of X_1 given $X_9 = 1$ are $E(X_1|X_9 = 1) = 0.70$ and $Var(X_1|X_9 = 1) = 12.371$. Note that after observing the key evidence $X_9 = 1$, the mean of X_1 has increased from 0 to 0.7 and the variance has decreased from 12.861 to 12.371. As can be seen in the last row of the table, when all the evidences are considered, $E(X_1|X_9 = 1, \ldots, X_{24} = 1) = 12.212$ and $Var(X_1|X_9 = 1, \ldots, X_{24} = 1) = 1.001$, an indication that the building is seriously damaged. In Figure 12.26 we show several of the conditional normal distributions of X_1 when a new evidence is added. The figure shows the increasing damage of the beam at different steps, as would be expected. Note that the mean increases and the variance decreases, an indication of decreasing uncertainty.

It can be seen from the above examples that any query posed by the engineer can be answered simply by propagating the evidence using the incremental algorithm.

12.5 Damage of Concrete Structures: The Gaussian Model

Step	Available Evidence	Damage Mean	Variance
0	None	0.000	12.861
1	$X_9 = 1$	0.700	12.371
2	$X_{10} = 1$	1.400	11.881
3	$X_{11} = 1$	1.900	11.631
4	$X_{12} = 1$	2.400	11.381
5	$X_{13} = 1$	3.950	8.979
6	$X_{14} = 1$	4.370	8.802
7	$X_{15} = 1$	4.790	8.626
8	$X_{16} = 1$	5.210	8.449
9	$X_{17} = 1$	5.630	8.273
10	$X_{18} = 1$	6.974	6.467
11	$X_{19} = 1$	8.318	4.660
12	$X_{20} = 1$	9.662	2.854
13	$X_{21} = 1$	10.162	2.604
14	$X_{22} = 1$	11.212	1.501
15	$X_{23} = 1$	11.712	1.251
16	$X_{24} = 1$	12.212	1.001

TABLE 12.16. Mean and variances of the damage, X_1, given the cumulative evidence of $x_9, x_{10}, \ldots, x_{24}$.

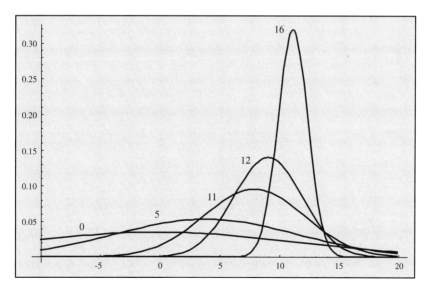

FIGURE 12.26. Conditional distributions of node X_1 corresponding to the cumulative evidence in Table 12.16. The step number is written near the curve.

566   12. Case Studies

| Available | Damage | |
Evidence	Mean	Variance
None	0	12.861
$X_9 = 1$	0.7	12.371
$X_{10} = 1$	$\dfrac{c - cm + 0.7v}{v}$	$\dfrac{-c^2 + 12.371v}{v}$
$X_{11} = x_{11}$	$\dfrac{c - cm + 0.7v + 0.5vx_{11}}{v}$	$\dfrac{-c^2 + 12.121v}{v}$
$X_{12} = 1$	$\dfrac{c - cm + 1.2v + 0.5vx_{11}}{v}$	$\dfrac{-c^2 + 11.871v}{v}$
$X_{13} = x_{13}$	$\dfrac{c - cm + 1.2v + 0.5vx_{11} + 1.55vx_{13}}{v}$	$\dfrac{-c^2 + 9.467v}{v}$
$X_{14} = 1$	$\dfrac{c - cm + 1.62v + 0.5vx_{11} + 1.55vx_{13}}{v}$	$\dfrac{-c^2 + 9.292v}{v}$

TABLE 12.17. Conditional means and variances of X_1, initially and after cumulative evidence.

12.5.3 Symbolic Computations

Suppose now that we are interested in the effect of the deflection of the beam, X_{10}, on the goal variable, X_1. Then, we consider X_{10} as a symbolic node. Let $E(X_{10}) = m, Var(X_{10}) = v, Cov(X_{10}, X_1) = Cov(X_1, X_{10}) = c$. The conditional means and variances of all nodes are calculated by applying the algorithm for symbolic propagation in Gaussian Bayesian networks introduced in Chapter 10. The conditional means and variances of X_1 given the sequential evidences $X_9 = 1, X_{10} = 1, X_{11} = x_{11}, X_{12} = 1, X_{13} = x_{13}, X_{14} = 1$, are shown in Table 12.17. Note that some of the evidences (X_{11}, X_{13}) are given in a symbolic form.

Note that the values in Table 12.16 are a special case of those in Table 12.17. They can be obtained by setting $m = 0$, $v = 1$, and $c = 0.7$ and considering the evidence values $X_{11} = 1, X_{13} = 1$. Thus the means and variances in Table 12.16 can actually be obtained from Table 12.17 by replacing the parameters by their values. For example, for the case of the evidence $X_9 = 1, X_{10} = 1, X_{11} = x_{11}$, the conditional mean of X_1 is $(c - cm + 0.7v + 0.5vx_{11})/v = 1.9$. Similarly, the conditional variance of X_1 is $(-c^2 + 12.121v)/v = 11.631$.

12.5 Damage of Concrete Structures: The Gaussian Model

D	F	$p(g\|D,F)$
d	f	θ_1
d	$\bar{f}$	θ_2
$\bar{d}$	f	1
$\bar{d}$	$\bar{f}$	0

E	G	$p(h\|E,G)$
e	g	1
e	$\bar{g}$	1
$\bar{e}$	g	θ_3
$\bar{e}$	$\bar{g}$	θ_4

TABLE 12.18. Modified conditional probabilities of failures of the variables G and H in the pressure tank system with a common cause of failure.

Exercises

12.1 Define the knowledge base for a rule-based expert system for the pressure tank example introduced in Section 12.2. Then

   (a) Use the rule chaining Algorithm 2.1 to draw new conclusions when the following new facts are known:

   i. $D = d$, $E = e$, and $F = f$.
   ii. $A = a$, $D = d$, $E = e$, and $F = f$.
   iii. $D = \bar{d}$, $E = e$, and $F = \bar{f}$.
   iv. $A = a$, $D = \bar{d}$, $E = e$, and $F = \bar{f}$.

   (b) Use the goal-oriented chain rule algorithm to obtain the causes that may influence the failure of the tank.

12.2 Use the clustering Algorithm 8.5 to propagate evidence in the pressure tank example given in Section 12.2. Verify the results shown in Figures 12.5, 12.6, and 12.7, which were obtained using the more efficient polytree Algorithm 8.1.

12.3 Replace the probabilities for nodes G and H in Table 12.2 by the ones given in Table 12.18. Use the symbolic methods to compute the conditional probabilities of the nodes given the evidence $F = f$.

12.4 Define the knowledge base for a rule-based expert system for the power distribution system example given in Section 12.3. Then

   (a) Use the rule chaining Algorithm 2.1 to draw new conclusions when the following new facts are known:

   i. $A = a$, $G = g$, and $M = m$.
   ii. $A = a$, $G = g$, $M = m$, and $F = f$.
   iii. $A = \bar{a}$, $G = \bar{g}$, and $M = m$.
   iv. $A = \bar{a}$, $G = \bar{g}$, $M = m$, and $F = f$.

   (b) Use the goal-oriented chain rule algorithm to obtain the causes that may influence the failure of motor 2.

I	H	$p(j\mid I,H)$
i	h	θ_1
i	$\bar{h}$	1
$\bar{i}$	h	θ_2
$\bar{i}$	$\bar{h}$	0

E	K	$p(l\mid E,K)$
e	k	θ_3
e	$\bar{k}$	θ_4
$\bar{e}$	k	0
$\bar{e}$	$\bar{k}$	0

TABLE 12.19. Modified conditional probabilities of failures of the variables J and L in the power distribution system (motor 2).

12.5 Replace the probabilities for nodes J and L in Table 12.7 by the ones given in Table 12.19. Use the symbolic methods to compute the conditional probabilities of the nodes given the evidence $K = k$.

12.6 Consider the power distribution system in Figure 12.14 and suppose that we are interested in knowing the operating status of motors 1 and 3. Repeat the steps described in Section 12.3 for either motor 1 or motor 3.

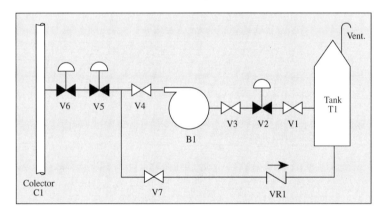

FIGURE 12.27. A simplified flow diagram of a standby water supply system.

12.7 Figure 12.27 shows a simplified flow diagram of a typical standby water-supply system. The system supplies water from tank $T1$ to collector $C1$. To operate correctly, the pump $B1$ is used to open the motorized valves $V2$, $V5$, and $V6$. Note that all valves shown in Figure 12.27 are in their normal positions (standby system). Pump $B1$ is tested once a month. During the test, we start the pump and let it work for ten minutes, allowing the water to flow through the manual valve $V7$ and the retention valve $VR1$, after opening the motorized valve $V2$, which returns the water to tank $T1$.

12.5 Damage of Concrete Structures: The Gaussian Model

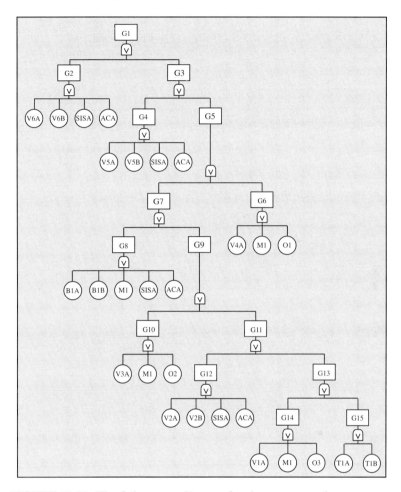

FIGURE 12.28. The failure tree diagram for the water supply system.

The system works correctly even when valve $V7$ is open. The maintenance to pump $B1$ is done two times a year. At maintenance time, valves $V3$ and $V4$ are closed, and the system becomes unavailable during the maintenance period, which takes seven hours. We are interested in the unavailability of the system. The relevant variables and their definitions are shown in Table 12.20.

Figure 12.28 shows the corresponding failure tree diagram. Note that some nodes are replicated (e.g., $M1$, $SISA$, and ACA) in order to keep the tree structure.

Figure 12.29 shows a DAG that avoids the replication of nodes of the fault tree and shows the corresponding dependency structure. We have assumed that there is a common cause of failure for valves $V2A$,

Variable	Definition
ACA	Electric power failure
$B1A$	Pump $B1$ fails to start
$B1B$	Pump $B1$ fails after starting
$G1$	Collector does not receive water flow
$G2$	Valve $V6$ fails to open
$G3$	Valve $V6$ does not receive water flow
$G4$	Valve $V5$ fails to open
$G5$	Valve $V5$ does not receive water flow
$G6$	Valve $V4$ is closed
$G7$	Valve $V4$ does not receive water flow
$G8$	Pump $B1$ failure
$G9$	Pump $B1$ does not receive water flow
$G10$	Valve $V3$ is closed
$G11$	Valve $V3$ does not receive water flow
$G12$	Valve $V2$ fails to open
$G13$	Valve $V2$ does not receive water flow
$G14$	Valve $V1$ is closed
$G15$	Valve $V1$ does not receive water flow
$M1$	Element out of service due to maintenance
$O1$	Operator forgets to open valve $V4$ after maintenance
$O2$	Operator forgets to open valve $V3$ after maintenance
$O3$	Operator forgets to open valve $V1$ after maintenance
$SISA$	Logic signal failure
$T1A$	Tank failure
$T1B$	Ventilation tank failure
$V1A$	Valve $V1$ is blocked
$V2A$	Valve $V2$ mechanical failure
$V2B$	Valve $V2$ is blocked
$V3A$	Valve $V3$ is blocked
$V4A$	Valve $V4$ is blocked
$V5A$	Valve $V5$ mechanical failure
$V5B$	Valve $V5$ is blocked
$V6A$	Valve $V6$ mechanical failure
$V6B$	Valve $V6$ is blocked

TABLE 12.20. Variables and their associated physical meanings.

$V5A$, and $V6A$. Calculate the reliability of collector $C1$, with and without a common cause of failure.

12.5 Damage of Concrete Structures: The Gaussian Model 571

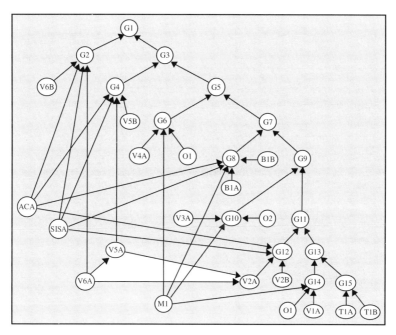

FIGURE 12.29. A directed graph for the water supply system assuming a common cause of failure for valves $V2A$, $V5A$, and $V6A$.

List of Notation

In this appendix we list the main notation used in this book. We have attempted to keep the notation consistent throughout the book as much as possible. The notation common to all chapters is listed first, followed by the notation mainly specific to each chapter.

Notation Common to All Chapters

$A \supseteq B$ .... A is a superset of or coincides with B
$A \subseteq B$ .... A is a subset of or coincides with B
$A \subset B$ .... A is a proper subset of B
A^T .... Transpose of matrix A
A^{-1} .... Inverse of a nonsingular matrix A
A_i .... Set $\{Y_{i+1}, \ldots, Y_n\}$ of variables after Y_i
B_i .... Set $\{Y_1, \ldots, Y_{i-1}\}$ of variables before Y_i
$card(A)$ .... Number of elements of set A
CIS .... Conditional independence statement
CPD .... Conditional probability distribution
D .... Directed graph
$det(A)$ .... Determinant of matrix A
$D(X,Y|Z)_G$ .... X is conditionally dependent on Y given Z in G
$D(X,Y|Z)_P$ .... X is conditionally dependent on Y given Z in P
DAG .... Directed acyclic graph
G .... Graph (directed or undirected)
GSM .... Graphically specified models
$I(X,Y|Z)_G$ ..... X is conditionally independent of Y given Z in G
$I(X,Y|Z)_P$ ..... X is conditionally independent of Y given Z in P
JPD .... Joint probability distribution
M .... Input list or a dependency model
MCS .... Maximum cardinality search
MPD .... Marginal probability distribution

574 List of Notation

$M_{ij}(s_{ij})$ .... Message clique C_i sends to clique C_j
n .... Number of variables or nodes
$p(c_i)$ .... JPD of clique C_i
$p(x_i|\pi_i)$ .... Conditional probability of $X_i = x_i$ given $\Pi_i = \pi_i$
$p(X = x|Y = y)$ Probability of $X = x$ given $Y = y$
$p(x_1, \ldots, x_n)$ .... Joint probability mass or density function
$p(x_i|x_k)$ .... Conditional probability of $X_i = x_i$ given $X_k = x_k$
$tr(A)$ .... Trace of matrix A
X .... Set of nodes or random variables
$X \setminus Y$ .... X minus Y (set difference)
$\{X_1, \ldots, X_n\}$ ... Set of nodes or random variables
X_i .... The ith random variable in the set $\{X_1, \ldots, X_n\}$
$X_i - X_j$ .... Undirected link between X_i and X_j
$X_i \notin S$ .... X_i is not in S
$X_i \to X_j$ .... Directed link between X_i and X_j
X_i .... Node, vertex, or variable
WWW .... World Wide Web
$\bar{A}$ .... Complement of A
$\cap$ .... Set intersection
$\cup$ .... Set union
ϕ .... Empty set
Π_i .... Parents of node X_i or Y_i
π_i .... An instantiation of the parents Π_i of X_i
$\Psi_i(c_i)$ .... Potential function of cluster or clique C_i
μ_X .... Mean vector of X
Σ_X .... Covariance matrix of X
σ_i^2 .... Variance of X_i
σ_{ij} .... Covariance of X_i and X_j
Σ_{XX} .... Covariance matrix of a set of variables X
Σ_{XY} .... Covariance of X and Y
$\exists$ .... Exists at least one
$\prod_{i=1}^n x_i$ .... $x_1 \times \ldots \times x_n$
$\sum_{i=1}^n x_i$ .... $x_1 + \ldots + x_n$
$\propto$ .... Proportional to
$\wedge$ .... and
$\vee$ .... or

Chapter 1

D_i .... Disease number i
NLP .... Natural language processing
S_i .... The ith symptom

List of Notation   575

Chapter 2

A, B, C	Objects used in rule-based expert systems
ATM	Automatic Teller Machine
F	False
GPA	Grade point average
PIN	Personal identification number
T	True

Chapter 3

A_i	Subset of a sample space S
D	Random variable containing all diseases including the state *healthy*
d	Value of D
D_i	The ith disease
d_i	Value of the ith component (disease) of D
$DRSM$	Dependent relevant symptoms model
DSM	Dependent symptoms model
$E(u)$	Expected value of u
G	Gastric adenocarcinoma
$IRSM$	Independent relevant symptoms model
ISM	Independent symptoms model
m	Number of diseases
lim	Limit
$\max_i (a_1, \ldots, a_k)$	Maximum value of the sequence $(a_1, \ldots, a_k)$
$\min_i (a_1, \ldots, a_k)$	Minimum value of the sequence $(a_1, \ldots, a_k)$
n	Number of symptoms
P	Symptom *pain*
p	Value of the symptom *pain*
S	Sample space
S_j	The jth symptom
s_j	Value of the jth symptom
$u_i(x)$	Utility function
V	Symptom vomiting
v	Value of the symptom vomiting
W	Symptom weight loss
w	Value of the symptom weight loss

Chapter 4

$A - D - H$	Undirected path between A and H
$AD(X_i)$	Ascending depth of node X_i

576 List of Notation

$Adj(X_i)$ .... Adjacency set of node X_i
AL_k .... The kth directed level of the ascending depth
$A \to D \to H$ .... Directed path between A and H
$Bnd(S)$ .... Boundary of the set S
C .... Set of clusters
C_i .... The ith clique or cluster
$D(X_i)$ .... Depth of node X_i
$DD(X_i)$ .... Descending depth of node X_i
DL_k .... The kth directed level of the descending depth
L .... Set of links
L_{ij} .... Link joining node X_i and X_j
$Nbr(X_i)$ .... Neighbors of node X_i
T .... Attainability matrix
$\alpha(i)$ .... The ith node in an ordering

Chapter 5

C .... Charitable contributions
E .... Employment status
H .... Health
IN .... Interaction
L .... Number of variables in U_i
m .... Number of conditional probabilities $p_i(u_i|v_i)$
P .... Happiness
SU .... Strong Union
U_i, V_i .... Disjoint subsets of X
V .... Investment income
W .... Wealth
X, Y, Z .... Disjoint sets of random variables

Chapter 6

$Bnd(S)$ .... Boundary of the set S
C_j .... The jth cluster or clique
L_{ij} .... Link between node X_i and X_j
R_i .... The ith residual set
S_i .... The ith separator set
v_A .... Variance of A
v_i .... Conditional variance of X_i given $\Pi_i = \pi_i$
$W = \Sigma^{-1}$ .... Inverse of the covariance matrix Σ
X, Y, Z .... Disjoint random variables
β_{CA} .... Coefficient of A in the regression of C on Π_C
β_{ij} .... Coefficient of X_j in the regression of X_i on Π_i

μ .... Mean of a normal random variable
μ_i .... Mean of X_i
μ_X .... Mean of X
Σ .... Covariance matrix of a normal random variable
Σ_{XX} .... Covariance matrix of X
Σ_{XY} .... Covariance of X and Y

Chapter 7

$B = (D, P)$ ..... Bayesian network
(G^ℓ, P^ℓ) .... The ℓth network model
C_1 .... All subsets of P to be checked for compatibility
L_{ij} .... Link joining nodes X_i and X_j
$MFPN$ .... Multifactorized probabilistic model
MNM .... Multinetwork model
$p^\ell(x_i^\ell | s_i^\ell)$ .... Conditional probabilities of the ℓth network model
Q_1 .... All subsets of P leading to uniqueness solutions
R_i .... Conditional probability $p(y_i | a_i)$
r_i^ℓ .... Cardinality of Y_i^ℓ
S_i .... Subset of $B_i = \{Y_1, \ldots, Y_{i-1}\}$
U_i .... Subset of $A_i = \{Y_{i+1}, \ldots, Y_n\}$
X, Y, Z .... Random variables
β_{ij} .... Coefficient of X_j in the regression of X_i on the rest of variables
ϵ_i .... Normal random variable, $N(0, \sigma_i^2)$
$\theta_{ij\pi}$ .... The parameter $p(X_i = j | \Pi_i = \pi)$
Θ^ℓ .... The set $\theta_{ij\pi}$ for the ℓth model
$\theta_{ij\pi}^\ell$ .... The parameter $\theta_{ij\pi}$ for the ℓth model

Chapter 8

C_i .... The ith cluster or clique
E .... Evidential set
$E_i^+, E_{X_i}^+$ .... Subset of E accessed from X_i through its parents
$E_i^-, E_{X_i}^-$ .... Subset of E accessed from X_i through its children
$E_{U_j X_i}^+$ .... Subset of E at the U_j-side of the link $U_j \to X_i$
$E_{X_i Y_j}^-$ .... Subset of E at the Y_j-side of the link $X_i \to Y_j$
e_i .... Value of E_i
M_{ij} .... A message clique C_i sends to its neighbor B_j
R_i .... The ith residual set
S_i .... The ith separator set
$S_{ij} = C_i \cap S_j$ ... Separator set of clusters C_i and C_j

578   List of Notation

v_i	Conditional variance of X_i given $\Pi_i = \pi_i$
$\beta_i(x_i)$	$\lambda_i(x_i)\rho_i(x_i)$
$\lambda_i(x_i)$	Conditional probability $p(e_i^-\|x_i)$
$\lambda_{Y_j X_i}(x_i)$	λ-message Y_j sends to its parent X_i
μ_X	Mean of X
$\psi_j(c_j)$	Potential function of cluster or clique C_j
$\psi_j^*(c_j)$	New potential function of cluster or clique C_j
$\rho_i(x_i)$	Joint probability $p(x_i, e_i^+)$
$\rho_{U_j X_i}(u_j)$	ρ-message U_j sends to its child X_i
Θ_i	Dummy parent of node X_i

Chapter 9

B	Set of nodes backward sampled
B_i^j	The ijth Branch
C_i	Set of children of X_i
E	Evidence set
F	Set of nodes forward sampled
f_j	Step equal to $(j - 0.5)/N$, $j = 1, \ldots, n$
$g(x)$	A function such that $0 \leq g(x) \leq 1$
$h(x)$	Simulation probability mass or density function
$H(x)$	Cumulative distribution associated with $h(x)$
I_J	Interval $[l_j, u_j] \subset [0, 1)$
l_j	Lower bound equal to $\sum_{x^i < x^j} p_e(x^i) \geq 0$
m	Number of all possible instantiations
N	Number of simulations (sample size)
$p(e)$	Probability of e
$p_e(x_i\|\pi_i)$	See Expression 9.14
$p_e(y)$	See Expression 9.7
r_i	Cardinality of X_i
$s(x)$	A score equal to $p(x)/h(x)$
u	A random value from $U(0, 1)$
$U(0, 1)$	Standard uniform distribution
u_j	Upper bound equal to $l_j + p_e(x^j) \leq 1$
x^j	The jth instantiation $\{x_1^j, \ldots x_n^j\}$
δ	Number of simulated identical instantiations
π_i^*	Set of uninstantiated parents of X_i

Chapter 10

Θ_i	Set of symbolic parameters associated with the symbolic node X_i
c_r	Numeric coefficient associated with monomial m_r

M .... Set of all monomials
E .... Set of evidential nodes
$q(e)$ .... Probability of a stochastic evidence e
$adj(\Sigma_z)$ .... Adjoint matrix of Σ_z

Chapter 11

$Desc_i$ .... The set of descendants of X_i
$Dim(B)$ .... Dimension of a Bayesian network model B (the number of free parameters, or degrees of freedom, required to specify the JPD of X)
D_c .... A complete directed graph
$f(N)$ .... A nonnegative penalizing function
$f(x, \theta)$ .... Likelihood function
$g(\theta, \eta)$ .... η-family of priors and posteriors for θ
$h(\eta; x)$ .... Posterior parameters as a function of the prior parameters and the sample x
iid .... Independently and identically distributed
m_i .... Intercept in the regression of X_i on Π_i
N .... Sample size
$N!$ .... N factorial $(1 \times 2 \times \ldots \times N)$
$\binom{N}{r}$ .... $\frac{N!}{r!(N-r)!}$
N_{ik} .... $\sum_{j=0}^{r_i} N_{ijk}$
N_{ijk} .... Number of cases in the sample data S with $X_i = k$ and Π_i equal to the jth instantiation of Π_i
$N_{x_1,\ldots,x_n}$ .... Number of cases in the sample data S with $X_i = x_i$
$p(B) = p(D, \theta)$ .. Prior probability assigned to the network model B
$Pred_i$ .... The set of predecessors of node X_i
$Q_B(B(\theta), S)$ .... Standard Bayesian quality measure of $(B(\theta), S)$
$Q_f(D', S)$ ...... Quality measure of (D', S) for a penalty function f
$Q_I(B, S)$ .... The information quality measure
$q_i(\Pi_i)$ .... Contribution of X_i with parent set Π_i to the quality of a network
$Q_{CH}(D, S)$ ..... The Cooper and Herskovits quality measure
$Q_{GH}(D, S)$ ..... The Geiger and Heckerman quality measure
$Q_{MDL}(B, S)$ .... The minimum description length quality measure
$Q_{SB}(D, S)$ ...... The standard Bayesian quality measure
r_i .... Cardinality of X_i
S .... Sample data
$W\Sigma^{-1}$ .... Precision matrix (inverse of covariance matrix)
$X_{(k)}$ .... All variables but X_k
$\bar{x}$ .... Arithmetic mean of $x_1, \ldots, x_N$
β_{ij} .... Regression coefficients when the variable X_i is

580   List of Notation

	regressed on the variables $X_1, \ldots, X_{i-1}$
η	Equivalent sample size for the prior distribution.
$\eta_{x_1,\ldots,x_n}$	Parameters of a Dirichlet distribution
η_{ik}	$\sum_{j=0}^{r_i} \eta_{ijk}$
η_{ijk}	Prior parameter associated with N_{ijk}
$\Gamma(.)$	Gamma function
$\hat{\theta}$	Posterior mode of θ
μ	Mean vector
Ω	Universal set

Chapter 12

$Nbr(X_i)$	Set of neighbors of X_i
$Nbr(x_i)$	A value of the set of neighbors of X_i
μ	Mean vector
θ_C	A parameter associated with node C

References

Abell, M. and Braselton, J. P. (1994), *Maple V by Example*. Academic Press, New York.

Akaike, H. (1974), A New Look at the Statistical Model Identification. *IEEE Transactions on Automatic Control*, 19:716–723.

Allen, J. (1995), *Natural Language Understanding, 2nd edition*. Addison-Wesley, Reading, MA.

Allen, J., Hendler, J., and Tate, A., editors (1990), *Readings in Planning*. Morgan Kaufmann Publishers, San Mateo, CA.

Almond, R. G. (1995), *Graphical Belief Modeling*. Chapman and Hall, New York.

Almulla, M. (1995), *Analysis of the Use of Semantic Trees in Automated Theorem Proving*. Ph.D. Thesis, McGill University.

Andersen, S. K., Olesen, K. G., Jensen, F. V., and Jensen, F. (1989), HUGIN: A Shell for Building Bayesian Belief Universes for Expert Systems. In *Proceedings of the 11th International Joint Conference on Artificial Intelligence (IJCAI-89)*. Morgan Kaufmann Publishers, San Mateo, CA, 1080–1085.

Anderson, T. W. (1984), *An Introduction to Multivariate Statistical Analysis, 2nd edition*. John Wiley and Sons, New York.

Arnold, B. C., Castillo, E., and Sarabia, J. M. (1992), *Conditionally Specified Distributions*. Springer-Verlag, New York.

Arnold, B. C., Castillo, E., and Sarabia, J. M. (1993), Conjugate Exponential Family Priors for Exponential Family Likelihoods. *Statistics*, 25:71–77.

Arnold, B. C., Castillo, E., and Sarabia, J. M. (1994), Priors with Convenient Posteriors. Technical Report, TR94-2, Department of Applied Mathematics, Cantabria University, Santander, Spain.

References

Arnold, B. C., Castillo, E., and Sarabia, J. M. (1996), Specification of Distributions by Combinations of Marginal and Conditional Distributions. *Statistics and Probability Letters*, 26:153–157.

Bachman, R. J., Levesque, H. J., and Reiter, R., editors (1991), *Artificial Intelligence, Special Volume on Knowledge Representation*, Volume 49.

Baker, M. and Boult, T. E. (1991), Pruning Bayesian Networks for Efficient Computation. In Bonissone, P. P., Henrion, M., Kanal, L. N., and Lemmer, J. F., editors, *Uncertainty in Artificial Intelligence 6*. North Holland, Amsterdam, 225–232.

Balas, E. and Yu, C. S. (1986), Finding a Maximum Clique in an Arbitrary Graph. *SIAM Journal on Computing*, 15(4):1054–1068.

Barr, A. and Feigenbaum, E. A. (1981), *The Handbook of Artificial Intelligence, Volume I*. William Kaufman, Los Altos, CA.

Barr, A. and Feigenbaum, E. A. (1982), *The Handbook of Artificial Intelligence, Volume II*. William Kaufman, Los Altos, CA.

Becker, A. and Geiger, D. (1994), Approximation Algorithms for the Loop Cutset Problem. In *Proceedings of the Tenth Conference on Uncertainty in Artificial Intelligence*. Morgan Kaufmann Publishers, San Francisco, CA, 60–68.

Beeri, C., Fagin, R., Maier, D., and Yannakis, M. (1993), On the Desirability of Acyclic Database Schemes. *Journal of the ACM*, 30:479–513.

Bench-Capon, T. J. M. (1990), *Knowledge Representation: An Approach to Artificial Intelligence*. Academic Press, San Diego.

Berge, C. (1973), *Graphs and Hypergraphs*. North-Holland, Amsterdam.

Bernardo, J. and Smith, A. (1994), *Bayesian Theory*. John Wiley and Sons, New York.

Bickel, P. J. and Doksum, K. A. (1977), *Mathematical Statistics: Basic Ideas and Selected Topics*. Holden-Day, Oakland, CA.

Biggs, N. L. (1989), *Discrete Mathematics, 2nd edition*. Oxford University Press, New York.

Billingsley, P. (1995), *Probability and Measure*. John Wiley and Sons, New York.

Bouckaert, R. R. (1994), A Stratified Simulation Scheme for Inference in Bayesian Belief Networks. In *Proceedings of the Tenth Conference on Uncertainty in Artificial Intelligence*. Morgan Kaufmann Publishers, San Francisco, CA, 110–117.

Bouckaert, R. R. (1995), *Bayesian Belief Networks: From Construction to Inference*. Ph.D. Thesis, Department of Computer Science, Utrecht University, Netherlands.

Bouckaert, R. R., Castillo, E., and Gutiérrez, J. M. (1996), A Modified Simulation Scheme for Inference in Bayesian Networks. *International Journal of Approximate Reasoning*, 14:55–80.

Bond, A. H. and Gasser, L., editors (1988), *Readings in Distributed Reasoning*. Morgan Kaufmann Publishers, San Mateo, CA.

Bondy, J. A. and Murty, U. S. R. (1976), *Graph Theory with Applications*. North Holland, New York.

Breese, J. S. and Fertig, K. W. (1991), Decision Making with Interval Influence Diagrams. In Bonissone, P. P., Henrion, M., Kanal, L. N., and Lemmer, J. F., editors, *Uncertainty in Artificial Intelligence 6*. North Holland, Amsterdam, 467–478.

Brown, L. D. (1986), *Fundamentals of Statistical Exponential Families: With Applications in Statistical Decision Theory*. Institute of Mathematical Statistics, Hayward, CA.

Brown, J. R. and Cunningham, S. (1989), *Programming the User Interface: Principles and Examples*. John Wiley and Sons, New York.

Buchanan, B. G. and Shortliffe, E. H. (1984), *Rule-Based Expert Systems: The MYCIN Experiments of the Stanford Heuristic Programming Project*. Addison-Wesley, Reading, MA.

Buckley, J. J., Siler, W., and Tucker, D. (1986), A Fuzzy Expert System. *Fuzzy Sets and Systems*, 20:1–16.

Bundy, A. (1983), *The Computer Modelling of Mathematical Reasoning*. Academic Press, New York.

Buntine, W. (1991), Theory Refinement on Bayesian Networks. In *Proceedings of the Seventh Conference on Uncertainty in Artificial Intelligence*. Morgan Kaufmann Publishers, San Mateo, CA, 52–60.

Campos, L. M. D. and Moral, S. (1995), Independence Concepts for Convex Sets of Probabilities. In *Proceedings of the Eleventh Conference on Uncertainty in Artificial Intelligence*. Morgan Kaufmann Publishers, San Francisco, CA, 108–115.

Cano, J., Delgado, M., and Moral, S. (1993), An Axiomatic Framework for Propagating Uncertainty in Directed Acyclic Networks. *International Journal of Approximate Reasoning*, 8:253–280.

Casella, G. and George, E. I. (1992), Explaining the Gibbs Sampler. *The American Statistician*, 46(3):167–174.

Castillo, E. and Alvarez, E. (1990), Uncertainty Methods in Expert Systems. *Microcomputers in Civil Engineering*, 5:43–58.

Castillo, E. and Alvarez, E. (1991), *Expert Systems: Uncertainty and Learning*. Computational Mechanics Publications and Elsevier Applied Science, London, U.K.

Castillo, E., Bouckaert, R., Sarabia, J. M., and Solares, C. (1995), Error Estimation in Approximate Bayesian Belief Network Inference. In *Proceedings of the Eleventh Conference on Uncertainty in Artificial Intelligence*. Morgan Kaufmann Publishers, San Francisco, CA, 55–62.

Castillo, E., Cobo, A., Gutiérrez, J. M., Iglesias, A., and Sagástegui, H. (1994), Causal Network Models in Expert Systems. *Microcomputers in Civil Engineering, Special Issue on Uncertainty in Expert Systems*, 9:55–60.

Castillo, E., Gutiérrez, J. M., and Hadi, A. S. (1995a), An Introduction to Expert Systems for Medical Diagnoses. *Biocybernetics and Biomedical Engineering*, 15:63–84.

Castillo, E., Gutiérrez, J. M., and Hadi, A. S. (1995b), Modelling Probabilistic Networks of Discrete and Continuous Variables. Technical Report, TR95-11, Department of Social Statistics, Cornell University, Ithaca, NY.

Castillo, E., Gutiérrez, J. M., and Hadi, A. S. (1995c), Parametric Structure of Probabilities in Bayesian Networks. *Lecture Notes in Artificial Intelligence*, 946:89–98.

Castillo, E., Gutiérrez, J. M., and Hadi, A. S. (1995d), Symbolic Propagation in Discrete and Continuous Bayesian Networks. In Keranen, V. and Mitic, P., editors, *Mathematics with Vision: Proceedings of the First International Mathematica Symposium*. Computational Mechanics Publications, Southampton, U.K., 77–84.

Castillo, E., Gutiérrez, J. M., and Hadi, A. S. (1996a), Goal Oriented Symbolic Propagation in Bayesian Networks. In *Proceedings of the Thirteenth National Conference on AI (AAAI-96)* AAAI Press/MIT Press, Menlo Park, CA, 1263–1268.

Castillo, E., Gutiérrez, J. M., and Hadi, A. S. (1996b), Multiply Factorized Bayesian Network Models. Technical Report, TR96-10, Department of Social Statistics, Cornell University, Ithaca, NY.

Castillo, E., Gutiérrez, J. M., and Hadi, A. S. (1996c), A New Method for Efficient Symbolic Propagation in Discrete Bayesian Networks. *Networks*, 28:31–43.

Castillo, E., Gutiérrez, J. M., and Hadi, A. S. (1996d), Sensitivity Analysis in Discrete Bayesian Networks. *IEEE Transactions on Systems, Man and Cybernetics*, 26. In press.

Castillo, E., Gutiérrez, J. M., and Hadi, A. S. (1996e), Constructing Probabilistic Models Using Conditional Probability Distributions. Technical Report, TR96-11, Department of Social Statistics, Cornell University, Ithaca, NY.

Castillo, E., Gutiérrez, J. M., Hadi, A. S, and Solares, C. (1996), Symbolic Propagation and Sensitivity Analysis in Gaussian Bayesian Networks with Application to Damage Assessment. *Artificial Intelligence in Engineering*. In press.

Castillo, E., Hadi, A. S., and Solares, C. (1996), Learning and Updating of Uncertainty in Dirichlet Models. *Machine Learning*. In press.

Castillo, E., Iglesias, A., Gutiérrez, J. M., Alvarez, E., and Cobo, A. (1993), *Mathematica*. Editorial Paraninfo, Madrid.

Castillo, E., Mora, E. and Alvarez, E. (1994), Log-Linear Models in Expert Systems. *Microcomputers in Civil Engineering, Special Issue on Uncertainty in Expert Systems*, 9:347–357.

Castillo, E., Solares, C., and Gómez, P. (1996a), Tail Sensitivity Analysis in Bayesian Networks. In *Proceedings of the Twelfth Conference on Uncertainty in Artificial Intelligence*. Morgan Kaufmann Publishers, San Francisco, CA, 133–140.

Castillo, E., Solares, C., and Gómez, P. (1996b), Estimating Extreme Probabilities Using Tail Simulated Data. *International Journal of Approximate Reasoning*. In press.

Chandrasekaran, B. (1988), On Evaluating AI Systems for Medical Diagnosis. *AI Magazine*, 4:34–37.

Chang, K.-C. and Fung, R. (1991), Symbolic Probabilistic Inference with Continuous Variables. In *Proceedings of the Seventh Conference on Uncertainty in Artificial Intelligence*. Morgan Kaufmann Publishers, San Mateo, CA, 77–85.

Char, B., Geddes, K., Gonnet, G., Leong, B., Monagan, M., and Watt, S. (1991), *Maple V Language Reference Manual*. Springer-Verlag.

Charniak, E. and McDermott, D. (1985), *Introduction to Artificial Intelligence*. Addison-Wesley, Reading, MA.

Chavez, R. and Cooper, G. (1990), A Randomized Approximation Algorithm for Probabilistic Inference on Bayesian Belief Networks. *SIAM Journal on Computing*, 20:661–685.

Cheeseman, P. (1985), In Defense of Probability. In *Proceedings of the 9th International Joint Conference on Artificial Intelligence (IJCAI-85)*. Morgan Kaufmann Publishers, San Mateo, CA, 1002–1009.

Chickering, D. M. (1995a), Search Operators for Learning Equivalence Classes of Bayesian Network Structures. Technical Report, R231, UCLA Cognitive Systems Laboratory, Los Angeles.

Chickering, D. M. (1995b), A Transformational Characterization of Equivalent Bayesian Network Structures. *Proceedings of the Eleventh Conference on Uncertainty in Artificial Intelligence*. Morgan Kaufmann Publishers, San Francisco, CA, 87–98.

Clancey, W. J. (1983), The Epistemology of Rule-Based Expert Systems. A Framework for Explanation. *Artificial Intelligence*, 20:215–251.

Cohen, P. R. and Feigenbaum, E. A. (1982), *The Handbook of Artificial Intelligence, Volume III*. William Kaufman, Los Altos, CA.

Cooper, G. F. (1984), *Nestor: A Computer-Based Medical Diagnostic Aid that Integrates Causal and Probabilistic Knowledge*. Ph.D. Thesis, Department of Computer Science, Stanford University.

Cooper, G. F. (1990), The Computational Complexity of Probabilistic Inference Using Bayesian Belief Networks. *Artificial Intelligence*, 42:393–405.

Cooper, G. F. and Herskovits, E. (1992), A Bayesian Method for the Induction of Probabilistic Networks from Data. *Machine Learning*, 9:309–347.

Cormen, T. H., Leiserson, C. E., and Rivest, R. L. (1990), *Introduction to Algorithms*. MIT Press, Boston, MA.

Dagum, P. and Luby, M. (1993), Approximating Probabilistic Inference in Bayesian Belief Networks is NP-hard. *Artificial Intelligence*, 60:141–153.

D'Ambrosio, B. (1994), SPI in Large BN2O Networks. In *Proceedings of the Tenth Conference on Uncertainty in Artificial Intelligence*. Morgan Kaufmann Publishers, San Francisco, CA, 128–135.

Darroch, J. N., Lauritzen, S. L., and Speed, T. P. (1980), Markov Fields and Log-linear Models for Contingency Tables. *Annals of Statistics*, 8:522–539.

Darwiche, A. (1995), Conditioning Algorithms for Exact and Approximate Inference in Causal Networks. In *Proceedings of the Eleventh Annual Conference on Uncertainty in Artificial Intelligence*. Morgan Kaufmann Publishers, San Francisco, CA, 99–107.

Davis, R. and Buchanan, B. G. (1977), Meta Level Knowledge. Overview and Applications. In *Proceedings of the Fifth International Joint Conference on Artificial Intelligence (IJCAI-77)*. Morgan Kaufmann Publishers, San Mateo, CA, 920–927.

Dawid, A. (1979), Conditional Independence in Statistical Theory. *Journal of the Royal Statistical Society, Series B*, 41:1–33.

Dawid, A. (1980), Conditional Independence for Statistical Operations. *Annals of Statistics*, 8:598–617.

Dawid, A. and Lauritzen, S. L. (1993), Hyper Markov Laws in the Statistical Analysis of Decomposable Graphical Models. *Annals of Statistics*, 21:1272–1317.

DeGroot, M. H. (1970), *Optimal Statistical Decisions*. Mc Graw Hill, New York.

DeGroot, M. H. (1987), *Probability and Statistics*. Addison-Wesley, Reading, MA.

Delcher, A. L., Grove, A., Kasif, S., and Pearl, J. (1995), Logarithmic-Time Updates and Queries in Probabilistic Networks. In *Proceedings of the Eleventh Conference on Uncertainty in Artificial Intelligence*. Morgan Kaufmann Publishers, San Francisco, CA, 116–124.

Dempster, A., Laird, N., and Rubin, D. (1977), Maximum Likelihood from Incomplete Data Via the EM Algorithm. *Journal of the Royal Statistical Society, B*, 39:1–38.

Devroye, L. (1986), *Non-Uniform Random Variate Generations*. Springer Verlag, New York.

Díez, F. J. (1994), *Sistema Experto Bayesiano para Ecocardiografía*. Ph.D. Thesis, Departamento de Informática y Automática, U.N.E.D., Madrid.

Díez, F. J. (1996), Local Conditioning in Bayesian Networks. *Artificial Intelligence*. In press.

Díez, F. J. and Mira, J. (1994), Distributed Inference in Bayesian Networks. *Cybernetics and Systems*, 25:39–61.

Dirac, G. A. (1961), On Rigid Circuit Graphs. *Abh. Math. Sem. Univ. Hamburg*, 25:71–76.

Duda, R. O., Gaschnig, J. G., and Hart, P. E. (1980), Model Design in the Prospector Consultant System for Mineral Exploration. In Michie, D., editor, *Expert Systems in the Microelectronic Age*. Edinburgh University Press, Edinburgh, 153–167.

Duda, R. O., Hart, P. E., and Nilsson, N. (1976), Subjective Bayesian Methods for Rule-Based Inference Systems. *Proceedings of the 1976 National Computer Conference (AFIPS)*, 45:1075–1082.

Durkin, J. (1994), *Expert Systems: Design and Development*. Maxwell Macmillan, New York.

Durrett, R. (1991), *Probability: Theory and Examples*. Wadsworth, Pacific Grove, CA.

Feller, W. (1968), *An Introduction to Probability Theory and Applications*. John Wiley and Sons, New York.

Fischler, M. A. and Firschein, O., editors (1987), *Readings in Computer Vision*. Morgan Kaufmann Publishers, San Mateo, CA.

Fisher, D. and Lenz, H. J., editors (1996), *Learning from Data: Artificial Intelligence and Statistics V*. Springer Verlag, New York.

Freeman, J. A. and Skapura, D. M. (1991), *Neural Networks: Algorithms, Applications, and Programming Techniques*. Addison-Wesley, Reading, MA.

Frydenberg, M. (1990), The Chain Graph Markov Property. *Scandinavian Journal of Statistics*, 17:333–353.

Fulkerson, D. R. and Gross, O. A. (1965), Incidence Matrices and Interval Graphs. *Pacific J. Math.*, 15:835–855.

Fung, R. and Chang, K. (1990), Weighing and Integrating Evidence for Stochastic Simulation in Bayesian Networks. In Henrion, M., Shachter, R. D., Kanal, L. N., and Lemmer, J. F., editors, *Uncertainty in Artificial Intelligence 5*. North Holland, Amsterdam, 209–219.

Fung, R. and Favero, B. D. (1994), Backward Simulation in Bayesian Networks. In *Proceedings of the Tenth Conference on Uncertainty in Artificial Intelligence*. Morgan Kaufmann Publishers, San Francisco, CA, 227–234.

García, O. N. and Chien, Y.-T., editors (1991), *Knowledge-Based Systems: Fundamentals and Tools*. IEEE Computer Society Press, Los Alamitos, CA.

Garey, M. R. and Johnson, D. S. (1979), *Computers and Intractability: A Guide to the Theory of NP-Completeness*. Freeman and Company, New York.

Gavril, T. (1972), Algorithms for Minimum Coloring, Maximum Clique, Minimum Covering by Cliques, and Maximum Indpendent Set of a Chordal Graph. *SIAM Journal of Computing*, 1:180–187.

Gavril, T. (1974), An Algorithm for Testing Chordality of Graphs. *Inform. Process. Lett.*, 3:110–112.

Geiger, D. (1987), Towards the Formalization of Informational Dependencies (M.S. Thesis). Technical Report, R-102, UCLA Cognitive Systems Laboratory, Los Angeles.

Geiger, D. (1990), *Graphoids: A Qualitative Framework for Probabilisitc Inference*. Ph.D. Thesis, Department of Computer Science, University of California.

Geiger, D. and Heckerman, D. (1994), Learning Gaussian Networks. In *Proceedings of the Tenth Conference on Uncertainty in Artificial Intelligence*. Morgan Kaufmann Publishers, San Francisco, CA, 235–243.

Geiger, D. and Heckerman, D. (1995), A Characterization of the Dirichlet Distribution with Application to Learning Bayesian Networks. In *Proceedings of the Eleventh Conference on Uncertainty in Artificial Intelligence*. Morgan Kaufmann Publishers, San Francisco, CA, 196–207.

Geiger, D. and Pearl, J. (1990), On the Logic of Causal Models. In Shachter, R. D., Levitt, T. S., Kanal, L. N., and Lemmer, J. F., editors, *Uncertainty in Artificial Intelligence 4*. North Holland, Amsterdam, 3–14.

Geiger, D., Verma, T., and Pearl, J. (1990a), D-separation: From Theorems to Algorithms. In Henrion, M., Shachter, R. D., Kanal, L. N., and Lemmer, J. F., editors, *Uncertainty in Artificial Intelligence 5*. North Holland, Amsterdam, 139–148.

Geiger, D., Verma, T., and Pearl, J. (1990b), Identifying Independence in Bayesian Networks. *Networks*, 20:507–534.

Gelfand, A. E. and Smith, A. F. (1990), Sampling-Based Approaches to Calculating Marginal Densities. *Journal of the American Statistical Association*, 85:398–409.

Gelman, A. and Speed, T. P. (1993), Characterizing a Joint Probability Distribution by Conditionals. *Journal of the Royal Statistical Society, Series B*, 55:185–188.

Gibbons, A. (1985), *Algorithmic Graph Theory*. Cambridge University Press, Cambridge.

Gilks, W. R. and Wild, P. (1992), Adaptive Rejection Sampling for the Gibbs Sampling. *Journal of the Royal Statistical Society, Series C*, 41:337–348.

Ginsberg, M. L. (1993), *Essentials of Artificial Intelligence*. Morgan Kaufmann, San Mateo, CA.

Golumbic, M. C. (1980), *Algorithmic Graph Theory and Perfect Graphs*. Academic Press, New York.

Gutiérrez, J. M. (1994), *Sistemas Expertos, Grafos y Redes Bayesianas*. Ph.D. Thesis, Departamento de Matemática Aplicada y Ciencias de la Computación, Universidad de Cantabria, Spain.

Gutiérrez, J. M. and Solares, C. (1995), Some Graph Algorithms for Causal Network Expert Systems. In Keranen, V. and Mitic, P., editors, *Mathematics with Vision: Proceedings of the First International Mathematica Symposium*. Computational Mechanics Publications, Southampton, U.K., 183–190

Hadi, A. S. (1996), *Matrix Algebra as a Tool*. Duxbury Press, Belmont, CA.

Harary, F., editor (1969), *Graph Theory*. Addison-Wesley, Reading, MA.

Hayes-Roth, F. (1985), Rule-Based Systems. *Comunications of the ACM*, 28:921–932.

Hayes-Roth, F., Waterman, D. A., and Lenat, D. B., editors (1983), *Building Expert Systems*. Addison-Wesley, Reading, MA.

Heckerman, D. (1990a), *Probabilistic Similarity Networks*. Ph.D. Thesis, Program in Medical Information Sciences, Stanford University.

Heckerman, D. (1990b), Probabilistic Similarity Networks. *Networks*, 20:607–636.

Heckerman, D. (1995), A Tutorial on Learning With Bayesian Networks. Technical Report, Msr TR-95-06, Microsoft Research, Redmond, WA.

Heckerman, D. and Geiger, D. (1995), Learning Bayesian Networks: A Unification for Discrete and Gaussian Domains. In *Proceedings of the Eleventh Conference on Uncertainty in Artificial Intelligence*. Morgan Kaufmann Publishers, San Francisco, CA, 274–284.

Heckerman, D., Geiger, D., and Chickering, D. M. (1994), Learning Bayesian Networks: The Combination of Knowledge and Statistical Data. In *Proceedings of the Tenth Conference on Uncertainty in Artificial Intelligence*. Morgan Kaufmann Publishers, San Francisco, CA, 293–301.

Henrion, M. (1988), Propagation of Uncertainty by Logic Sampling in Bayes' Networks. In Lemmer, J. F. and Kanal, L. N., editors, *Uncertainty in Artificial Intelligence 2*. North Holland, Amsterdam, 149–164.

Henrion, M. (1991), Search-Based Methods to Bound Diagnostic Probabilities in Very Large Belief Nets. In *Proceedings of the Seventh Conference on Uncertainty in Artificial Intelligence*. Morgan Kaufmann Publishers, San Mateo, CA, 142–150.

Hernández, L. D. (1995), *Diseño y Validación de Nuevos Algoritmos para el Tratamiento de Grafos de Dependencias.* Ph.D. Thesis, Departamento de Ciencias de la Computación e Inteligencia Artificial, Universidad de Granada. Spain.

Hogg, R. V. (1993), *Probability and Statistical Inference.* Maxwell Macmillan International, New York.

Horvitz, E., Breese, J., and Henrion, M. (1988), Decision Theory in Expert Systems and Artificial Intelligence. *International Journal of Approximate Reasoning,* 2:247–302.

Horvitz, E., Heckerman, D., and Langlotz, C. (1986), A Framework for Comparing Alternative Formalisms for Plausible Reasoning. In *Fifth National Conference on Artificial Intelligence (AAAI-86).* AAAI Press/MIT Press, Menlo Park, CA, 210–214.

Isham, B. (1981), An Introduction to Spatial Point Processes and Markov Random Fields. *International Statistical Review,* 49:21–43.

Jackson, P. (1990), *Introduction to Expert Systems, 2nd edition.* Addison-Wesley, Reading, MA.

Jensen, F. V. (1988), *Junction Trees and Decomposable Hypergraphs.* JUDEX Research Report, Aalborg, Denmark.

Jensen, F. V. (1996), *An Introduction to Bayesian Networks.* Springer-Verlag, New York.

Jensen, F., Lauritzen, S., and Olesen, K. (1990), Bayesian Updating in Causal Probabilistic Networks by Local Computations. *Computational Statistics Quarterly,* 4:269–282.

Jensen, F. V., Olesen, K. G., and Andersen, S. K. (1990), An Algebra of Bayesian Belief Universes for Knowledge-Based Systems. *Networks,* 20:637–660.

Johnson, L. and Keravnou, E. T. (1988), *Expert System Architectures.* Kogan Page Limited, London, U.K.

Johnson, R. A. and Wichern, D. W. (1988), *Applied Multivariate Analysis, 2nd edition.* Prentice Hall, Englewood Cliffs, NJ.

Jones, J. L. and Flynn, A. M. (1993), *Mobile Robots: Inspiration to Implementation.* A. K. Peters, Wellesley, MA.

Jordan, M. and Jacobs, R. (1993), Supervised Learning and Divide-And-Conquer: A Statistical Approach. In *Machine Learning: Proceedings of the Tenth International Conference.* Morgan Kaufmann Publishers, San Mateo, CA, 159–166.

Jubete, F. and Castillo, E. (1994), Linear Programming and Expert Systems. *Microcomputers in Civil Engineering, Special Issue on Uncertainty in Expert Systems,* 9:335–345.

Kenley, C. R. (1986), *Influence Diagram Models with Continuous Variables.* Ph.D. Thesis, Stanford, Stanford University.

Kim, J. H. (1983), *CONVINCE: A Conversation Inference Consolidation Engine.* Ph.D. Thesis, Department of Computer Science, University of California.

Kim, J. H. and Pearl, J. (1983), A Computational Model for Combined Causal and Diagnostic Reasoning in Inference Systems. In *Proceedings of the 8th*

International Joint Conference on Artificial Intelligence (IJCAI-83). Morgan Kaufmann Publishers, San Mateo, CA, 190–193.

Larrañaga, P. (1995), *Aprendizaje Estructural y Descomposición de Redes Bayesianas Via Algoritmos Genéticos*. Ph.D. Thesis, Departamento de Ciencias de la Computación e Inteligencia Artificial, Universidad del País Vasco. Spain.

Larrañaga, P., Kuijpers, C., Murga, R., and Yurramendi, Y. (1996), Searching for the Best Ordering in the Structure Learning of Bayesian Networks. *IEEE Transactions on Systems, Man and Cybernetics*, 26. In press.

Laskey, K. B. (1995), Sensitivity Analysis for Probability Assessments in Bayesian Networks. *IEEE Transactions on Systems, Man and Cybernetics*, 25:901–909.

Lauritzen, S. L. (1974), Sufficiency, Prediction and Extreme Models. *Scandinavian Journal of Statistics*, 1:128–134.

Lauritzen, S. L. (1982), *Lectures on Contingency Tables, 2nd edition*. Aalborg University Press, Aalborg, Denmark.

Lauritzen, S. L. (1992), Propagation of Probabilities, Means, and Variances in Mixed Graphical Association Models. *Journal of the American Statistical Association*, 87:1098–1108.

Lauritzen, S. L., Dawid, A. P., Larsen, B. N., and Leimer, H. G. (1990), Independence Properties of Directed Markov Fields. *Networks*, 20:491–505.

Lauritzen, S. L., Speed, T. P., and Vijayan, K. (1984), Decomposable Graphs and Hypergraphs. *Journal of the Australian Mathematical Society, Series A*, 36:12–29.

Lauritzen, S. L. and Spiegelhalter, D. J. (1988), Local Computations with Probabilities on Graphical Structures and Their Application to Expert Systems. *Journal of the Royal Statistical Society, Series B*, 50:157–224.

Lauritzen, S. L. and Wermuth, N. (1989), Graphical Models for Association Between Variables, Some of Which are Qualitative and Some Quantitative. *Annals of Statistics*, 17:31–54.

Levy, D. N. L., editor (1988), *Computer Games*. Springer Verlag, New York.

Li, Z. and D'Ambrosio, B. (1994), Efficient Inference in Bayes Nets as a Combinatorial Optimization Problem. *International Journal of Approximate Reasoning*, 11:55–81.

Lindley, D. V. (1987), The Probability Approach to the Treatment of Uncertainty in Artificial Intelligence. *Statistical Science*, 2:17–24.

Lisboa, P. G. L., editor (1992), *Neural Networks: Current Applications*. Chapman and Hall, New York.

Liu, C. (1985), *Elements of Discrete Mathematics*. McGraw-Hill, New York.

Liu, X. and Li, Z. (1994), A Reasoning Method in Damage Assessment of Buildings. *Microcomputers in Civil Engineering, Special Issue on Uncertainty in Expert Systems*, 9:329–334.

Luger, G. F. and Stubblefield, W. A. (1989), *Artificial Intelligence and the Design of Expert Systems*. Benjamin/Cummings, Redwood City, CA.

Madigan, D. and Raftery, A. (1994), Model Selection and Accounting for Model Uncertainty in Graphical Models Using Occam's Window. *Journal of the American Statistical Association*, 89:1535–1546.

Martos, B. (1964), Hyperbolic Programming. *Naval Research Logistic Quarterly*, 11:135–156.

McHugh, J. A. (1990), *Algorithmic Graph Theory*. Prentice Hall, Englewood Cliffs, NJ.

McKeown, K. R. (1985), *Text Generation: Using Discourse Strategies and Focus Constraints to Generate Natural Language Text*. Cambridge University Press, New York.

McKerrow, P. (1991), *Introduction to Robotics*. Addison-Wesley, Reading, MA.

Moore, J. D. and Swartout, W. R. (1990), Pointing: A Way Toward Explanation Dialogue. In *Proceedings of the 8th National Conference on AI (AAAI-90)*. AAAI Press/MIT Press, Menlo Park, CA, 457–464.

Naylor, C. (1983), *Build Your Own Expert System*. Sigma Press, Wilmslow, U.K.

Neapolitan, R. E. (1990), *Probabilistic Reasoning in Expert Systems: Theory and Algorithms*. Wiley-Interscience, New York.

Newborn, M. (1994), *The Great Theorem Prover, Version 2.0*. NewBorn Software, Westmount, Quebec.

Newell, A., Shaw, J. C., and Simon, H. A. (1963), Chess-Playing Programs and the Problem of Complexity. In Feigenbaum, E. and Feldman, J., editors, *Computers and Thought*, McGraw-Hill, New York.

Niemann, H. (1990), *Pattern Analysis and Understanding, 3rd edition*. Series in Information Sciences. Springer-Verlag, Berlin.

Normand, S.-L. and Tritchler, D. (1992), Parameter Updating in Bayes Network. *Journal of the American Statistical Association*, 87:1109–1115.

Norton, S. W. (1988), An Explanation Mechanism for Bayesian Inference Systems. In Lemmer, J. F. and Kanal, L. N., editors, *Uncertainty in Artificial Intelligence 2*. North Holland, Amsterdam, 165–174.

O'Keefe, R. M., Balci, O., and Smith, E. P. (1987), Validating Expert System Performance. *IEEE Expert*, 2:81–90.

Olesen, K. G. (1993), Causal Probabilistic Networks with Both Discrete and Continuous Variables. *IEEE Transactions on Pattern Analysis and Machine Intelligence*, 3:275–279.

Patrick, E. A. and Fattu, J. M. (1984), *Artificial Intelligence with Statistical Pattern Recognition*. Prentice-Hall, Englewood Cliffs, N.J.

Paz, A. (1987), A Full Characterization of Pseudographoids in Terms of Families of Undirected Graphs. Technical Report, R-95, UCLA Cognitive Systems Laboratory, Los Angeles.

Paz, A. and Schulhoff, R. (1988), Closure Algorithms and Decision Problems for Graphoids Generated by Two Undirected Graphs. Technical Report, R-118, UCLA Cognitive Systems Laboratory, Los Angeles.

Pearl, J. (1984), *Heuristics*. Addison–Wesley, Reading, MA.

Pearl, J. (1986a), A Constraint-Propagation Approach to Probabilistic Reasoning. In Kanal, L. N. and Lemmer, J. F., editors, *Uncertainty in Artificial Intelligence*. North Holland, Amsterdam, 357–369.

Pearl, J. (1986b), Fusion, Propagation and Structuring in Belief Networks. *Artificial Intelligence*, 29:241–288.

Pearl, J. (1987a), Distributed Revision of Compatible Beliefs. *Artificial Intelligence*, 33:173–215.

Pearl, J. (1987b), Evidential Reasoning Using Stochastic Simulation of Causal Models. *Artificial Intelligence*, 32:245–257.

Pearl, J. (1988), *Probabilistic Reasoning in Intelligent Systems: Networks of Plausible Inference*. Morgan Kaufmann, San Mateo, CA.

Pearl, J., Geiger, D., and Verma, T. (1989), The Logic of Influence Diagrams. In Oliver, R. M. and Smith, J. D., editors, *Influence Diagrams, Belief Networks and Decision Analysis*. John Wiley and Sons, New York, 67–87.

Pearl, J. and Paz, A. (1987), Graphoids: A Graph-Based Logic for Reasoning about Relevance Relations. In Boulay, B. D., Hogg, D., and Steels, L., editors, *Advances in Artificial Intelligence-II*. North Holland, Amsterdam, 357–363.

Pearl, J. and Verma, T. (1987), The Logic of Representing Dependencies by Directed Graphs. In *Proceedings of the National Conference on AI (AAAI-87)*. AAAI Press/MIT Press, Menlo Park, CA, 374–379.

Pedersen, K. (1989), *Expert Systems Programming: Practical Techniques for Rule-Based Expert Systems*. John Wiley and Sons, New York.

Peot, M. A. and Shachter, R. D. (1991), Fusion and Propagation With Multiple Observations in Belief Networks. *Artificial Intelligence*, 48:299–318.

Poole, D. (1993a), Average-Case Analysis of a Search Algorithm for Estimating Prior and Posterior Probabilities in Bayesian Networks with Extreme Probabilities. In *Proceedings of the 13th International Joint Conference on Artificial Intelligence (IJCAI-93)*. Morgan Kaufmann Publishers, San Mateo, CA, 606–612.

Poole, D. (1993b), The Use of Conflicts in Searching Bayesian Networks. In *Proceedings of the Ninth Conference on Uncertainty in Artificial Intelligence*. Morgan Kaufmann Publishers, San Mateo, CA, 359–367.

Preece, A. D. (1990), Towards a Methodology for Evaluating Expert Systems. *Expert Systems*, 7:215–293.

Preparata, F. P. and Shamos, M. I. (1985), *Computational Geometry*. Springer-Verlag, New York.

Press, S. J. (1992), *Bayesian Statistics: Principles, Models, and Applications*. John Wiley and Sons, New York.

Press, W. H., Teulosky, S. A., Vetterling, W. T., and Flannery, B. P. (1992), *Numerical Recipies, 2nd edition*. Cambridge University Press, Cambridge.

Quinlan, J., editor (1987), *Applications of Expert Systems, Volume 1*. Addison-Wesley, Reading, MA.

Quinlan, J., editor (1989), *Applications of Expert Systems, Volume 2*. Addison-Wesley, Reading, MA.

Rabiner, L. and Juang, B. H. (1993), *Fundamentals of Speech Recognition.* Prentice Hall, Englewood Cliffs, N.J.

Raiffa, H. and Schlaifer, R. (1961), *Applied Statistical Decision Theory.* Division of Research, Harvard Business School, Boston.

Rencher, A. C. (1995), *Methods of Multivariate Analysis.* John Wiley and Sons, New York.

Ripley, B. D. (1987), *Stochastic Simulation.* John Wiley and Sons, New York.

Rissanen, J. (1983), A Universal Data Compression System. *IEEE Transactions on Information Theory*, IT-29:656–664.

Rissanen, J. (1986), Stochastic Complexity and Modeling. *Annals of Statistics*, 14:1080–1100.

Rich, E. and Knight, K. (1991), *Artificial Intelligence, 2nd edition.* McGraw-Hill, New York.

Rose, D. J., Tarjan, R. E., and Leuker, G. S. (1976), Algorithmic Aspects of Vertex Elimination on Graphs. *SIAM Journal of Computing*, 5:266–283.

Ross, K. A. and Wright, C. R. (1988), *Discrete Mathematics.* Prentice Hall, Englewood Cliffs, N.J.

Rubinstein, R. Y. (1981), *Simulation and the Monte Carlo Method.* John Wiley and Sons, New York.

Russell, S. J. and Norvig, P. (1995), *Artificial Intelligence: A Modern Approach.* Prentice Hall, Englewood Cliffs, N.J.

Schank, R. C. and Abelson, R. P. (1977), *Scripts, Plans, Goals and Understanding. An Inquiry into Human Knowledge Structures.* Lawrence Erlbaum Associates, Hillsdale, N.J.

Schwarz, G. (1978), Estimation the Dimension of a Model. *Annals of Statistics*, 17(2):461–464.

Shachter, R. D. (1986), Evaluating Influence Diagrams. *Operations Research*, 34:871–882.

Shachter, R. D. (1988), Probabilistic Inference and Influence Diagrams. *Operations Research*, 36:589–605.

Shachter, R. D. (1990a), Evidence Absorption and Propagation Through Evidence Reversals. In Henrion, M., Shachter, R. D., Kanal, L. N., and Lemmer, J. F., editors, *Uncertainty in Artificial Intelligence 5.* North Holland, Amsterdam, 173–190.

Shachter, R. D. (1990b), An Ordered Examination of Influence Diagrams. *Networks*, 20:535–563.

Shachter, R. D., Andersen, S. K., and Szolovits, P. (1994), Global Conditioning for Probabilistic Inference in Belief Networks. In *Proceedings of the Tenth Conference on Uncertainty in Artificial Intelligence.* Morgan Kaufmann Publishers, San Francisco, CA, 514–522.

Shachter, R., D'Ambrosio, B., and DelFavero, B. (1990), Symbolic Probabilistic Inference in Belief Networks. In *Proceedings of the 8th National Conference on AI (AAAI-90).* AAAI Press/MIT Press, Menlo Park, CA, 126–131.

Shachter, R. and Kenley, C. (1989), Gaussian Influence Diagrams. *Management Science*, 35(5):527–550.

Shachter, R. and Peot, M. (1990), Simulation Approaches to General Probabilistic Inference on Belief Networks. In Henrion, M., Shachter, R. D., Kanal, L. N., and Lemmer, J. F., *Uncertainty in Artificial Intelligence 5*. North Holland, Amsterdam, 221–231.

Shafer, G. (1976), *A Mathematical Theory of Evidence*. Princeton University Press, Princeton, NJ.

Shapiro, L. G. and Rosenfeld, A. (1992), *Computer Vision and Image Processing*. Academic Press, Boston, MA.

Shapiro, S. C., editor (1987), *Encyclopedia of Artificial Intelligence*, John Wiley and Sons, New York.

Shneiderman, B. (1987), *Designing the Human Interface*. Addison-Wesley, Reading, MA.

Shwe, M. and Cooper, G. (1991), An Empirical Analysis of Likelihood-Weighting Simulation on a Large Multiply Connected Medical Belief Network. *Computers and Biomedical Research*, 24:453–475.

Simons, G. L. (1985), *Introducing Artificial Intelligence*. John Wiley and Sons, New York.

Sing-Tze, B. (1984), *Pattern Recognition*. Marcel Dekker, New York.

Skiena, S. S. (1990), *Implementing Discrete Mathematics: Combinatorics and Graph Theory with Mathematica*. Addison-Wesley, Reading, MA.

Smith, C. A. B. (1961), Consistency in Statistical Inference and Decision. *Journal of the Royal Statistical Society, Series B*, 23:1–37.

Spirtes, P. and Meek, C. (1995), Learning Bayesian Netwoks with Discrete Variables from Data. In *Proceedings of the First International Conference on Knowledge Discovery and Data Mining*. AAAI Press, Menlo Park, CA, 294–300.

Srinivas, S. and Nayak, P. (1996), Efficient Enumeration of Instantiations in Bayesian Networks. In *Proceedings of the Twelfth Conference on Uncertainty in Artificial Intelligence*. Morgan Kaufmann Publishers, San Francisco, CA, 500–508.

Stevens, L. (1984), *Artificial Intelligence. The Search for the Perfect Machine*. Hayden Book Company, Hasbrouck Heights, N.J.

Stillman, J. (1991), On Heuristics for Finding Loop Cutsets in Multiply Connected Belief Networks. In Bonissone, P. P., Henrion, M., Kanal, L. N., and Lemmer, J. F., editors, *Uncertainty in Artificial Intelligence 6*. North Holland, Amsterdam, 233–343.

Strat, T. M. and Lowrance, J. D. (1989), Explaining Evidential Analysis. *International Journal of Approximate Reasoning*, 3:299–353.

Studený, M. (1989), Multiinformation and the Problem of Characterization of Conditional Independence Relations. *Problems of Control and Information Theory*, 18:3–16.

Studený, M. (1992), Conditional Independence Relations Have No Finite Complete Characterization. In Kubíck, S. and Vísek, J. A., editors, *Information Theory, Statistical Decision Functions and Random Processes: Transactions of 11th Prague Conference B*. Kluwer, Dordrecht, 377–396.

Studený, M. (1994), Semigraphoids are Two-Antecedental Approximations of Stochastic Conditional Independence Models. In *Proceedings of the Tenth Conference on Uncertainty in Artificial Intelligence*. Morgan Kaufmann Publishers, San Francisco, CA, 546–552.

Suermondt, H. J. (1992), *Explanation in Bayesian Belief Networks*. Ph.D. Thesis, Department of Computer Science, Stanford University.

Suermondt, H. J. and Cooper, G. F. (1990), Probabilistic Inference in Multiply Connected Belief Networks Using Loop Cutsets. *International Journal of Approximate Reasoning*, 4:283–306.

Suermondt, H. J. and Cooper, G. F. (1991a), A Combination of Exact Algorithms for Inference on Bayesian Belief Networks. *International Journal of Approximate Reasoning*, 5:521–542.

Suermondt, H. J. and Cooper, G. F. (1991b), Initialization for the Method of Conditioning in Bayesian Belief Networks. *Artificial Intelligence*, 50:83–94.

Suermondt, H., Cooper, G., and Heckerman, D. (1991), A Combination of Cutset Conditioning with Clique-Tree Propagation in the Pathfinder System. In Bonissone, P. P., Henrion, M., Kanal, L. N., and Lemmer, J. F., editors, *Uncertainty in Artificial Intelligence 6*. North-Holland, Amsterdam, 245–253.

Tamassia, R. and Tollis, I. G., editors (1995), *Graph Drawing (Proceedings of GD'94)*, Lecture Notes in Computer Science. Springer-Verlag, New York.

Tarjan, R. E. (1983), *Data Structures and Network Algorithms*. SIAM (Society for Industrial and Applied Mathematics), Philadelphia, PA.

Tarjan, R. E. and Yannakakis, M. (1984), Simple Linear-Time Algorithms to Test Chordality of Graphs, Test Acyclity of Hypergraphs and Selectively Reduce Acyclic Hypergraphs. *SIAM Journal of Computing*, 13:566–579.

Tessem, B. (1992), Interval Probability Propagation. *International Journal of Approximate Reasoning*, 7:95–120.

Ur, S. and Paz, A. (1994), The Representation Power of Probabilistic Knowledge by Undirected Graphs and Directed Acyclic Graphs: A Comparasion. *International Journal of General Systems*, 22:219–231.

Verma, T. S. (1987), Some Mathematical Properties of Dependency Models. Technical Report, R-103, UCLA Cognitive Systems Laboratory, Los Angeles.

Verma, T. and Pearl, J. (1990), Causal Networks: Semantics and Expressiveness. In Shachter, R. D., Levitt, T. S., Kanal, L. N., and Lemmer, J. F., editors, *Uncertainty in Artificial Intelligence 4*. North Holland, Amsterdam, 69–76.

Verma, T. and Pearl, J. (1991), Equivalence and Synthesis of Causal Models. In Bonissone, P. P., Henrion, M., Kanal, L. N., and Lemmer, J. F., editors, *Uncertainty in Artificial Intelligence 6*. North Holland, Amsterdam, 255–268.

Von Neumann, J. (1951), Various Techniques Used in Connection with Random Numbers. *U.S. Nat. Bur. Stand. Appl. Math. Ser.*, 12:36–38.

Waterman, D. A. (1985), *A Guide to Expert Systems*. Addison-Wesley, Reading, MA.

Weiss, S. M. and Kulikowski, C. A. (1984), *A Practical Guide to Designing Expert Systems*. Rowman and Allanheld, Totowa, N.J.

Wermuth, N. and Lauritzen, S. L. (1983), Graphical and Recursive Models for Contingency Tables. *Biometrika*, 70:537–552.

Whittaker, J. (1990), *Graphical Models in Applied Mathematical Multivariate Statistics*. John Wiley and Sons, New York.

Wick, M. R. and Thompson, W. B. (1992), Reconstructive Expert System Explanation. *Artificial Intelligence*, 54:33–70.

Winston, P. H. (1992), *Artificial Intelligence, 3rd edition*. Addison-Wesley, Reading, MA.

Wolfram, S. (1991), *Mathematica: A System for Doing Mathematics by Computer*. Addison-Wesley, Reading, MA.

Wos, L., Overbeek, R., Lusk, E., and Boyle, J. (1984), *Automated Reasoning. Introduction and Applications*. Prentice-Hall, Englewood Cliffs, N.J.

Xu, H. (1995), Computing Marginals for Arbitrary Subsets from Marginal Representation in Markov Trees. *Artificial Intelligence*, 74:177–189.

Xu, L. and Pearl, J. (1989), Structuring Causal Tree Models with Continuous Variables. In Kanal, L. N., Levitt, T. S., and Lemmer, J. F., editors, *Uncertainty in Artificial Intelligence 3*. North Holland, Amsterdam, 209–219.

Yager, R. R., Ovchinnikov, S., Yong, R. M., and Nguyen, H. T., editors (1987), *Fuzzy Sets and Applications: Selected Papers by L. A. Zadeh*. John Wiley and Sons, New York.

Yannakakis, M. (1981), Computing the Minimal Fill-in is NP-Complete. *SIAM Journal of Algebraic Discrete Methods*, 2:77–79.

Yu, Q., Almulla, M., and Newborn, M. (1996), Heuristics for a Semantic Tree Theorem Prover. In *Proceedings of the Fourth International Symposium on Artificial Intelligence and Mathematics (AI/MATH-96)*, Fort Lauderdale, Florida, 162–165.

Zadeh, L. A. (1983), The Role of Fuzzy Logic in the Management of Uncertainty in Expert Systems. *Fuzzy Sets and Systems*, 11:199–227.

Index

A-separation, 183
Absorption of evidence, 343, 354
Abstract knowledge, 11, 22
Acceptance-rejection method, 397
   algorithm, 399
   pseudocode, 407
Action execution, 13
Acyclic graph, 128, 155
Adjacency
   matrix, 155, 156
   set, 115, 119, 122, 161
Algebraic structure of probabilities, 454
Ancestral
   numbering, 125, 153
   ordering, 275, 407, 411
   set, 124
Approximate propagation, 317, 393
   backward-forward sampling, 413
   likelihood weighing, 411
   logic sampling, 407
   Markov sampling, 415
   maximum probability search, 429
   search-based methods, 394, 429
   simulation methods, 394
   systematic sampling, 419
   uniform sampling, 409

Artificial intelligence, 1, 16–20
Artificial vision, 19
Automatic
   elimination of feasible values, 51
   game playing, 18
   language translation, 19
   teller machine, 3
   theorem proving, 18
Automatic teller machine, 23
Axiom
   additivity, 71
   boundary, 71
   continuity-consistency, 72
   monotonicity, 72

B-algorithm, 511
Backward chaining, 37
Backward-forward sampling, 413
   pseudocode, 414
Banking transactions example, 3
Bayes' theorem, 9, 71, 79–82, 104, 108, 109, 111
Bayesian measure, 486
   for multinomial network, 490
Bayesian network, 9, 70, 102, 128, 201, 246, 270, 285, 287, 289,

Bayesian (*cont.*)
    296, 302, 342, 351, 364, 373,
    377, 393, 405, 445, 481–484,
    487, 489, 490, 508, 511, 525
  definition, 248
  dimension, 486
  example, 249, 251
  Gaussian (normal), 249, 250
  learning, 482
  multinomial, 249
Bernoulli
  conjugate, 517
  process, 517
Beta distribution, 518
Binomial distribution, 517
Bounds, 434
Breadth-first search, 160, 161
  algorithm, 165

Canonical CPD, 197
Canonical form, 198
  algorithm, 198
Causal list, 248, 276, 277
  closure, 247
  completeness, 248
  soundness, 247
Causal models, 245, 246, 248
Certainty factors, 9
Chain graph, 213
Chain of cliques, 138, 352
  algorithm, 138
  existence of, 138
Chain rule factorization, 195, 196
Chaining
  backward, 37
  forward, 37
  goal-oriented, 28, 37
  rule, 28, 35, 36
Chord of a loop, 129
Chordal graph, 129, 261
Chordality property, 191, 238
CIS, 78, 177
Cliques, 118
  chain of, 138, 352
  graph of, 140
Cluster, 139
  graph, 139
Clustering, 139

Clustering algorithm, 318, 342, 351
  in Bayesian networks, 364
  in Markov networks, 351
Coherence
  control, 11, 48, 104
  control subsystem, 11
  of facts, 51
  of rules, 48
Compatibility, 279
  of CSPM, 301
  of multifactorized models, 284
  of multigraph models, 273
Compiling rules, 28, 47
Complexity
  approximate propagation, 439
  clustering algorithm, 342
  conditioning algorithm, 342, 344
  of a network, 486
  propagation in polytrees, 331
Composition property, 187, 238
Conclusion, 23
  mixed, 28
  simple, 28
Concrete knowledge, 11, 22
Conditional
  dependence, 78
  dependence statement, 78
  independence, 78, 185, 192, 201
  independence properties, 212
  independence statement, 78, 177
  probability distribution, 73
Conditionally specified models, 102,
    201, 212, 298
Conditioning algorithm, 318, 342
Conjugate
  Bernoulli, 517
  distributions, 516
  exponential family, 517
  multinomial, 519, 520
  multivariate normal, 521
  normal, 519
  Poisson, 518
Connected
  components, 121, 168
  graph, 120, 128, 167
Contraction property, 187, 238
Convenient posterior, 524
Cooper-Herskovits measure, 494
CPD, 73

Index 599

CPD (cont.)
  canonical, 197
  standard canonical, 199
CSPM, 298
  assessment of, 308
  consistency, 301
  uniqueness, 300
Cutset, 178
  loop, 343
Cycle, 127, 155
  finding algorithm, 169
Cyclic graph, 128

D-separation, 180–182, 188, 207, 247, 248, 266, 267, 296, 312, 378, 380, 386
  algorithm, 183
DAG, 128, 556
Damage of structures example, 550, 562
Data
  complete, 482
  incomplete, 482, 513
  set, 486
Decomposable Markov network, 355, 445
Decomposable model, 129, 229
  factorization, 230
Decomposition property, 185, 219, 238
Dependence, 74
  conditional, 78
Dependency map, 216
Dependency models, 195, 211
  directed graph, 237
  probabilistic, 212
  undirected graph, 218
Depth
  algorithm, 154
  ascending, 153
  descending, 153
  level, 153
Depth-first search, 160, 161
  algorithm, 162
Deterministic
  expert systems, 8
  knowledge, 12

Dimension of a Bayesian network, 486
Directed acyclic graph, 128, 556
Directed graphs, 113
Dirichlet distribution, 520
Disconnected subset, 147
Distributed reasoning, 17
Distribution
  beta, 554
  binomial, 264, 554
  conjugate, 516
  multinomial, 249
  multivariate normal, 220, 250, 292, 383, 476, 501
  natural conjugate, 516
  normal-Wishart, 501
  of conditional parameters, 492
  population, 395
  posterior, 103
  simulation, 395
  uniform, 394
Distribution equivalent Bayesian networks, 258

EM-algorithm, 514
Equivalent
  Bayesian networks, 253, 482
  graphical models, 252
  Markov networks, 253
Estimation, 483
  Bayes, 485
  maximum likelihood, 485
Evidence, 318, 400
  absorption, 343, 354
  collecting, 366
  distributing, 366
  propagation, 104, 317
Exact propagation, 317
  clustering algorithm, 351
  conditioning algorithm, 342
  goal-oriented, 377
  in Gaussian networks, 382
  in join trees, 366
  in Markov networks, 351
  in polytrees, 321
Experiment, 394
Expert systems, 2
  comparing, 106

Expert (cont.)
  components of, 10
  deterministic, 8
  development of, 14
  HUGIN, 9
  human component, 11
  MYCIN, 9
  probabilistic, 9, 65, 69, 70, 86
  PROSPECTOR, 9, 85
  rule-based, 9, 21, 70
  stochastic, 8
  types of, 8
  why?, 7
  X-pert, 9, 341, 448, 530, 533
Explanation subsystem, 13, 53
Exponential distribution, 517
Exponential family, 517
  conjugate, 517

Factorization, 195
  according to a DAG, 244
  according to an undirected graph, 228
  by potentials, 195
  canonical chain rule, 196
  chain rule, 195
Failure tree, 532
False negative decision, 83
False positive decision, 83
Family tree, 143, 367
  algorithm, 143
Famous people example, 24
Filling-in, 131
  minimal, 131
Forward chaining, 37
Forward sampling, 407
Fuzzy logic, 9

Game playing, 18
Gamma distribution, 519
Gamma function, 493, 522
Gaussian
  Bayesian network, 250
Gaussian Bayesian network, 391, 444, 474
Geiger-Heckerman quality measure, 493, 503

Generalized rule, 70, 85
Gibbs' sampling, 513
Goal-oriented
  propagation of evidence, 377
  rule chaining, 37
Graph, 114
  acyclic, 128
  algorithms, 158
  chain, 213
  chordal, 129
  clique, 140
  cluster, 139
  complete, 118
  connected, 120, 128, 167
  cyclic, 128
  DAG, 128
  directed, 115
  equivalent, 253
  expressiveness of, 259
  from directed to undirected, 125
  join, 140
  junction, 140
  moral, 126, 260
  multiply-connected, 121, 128
  numerical representation of, 155
  polytree, 128
  representation, 144, 146
  separation, 177, 212, 217, 268
  simple tree, 128
  tree, 121, 128
  triangulated, 129, 133
  undirected, 115, 118, 177
Graphical independence, 178
Graphically specified models, 177, 201, 203, 211, 212
Graphoid, 192, 220, 222
  algorithm, 193

HUGIN, 9

I-map, 270
Image recognition, 18
Independence, 74
  conditional, 78, 185, 192, 201
  graphical, 178
  parameter, 488
Independency map, 216

Independency (cont.)
  existence of, 222
  minimal, 217
  minimal directed, 239
  minimal undirected, 222
Infeasible values, 50
  elimination of, 51
Inference engine, 12
  probabilistic expert systems, 102
  rule-based expert systems, 28
Information
  acquisition, 12
  measures, 509
  prior, 485
Input list, 185, 201, 212, 248
  causal, 245
  compatible with a JPD, 192
  graphoid, 192
  nonextreme probabilistic, 192
  probabilistic, 192
  semigraphoid, 192
  types of, 192
Instantiation, 403
  ordering, 419
Intersection property, 188, 219, 238

Join graph, 140
Join tree, 140, 342, 366, 373, 448
  algorithm, 141, 318
  existence of, 141
  propagation algorithm, 367
Joint probability distribution, 73
  factorization of, 195
Junction
  graph, 140
  tree, 140

K2-algorithm, 511
Knowledge
  abstract, 11, 22, 28, 50
  acquisition subsystem, 11
  base, 11, 22, 91
  concrete, 11, 22, 28, 51
  deterministic, 12
  engineer, 11
  probabilistic, 12
  representation, 17

Language
  generating, 19
  translation, 19
Learning, 202, 481
  Bayesian networks, 482
  parametric, 14, 483
  structural, 14, 483
  subsystem, 14
Likelihood, 80
Likelihood weighing, 411
  pseudocode, 412
Link, 114
  directed, 115
  irreversible, 255
  reversible, 255
  undirected, 115
Local Markov property, 228
Logic sampling method, 407
Logical
  operator, 22
  reasoning, 9
Logical expression, 22
  complex, 23
  simple, 23
Loop, 119, 127
  chord of, 129
  finding algorithm, 169
Loop-cutset, 343
Lower Bounds, 434

Map
  dependency, 216
  independency, 216
  perfect, 213
Maple, 444, 448
Marginal probability distribution, 73
Markov
  blanket, 416
  field, 178
Markov network, 9, 70, 102, 137, 201, 219, 270, 342, 351, 366, 372, 377, 393, 445, 481
  definition, 234
  example, 234
Markov sampling, 415
  pseudocode, 416

Mathematica, 199, 225, 285, 295, 309, 444, 447, 448, 452
  program, 199, 225, 242, 285, 309, 311, 312, 385, 448, 449, 475, 479, 496, 504, 505
Matrix
  adjacency, 155, 156
  attainability, 157
  canonical, 457
Maximum cardinality search, 132
  algorithm, 133
  fill-in, 135
Maximum probability search, 429
Medical diagnosis example, 4, 87
Minimal directed I-map, 240, 248
  example, 239, 241
Minimal filling-in, 131
Minimal undirected I-map, 222
  algorithm, 224
  example, 223, 225
  factorization, 228
Minimum description length measure, 506
Missing values, 482
Models
  Bayesian network, 248
  conditionally specified, 200, 298
  decomposable, 229
  dependent relevant symptoms, 100
  dependent symptoms, 92
  graphically specified, 177, 211
  independent relevant symptoms, 98
  independent symptoms, 95
  Markov network, 219, 234
  multifactorized, 275
  probabilistic, 211
  specified by input lists, 102, 177, 185, 200, 275
  specified by multiple graphs, 269
Modus Ponens, 29
Modus Tollens, 29
Moral graph, 126, 260
Multifactorized
  models, 275
  multinomial models, 279
  normal models, 292
  probabilistic models, 279

Multigraph, 269
  compatibility, 273
  models, 269
  reduction of, 271
  redundancy in, 271
Multilevel representation, 147, 148, 150, 153
  algorithm, 149
  existence of, 152
Multinomial
  Bayesian network, 249, 509, 514
  conjugate, 519, 520
  distribution, 249, 519
Multinormal Bayesian network, 499
Multiply-connected graph, 128
Multivariate normal
  conjugate, 521
  distribution, 250, 292, 501
MYCIN, 9

Natural language processing, 19
Network, 114
  Bayesian, 9, 70, 102, 128, 201, 246, 248
  Markov, 9, 70, 102, 201, 234
  neural, 19
Neural networks, 19
Node, 114
  ancestors of, 124
  boundary of, 119
  children of, 122
  descendants of, 124
  evidential, 318, 400
  family of, 123
  goal, 378
  head-to-head, 180
  instantiation, 403
  neighbors of, 119
  nonevidential, 319, 400
  parent of, 122
  root, 148
  symbolic, 445
  uncoupled head-to-head, 253
Normal
  Bayesian network, 249, 250, 382
  conjugate, 519
  distribution, 220, 476, 517

Normal-Wishart distribution, 501, 521
NP-hard problem, 331, 344, 439
Numbering, 124
  ancestral, 125, 153
  perfect, 132, 133

Ordering
  ancestral, 275
  instantiation, 419
  links in a graph, 255
  valid, 413

Parallel computing, 17
  propagation, 329, 330, 374
Parameter
  conditional, 492
  independence, 488
  modularity, 489
Parameter independence
  global, 488
  local, 489
Parametric learning, 483
Partial correlation, 293
Path, 116, 157
  closed, 117
  cycle, 127, 169
  loop, 119, 127, 169
Pathfinding algorithms, 161
Pattern recognition, 18
Perfect map, 213
  directed, 238
  existence of, 219, 238
Perfect numbering, 132
  existence of, 133
Perfect sample, 495
Planar graph, 146
Planning, 17
Poisson
  conjugate, 518
  distribution, 517, 518
  process, 518, 519
Polytree, 318, 342
Population distribution, 395
Posterior
  convenient, 524
  distribution, 103, 516

Potential
  factorization, 195
  representation, 228, 351
Power distribution system example, 542
Premise, 22
Pressure tank example, 530
Prior
  assessment of normal networks, 502
  distribution, 103, 515
  information, 485
Prisoners' problem example, 6, 61
Probabilistic
  dependency model, 212
  inference, 9
  input list, 192
  knowledge, 12
  models, 211
  network models, 9
  reasoning, 9
Probabilistic expert systems, 9, 65, 69, 70, 86
Probabilistic network model, 454
Probability
  axioms, 71
  conditional, 73, 80
  density function, 73
  distributions, 72
  mass function, 73
  measure, 70, 71
  posterior, 80
  prior, 80
  theory, 71
Probability bounds, 473
Prolog, 18
Propagation of evidence, 317
  approximate, 317, 393
  clustering algorithm, 342, 351, 355, 364
  conditioning algorithm, 342
  exact, 317
  goal-oriented, 377
  in Gaussian networks, 382
  in join trees, 366, 372
  in polytrees, 321
  search-based methods, 430
  simulation methods, 394
  symbolic, 283, 317, 443

Property
  chordality, 191, 238
  composition, 187, 238
  contraction, 187, 238
  decomposition, 185, 219, 238
  intersection, 188, 219, 238
  local Markov, 228
  of conditional independence, 184
  running intersection, 138, 141, 229, 352
  strong transitivity, 190
  strong union, 188, 219
  symmetry, 74, 185, 219, 238
  transitivity, 219
  weak transitivity, 190, 238
  weak union, 187, 238
PROSPECTOR, 9, 85

Qualitative structure, 201, 202, 211
Quality measure, 483
  Bayesian, 486
  Cooper-Herskovits, 494
  for normal network, 499
  Geiger and Heckerman, 493, 503
  information, 509
  minimum description length, 506
Quantitative structure, 202, 213

Reasoning
  logical, 9
  parallel, 17
  probabilistic, 9
Recognition
  image, 18
  pattern, 18
  signal, 18
  speech, 18
Resolution mechanism, 28, 31
Robotics, 19
Root node, 148
Rule, 23
  chaining, 28, 35, 36
  chaining algorithm, 35
  coherence of, 48
  compiling, 28, 47
  complex, 23
  conclusion of, 23

Rule (*cont.*)
  deterministic, 21
  generalized, 70, 85
  premise of, 22
  simple, 23
  substitution, 26, 27, 55
Rule-based expert systems, 9, 21
Running intersection property, 138, 141, 229, 352

Sample, 395
Sampling
  acceptance-rejection, 406
  backward-forward, 413
  forward, 407
  Gibbs', 513
  likelihood weighing, 411
  logic, 407
  Markov, 415
  maximum probability search, 429
  systematic, 419
  uniform, 409
Scheduling problems example, 4
Score
  equivalence, 484
  function, 399, 406
Search, 159
  algorithm, 483
  based propagation, 430
  Bayesian network algorithm, 511
  breadth-first, 160, 161
  depth-first, 160, 161
  maximum cardinality, 132
Secret agents example, 5, 58
Semigraphoid, 192, 220
Separation
  A-, 183
  D-, 180–182, 188, 207, 247, 248, 267, 296, 312, 378, 380, 386
  in directed graphs, 180
  in undirected graphs, 177
  U-, 178, 206, 214
Signal recognition, 18
Simulation, 394
  distribution, 395, 406
  general algorithm, 401
Speech recognition, 18
Standard Bayesian measure

Index   605

Standard (*cont.*)
  multinomial networks, 494
  multinormal networks, 504
Standard canonical CPD, 199, 308
Stochastic expert systems, 8
Stratified sampling, 422
Strong transitivity property, 190
  violation, 220, 221
Strong union property, 188, 219
  violation, 220, 221
Structural learning, 483
Symbolic node, 445
Symbolic propagation, 283, 317, 443
  code generation, 447
  in normal networks, 474
  numeric canonical components, 456
Symmetry property, 74, 185, 219, 238
Systematic sampling, 419

Theorem proving, 18
Theory of evidence, 9
Totally disconnected subset, 147
Traffic control example, 4, 53
Transitivity property, 219
Tree, 121, 128
  failure, 532
  family, 143, 367
  join, 140, 366, 448
  junction, 140
  polytree, 128
  simple tree, 128
Trial, 394
Triangulated graph, 129, 133
Triangulation, 131
Truth table, 27, 32
Type I error, 83
Type II error, 83

U-separation, 178, 206, 214
Uncertainty, 65
  measure, 70
  propagation, 12, 104, 317
Undirected graphs, 113
Uniform sampling, 409

Uniform (*cont.*)
  pseudocode, 410
Uniqueness of CSPM, 300
Upper Bounds, 434
User interface, 12, 13
Utility function, 84

V-structure, 253
Valid ordering, 413

Weak transitivity property, 190, 238
Weak union property, 187, 238
Working memory, 11, 22
World Wide Web site, 2, 17, 56, 193, 341, 448, 530, 533

X-pert, 341, 448, 530, 533
X-pert Maps, 193
X-pert Nets, 9
X-pert Rules, 56